The Microwave Processing of Foods

Related Titles Page
Regier, Knoerzer, Schubert/The Microwave Processing of Foods, Second Edition

Related Titles

Innovative Food Processing Technologies
(ISBN 978-0-08100-294-0)

Thermal Technologies in Food Processing
(ISBN 978-1-85573-558-3)

Improving the Thermal Processing of Foods
(ISBN 978-1-85573-730-3)

Food Preservation Techniques
(ISBN 978-1-85573-530-9)

Woodhead Publishing Series in Food Science, Technology and Nutrition

The Microwave Processing of Foods

Second Edition

Edited by

Marc Regier

Kai Knoerzer

Helmar Schubert

ELSEVIER

AMSTERDAM • BOSTON • HEIDELBERG • LONDON
NEW YORK • OXFORD • PARIS • SAN DIEGO
SAN FRANCISCO • SINGAPORE • SYDNEY • TOKYO
Woodhead Publishing is an imprint of Elsevier

WP WOODHEAD PUBLISHING

Woodhead Publishing is an imprint of Elsevier
The Officers' Mess Business Centre, Royston Road, Duxford, CB22 4QH, United Kingdom
50 Hampshire Street, 5th Floor, Cambridge, MA 02139, United States
The Boulevard, Langford Lane, Kidlington, OX5 1GB, United Kingdom

Second Edition 2017

First Edition 2005

British Library Cataloguing-in-Publication Data
A catalogue record for this book is available from the British Library

Library of Congress Cataloging-in-Publication Data
A catalog record for this book is available from the Library of Congress

ISBN: 978-0-08-100528-6 (print)
ISBN: 978-0-08-100531-6 (online)

For information on all Woodhead Publishing visit our website at https://www.elsevier.com

Publisher: Nikki Levy
Acquisition Editor: Nina D. Bandeira
Editorial Project Manager: Ana Claudia Garcia
Production Project Manager: Julie-Ann Stansfield
Designer: Mark Rogers

Typeset by MPS Limited, Chennai, India

Contents

List of contributors

M. Benlloch-Tinoco University of Nottingham, Loughborough, United Kingdom

S.L. Birla ConAgra Foods, Omaha, NE, United States

H. Bozkurt-Cekmer University of Tennessee, Knoxville, TN, United States

C. Contreras Universitat Politècnica de València, Valencia, Spain

P.M. Davidson University of Tennessee, Knoxville, TN, United States

L. Dorantes-Alvarez Instituto Politécnico Nacional, Mexico City, Mexico

U. Erle Nestlé R&D Center, Solon, OH, United States

V. Gaukel Karlsruhe Institute of Technology (KIT), Karlsruhe, Germany

R. Guzmán-Gerónimo Universidad Veracruzana, Veracruz, Mexico

S. Isaksson SP Food and Bioscience, RISE Research Institutes of Sweden, Gothenburg, Sweden

C. James Food Refrigeration and Process Engineering Research Centre, Grimsby, United Kingdom

S.J. James Food Refrigeration and Process Engineering Research Centre, Grimsby, United Kingdom

K. Knoerzer CSIRO Agriculture and Food, Melbourne, VIC, Australia

K. Krishnaswamy McGill University, Montreal, QC, Canada

N. Martínez-Navarrete Universitat Politècnica de València, Valencia, Spain

V. Meda University of Saskatchewan, Saskatoon, SK, Canada

V. Orsat McGill University, Montreal, QC, Canada

A. Ortiz-Moreno Instituto Politécnico Nacional, Mexico City, Mexico

M. Ozcelik Ankara University, Ankara, Turkey

L. Parada-Dorantes Universidad del Caribe, Cancún, Mexico

K. Petersen 915 Labs, Denver, CO, United States

K. Pitchai University of Nebraska-Lincoln, Lincoln, NE, United States

G. Purnell Food Refrigeration and Process Engineering Research Centre, Grimsby, United Kingdom

P.-A. Püschner Püschner GmbH & Co KG, Schwanewede, Germany

G.S.V. Raghavan McGill University, Montreal, QC, Canada

V. Raghavan McGill University, Montreal, QC, Canada

M. Regier Trier University of Applied Sciences, Trier, Germany

D. Rodrigo Instituto de Agroquímica y Tecnología de Alimentos (IATA-CSIC), Valencia, Spain

S. Sahin Middle East Technical University, Ankara, Turkey

R. Schiffmann R.F. Schiffmann Associates, Inc., New York, NY, United States

H. Schubert Karlsruhe Institute of Technology (KIT), Karlsruhe, Germany

T. Siebert Karlsruhe Institute of Technology (KIT), Karlsruhe, Germany

R.A. Stanley University of Tasmania, Hobart, TAS, Australia

G. Sumnu Middle East Technical University, Ankara, Turkey

B. Wäppling Raaholt SP Food and Bioscience, RISE Research Institutes of Sweden, Gothenburg, Sweden

E. Turabi Yolacaner Hacettepe University, Ankara, Turkey

About the editors

Marc Regier studied Physics at the University of Karlsruhe. He received his PhD in Food Process Engineering in 2003 (University of Karlsruhe) and worked further on the topic of food drying amongst others at the Max Rubner Institute (German Federal Research Institute of Nutrition and Food). After becoming Professor in 2007 at Beuth University of Applied Sciences in Berlin, he moved to Trier University of Applied Sciences in 2010, being now Professor of Food Process Engineering and Vice Dean of the faculty. His research interests are innovative as well as conventional thermal food processes and new electromagnetic in- and online process analytical technologies. He holds four patents, has published more than 80 technical papers and book chapters, and edited 2 books in this area. He is elected member of ProcessNet Subject Divisions of drying and particle measurement.

Kai Knoerzer received his MSc in Chemical Engineering and PhD in Food Process Engineering from the University of Karlsruhe (Germany) in 2002 and 2006, respectively. He commenced a Postdoctoral Fellowship at Food Science Australia (Melbourne, Australia), a joint venture of the CSIRO (Commonwealth Scientific and Industrial Research Organisation) and the Victorian Government in 2006, and was appointed Research Scientist at CSIRO Food and Nutritional Sciences (Melbourne, Australia) in 2008. Since 2015 he has been Principal Research Scientist at CSIRO Agriculture and Food, and is leading the Food Process Engineering Team in CSIRO's Werribee facilities. His main research interests are in innovative food processing technologies, such as high pressure and high pressure thermal, pulsed electric fields, ultrasonic and megasonic and microwave processing, and he has published more than 90 technical papers, edited 4 books, presented over 140 oral and poster presentations, and has 5 published patents in this area. Dr Knoerzer is past chair of the International Division of the Institute of Food Technologists (USA) and he serves as section editor of Elsevier's Food Science Reference Module.

Helmar Schubert studied Mechanical and Chemical Engineering at the Technical University of Berlin and the University of Karlsruhe (Germany) and received his MSc in Chemical Engineering, PhD, and Habilitation from the University of Karlsruhe in 1966, 1972, and 1981, respectively. He was Director and Professor at the German Federal Research Centre for Nutrition, Karlsruhe (1977–1991) and since 1986 has been Professor and head of the Institute of Food Process Engineering at the University of Karlsruhe. He has been Professor Emeritus since

2004. His main research interests are emulsification, powder technology, microwave heating, and water processing, and he has published more than 350 technical papers and 3 books on these topics. Professor Schubert has been elected a member of the International Academy of Food Science and Technology, the Berlin-Brandenburg Academy of Sciences and Humanities (formerly Prussian Academy of Sciences), and the German National Academy of Science and Engineering.

Preface

Over ten years ago, in the first edition of this book published in 2005, we reported some remarkable market data of microwave ovens and industrial applications. Concerning these numbers, what has changed over the last decade?

In the early years of the new millennium, the total number of sales of microwave ovens in the United States was at a more or less constant level of approximately 10–15 million per year. The corresponding number in Europe was in the same range. A further increase in these developed countries was not expected, as microwave ovens were commonplace in nearly every home (penetration into private households was assumed to be at approximately 90%).

On the other hand, the development of emerging markets, such as China and Brazil, led to a significant increase of the total global sales of up to 70 to 75 millions (Osepchuk, 2013). Therefore, it is estimated that the number of microwave ovens globally is peaking at an incredible order of 1 billion (Osepchuk, 2013).

Using this estimation, and updating the calculation published in the chapter in the first edition (average power of an oven 0.8 kW), the power of all installed household microwaves globally reaches a value of 800 GW.

Compared to these numbers, the number of microwave systems in the food industry is significantly lower, estimated to be less than 1000 in 2001 (Datta, 2001). Suggesting an average power of 100 kW (Datta, 2001) for an industrial microwave installation, the total installed power of all microwave systems in the food industry was approximately 100 MW in 2001, being almost four orders of magnitude lower than the installed power of household microwave ovens. Unfortunately, a more recent estimate on the number of installed microwave systems in the food industry could not be found in the literature. However, from interactions with the food industry on different continents, it is the editors' belief that after some stagnation, the progress in microwave control and better understanding of the fundamentals and interactions with food products, has led to a strong increase in the number of industrial microwave installation, possibly by a factor of 5 to 10; which means that the installed power of all microwave systems in the food industry will still be more than a factor of 1000 lower than the installed power of all household microwave ovens.

Assuming the microwave in a household is operating for approximately 15 min every day, whereas an industrial installation may run in three shifts 24/7, possibly with downtime for repair and maintenance etc., ~15 h a day, this corresponds to a daily energy consumption of household microwaves of "only" 200 GWh, and of 1000 industrial systems ~2 GWh per day; i.e., a two order of magnitude difference.

Estimating the number of installations in the food industry being in the order of 10,000 now, and provided the earlier assumptions regarding time of use in both household and industry, we still end up one order of magnitude lower in the food processing industry compared to energy consumption in the domestic use. Furthermore, it is the editors' belief, that mainly due to financial reasons, probably still only less than 1% of food industrial heat treatments is done by the help of microwaves (and the technology can, therefore, still be regarded as a "niche technology").

Nonetheless, microwave processing still becomes more and more attractive, but also challenging for food engineers, due to its unique potential for volumetric (instead of surface) heating.

The fundamental part of this book deals with basic principles of microwaves, dielectric properties, and specialized measuring approaches. This is followed by a section on applications, classified in unit operations as (re)heating, blanching, baking, frying, tempering, microbial inactivation, and dehydration, emphasizing the special advantages of the microwave technology over conventional or alternative technologies without ignoring the drawbacks.

More general chapters treat modeling/simulation and optimization techniques and nutritional and sensorial quality of microwave processed foods. Furthermore, new developments in microwave processing and technology, microwave control, and food and packaging developments for household and gastronomy microwave ovens are presented.

The overarching objective of this book is to give an overview on up-to-date microwave applications and developments in food processing and to become a valuable resource for academics and students of any discipline related to food, and for the food industry alike.

References

Datta, A. K., & Anantheswaran, R. (2001). *Handbook of microwave technology for food applications*. Marcel Dekker.

Osepchuk, J. M. (2013). *Revision of "microwave ovens: An overview: 2013"*. Website of IMPI <http://impi.org/microwave-ovens-an-overview-2013/> Visited 22.10.15.

Woodhead publishing series in food science, technology and nutrition

1. **Chilled foods: A comprehensive guide**
   Edited by C. Dennis and M. Stringer
2. **Yoghurt: Science and technology**
   A. Y. Tamime and R. K. Robinson
3. **Food processing technology: Principles and practice**
   P. J. Fellows
4. **Bender's dictionary of nutrition and food technology Sixth Edition**
   D. A. Bender
5. **Determination of veterinary residues in food**
   Edited by N. T. Crosby
6. **Food contaminants: Sources and surveillance**
   Edited by C. Creaser and R. Purchase
7. **Nitrates and nitrites in food and water**
   Edited by M. J. Hill
8. **Pesticide chemistry and bioscience: The food-environment challenge**
   Edited by G. T. Brooks and T. Roberts
9. **Pesticides: Developments, impacts and controls**
   Edited by G. A. Best and A. D. Ruthven
10. **Dietary fibre: Chemical and biological aspects**
    Edited by D. A. T. Southgate, K. W. Waldron, I. T. Johnson and G. R. Fenwick
11. **Vitamins and minerals in health and nutrition**
    M. Tolonen
12. **Technology of biscuits, crackers and cookies Second Edition**
    D. Manley
13. **Instrumentation and sensors for the food industry**
    Edited by E. Kress-Rogers

14. **Food and cancer prevention: Chemical and biological aspects**
Edited by K. W. Waldron, I. T. Johnson and G. R. Fenwick
15. **Food colloids: Proteins, lipids and polysaccharides**
Edited by E. Dickinson and B. Bergenstahl
16. **Food emulsions and foams**
Edited by E. Dickinson
17. **Maillard reactions in chemistry, food and health**
Edited by T. P. Labuza, V. Monnier, J. Baynes and J. O'Brien
18. **The Maillard reaction in foods and medicine**
Edited by J. O'Brien, H. E. Nursten, M. J. Crabbe and J. M. Ames
19. **Encapsulation and controlled release**
Edited by D. R. Karsa and R. A. Stephenson
20. **Flavours and fragrances**
Edited by A. D. Swift
21. **Feta and related cheeses**
Edited by A. Y. Tamime and R. K. Robinson
22. **Biochemistry of milk products**
Edited by A. T. Andrews and J. R. Varley
23. **Physical properties of foods and food processing systems**
M. J. Lewis
24. **Food irradiation: A reference guide**
V. M. Wilkinson and G. Gould
25. **Kent's technology of cereals: An introduction for students of food science and agriculture Fourth Edition**
N. L. Kent and A. D. Evers
26. **Biosensors for food analysis**
Edited by A. O. Scott
27. **Separation processes in the food and biotechnology industries: Principles and applications**
Edited by A. S. Grandison and M. J. Lewis
28. **Handbook of indices of food quality and authenticity**
R. S. Singhal, P. K. Kulkarni and D. V. Rege
29. **Principles and practices for the safe processing of foods**
D. A. Shapton and N. F. Shapton
30. **Biscuit, cookie and cracker manufacturing manuals Volume 1: Ingredients**
D. Manley

31. **Biscuit, cookie and cracker manufacturing manuals Volume 2: Biscuit doughs**
    D. Manley
32. **Biscuit, cookie and cracker manufacturing manuals Volume 3: Biscuit dough piece forming**
    D. Manley
33. **Biscuit, cookie and cracker manufacturing manuals Volume 4: Baking and cooling of biscuits**
    D. Manley
34. **Biscuit, cookie and cracker manufacturing manuals Volume 5: Secondary processing in biscuit manufacturing**
    D. Manley
35. **Biscuit, cookie and cracker manufacturing manuals Volume 6: Biscuit packaging and storage**
    D. Manley
36. **Practical dehydration Second Edition**
    M. Greensmith
37. **Lawrie's meat science Sixth Edition**
    R. A. Lawrie
38. **Yoghurt: Science and technology Second Edition**
    A. Y. Tamime and R. K. Robinson
39. **New ingredients in food processing: Biochemistry and agriculture**
    G. Linden and D. Lorient
40. **Benders' dictionary of nutrition and food technology Seventh Edition**
    D. A. Bender and A. E. Bender
41. **Technology of biscuits, crackers and cookies Third Edition**
    D. Manley
42. **Food processing technology: Principles and practice Second Edition**
    P. J. Fellows
43. **Managing frozen foods**
    Edited by C. J. Kennedy
44. **Handbook of hydrocolloids**
    Edited by G. O. Phillips and P. A. Williams
45. **Food labeling**
    Edited by J. R. Blanchfield
46. **Cereal biotechnology**
    Edited by P. C. Morris and J. H. Bryce

47. **Food intolerance and the food industry**
Edited by T. Dean

48. **The stability and shelf-life of food**
Edited by D. Kilcast and P. Subramaniam

49. **Functional foods: Concept to product**
Edited by G. R. Gibson and C. M. Williams

50. **Chilled foods: A comprehensive guide Second Edition**
Edited by M. Stringer and C. Dennis

51. **HACCP in the meat industry**
Edited by M. Brown

52. **Biscuit, cracker and cookie recipes for the food industry**
D. Manley

53. **Cereals processing technology**
Edited by G. Owens

54. **Baking problems solved**
S. P. Cauvain and L. S. Young

55. **Thermal technologies in food processing**
Edited by P. Richardson

56. **Frying: Improving quality**
Edited by J. B. Rossell

57. **Food chemical safety Volume 1: Contaminants**
Edited by D. Watson

58. **Making the most of HACCP: Learning from others' experience**
Edited by T. Mayes and S. Mortimore

59. **Food process modeling**
Edited by L. M. M. Tijskens, M. L. A. T. M. Hertog and B. M. Nicolaï

60. **EU food law: A practical guide**
Edited by K. Goodburn

61. **Extrusion cooking: Technologies and applications**
Edited by R. Guy

62. **Auditing in the food industry: From safety and quality to environmental and other audits**
Edited by M. Dillon and C. Griffith

63. **Handbook of herbs and spices Volume 1**
Edited by K. V. Peter

64. **Food product development: Maximising success**
M. Earle, R. Earle and A. Anderson
65. **Instrumentation and sensors for the food industry Second Edition**
Edited by E. Kress-Rogers and C. J. B. Brimelow
66. **Food chemical safety Volume 2: Additives**
Edited by D. Watson
67. **Fruit and vegetable biotechnology**
Edited by V. Valpuesta
68. **Foodborne pathogens: Hazards, risk analysis and control**
Edited by C. de W. Blackburn and P. J. McClure
69. **Meat refrigeration**
S. J. James and C. James
70. **Lockhart and Wiseman's crop husbandry Eighth Edition**
H. J. S. Finch, A. M. Samuel and G. P. F. Lane
71. **Safety and quality issues in fish processing**
Edited by H. A. Bremner
72. **Minimal processing technologies in the food industries**
Edited by T. Ohlsson and N. Bengtsson
73. **Fruit and vegetable processing: Improving quality**
Edited by W. Jongen
74. **The nutrition handbook for food processors**
Edited by C. J. K. Henry and C. Chapman
75. **Colour in food: Improving quality**
Edited by D. MacDougall
76. **Meat processing: Improving quality**
Edited by J. P. Kerry, J. F. Kerry and D. A. Ledward
77. **Microbiological risk assessment in food processing**
Edited by M. Brown and M. Stringer
78. **Performance functional foods**
Edited by D. Watson
79. **Functional dairy products Volume 1**
Edited by T. Mattila-Sandholm and M. Saarela
80. **Taints and off-flavours in foods**
Edited by B. Baigrie
81. **Yeasts in food**
Edited by T. Boekhout and V. Robert

82. **Phytochemical functional foods**
Edited by I. T. Johnson and G. Williamson

83. **Novel food packaging techniques**
Edited by R. Ahvenainen

84. **Detecting pathogens in food**
Edited by T. A. McMeekin

85. **Natural antimicrobials for the minimal processing of foods**
Edited by S. Roller

86. **Texture in food Volume 1: Semi-solid foods**
Edited by B. M. McKenna

87. **Dairy processing: Improving quality**
Edited by G. Smit

88. **Hygiene in food processing: Principles and practice**
Edited by H. L. M. Lelieveld, M. A. Mostert, B. White and J. Holah

89. **Rapid and on-line instrumentation for food quality assurance**
Edited by I. Tothill

90. **Sausage manufacture: Principles and practice**
E. Essien

91. **Environmentally-friendly food processing**
Edited by B. Mattsson and U. Sonesson

92. **Bread making: Improving quality**
Edited by S. P. Cauvain

93. **Food preservation techniques**
Edited by P. Zeuthen and L. Bøgh-Sørensen

94. **Food authenticity and traceability**
Edited by M. Lees

95. **Analytical methods for food additives**
R. Wood, L. Foster, A. Damant and P. Key

96. **Handbook of herbs and spices Volume 2**
Edited by K. V. Peter

97. **Texture in food Volume 2: Solid foods**
Edited by D. Kilcast

98. **Proteins in food processing**
Edited by R. Yada

99. **Detecting foreign bodies in food**
Edited by M. Edwards

100. Understanding and measuring the shelf-life of food
Edited by R. Steele

101. Poultry meat processing and quality
Edited by G. Mead

102. Functional foods, ageing and degenerative disease
Edited by C. Remacle and B. Reusens

103. Mycotoxins in food: Detection and control
Edited by N. Magan and M. Olsen

104. Improving the thermal processing of foods
Edited by P. Richardson

105. Pesticide, veterinary and other residues in food
Edited by D. Watson

106. Starch in food: Structure, functions and applications
Edited by A.-C. Eliasson

107. Functional foods, cardiovascular disease and diabetes
Edited by A. Arnoldi

108. Brewing: Science and practice
D. E. Briggs, P. A. Brookes, R. Stevens and C. A. Boulton

109. Using cereal science and technology for the benefit of consumers: Proceedings of the 12P^{thP} International ICC Cereal and Bread Congress, 24 – 26P^{thP} May, 2004, Harrogate, UK
Edited by S. P. Cauvain, L. S. Young and S. Salmon

110. Improving the safety of fresh meat
Edited by J. Sofos

111. Understanding pathogen behaviour: Virulence, stress response and resistance
Edited by M. Griffiths

112. The microwave processing of foods
Edited by H. Schubert and M. Regier

113. Food safety control in the poultry industry
Edited by G. Mead

114. Improving the safety of fresh fruit and vegetables
Edited by W. Jongen

115. Food, diet and obesity
Edited by D. Mela

116. Handbook of hygiene control in the food industry
Edited by H. L. M. Lelieveld, M. A. Mostert and J. Holah

117. **Detecting allergens in food**
Edited by S. Koppelman and S. Hefle

118. **Improving the fat content of foods**
Edited by C. Williams and J. Buttriss

119. **Improving traceability in food processing and distribution**
Edited by I. Smith and A. Furness

120. **Flavour in food**
Edited by A. Voilley and P. Etievant

121. **The Chorleywood bread process**
S. P. Cauvain and L. S. Young

122. **Food spoilage microorganisms**
Edited by C. de W. Blackburn

123. **Emerging foodborne pathogens**
Edited by Y. Motarjemi and M. Adams

124. **Benders' dictionary of nutrition and food technology Eighth Edition**
D. A. Bender

125. **Optimising sweet taste in foods**
Edited by W. J. Spillane

126. **Brewing: New technologies**
Edited by C. Bamforth

127. **Handbook of herbs and spices Volume 3**
Edited by K. V. Peter

128. **Lawrie's meat science Seventh Edition**
R. A. Lawrie in collaboration with D. A. Ledward

129. **Modifying lipids for use in food**
Edited by F. Gunstone

130. **Meat products handbook: Practical science and technology**
G. Feiner

131. **Food consumption and disease risk: Consumer–pathogen interactions**
Edited by M. Potter

132. **Acrylamide and other hazardous compounds in heat-treated foods**
Edited by K. Skog and J. Alexander

133. **Managing allergens in food**
Edited by C. Mills, H. Wichers and K. Hoffman-Sommergruber

134. **Microbiological analysis of red meat, poultry and eggs**
Edited by G. Mead

135. **Maximising the value of marine by-products**
Edited by F. Shahidi

136. **Chemical migration and food contact materials**
Edited by K. Barnes, R. Sinclair and D. Watson

137. **Understanding consumers of food products**
Edited by L. Frewer and H. van Trijp

138. **Reducing salt in foods: Practical strategies**
Edited by D. Kilcast and F. Angus

139. **Modelling microorganisms in food**
Edited by S. Brul, S. Van Gerwen and M. Zwietering

140. **Tamime and Robinson's Yoghurt: Science and technology Third Edition**
A. Y. Tamime and R. K. Robinson

141. **Handbook of waste management and co-product recovery in food processing Volume 1**
Edited by K. W. Waldron

142. **Improving the flavour of cheese**
Edited by B. Weimer

143. **Novel food ingredients for weight control**
Edited by C. J. K. Henry

144. **Consumer-led food product development**
Edited by H. MacFie

145. **Functional dairy products Volume 2**
Edited by M. Saarela

146. **Modifying flavour in food**
Edited by A. J. Taylor and J. Hort

147. **Cheese problems solved**
Edited by P. L. H. McSweeney

148. **Handbook of organic food safety and quality**
Edited by J. Cooper, C. Leifert and U. Niggli

149. **Understanding and controlling the microstructure of complex foods**
Edited by D. J. McClements

150. **Novel enzyme technology for food applications**
Edited by R. Rastall

151. **Food preservation by pulsed electric fields: From research to application**
Edited by H. L. M. Lelieveld and S. W. H. de Haan

152. **Technology of functional cereal products**
Edited by B. R. Hamaker

153. **Case studies in food product development**
Edited by M. Earle and R. Earle

154. **Delivery and controlled release of bioactives in foods and nutraceuticals**
Edited by N. Garti

155. **Fruit and vegetable flavour: Recent advances and future prospects**
Edited by B. Brückner and S. G. Wyllie

156. **Food fortification and supplementation: Technological, safety and regulatory aspects**
Edited by P. Berry Ottaway

157. **Improving the health-promoting properties of fruit and vegetable products**
Edited by F. A. Tomás-Barberán and M. I. Gil

158. **Improving seafood products for the consumer**
Edited by T. Børresen

159. **In-pack processed foods: Improving quality**
Edited by P. Richardson

160. **Handbook of water and energy management in food processing**
Edited by J. Klemeš, R.. Smith and J.-K. Kim

161. **Environmentally compatible food packaging**
Edited by E. Chiellini

162. **Improving farmed fish quality and safety**
Edited by Ø. Lie

163. **Carbohydrate-active enzymes**
Edited by K.-H. Park

164. **Chilled foods: A comprehensive guide Third Edition**
Edited by M. Brown

165. **Food for the ageing population**
Edited by M. M. Raats, C. P. G. M. de Groot and W. A Van Staveren

166. **Improving the sensory and nutritional quality of fresh meat**
Edited by J. P. Kerry and D. A. Ledward

167. **Shellfish safety and quality**
Edited by S. E. Shumway and G. E. Rodrick

168. **Functional and speciality beverage technology**
Edited by P. Paquin

169. **Functional foods: Principles and technology**
M. Guo

170. **Endocrine-disrupting chemicals in food**
Edited by I. Shaw

171. **Meals in science and practice: Interdisciplinary research and business applications**
Edited by H. L. Meiselman

172. **Food constituents and oral health: Current status and future prospects**
Edited by M. Wilson

173. **Handbook of hydrocolloids Second Edition**
Edited by G. O. Phillips and P. A. Williams

174. **Food processing technology: Principles and practice Third Edition**
P. J. Fellows

175. **Science and technology of enrobed and filled chocolate, confectionery and bakery products**
Edited by G. Talbot

176. **Foodborne pathogens: Hazards, risk analysis and control Second Edition**
Edited by C. de W. Blackburn and P. J. McClure

177. **Designing functional foods: Measuring and controlling food structure breakdown and absorption**
Edited by D. J. McClements and E. A. Decker

178. **New technologies in aquaculture: Improving production efficiency, quality and environmental management**
Edited by G. Burnell and G. Allan

179. **More baking problems solved**
S. P. Cauvain and L. S. Young

180. **Soft drink and fruit juice problems solved**
P. Ashurst and R. Hargitt

181. **Biofilms in the food and beverage industries**
Edited by P. M. Fratamico, B. A. Annous and N. W. Gunther

182. **Dairy-derived ingredients: Food and neutraceutical uses**
Edited by M. Corredig

183. **Handbook of waste management and co-product recovery in food processing Volume 2**
Edited by K. W. Waldron

184. **Innovations in food labeling**
Edited by J. Albert

185. **Delivering performance in food supply chains**
Edited by C. Mena and G. Stevens

186. **Chemical deterioration and physical instability of food and beverages**
Edited by L. H. Skibsted, J. Risbo and M. L. Andersen

187. **Managing wine quality Volume 1: Viticulture and wine quality**
Edited by A. G. Reynolds

188. **Improving the safety and quality of milk Volume 1: Milk production and processing**
Edited by M. Griffiths

189. **Improving the safety and quality of milk Volume 2: Improving quality in milk products**
Edited by M. Griffiths

190. **Cereal grains: Assessing and managing quality**
Edited by C. Wrigley and I. Batey

191. **Sensory analysis for food and beverage quality control: A practical guide**
Edited by D. Kilcast

192. **Managing wine quality Volume 2: Oenology and wine quality**
Edited by A. G. Reynolds

193. **Winemaking problems solved**
Edited by C. E. Butzke

194. **Environmental assessment and management in the food industry**
Edited by U. Sonesson, J. Berlin and F. Ziegler

195. **Consumer-driven innovation in food and personal care products**
Edited by S. R. Jaeger and H. MacFie

196. **Tracing pathogens in the food chain**
Edited by S. Brul, P. M. Fratamico and T. A. McMeekin

197. **Case studies in novel food processing technologies: Innovations in processing, packaging, and predictive modeling**
Edited by C. J. Doona, K. Kustin and F. E. Feeherry

198. **Freeze-drying of pharmaceutical and food products**
T.-C. Hua, B.-L. Liu and H. Zhang

199. **Oxidation in foods and beverages and antioxidant applications Volume 1: Understanding mechanisms of oxidation and antioxidant activity**
Edited by E. A. Decker, R. J. Elias and D. J. McClements

200. **Oxidation in foods and beverages and antioxidant applications Volume 2: Management in different industry sectors**
Edited by E. A. Decker, R. J. Elias and D. J. McClements

201. **Protective cultures, antimicrobial metabolites and bacteriophages for food and beverage biopreservation**
Edited by C. Lacroix

202. **Separation, extraction and concentration processes in the food, beverage and nutraceutical industries**
Edited by S. S. H. Rizvi

203. **Determining mycotoxins and mycotoxigenic fungi in food and feed**
Edited by S. De Saeger

204. **Developing children's food products**
Edited by D. Kilcast and F. Angus

205. **Functional foods: Concept to product Second Edition**
Edited by M. Saarela

206. **Postharvest biology and technology of tropical and subtropical fruits Volume 1: Fundamental issues**
Edited by E. M. Yahia

207. **Postharvest biology and technology of tropical and subtropical fruits Volume 2: Açai to citrus**
Edited by E. M. Yahia

208. **Postharvest biology and technology of tropical and subtropical fruits Volume 3: Cocona to mango**
Edited by E. M. Yahia

209. **Postharvest biology and technology of tropical and subtropical fruits Volume 4: Mangosteen to white sapote**
Edited by E. M. Yahia

210. **Food and beverage stability and shelf life**
Edited by D. Kilcast and P. Subramaniam

211. **Processed Meats: Improving safety, nutrition and quality**
Edited by J. P. Kerry and J. F. Kerry

212. **Food chain integrity: A holistic approach to food traceability, safety, quality and authenticity**
Edited by J. Hoorfar, K. Jordan, F. Butler and R. Prugger

213. **Improving the safety and quality of eggs and egg products Volume 1**
Edited by Y. Nys, M. Bain and F. Van Immerseel

214. **Improving the safety and quality of eggs and egg products Volume 2**
Edited by F. Van Immerseel, Y. Nys and M. Bain

215. **Animal feed contamination: Effects on livestock and food safety**
Edited by J. Fink-Gremmels

216. **Hygienic design of food factories**
Edited by J. Holah and H. L. M. Lelieveld

217. **Manley's technology of biscuits, crackers and cookies Fourth Edition**
Edited by D. Manley

218. **Nanotechnology in the food, beverage and nutraceutical industries**
Edited by Q. Huang

219. **Rice quality: A guide to rice properties and analysis**
K. R. Bhattacharya

220. **Advances in meat, poultry and seafood packaging**
Edited by J. P. Kerry

221. **Reducing saturated fats in foods**
Edited by G. Talbot

222. **Handbook of food proteins**
Edited by G. O. Phillips and P. A. Williams

223. **Lifetime nutritional influences on cognition, behaviour and psychiatric illness**
Edited by D. Benton

224. **Food machinery for the production of cereal foods, snack foods and confectionery**
L.-M. Cheng

225. **Alcoholic beverages: Sensory evaluation and consumer research**
Edited by J. Piggott

226. **Extrusion problems solved: Food, pet food and feed**
M. N. Riaz and G. J. Rokey

227. **Handbook of herbs and spices Second Edition Volume 1**
Edited by K. V. Peter

228. **Handbook of herbs and spices Second Edition Volume 2**
Edited by K. V. Peter

229. **Breadmaking: Improving quality Second Edition**
Edited by S. P. Cauvain

230. **Emerging food packaging technologies: Principles and practice**
Edited by K. L. Yam and D. S. Lee

231. **Infectious disease in aquaculture: Prevention and control**
Edited by B. Austin

232. **Diet, immunity and inflammation**
Edited by P. C. Calder and P. Yaqoob

233. **Natural food additives, ingredients and flavourings**
Edited by D. Baines and R. Seal

234. **Microbial decontamination in the food industry: Novel methods and applications**
Edited by A. Demirci and M.O. Ngadi

235. **Chemical contaminants and residues in foods**
Edited by D. Schrenk

236. **Robotics and automation in the food industry: Current and future technologies**
Edited by D. G. Caldwell

237. **Fibre-rich and wholegrain foods: Improving quality**
Edited by J. A. Delcour and K. Poutanen

238. **Computer vision technology in the food and beverage industries**
Edited by D.-W. Sun

239. **Encapsulation technologies and delivery systems for food ingredients and nutraceuticals**
Edited by N. Garti and D. J. McClements

240. **Case studies in food safety and authenticity**
Edited by J. Hoorfar

241. **Heat treatment for insect control: Developments and applications**
D. Hammond

242. **Advances in aquaculture hatchery technology**
Edited by G. Allan and G. Burnell

243. **Open innovation in the food and beverage industry**
Edited by M. Garcia Martinez

244. **Trends in packaging of food, beverages and other fast-moving consumer goods (FMCG)**
Edited by N. Farmer

245. **New analytical approaches for verifying the origin of food**
Edited by P. Brereton

246. **Microbial production of food ingredients, enzymes and nutraceuticals**
Edited by B. McNeil, D. Archer, I. Giavasis and L. Harvey

247. **Persistent organic pollutants and toxic metals in foods**
Edited by M. Rose and A. Fernandes

248. **Cereal grains for the food and beverage industries**
E. Arendt and E. Zannini

249. Viruses in food and water: Risks, surveillance and control
Edited by N. Cook

250. Improving the safety and quality of nuts
Edited by L. J. Harris

251. Metabolomics in food and nutrition
Edited by B. C. Weimer and C. Slupsky

252. Food enrichment with omega-3 fatty acids
Edited by C. Jacobsen, N. S. Nielsen, A. F. Horn and A.-D. M. Sørensen

253. Instrumental assessment of food sensory quality: A practical guide
Edited by D. Kilcast

254. Food microstructures: Microscopy, measurement and modeling
Edited by V. J. Morris and K. Groves

255. Handbook of food powders: Processes and properties
Edited by B. R. Bhandari, N. Bansal, M. Zhang and P. Schuck

256. Functional ingredients from algae for foods and nutraceuticals
Edited by H. Domínguez

257. Satiation, satiety and the control of food intake: Theory and practice
Edited by J. E. Blundell and F. Bellisle

258. Hygiene in food processing: Principles and practice Second Edition
Edited by H. L. M. Lelieveld, J. Holah and D. Napper

259. Advances in microbial food safety Volume 1
Edited by J. Sofos

260. Global safety of fresh produce: A handbook of best practice, innovative commercial solutions and case studies
Edited by J. Hoorfar

261. Human milk biochemistry and infant formula manufacturing technology
Edited by M. Guo

262. High throughput screening for food safety assessment: Biosensor technologies, hyperspectral imaging and practical applications
Edited by A. K. Bhunia, M. S. Kim and C. R. Taitt

263. Foods, nutrients and food ingredients with authorised EU health claims: Volume 1
Edited by M. J. Sadler

264. Handbook of food allergen detection and control
Edited by S. Flanagan

265. Advances in fermented foods and beverages: Improving quality, technologies and health benefits
Edited by W. Holzapfel

266. Metabolomics as a tool in nutrition research
Edited by J.-L. Sébédio and L. Brennan

267. Dietary supplements: Safety, efficacy and quality
Edited by K. Berginc and S. Kreft

268. Grapevine breeding programs for the wine industry
Edited by A. G. Reynolds

269. Handbook of antimicrobials for food safety and quality
Edited by T. M. Taylor

270. Managing and preventing obesity: Behavioural factors and dietary interventions
Edited by T. P. Gill

271. Electron beam pasteurization and complementary food processing technologies
Edited by S. D. Pillai and S. Shayanfar

272. Advances in food and beverage labelling: Information and regulations
Edited by P. Berryman

273. Flavour development, analysis and perception in food and beverages
Edited by J. K. Parker, S. Elmore and L. Methven

274. Rapid sensory profiling techniques and related methods: Applications in new product development and consumer research
Edited by J. Delarue, J. B. Lawlor and M. Rogeaux

275. Advances in microbial food safety: Volume 2
Edited by J. Sofos

276. Handbook of antioxidants for food preservation
Edited by F. Shahidi

277. Lockhart and Wiseman's crop husbandry including grassland: Ninth Edition
H. J. S. Finch, A. M. Samuel and G. P. F. Lane

278. Global legislation for food contact materials
Edited by J. S. Baughan

279. Colour additives for food and beverages
Edited by M. Scotter

280. A complete course in canning and related processes 14th Edition: Volume 1
Revised by S. Featherstone

281. **A complete course in canning and related processes 14th Edition: Volume 2**
Revised by S. Featherstone

282. **A complete course in canning and related processes 14th Edition: Volume 3**
Revised by S. Featherstone

283. **Modifying food texture: Volume 1: Novel ingredients and processing techniques**
Edited by J. Chen and A. Rosenthal

284. **Modifying food texture: Volume 2: Sensory analysis, consumer requirements and preferences**
Edited by J. Chen and A. Rosenthal

285. **Modeling food processing operations**
Edited by S. Bakalis, K. Knoerzer and P. J. Fryer

286. **Foods, nutrients and food ingredients with authorised EU health claims Volume 2**
Edited by M. J. Sadler

287. **Feed and feeding practices in aquaculture**
Edited by D. Allen Davis

288. **Foodborne parasites in the food supply web: Occurrence and control**
Edited by A. Gajadhar

289. **Brewing microbiology: design and technology applications for spoilage management, sensory quality and waste valorisation**
Edited by A. E. Hill

290. **Specialty oils and fats in food and nutrition: Properties, processing and applications**
Edited by G. Talbot

291. **Improving and tailoring enzymes for food quality and functionality**
Edited by R. Yada

292. **Emerging Technologies for Promoting Food Security: Overcoming the World Food Crisis**
Edited by C. Madramootoo

293. **Innovation and Future Trends in Food Manufacturing and Supply Chain Technologies**
Edited by C. E. Leadley

294. **Functional Dietary Lipids: Food Formulation, Consumer Issues and Innovation for Health**
Edited by T. Sanders

295. Handbook on Natural Pigments in Food and Beverages: Industrial Applications for Improving Color
Edited by R. Carle and R. M. Schweiggert

296. Integrating the Packaging and Product Experience in Food and Beverages: A Road-Map to Consumer Satisfaction
Edited by P. Burgess

297. The Stability and Shelf Life of Food Second Edition
Edited by Persis Subramaniam and Peter Wareing

298. Multisensory Flavor Perception: From Fundamental Neuroscience through to the Marketplace
Edited by Betina Piqueras-Fiszman and Charles Spence

299. Flavor: From food to behaviors, wellbeing and health
Edited by Andrée Voilley, Christian Salles, Elisabeth Guichard and Patrick Etiévant

300. Developing Food Products for Consumers with Specific Dietary Needs
Edited by Wayne Morley and Steve Osborn

301. Advances in Food Traceability Techniques and Technologies: Improving Quality Throughout the Food Chain
Edited by Montserrat Espiñeira and J. Francisco Santaclara

302. Innovative Food Processing Technologies: Extraction, Separation, Component Modification and Process Intensification
Edited by Kai Knoerzer, Pablo Juliano, and Geoffrey Smithers

303. Steamed Breads: Ingredients, Process and Quality
Sidi Huang and *Diane Miskelly*

304. Handbook of Hygiene Control in the Food Industry, Second Edition
Edited by Huub Lelieveld, Domagoj Gabrić, and John Holah

305. Handbook for Sensory and Consumer-Driven New Product Development: Innovative Technologies for the Food and Beverage Industry
Maurice O'Sullivan

306. Early Nutrition and Long-Term Health: Mechanisms, Consequences and Opportunities
Edited by Jose M. Saavedra and Anne M. Dattilo

307. Baking Problems Solved, Second Edition
S. P. Cauvain

308. Non-Equilibrium States and Glass Transitions in Food
Edited by Bhesh Bhandari and Yrjö Roos

309. New Aspects of Meat Quality: From Genes to Ethics
Edited by Peter Purslow

310. **Advances in Food Authenticity Testing**
Edited by Gerald Downey

311. **Food for the Ageing Population, Second Edition**
Edited by Monique Raats, Lisette De Groot and Dieneke van Asselt

312. **Microbiological Quality of Food**
Edited by Antonio Bevilacqua, Maria Rosaria Corbo and Milena Sinigaglia

313. **Swainson's Handbook of Technical and Quality Management for the Food Manufacturing Sector**
Edited by Mark Swainson

314. **Food Protection and Security**
Edited by Shaun Kennedy

315. **Cereal Grains: Assessing and Managing Quality, Second Edition**
Edited by Colin Wrigley, Ian Bate and Diane Miskelly

316. **Advances in Food Rheology and Its Applications**
Edited by Jasim Ahmed, Pawel Ptazsek and Santanu Basu

317. **Improving Food Quality through Advances in the Cold Food Chain**
Edited by Steven James and Christian James

318. **A Practical Guide to Sensory and Consumer Evaluation**
Christopher Findlay

319. **The Microwave Processing of Foods, Second Edition**
Marc Regier, Kai Knoerzer and Helmar Schubert

Introducing microwave-assisted processing of food: fundamentals of the technology

1

M. Regier[1], K. Knoerzer[2] and H. Schubert[3]
[1]Trier University of Applied Sciences, Trier, Germany, [2]CSIRO Agriculture and Food, Melbourne, VIC, Australia, [3]Karlsruhe Institute of Technology (KIT), Karlsruhe, Germany

1.1 Introduction

The following chapter covers the fundamentals of microwaves, the corresponding physical theory, as well as some general remarks and considerations on the setup of microwave applications. First, the definition of frequency and wavelength ranges, and legislative regulations of frequency bands are discussed. Then, the basic equations, i.e., Maxwell's equations and the equations describing the interaction of electromagnetic waves with matter, are introduced. Starting with these basics, the wave equation and some exemplary solutions are derived, and the important concept of the penetration depth and power absorption, which is useful for the estimation of thermal microwave–matter interaction, can be introduced. After covering the general setup and configuration of microwave applications, consisting of microwave sources, waveguides, and applicators, the chapter is completed by useful links to further literature on this topic.

1.2 Physical definitions and legislative regulations

1.2.1 Definition

Microwaves are electromagnetic waves within a frequency band of 300 MHz to 300 GHz. In the electromagnetic spectrum (Fig. 1.1), the microwaves are located between the radio frequency range at lower frequencies and by infrared and visible light at higher frequencies. Thus, microwaves are part of the spectrum covering nonionizing radiation.

Through Eq. (1.1), the frequency f is linked with the speed of light c to a corresponding wavelength λ.

$$c = \lambda \cdot f \tag{1.1}$$

In this case, the speed of light, as well as the wavelength within matter, is dependent on the material. For the speed of light in vacuum ($c_0 \approx 3 \times 10^8$ m/s), the corresponding wavelength of microwaves is in the range between 1 m and 1 mm,

The Microwave Processing of Foods. DOI: http://dx.doi.org/10.1016/B978-0-08-100528-6.00001-2

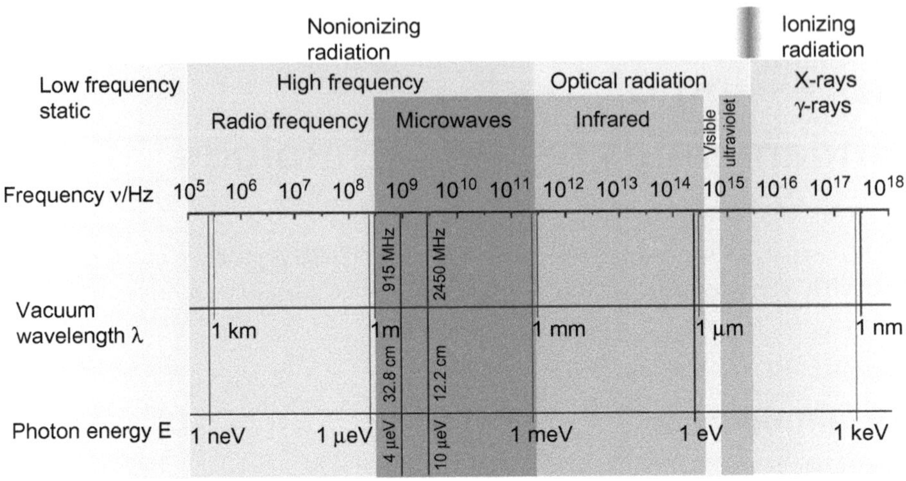

Figure 1.1 Electromagnetic spectrum. Additionally, the two most frequently used microwave frequency bands (at 915 and 2450 MHz) are depicted.

hence, the term microwaves is somewhat misleading. Their name rather refers to their wavelength within the matter, where it can be in the micrometer range, indeed.

1.2.2 Regulations—ISM bands

1.2.2.1 ISM bands

As already shown in Fig. 1.1, the frequency range of microwaves is adjacent to the range of radio frequencies, which is primarily used for television and radio broadcast. However, also the microwave frequency range is used for telecommunication, such as mobile phones, wireless networks, and radar. In order to prevent interference problems, certain frequency bands are reserved for industrial, scientific, and medical (so-called ISM) applications, where a certain radiation level has to be tolerated by other applications (e.g., communication devices). In the range of microwaves, the ISM bands are located at 433, 915, 2450, and 5800 MHz, whereby the first is not commonly used, the second is not generally permitted in continental Europe and Australia. Outside of the permitted frequency range, leakage is very restricted.

Whereas 915 MHz has some considerable advantages for industrial applications, domestic microwave ovens only utilize a frequency of 2450 MHz.

1.2.2.2 Safety regulations

Apart from the regulations concerning interference, there are two further types of safety regulations:

1. the regulation concerning the maximum exposure or absorption of a human, working in a microwave environment;
2. the regulation concerning the maximum emission or leakage of the microwave equipment.

Exposure

The exposure limits for humans is based on the estimation of thermal effects that microwaves can cause in the human body. Particularly sensitive organs, such as the eye, with a limited thermal balancing possibility and/or geometric focusing effects are taken into account.

Thus, the limit for human exposure, which is generally considered to be safe, is at a level of 1 mW/cm^2 body surface in most countries. As is the case for ionizing radiation, also for microwaves it becomes common to express the exposure or absorption by humans through the value of the specific absorption rate (SAR), which is defined as the quotient of the incident power and body weight. For microwaves, the International Commisson on Non-Ionizing Radiation Protection (ICNIRP, 1998; IRPA, 1988) recommends a value for the SAR to be set to 0.4 W/kg.

Emission

The maximum emission of a microwave equipment is limited to a value of 5 mW/cm^2 measured at a distance of 5 cm from the point where the leakage has the maximum level. Thus, the permissible leakage level is higher than the maximum exposure limit. However, the power density of nonfocused radiation, which is generally the case for leakage, decreases proportional to the inversed square of the distance from the source. So a leakage, which just manages to stay in the limits of 5 mW/cm^2 at a distance of 5 cm, is below the maximum exposure limit of 1 mW/cm^2 already at a distance of 11.2 cm.

1.3 Electromagnetic theory

1.3.1 Maxwell's and constitutive equations

As already mentioned above, microwaves are electromagnetic waves, which can be described by Maxwell's Eqs. (1.2–1.5):

$$\nabla \cdot \vec{D} = \rho \tag{1.2}$$

$$\nabla \times \vec{E} = -\frac{\partial \vec{B}}{\partial t} \tag{1.3}$$

$$\nabla \cdot \vec{B} = 0 \tag{1.4}$$

$$\nabla \times \vec{H} = \vec{j} + \frac{\partial \vec{D}}{\partial t} \tag{1.5}$$

The first and third Eqs. (1.2 and 1.4) describe the source of an electric field (ρ) and that no magnetic monopole as source for the magnetic field can exist.

On the other hand, Eqs. (1.3 and 1.5) show the coupling between electric and magnetic fields.

The interaction of electromagnetism with matter is included through the material equations, also referred to as constitutive relations (Eqs. 1.6–1.8), where the permittivity or dielectric constant ε (interaction of nonconducting matter with an electric field $\vec{E}$), the conductivity σ, and the permeability μ (interaction with a magnetic field $\vec{H}$) describe their behavior (see also Chapter 2: Microwave heating and the dielectric properties of foods). The zero indexed values describe the respective values in vacuum, so that ε and μ are relative values.

$$\vec{D} = \varepsilon_0 \varepsilon \cdot \vec{E} \tag{1.6}$$

$$\vec{B} = \mu_0 \mu \cdot \vec{H} \tag{1.7}$$

$$\vec{j} = \sigma \cdot \vec{E} \tag{1.8}$$

In general, all these material parameters can be complex tensors (with directional dependent behavior). In the case of food substances, some simplifications are possible for most practical uses: since food behaves nonmagnetically, the relative permeability can be set to $\mu = 1$ and the permittivity tensor can be reduced to a complex constant with real (ε') and imaginary part (ε''), which may include the conductivity σ (see Chapter 2: Microwave heating and the dielectric properties of foods).

1.3.2 Wave equations and boundary conditions

The Maxwell's equations cover all aspects of electromagnetism. In order to describe the more specific theme of electromagnetic waves, the corresponding wave equations (for the electric or the magnetic field) can be easily derived, starting from Maxwell's equations, with the simplifications of no charge ($\rho = 0$) and no current density ($\vec{j} = 0$). The derivation is shown here only for the electric field; however, it can be easily transferred to the magnetic field:

Applying the curl-operator ($\nabla \times$) on Eq. (1.3) yields Eq. (1.9):

$$\nabla \times \left(\nabla \times \vec{E}\right) = -\nabla \times \frac{\partial \vec{B}}{\partial t} = -\frac{\partial}{\partial t}\left(\nabla \times \vec{B}\right) \tag{1.9}$$

Using the constitutive equation for the magnetic field Eq. (1.7), this can be transformed to Eq. (1.10), assuming the permeability μ to be constant and introducing Eq. (1.5):

$$\nabla \times \left(\nabla \times \vec{E}\right) = -\mu_0 \mu \frac{\partial}{\partial t}\left(\frac{\partial \vec{D}}{\partial t}\right) \tag{1.10}$$

Utilizing the material equation for the electric field Eq. (1.6), first Maxwell's Eq. (1.2) and the vector identity $\nabla\times(\nabla\times\vec{X}) = \nabla(\nabla\cdot\vec{X}) - \Delta\vec{X}$, one gets the following well-known wave Eq. (1.11):

$$\Delta\vec{E} - \mu_0\mu\varepsilon_0\varepsilon\frac{\partial^2\vec{E}}{\partial t^2} = 0 \tag{1.11}$$

The corresponding wave equation for the magnetic component $\vec{B}$ can be similarly derived, yielding the same equation, where $\vec{E}$ is replaced by $\vec{B}$.

Comparing this wave Eq. (1.11) with the standard wave equation, one can infer that here the wave velocity is defined by Eq. (1.12):

$$c = \frac{1}{\sqrt{\mu_0\varepsilon_0\mu\varepsilon}} = \frac{c_0}{\sqrt{\mu\varepsilon}} \tag{1.12}$$

The nature of possible solutions of Eq. (1.11) can be illustrated, considering the case of a so-called linearly polarized plane wave. Linearly polarized means that, for example, the electric field consists of only one component, e.g., in the z-direction E_z. If this component only depends on the one local coordinate, e.g., x (and the time), the wave is called a plane wave. If additionally the material parameters are frequency independent, Eq. (1.11) then reduces to

$$\frac{\partial^2 E_z}{\partial x^2} - \frac{1}{c^2}\frac{\partial^2 E_z}{\partial t^2} = 0 \tag{1.13}$$

It can be shown that all functions of the form $f(kx \pm \omega t)$ solve this equation. Often used as solutions also for the more complex case Eq. (1.11) are time harmonic functions Eq. (1.14):

$$\begin{aligned}\vec{E} &= \vec{E}_0\cdot\cos\left(\vec{k}\vec{x} - \omega t\right)\\ \vec{E} &= \vec{E}_0\cdot\sin\left(\vec{k}\vec{x} - \omega t\right)\\ \vec{E} &= \Re\left[\vec{E}_0\cdot\exp\left\{i\left(\vec{k}\vec{x} - \omega t\right)\right\}\right]\end{aligned} \tag{1.14}$$

Here $\vec{k}$ is the wave vector pointing to the direction of propagation with its absolute value defined as

$$\vec{k}^2 = \frac{\omega^2}{c^2} \tag{1.15}$$

and $\omega = 2\pi\cdot f$, the circular frequency of the wave.

Table 1.1 Dependency between magnetic and electric fields

Transversality	Correlation of electric and magnetic field
$\vec{k} \cdot \vec{E}_0 = 0$ $\vec{k} \cdot \vec{B}_0 = 0$	$\vec{k} \times \vec{E}_0 = \omega \cdot \vec{B}_0$ $\vec{k} \times \vec{B}_0 = -\omega \cdot \mu_0\mu \cdot \varepsilon_0\varepsilon \cdot \vec{E}_0$

It should be noted that the separate wave equations for the electric and the magnetic field cannot completely replace Maxwell's equations. Instead, further conditions, listed in Table 1.1, show the dependency between the magnetic and the electric field, which are not independent of each other.

In this theory, the dispersion (the dependence of the speed of light on the frequency ω in materials) is included. To include absorption within matter, a complex permittivity and with this a complex wave vector has to be introduced.

When additionally a finite conductivity σ in Eq. (1.10) is allowed, so that a current density $\vec{j} = \sigma \cdot \vec{E}$ flows, instead of the simple wave Eq. (1.11), the expanded Eq. (1.11a) has to be used:

$$\Delta\vec{E} - \mu_0\mu\sigma\frac{\partial\vec{E}}{\partial t} - \mu_0\mu\varepsilon_0\varepsilon\frac{\partial^2\vec{E}}{\partial t^2} = 0 \tag{1.11a}$$

Taking time harmonic functions for the electric field as solutions as above, Eq. (1.11a) reduces to:

$$\Delta\vec{E} + \omega^2\mu_0\mu\varepsilon_0\left(\varepsilon - i\frac{\sigma}{\varepsilon_0\omega}\right)\vec{E} = 0 \tag{1.11b}$$

This equation shows that a finite conductivity σ is equivalent to an imaginary term in the permittivity ε.

1.3.3 *Exemplary solutions, the exponentially damped plane wave*

Coming back to an exemplary solution in the case of an absorbing material, where the permittivity ε has an imaginary part $\varepsilon = \varepsilon' - i\varepsilon''_{\text{total}}$

$$\varepsilon''_{\text{total}} = \varepsilon'' + \frac{\sigma}{\varepsilon_0\omega} \tag{1.16}$$

Then the time harmonic plane wave has to be a solution of Eq. (1.11c).

$$\frac{\partial^2 E_z}{\partial x^2} + \omega^2\mu_0\mu\varepsilon_0\left(\varepsilon' - i\varepsilon''_{\text{total}}\right)E_z = 0 \tag{1.11c}$$

Table 1.2 Boundary conditions for electric and magnetic field components

Prerequisite	Boundary condition
No surface charge	Continuity of $D_\perp$
—	Continuity of $B_\perp$
No surface current	Continuity of $H_\parallel$
—	Continuity of $E_\parallel$
Ideally conducting wall (metallic)	$E_\parallel = 0$
Ideally conducting wall (metallic)	$B_\perp = 0$

For the magnetic component of the plane wave H_y (which has to be orthogonal to the electric field E_z), a similar equation can be derived, leading to a general solution with g, h, m, and n constants to satisfy the boundary conditions (see Table 1.2):

$$\begin{aligned} E_z &= g \cdot \exp\{(ik + \kappa)x\} + h \cdot \exp\{-(ik + \kappa)x\} \\ H_y &= m \cdot \exp\{(ik + \kappa)x\} + n \cdot \exp\{-(ik + \kappa)x\} \end{aligned} \tag{1.17}$$

The continuity of $E_\parallel$ (which is one boundary condition of Table 1.2) should be emphasized, as it can explain the often observed effect of edge or corner overheating. Later it will be shown that the power dissipation in a sample volume is proportional to the squared electric field Eq. (1.23). At edges, or especially at corners, the microwaves can not only intrude from two or three directions, respectively, but also electric fields of two or three polarizations have a parallel surface to intrude continuously without amplitude decrease. Therefore, heat generation there will be high.

The solution approach Eq. (1.17) describes an exponentially damped wave, with wave number k and damping constant κ, both dependent on the permittivity ε. The comparison of coefficients yields

$$\omega^2 \mu_0 \mu \varepsilon_0 (\varepsilon' - i\varepsilon''^*) = (\kappa + ik)^2 \tag{1.18}$$

leading to

$$k = \omega \sqrt{\frac{\mu_0 \mu \varepsilon_0 \varepsilon'}{2} \left(\sqrt{1 + \frac{\varepsilon''^{*2}}{\varepsilon'^2}} + 1 \right)} \tag{1.19}$$

and

$$\kappa = \omega \sqrt{\frac{\mu_0 \mu \varepsilon_0 \varepsilon'}{2} \left(\sqrt{1 + \frac{\varepsilon''^{*2}}{\varepsilon'^2}} - 1 \right)}. \tag{1.20}$$

The corresponding electric field penetration depth, the distance in which the electric field is reduced to $1/e$ is defined by Eq. (1.21).

$$\delta_E = \frac{1}{\kappa} = \frac{1}{\omega} \cdot \sqrt{\frac{2}{\mu_0 \mu \varepsilon_0 \varepsilon' \cdot \left(\sqrt{\sqrt{1 + \frac{\varepsilon''^{*2}}{\varepsilon'^2}} - 1}\right)}} \tag{1.21}$$

An important consequence of the frequency dependence of κ is that microwaves of 915 MHz do have an approximately 2.5 times larger penetration depth than waves of 2450 MHz, when similar permittivities at both frequencies are assumed. This larger penetration depth helps to heat larger pieces at an industrial level more homogeneously.

With the assumption of the excitation and the propagation of a plane wave, which satisfies the boundary conditions, first estimations of the field configurations are possible. This yields, for example, the laws of the geometric optics, which are also valid for microwaves, when a typical object size is much larger than the wavelength.

1.3.4 Geometric optics: reflection and refraction

1.3.4.1 Angles of reflection and refraction

Taking again a plane wave, as shown in Fig. 1.2, traveling from a semi-infinite non-absorbing medium I ($n_1 = \sqrt{\varepsilon_1}$) into a semi-infinite absorbing medium II ($n_2 = \sqrt{\varepsilon_2} = n_{2r} + in_{2i}$). Both permeabilities should be $\mu = 1$ and the boundary plane between the two media should be the x-y-plane at $z = 0$. The electric field of the plane wave can be written as (neglecting the explicit writing of the real part):

$$\vec{E} = \vec{E}_0 \cdot \exp[i(\vec{k}\,\vec{x} - \omega t)] \tag{1.22a}$$

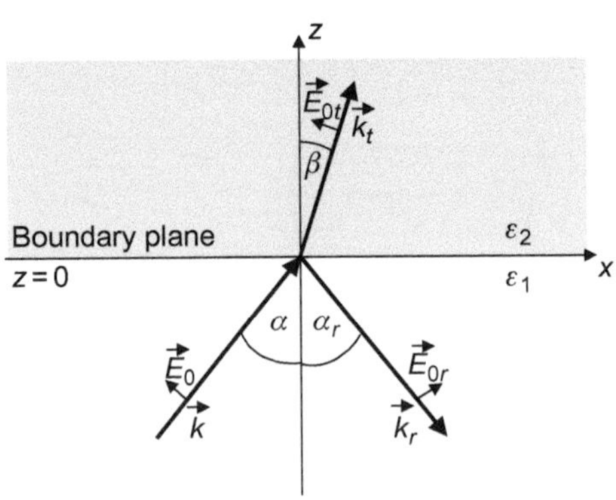

Figure 1.2 Reflection and refraction of a plane wave incident on a plane dielectric boundary.

As per Table 1.1 the corresponding magnetic field is defined by:

$$\vec{B} = \frac{1}{\omega} \cdot \vec{k} \times \vec{E} \tag{1.22b}$$

This wave transports energy in the direction of the wave vector $\vec{k}$ depicted as arrows in Fig. 1.2. Also in this case, the boundary conditions (with no surface charge and current) of Table 1.2 are valid; therefore, a reflected Eq. (1.23) and a refracted (transmitted) wave Eq. (1.24) with the same time dependence have to be present.

$$\begin{aligned} \vec{E}_r &= \vec{E}_{0r} \cdot \exp\left[i\left(\vec{k}_r \vec{x} - \omega t\right)\right] \\ \vec{B}_r &= \frac{1}{\omega} \cdot \vec{k}_r \times \vec{E}_r \end{aligned} \qquad (z < 0) \tag{1.23}$$

$$\begin{aligned} \vec{E}_t &= \vec{E}_{0t} \cdot \exp\left[i\left(\vec{k}_t \vec{x} - \omega t\right)\right] \\ \vec{B}_t &= \frac{1}{\omega} \cdot \vec{k}_t \times \vec{E}_t \end{aligned} \qquad (z > 0) \tag{1.24}$$

The wave vectors obey Eq. (1.25), so that $\vec{k}$ and $\vec{k}_r$ are real but $\vec{k}_t$ is generally complex:

$$\frac{\vec{k}^2}{n_1{}^2} = \frac{\vec{k}_r{}^2}{n_1{}^2} = \frac{\vec{k}_t{}^2}{n_2{}^2} = \frac{\omega^2}{c^2} \tag{1.25}$$

Recapitulating, for $z > 0$, $\vec{E}_t$ is the solution, whereas for $z < 0$ the solution consists of the sum $\vec{E} + \vec{E}_r$. By taking the incident wave with $\vec{E}$ and $\vec{k}$ as starting points, the remaining variables $\vec{E}_r, \vec{E}_t, \vec{k}_r, \vec{k}_t$ can be determined by the boundary conditions of Table 1.2.

At the plane $z = 0$, the local dependency of all waves $\vec{E}, \vec{E}_r, \vec{E}_t$ has to coincide, so that

$$k_x x + k_y y = k_{r,x} x + k_{r,y} y = k_{t,x} x + k_{t,y} \tag{1.26}$$

Without constraining universality, the y-component can be chosen to vanish $k_y = 0$, which yields:

$$k_y = k_{r,y} = k_{t,y} = 0 \tag{1.27}$$

and

$$k_x = k_{r,x} = k_{t,x} \tag{1.28}$$

Eq. (1.27) shows that the incident, the reflected, and the diffracted wave vector are in the same plane (this is the plane depicted in Fig. 1.2). In order to state equations for the angles, shown in Fig. 1.2, they are defined by the following equation (which is even more general, since $\vec{k}_t$, and with it β, may be complex).

$$\begin{aligned} k_x &= k \cdot \sin\alpha \\ k_{t,x} &= k_t \cdot \sin\beta \\ k_{r,x} &= k_r \cdot \sin\alpha_r \end{aligned} \tag{1.29}$$

Eqs. (1.25, 1.28, and 1.29) directly yield the law of reflection Eq. (1.30):

$$\alpha = \alpha_r \tag{1.30}$$

and the law of refraction Eq. (1.31), taking weak damping ($n_{2i} << n_{2r}$) into account:

$$\frac{\sin\beta}{\sin\alpha} = \frac{n_1}{n_2} \overset{n_{2i} << n_{2r}}{\approx} \frac{n_1}{n_{2r}} \tag{1.31}$$

1.3.4.2 Intensities

In order to determine the intensities of the reflected and the transmitted waves, the boundary conditions of Table 1.2 have to be used. Again, all permeabilities are set to $\mu = 1$, so that the following equations are derived:

$$\left[\varepsilon_1 \cdot \left(\vec{E}_0 + \vec{E}_{0r}\right) - \varepsilon_2 \cdot \vec{E}_{0t}\right] \cdot \hat{e}_z = 0 \tag{1.32}$$

$$\left[\vec{k} \times \vec{E}_0 + \vec{k}_r \times \vec{E}_{0r} - \vec{k}_t \times \vec{E}_{0t}\right] \cdot \hat{e}_z = 0 \tag{1.33}$$

$$\left[\vec{E}_0 + \vec{E}_{0r} - \vec{E}_{0t}\right] \times \hat{e}_z = 0 \tag{1.34}$$

$$\left[\vec{k} \times \vec{E}_0 + \vec{k}_r \times \vec{E}_{0r} - \vec{k}_t \times \vec{E}_{0t}\right] \times \hat{e}_z = 0 \tag{1.35}$$

Two orthogonal cases of linear polarizations have to be distinguished, with which any kind of polarization can be formed: The first case, where $\vec{E}_0$ (and with it also the fields of the transmitted and reflected wave) is parallel to the incident plane, $\vec{E}_0 \cdot \hat{e}_y = 0$, should be covered here in detail; for the second polarization, orthogonal to the incident plane $\vec{E}_0 = E_0 \cdot \hat{e}_y$, only the results are presented.

Due to the fact that all wave vectors as well as all electric field vectors are in the incident plane, which is parallel to the z-axis, Eq. (1.33) is trivially fulfilled. With the angles defined in Eq. (1.29), the remaining equations yield:

$$\varepsilon_1 \cdot (E_0 + E_{0r}) \cdot \sin\alpha - \varepsilon_2 \cdot E_{0t} \cdot \sin\beta = 0 \tag{1.36}$$

$$(E_0 - E_{0r}) \cdot \cos\alpha - E_{0t} \cdot \cos\beta = 0 \tag{1.37}$$

$$\sqrt{\varepsilon_1} \cdot (E_0 + E_{0r}) - \sqrt{\varepsilon_2} \cdot E_{0t} = 0 \tag{1.38}$$

Eqs. (1.38 and 1.36) are equivalent, as long as the law of refraction Eq. (1.31) and $n = \sqrt{\varepsilon}$ are taken into account, so that one of them can be neglected. The remaining equations can be solved for E_{0r} and E_{0t}, yielding Fresnel's equations.

$$\frac{E_{0t}}{E_0} = \frac{2\sqrt{\varepsilon_1\varepsilon_2} \cdot \cos\alpha}{\varepsilon_2 \cdot \cos\alpha + \sqrt{\varepsilon_1 \cdot \left(\varepsilon_2 - \varepsilon_1 \cdot \sin^2\alpha\right)}} \tag{1.39a}$$

$$\frac{E_{0r}}{E_0} = -\frac{\varepsilon_2 \cdot \cos\alpha - \sqrt{\varepsilon_1\left(\varepsilon_2 - \varepsilon_1 \cdot \sin^2\alpha\right)}}{\varepsilon_2 \cdot \cos\alpha + \sqrt{\varepsilon_1 \cdot \left(\varepsilon_2 - \varepsilon_1 \cdot \sin^2\alpha\right)}} \tag{1.40a}$$

The squared field ratios correspond to the reflection and transmission coefficient, respectively, so that the sum of both equals 1.

If the electric field is orthogonal to the incident plane, a very similar derivation yields the corresponding Fresnel's Eqs. (1.39b and 1.40b):

$$\frac{E_{0t}}{E_0} = \frac{2\cos\alpha}{\cos\alpha + \sqrt{\frac{\varepsilon_2}{\varepsilon_1} - \sin^2\alpha}} \tag{1.39b}$$

$$\frac{E_{0r}}{E_0} = \frac{\cos\alpha - \sqrt{\frac{\varepsilon_2}{\varepsilon_1} - \sin^2\alpha}}{\cos\alpha + \sqrt{\frac{\varepsilon_2}{\varepsilon_1} - \sin^2\alpha}} \tag{1.40b}$$

Both cases are depicted in Fig. 1.3 for the transition from air to a nonabsorbing dielectric ($\varepsilon_2 = 80$). The case where the electric field is parallel to the incident plane is interesting; an angle (the so-called Brewster's angle) exists, where $E_{0r} = 0$; i.e., no reflection occurs.

With this approach (especially of Eq. (1.31)), the particular center heating of objects of cm-dimensions with convex surfaces (e.g., eggs) can be easily understood, as at the convex surface the microwave "rays" are refracted and focused to the center.

For objects of the same size as the wavelength or smaller, historically, the theory of Mie has been used to determine the microwave absorption, nowadays direct field

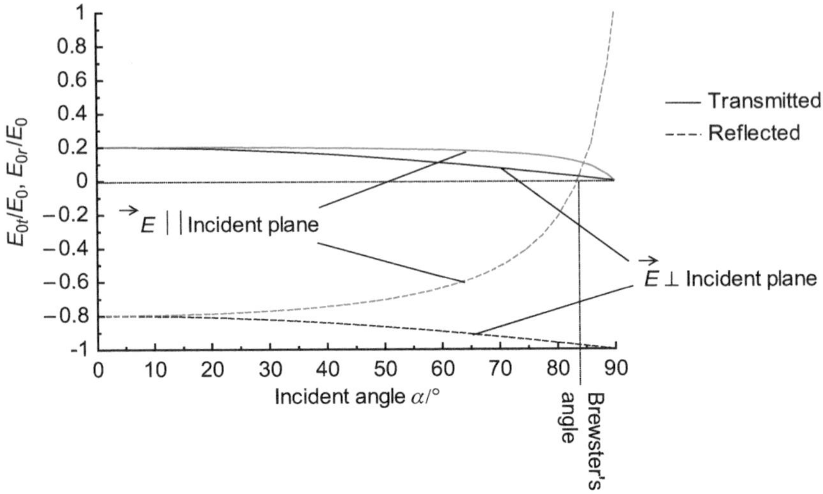

Figure 1.3 Reflected and transmitted parts of the electric field of a plane electromagnetic wave hitting a half space of a dielectric ($\varepsilon_2 = 80$) with incident angle α.

modeling by numerical solutions of Maxwell's equations (see Chapter 18: Simulation of microwave processes) become more and more important.

In order to calculate temperature changes within an object by microwave heating, it is important to determine the power density, starting from the electromagnetic field configuration. Since normal food substances are not significantly magnetically different from vacuum ($\mu = 1$), in most cases the knowledge of the electric field is enough to calculate the heat generation caused by power dissipation. This power dissipation (per unit volume) p_v is determined by ohmic losses which can be calculated by:

$$p_v = \frac{1}{2} \cdot \Re\left(\vec{E} \cdot \vec{j}^*\right) \tag{1.41}$$

The current density $\vec{j}$ is determined by the conductivity and the electric field by Eq. (1.8). The equivalence of an imaginary part of the permittivity and a conductivity Eq. (1.16) can also be described as

$$\sigma_{\text{total}} = \sigma + \omega\varepsilon_0\varepsilon'' \tag{1.42}$$

The resulting power dissipation can be written in terms of the total conductivity or the total imaginary part of the permittivity, the so-called loss factor:

$$p_v = \frac{1}{2}\sigma_{\text{total}} \cdot \left|\vec{E}\right|^2 = \frac{1}{2}\omega\varepsilon_0\varepsilon_{\text{total}} \cdot \left|\vec{E}\right|^2 \tag{1.43}$$

The dependence on the squared electric field magnitude determines that the power dissipation penetration depth δ_{p} is only half the value of the electric field penetration depth δ_{E} Eq. (1.25).

$$\delta_{\mathrm{p}} = \frac{1}{\omega} \cdot \sqrt{\frac{1}{2 \cdot \mu_0 \mu \varepsilon_0 \varepsilon' \cdot \left(\sqrt{\sqrt{1 + \frac{\varepsilon''^{*2}}{\varepsilon'^2}} - 1} \right)}} \tag{1.44}$$

1.3.5 The effect of utilizing microwaves in short time thermal processing

For short time thermal processes, the importance of heating rate increases, while the impact of holding time becomes less significant. Thus, short time processes rely on rapid heat transfer mechanisms: Whereas for pure liquids conventional heating techniques in plate or scraped surface heat exchangers may be sufficient to handle this challenge, already particulates of some millimeters may suffer from insufficient heating of the particle centers. For solid foods, this problem becomes even more dominant (Regier et al., 2009).

Assuming a typical food particle size (characteristic length) of X, the thermal conductivity of the food product λ_{p}, and the heat transfer coefficient α, the Biot number $Bi = \alpha^* X / \lambda$ can be calculated. This dimensionless number describes the ratio of the (outer) heat transfer from the surrounding medium to the product surface to the heat transfer by conduction within the product. For Biot numbers much greater than 1 (the heat transport resistance is dominated by the inner heat conduction) a further enhancement of the outer heat transfer (e.g., by higher temperatures or velocities) does not accelerate the inner heating significantly. Therefore, only a reduction in particle/product size or a change of the inner heat transfer mechanism will lead to a decrease in Biot number and, therefore, an improvement of heating rate and temperature uniformity in short time thermal processing. One such change of inner heat transfer is the utilization of microwaves, which are an interesting alternative for such processes. The resistance to heat conduction may be circumvented by the penetration of microwaves, generating the heat through dissipation of the electromagnetic energy within the product. Nonetheless, it has to be noted that the penetration depths (as discussed in the Section 1.3.4.2) in comparison with the characteristic product size have to be taken into consideration.

1.4 General configuration of microwave systems

Each microwave system generally consists of three basic parts: the microwave source, the waveguide, and the actual applicator. In the following section, these parts will be described in more detail.

1.4.1 Microwave sources

The by far most commonly used microwave source for industrial and domestic applications ((Metaxas, 1996) names a number of 98%) is the magnetron tube. Therefore, the main part of this chapter should be limited to the description of a magnetron and only from a phenomenological point of view. More detailed descriptions can be found, for example, in Metaxas & Meredith (1983) and Püschner (1966). Over the last few years, new development of solid state microwave generators with powers suitable also for heating application appeared. These solid state transistors show some potential for advantages over magnetron tube microwave generation. Therefore, these solid state devices will also briefly be covered here.

1.4.1.1 Magnetrons

A magnetron consists of a vacuum tube with a central electron emitting cathode of highly negative potential (see Fig. 1.4). This cathode is surrounded by a structured anode, consisting of open metal cavities (also referred to as cavity resonators). Electrons are accelerated from the cathode radially; however, since an orthogonal magnetic field is applied, they are deflected, yielding a spiral motion. Electrons passing by the openings of the cavities excite electromagnetic wave oscillations in the cavity. The electric and the magnetic field strength are chosen appropriately, so that the resonant cavities take energy from the electrons. This phenomenon can be compared to the excitation of the resonance by whistling on a bottle. The stored electromagnetic energy can be coupled out by a circular loop antenna in one of the cavities into a waveguide or a coaxial line.

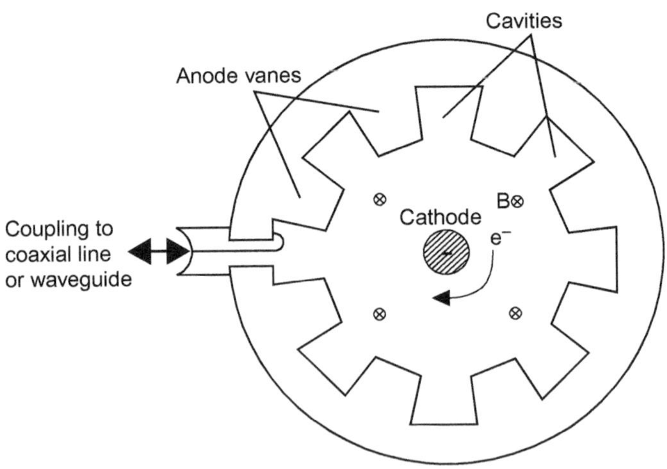

Figure 1.4 Schematic view of a magnetron tube.
Source: Adapted from Regier, M., & Schubert, H. (2001). Microwave processing. In S. Richardson (Ed.), *Thermal technologies in food processing*. Cambridge: Woodhead Publishing Limited (Regier & Schubert, 2001).

The power output of a magnetron can be controlled by the tube current or the magnetic field strength. Its maximum power is generally limited by the temperature of the anode, which has to be prevented from melting. Practical limits at 2.45 GHz are approximately 1.5 and 25 kW for air or water cooled anodes, respectively (Roussy & Pearce, 1995). The 915 MHz magnetrons have larger cavities (lower resonant frequency means larger wavelength) and thus can achieve higher powers per unit. The efficiencies of modern 2.45 GHz magnetrons range at approximately 70%, mainly limited by the magnetic flux of the utilized economic ferrite magnets (Yokoyama & Yamada, 1996), whereas the total efficiency of microwave heating applications are often lower due to unmatched loads.

1.4.1.2 Solid state microwave generators

Over the last few years, with advances in semiconductors, especially the GaN (Gallium-nitride) technology and advanced silicon LDMOS (laterally diffused metal oxide semiconductor) transistors, it has become possible to generate microwave power of up to some hundred watts. Therefore, the power required for heating applications can now be achieved also with this solid state (instead of vacuum tube) technology. One prominent example for such a device is offered by NXP Semiconductors (Toulouse, France). For example, NXP Semiconductors (former Freescale Semiconductors) offer an LDMOS transistor with a nominal power of 250W, operating at a frequency of 2450 MHz, as well as a system with a nominal power of 350W, and an operating frequency of 915 MHz; they report efficiencies of up to 67% (Freescale, 2015). Considering that the maximum power of such solid state devices has increased by more than 100% and the efficiency by more than 40%, both in the last 4 years (Viza, 2015), their application becomes more and more attractive. Further claimed advantages of solid state devices are their small dimensions, the use of low voltages (<50V) in contrast to some kV for magnetrons, a lifetime of up to 20 years without degradation, and better control characteristics; e.g., the frequency, its phase, and even the power is continuously variable in contrast to the typical on–off-control for magnetrons (Freescale, 2015). With multiple transistors of variable frequencies and phases and a feedback control in which the reflected signals can be detected and analyzed, microwave-assisted processing can be revolutionized through homogenization of the undesired uneven heating, which is often a problem in conventional microwave-assisted processing.

1.4.2 Waveguides

For guiding an electromagnetic wave, transmission lines (e.g., coaxial lines) and waveguides can be used. Due to lower losses of waveguides at higher frequencies, e.g., microwave frequencies, these parts are used for microwave power applications. Basically, waveguides are hollow conductors of, in most cases, constant cross-section, whereby rectangular and circular forms are of most practical use. The internal size defines a minimum frequency f_c (the so-called cut-off frequency) through the solution of the wave equations (Eq. (1.11) and the corresponding one for the

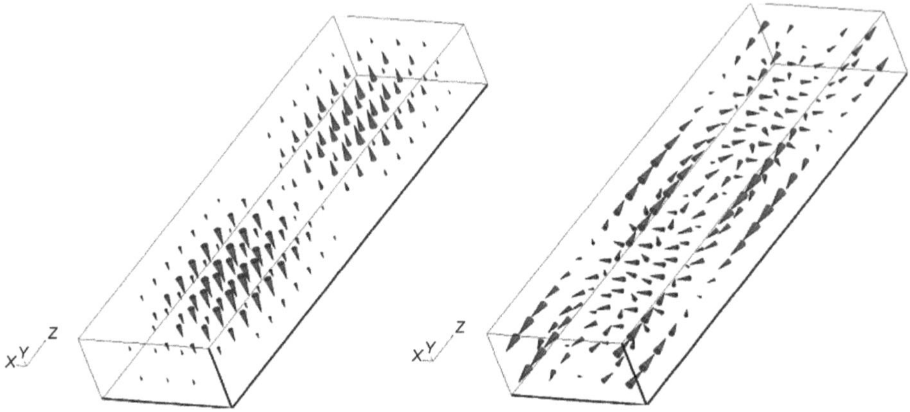

Figure 1.5 Electric and magnetic field configurations in a TE10 rectangular waveguide.
Source: Adapted from Regier, M., & Schubert, H. (2001). Microwave processing. In S. Richardson (Ed.), *Thermal technologies in food processing*. Cambridge: Woodhead Publishing Limited.

magnetic field) and appropriate boundary conditions (Table 1.2), below which waves do not propagate. For rectangular waveguides with width a and height b, the following Eq. (1.45) can be derived for the cut-off frequency f_c.

$$f \geq \frac{\sqrt{\left(\frac{m}{a}\right)^2 + \left(\frac{n}{b}\right)^2}}{2\sqrt{\mu\mu_0\varepsilon\varepsilon_0}} = \text{minimum}\left\{\begin{array}{ll} \dfrac{1}{2a\sqrt{\mu\mu_0\varepsilon\varepsilon_0}}, & a \geq b \\ \dfrac{1}{2b\sqrt{\mu\mu_0\varepsilon\varepsilon_0}}, & a \leq b \end{array}\right\} = f_c \qquad (1.45)$$

Within the waveguide, the wave may spread out in so-called modes, which define the electromagnetic field distribution within the waveguide. These modes can be split into transversal electric (TE) and transversal magnetic (TM) ones, describing the direction of the electric and the magnetic field, respectively, toward the propagation direction. The most commonly used waveguide is of rectangular cross-section with a width a equals to double the height b and is used in TE10 mode, which is depicted in Fig. 1.5.

1.4.3 Microwave applicators and tuners

The waveguide itself can be used as an applicator for microwave heating when the material to be heated is introduced by wall slots and the waveguide is terminated by a matched load (Fig. 1.6). This configuration is then called traveling wave device, since the location of the field maxima change with time. A radiation through the slots only occurs if wall current lines are cut and the slots exceed a certain dimension, which can be avoided (Roussy & Pearce, 1995).

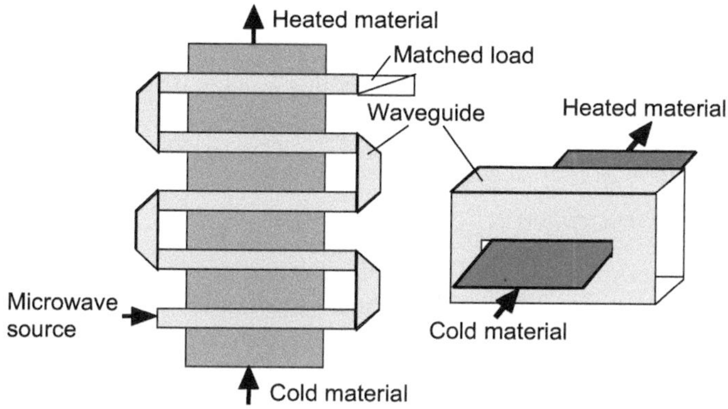

Figure 1.6 Example for a traveling wave device.
Source: Adapted from Roussy, G., & Pearce, J. A. (1995). *Foundations and industrial applications of microwaves and radio frequency fields*. Chichester: John Wiley & Sons.

More common in the food industry and domestic application are standing wave devices described in the next section, where the microwaves irradiate by slot arrays (that cut wall currents) or horn antennas (specially formed open ends) of waveguides.

For receiving high power absorption and little back-reflection of microwaves from the applicator to the source, the impedance of the load containing applicator has to be matched with the corresponding impedance of the source and the waveguide. In order to achieve such a situation, tuners are introduced. Tuners are waveguide components used to match the load impedance to the impedance of the waveguide. Tuners minimize the amount of reflected power, resulting in the most efficient coupling of power to the load.

Due to changes of the load during processing, this matching has to be controlled continuously or optimized to provide an average load. The remainder of the reflected power has to be prevented from coming back to and overheating the microwave source. Therefore, circulators—directional dependent microwave traveling devices—are used, which let the incident wave pass and guide the reflected wave into an additional load (in most cases water). As a side effect, by heating this load, the reflected power can be determined too.

Common applicators can be classified by the type of field configuration into three types.

1.4.3.1 Near-field applicators

In the case of near-field applicators, the microwaves originating from a horn antenna or slot array directly "hit" the product to be heated. The power should be set to a level that can be practically completely absorbed by the product, so that only minor parts of the power is transmitted and transformed into heat in dielectric loads (usually water) behind the product.

Similar to the traveling wave device, in this case standing waves do not exist. Consequently, a relatively homogeneous electrical field distribution (depending on the mode irradiated from the waveguide) within a plane orthogonal to the direction of propagation of the wave can be achieved.

1.4.3.2 Single mode applicators

The near-field applicators as well as the traveling wave devices work best with materials with high losses. In order to heat substances with low dielectric losses effectively by microwaves, applicators with resonant modes, which enhance the electric field at certain positions, are better suited. At these positions, where the electric field is concentrated, the material to be heated should be located. Single mode applicators consist generally of one feeding waveguide and a tuning aperture and a relatively small microwave resonator with dimensions in the range of the wavelength. As in the case of the dielectric measurements by resonators (see Chapter 3: Determination of the dielectric properties of foods), a standing wave (resonance) exists within the cavity at a certain frequency. The standing wave yields a defined electric field pattern, which can then be used to heat the product. It has to be noted that these types of applicators have to be well matched to the load, since the insertion of the dielectric material naturally shifts the resonant modes. An example of such a system is shown in Fig. 1.7, where a cylindrical TM010 field configuration, with high electric field strength at the center, is used to heat a cylindrical product, which could correspond to a pumpable product being transported through tubes (see also Chapter 16: Microwave plant requirements and process control for advanced applications).

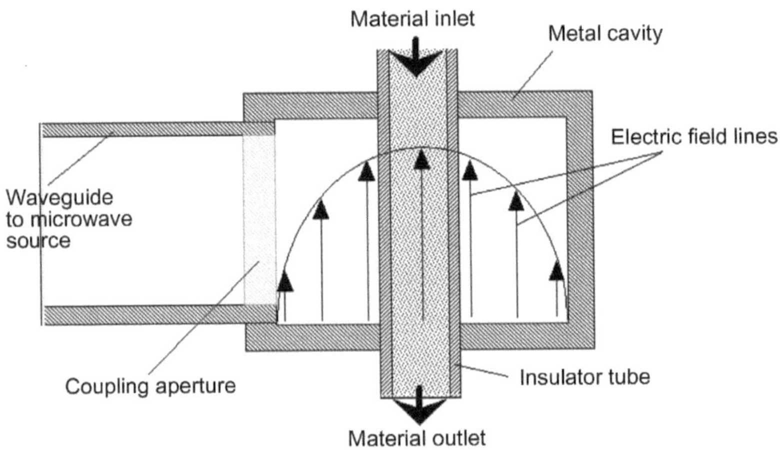

Figure 1.7 A TM010 flow applicator schematically as an example for a single mode device. *Source*: Adapted from Regier, M., & Schubert, H. (2001). Microwave processing. In S. Richardson (Ed.), *Thermal technologies in food processing*. Cambridge: Woodhead Publishing Limited.

The small dimensions of the applicator are needed to avoid different modes from the used one, since the number of modes per frequency range grows very fast with cavity dimensions.

1.4.3.3 Multimode applicators

By increasing the dimensions of the cavity, a fast transition from the single mode to the multimode applicator occurs, due to the strongly increasing mode density with applicator size. Additionally, it has to be taken into account that common microwave power generators like magnetrons do not emit a single frequency but rather a frequency band.

In industrial as well as in domestic applications, the multimode applicators play by far the most important role, since most of the conveyor-belt-tunnel applicators employed in industry, as well as the microwave ovens at home, are of the multimode type due to their typical dimensions. Despite the high number of stimulated modes, often a nonhomogeneous field distribution, which is stationary over time, will develop. This field distribution mainly depends on the cavity and the product geometry and the dielectric properties of the material to be processed.

In contrast to the case of the single mode application, this nonuniform field distribution, which would result in nonuniform heating patterns, is generally not desired, as it is difficult to control. An undesired nonuniform heating pattern can be prevented by changing the field configuration through variations in the cavity geometries (e.g., mode stirrer) or by moving of the product (e.g., on a conveyor belt, or turn table), which also impacts on the field distribution.

Industrial applications are often operated in continuous mode due to the high throughputs required. Therefore, continuous microwave applicators were developed as early as 1952 with the first conveyor belt oven patent (Spencer, 1952). Nevertheless, due to the lack of high power microwave generators, industrial use was not realized until nearly 10 years later.

Today's industrial ovens (a more complete overview can be found in the corresponding chapters) may be differentiated into two groups by the number and power of microwave sources: high power single magnetron and low power multimagnetron devices.

Whereas for a single mode unit only a single source is possible, in all other systems (multimode, near-field, or traveling wave system) the microwave energy can be irradiated optionally by one high power magnetron or several low power magnetrons. Industrial high power magnetrons have often longer operating lifetimes; low power magnetrons have the advantage of lower capital investment due to the high production numbers for the domestic market, as well as reduced downtime for repair and maintenance.

As mentioned above, an important hurdle for all microwave systems, but in particular for continuously operated ones, is the prevention of leakage radiation through the product in- and outlet. For fluids or granular products with small dimensions (cm-range), the legislative limits can be guaranteed by the small in- and outlet sizes, together with the absorption in the entering product, sometimes with

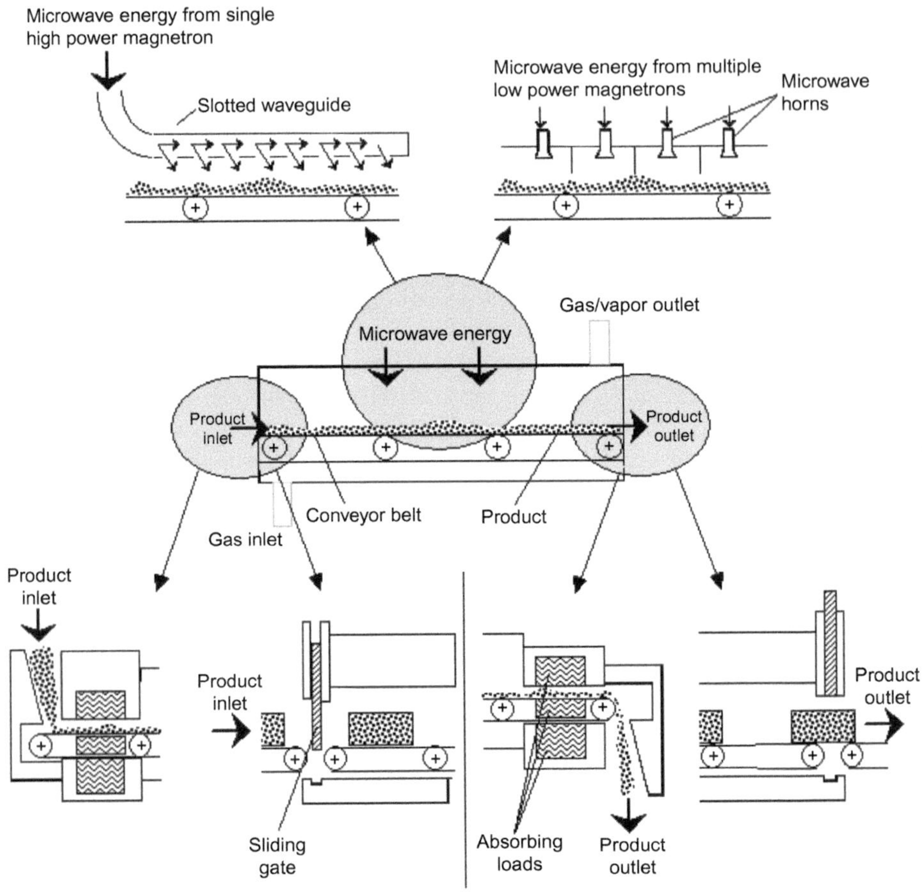

Figure 1.8 Continuous conveyor belt device, with different product input and output systems and various microwave energy inputs.
Source: Adapted from Regier, M., & Schubert, H. (2001). Microwave processing. In S. Richardson (Ed.), *Thermal technologies in food processing*. Cambridge: Woodhead Publishing Limited.

additional dielectric loads just in front of the openings. However, especially in the case of larger product pieces, inlet and outlet gates, which completely close the microwave application device, have to be used. Schematically, a microwave system with conveyor belt and its power sources and openings is shown in Fig. 1.8.

1.5 Summary

In this chapter, microwaves are introduced as a type of electromagnetic wave of frequencies between 300 MHz and 300 GHz. The "technical" microwaves used for processing are regulated by the ISM bands and by certain maximum emission levels and exposure limits of humans. Theoretical aspects of the electromagnetic theory, starting from Maxwell's and the constitutive material equations, over the general wave equations to exemplary solutions like the plane wave, the exponentially damped wave, and Fresnel's reflection formulas were discussed. Finally, the general configuration and operation of microwave-assisted systems, consisting of a microwave source (still in most cases the magnetron), a waveguide, and an applicator was presented.

1.6 Nomenclature

Due to the used nomenclature that are common in the particular fields, some of the variable names are ambiguous, but their meaning should become clear in the equations context.

a	width of waveguide
b	height of waveguide
$\vec{B}$	magnetic flux density
c, c_0	velocity of light, in vacuum
$\vec{D}$	electric flux density
$\hat{e}_i$	unit vector in the direction of i
$\vec{E}$	electric field
f	frequency, constant
g	constant
$\vec{H}$	magnetic field
h	constant
h	Planck's constant
i	imaginary unit
$\vec{j}$	electric current density
$\vec{k}$, k	wave vector, absolute value
m	constant
n	refractive index, constant
$\Re(x)$	real part of x
t	time
$\vec{x} = (x, y, z)$	local vector
α, α_r	incident-, reflection angle
β	refraction angle
δ_E	electric field attenuation length
δ_P	power attenuation length
ε_0	dielectric constant of vacuum
$\varepsilon = \varepsilon' - i\varepsilon''$	relative permittivity
κ	damping constant
λ	wavelength
μ	relative permeability
μ_0	permeability of vacuum
ρ	charge density
σ	conductivity
ω	circular frequency

References

Freescale. (2015). *Websites of Freescale Semiconductor, Inc.* http://www.freescale.com/products/rf/rf-cooking/rf-and-microwave-power-solutions-for-heating:RF_HEATING_PG, and http://www.freescale.com/products/rf/rf-cooking:RF-COOKING?cof=0&am=0 Visited 22.10.15.

ICNIRP. (1998). Guidelines on limits of exposure to time-varying electric, magnetic and electromagnetic fields (up to 300 GHz). *Health Physics*, *74*, 494–522.

IRPA. (1988). Guidelines on limits of exposure to radiofrequency electromagnetic fields in the frequency range from 100 kHz to 300 GHz. *Health Physics*, *54*(1), 115–123.

Metaxas, A. C. (1996). *Foundations of electroheat*. Chichester: John Wiley & Sons.

Metaxas, A. C., & Meredith, R. J. (1983). *Industrial microwave heating*. London: Polu Pelegrinus Ltd.

Püschner, H. A. (1966). *Heating with microwaves*. Berlin: Philips Technical Library.

Regier, M., Rother, M., & Schuchmann, H. P. (2009). Alternative heating technologies. In E. Ortega-Rivas (Ed.), *Processing effects on safety and quality of foods*. Boca Raton, FL: CRC Press.

Regier, M., & Schubert, H. (2001). Microwave processing. In S. Richardson (Ed.), *Thermal technologies in food processing*. Cambridge: Woodhead Publishing Limited.

Roussy, G., & Pearce, J. A. (1995). *Foundations and industrial applications of microwaves and radio frequency fields*. Chichester: John Wiley & Sons.

Spencer, P. (1952). Means for Treating Foodstuffs, U. S. Patent 2,605,383, 605, 383.

Viza, D. (2015, Sept 9). *RF cooking information for Prof. Regier*, personal communication with Dan Viza, Global Marketing Manager, RF Cooking, Freescale Semiconductor, Inc.

Yokoyama, R., & Yamada, A. (1996). Development status of magnetrons for microwave ovens. *Proceedings of 31st microwave power symposium*, 132–135.

Microwave heating and the dielectric properties of foods

2

V. Meda[1], *V. Orsat*[2] *and V. Raghavan*[2]
[1]University of Saskatchewan, Saskatoon, SK, Canada, [2]McGill University, Montreal, QC, Canada

2.1 Introduction

Microwave heating is highly popular in a variety of industrial and domestic applications. The major factor supporting the popularity of microwave heating is its rapidness which matches very well the current fast lifestyle of people. Other advantages of microwave heating are its ease of operation, low maintenance, and instant heating. The combination of traditional means of drying (such as hot air) with microwave technology can reduce processing times, produce higher quality foods, and provide flexibility in the selections of foods to be dried, thus it is widely used in the food industry (Datta & Rakesh, 2013; Gowen, Abu Ghanam, Frias, & Oliviera, 2006; Zhang, Tang, Mujumdar, & Wang, 2006). However, the occurrence of uneven temperatures during microwave food heating remains a major concern regarding the use of microwave energy (Vadivambal & Jayas, 2010).

Thermal technology efficiency dictates the quality, economics, and environmental impact of most processing plants. It is by far the most sensitive aspect of food processing. Green engineering regulations call for more efficient energy usage and more environment-friendly raw materials as well as effluents. Efficiency demands minimal processing of materials, especially for highly nutritious foods. Hence, electric heating technologies, such as radio frequency, microwave, ohmic, and infrared are being considered; among them, microwave processing shows a continued promising future (Chan & Reader, 2000; Torrealba-Meléndez, Sosa-Morales, Olvera-Cervantes, & Corona-Chávez, 2015; Zhang et al., 2006). Microwaves are energy efficient and can operate in combination modes.

Microwaves are electromagnetic waves with frequencies ranging from 300 MHz to 300 GHz (Singh & Heldman, 2009). Certain frequencies within this range of the electromagnetic spectrum are set aside by the International Telecommunications Union for industrial, scientific, medical, and domestic use. These so-called ISM & D bands in North America are 2450 MHz, 915 MHz, and a few other frequencies according to the geographic location. By far the most common in North America and Europe is 2450 MHz, which is the frequency at which most domestic microwave ovens operate. These waves (i.e., in the frequency range of 300 MHz to 300 GHz) propagate with "a time interval between peaks" during oscillation, ranging from 3×10^{-8} to 3×10^{-11} seconds (Venkatesh & Raghavan, 2004).

The Microwave Processing of Foods. DOI: http://dx.doi.org/10.1016/B978-0-08-100528-6.00002-4

This range coincides with the temporal sequence of events at atomic and molecular transitions, such as reactions in water, molecular dissociation and, most importantly (for industrial microwave technology), dielectric relaxation of water. The dielectric relaxation of water may vary from 100 MHz for bound water to 18 GHz for pure water (Miura, Yagihara, & Mashimo, 2003), and this is the property that is studied extensively when heating effects of microwaves are investigated. The most effective conversion of microwave energy to thermal energy in biological materials (or otherwise moist materials) will occur in this frequency range.

2.2 Microwave heating and the dielectric properties of foods (see also Chapter 1: Introducing microwave-assisted processing of food: fundamentals of the technology)

As mentioned earlier, microwave energy is transported as an electromagnetic wave in certain frequency bands in the range between about 0.3 GHz and 300 GHz. When microwaves impinge on a dielectric material, part of the energy is transmitted, part is reflected, and part is absorbed by the material, where it is dissipated as heat. Heating is due to an increase in molecular kinetic energy (i.e., an increased movement of permanent dipoles within the material as they try to reorient themselves with the oscillating (electrical) field of the incident wave), as well as associated "molecular friction." The power dissipated in a material is proportional to the frequency of the source, the dielectric loss of the material, and the square of the field strength within it (see Eq. 2.2). A material is subjected to microwave energy in a device known as an applicator or cavity. Considering all these features, it is possible to identify those candidate materials and processes that can use microwave heating effectively and understand microwave ingredient interaction mechanisms. Only after such a step is taken can microwave heating be exploited fully in terms of its unique characteristics, which include the facts that no contact is required between the energy source and the target and that heating is volumetric, rapid, and highly specific in nature.

As mentioned before, there are only a few frequencies that are permitted by an international convention to be used in industrial, scientific, and medical applications, the most favored being 2.45 GHz. At this frequency, the electric field swings the orientation of water molecules 2.45×10^9 (2,450,000,000) times every second, generating an intense heat that can escalate as quickly as 10°C per second (Lew, Krutzik, Hart, & Chamberlin, 2002). Water being the predominant component of biological materials, its content directly influences heating. However, there are minor contributions from a host of other factors (Chandrasekaran, Ramanathan, & Basak, 2013; Schiffmann, 1986): heating is accelerated by ionic effects (mostly salt content) and specific heat of the composite material (Decareau, 1992). Specific heat is an important property in the thermal behavior of a food subjected to microwaves. Products with low specific heat may heat very rapidly, and even faster than water of the same weight, one example being oil (Schiffmann, 1986). Hence, for oily materials, the influence of specific heat becomes the determining factor in microwave heating, owing to the low specific heat of oils—often less than half that of water (Ohlsson, 1983).

2.3 Microwave interactions with dielectric materials

Microwave heating is a result of interactions among dielectric materials and the electromagnetic waves. Dielectric properties govern the efficiency and quality of the heating process (Curet, Rouaud, & Boillereaux, 2014). Occurrence of dipolar interactions due to penetration of microwaves in food generates molecular motion and intermolecular friction (Salazar-Gonzalez, San Martin-Gonzalez, Lopez Malo, & Sosa-Morales, 2012). The heating effect occurs by the dissipation of energy from the microwave field. This brings a temperature gradient within the food material which leads to the propagation of heat in all directions (Hossan, Byun, & Dutta, 2010). Solutions with high concentrations of ions may get heated more quickly with increasing frequency of operation (Singh & Heldman, 2009; Suarez, Viollaz, Rovedo, Tolaba, & Haros, 2000).

When an oscillating electrical field is applied to a polar dielectric material, the dipoles within the material attempt to align themselves (polarize) with the field. The rate of change of polarization represents a displacement current in the dielectric material, and the product of this and the applied field gives the power converted into heat. Averaged over a cycle, the power "lost" in the material (i.e., dissipated as heat) depends on the phase angle between the applied field and the polarization. For most dielectric materials, the lag depends on the flexibility and mass of the molecules that house the dipoles, and the randomization effect of temperature.

2.3.1 *Power dissipated*

Absorption of microwaves by the food material leads to attenuation of the microwaves. The distance at which the power of microwaves is reduced by $1/e \approx 36.8\%$ of its initial value is called the penetration depth (Yam & Lai, 2006) and is presented in the next section.

The fundamental dielectric property of a material is the complex relative permittivity ϵ^* where

$$\epsilon^* = \epsilon' - j\epsilon'' \tag{2.1}$$

The real part of ϵ^*, ϵ', is called the permittivity, and the imaginary part, ϵ'' is called the loss factor. The ratio ϵ''/ϵ' is called the loss tangent, written as $\tan\delta$, with δ being the phase lag. Selected values of dielectric properties are presented in Tables 2.1 and 2.2.

Table 2.1 **The dielectric properties and penetration depth of selected foods (Tang, 2005)**

Foods	915 MHz				2450 MHz		
	Temp. (°C)	ϵ'	ϵ''	D_p (mm)	ϵ'	ϵ''	D_p (mm)
Air		1.0	0		1.0	0	
Water							
Distilled/deionized	20	79.5	3.8	122.4	78.2	10.3	16.8
0.5% salt	23	77.2	20.8	22.2	75.8	15.6	10.9
Ice	− 12	—	—	—	3.2	0.003	11615
Corn oil	25	2.6	0.18	467	2.5	0.14	220
Fresh fruit and vegetables							
Apples (red delicious)	22	60	9.5	42.6	57	12	12.3
Potato	25	65	20	21.3	54	16	9.0
Asparagus	21	74	21	21.5	71	16	10.3

Table 2.2 **Dielectric properties of egg whites and whole eggs (Sosa-Morales, Valerio-Junco, Lopez-Malo, & Garcia 2010; Wang, Tang, Wang & Swanson, 2009)**

Egg product	State	Temperature (°C)	Dielectric constant		Dielectric loss factor	
			Frequency		Frequency	
			27.12 MHz	915 MHz	27.12 MHz	915 MHz
Egg white	Liquid	20	84.6	64.0	427.0	18.7
		80	98.3	50.5	866.5	33.3
		120	135.1	53.2	1665.8	56.9
	Precooked	20	89.3	64.5	411.8	18.9
		80	99.5	53.0	937.1	34.6
		120	124.4	50.1	1480.5	52.2
Whole egg	Liquid	20	76.3	55.5	335.9	15.8
		80	87.5	48.9	801.8	30.5
		120	106.1	44.7	1132.7	42.3
	Precooked	20	79.6	56.5	336.8	16.3
		80	89.0	48.5	745.8	29.0
		120	104.8	44.3	1020.0	39.5

The power dissipated per unit volume can be expressed as:

$$P = \omega\epsilon_0\epsilon'(\tan\delta)\, E^2_{\text{rms}} = 2\pi f\, \epsilon_0\epsilon'' E^2_{\text{rms}} \tag{2.2}$$

where ϵ_0 is the permittivity of free space (8.85×10^{-12} F/m), f is the microwave frequency (Hz), and E is the electrical field strength (V/m) in the material.

For completeness, the total loss in a material is made up of the dielectric loss (from polarization) and the conductive (ohmic loss), as follows:

$$\epsilon_t'' = \epsilon'' + \sigma/\omega\epsilon_0 \tag{2.3}$$

where σ is the electrical conductivity of the material.

For metals, the polarization loss is zero at microwave frequencies while for dielectrics having no conductive losses the conductivity term is zero. In the absence of any other energy processes, the conservation of energy equation for heat transfer in one direction (x), through a plane slab of material exposed to microwave energy, may be written as:

$$dT/dt = d(\alpha\, dT/dx)/dx + (2\pi\epsilon_0\epsilon'' E^2)/C_{\text{p}} \tag{2.4}$$

where C_{p} is the specific heat capacity of the material and α the temperature conductivity. If all the parameters in the microwave source term are independent of location (x), this term is constant, and its contribution is to cause the temperature of the material to increase linearly and uniformly with time. It is this that gives rise to the notion of volumetric heating. If, in addition, the boundary condition at the surfaces is represented by a convective heat transfer coefficient and the surrounding medium is at the original temperature of the material, the material will be hotter in the middle than at the edges. This gives rise to the notion that microwave energy heats from the inside out.

Thus, the initial increase in temperature depends on the size of the microwave source term, and, in order to maximize this, it would seem desirable to plan that the electrical field, the frequency, and the loss factor of the material should all be as large as possible. However, for all practical (commercial) purposes the frequency is fixed within the ISM & D range, the upper limit of the field is determined by the dielectric strength of the substrate or the surrounding air (20–30 kV/m), and the upper limit of the loss factor is determined by the rate of absorption of microwave energy through the material.

2.3.2 Field strength within a material

Field strength E varies with the position within the material. For a plane wave, this relationship can be expressed in terms of the depth of material, D_{p}, over which the incident power is reduced by a factor of $1/e$:

$$D_{\text{p}} = \frac{1}{\omega}\left[2\mu_0\mu'\,\epsilon_0\,\epsilon'\left(\sqrt{1+(\tan\,\delta)^2}-1\right)\right]^{-1/2} \approx \lambda_0\sqrt{\epsilon'}/2\pi\epsilon''$$

Here μ_0 is the magnetic permeability of free space (1.25×10^{-6} H/m) and μ' is the magnetic permeability of the material which typically (for nonmagnetic materials) is taken as 1 and λ_0 is the free space wavelength.

This approximation applies to a low loss dielectric material. Note that the attenuation of an incident wave is more commonly expressed as the depth over which the incident electrical field falls by a factor of $1/e$. This so-called skin depth, d, is inversely proportional to the attenuation constant of the wave, a. Numerically, $D_p = d/2$.

2.3.3 Using microwave heating effectively

High surface heating can be obtained by higher frequencies (Wang & Wang, 2009), while greater penetration depth is typically obtained at lower frequencies. For even quasivolumetric heating, the size of a target material should be of the same order as D_p and, for preference, considerably less.

A relatively simple measure of the effectiveness of microwave heating is to compare the rate of heat conduction with the rate of heat generation. For a plane slab, this gives a value m, where

$$m = (\omega \epsilon_0 \epsilon'' E^2 x^2)/k\Delta T \tag{2.6}$$

That is to say, microwave heating is favored for thick materials (large x) with high dielectric loss factor and low thermal conductivity (k) and where a large temperature gradient is undesirable. Here, it is assumed that $D_p \gg x$ so that the field is substantially constant throughout.

2.4 Measuring microwave heating

Industrial microwaves are high frequency applications. At these frequencies, electronic components do not remain inert. The electronic circuits are therefore visualized as guides for waves. Measurements are often made in terms of power transmitted and reflected and not voltage and current.

Even as accurate measurements challenge the microwave engineer, there are basic areas such as efficient conversion of microwave energy to thermal energy and its numerical modeling that leave many experts still guessing. Quoting Meredith (1998), "numerical analysis ... is a topic requiring great professional skill and generally remains within the ambit of university/industrial research. It is a potential trap for the unwary, readily producing incorrect but plausible results." What follows is hence a summarized introduction of governing equations. Nevertheless in the meantime, there was much progress in that and further details can be found in Chapter 18, Simulation of Microwave Processes.

The heat generated by microwaves is represented in a conduction equation as given in Eq. (2.7):

$$\rho C_p \frac{\partial T}{\partial t} = \lambda \nabla^2 T + Q \tag{2.7}$$

This simplistic case has to be expanded with terms for convection and transport in order to get an accurate numerical solution.

The source term Q for heat generation being in the form of an electromagnetic field (microwave) is a function of field frequency and absorbed power by the material.

The two equations used for deriving the field equations for microwaves are Ampere's law (Eq. 2.8) and Faraday's Law (Eq. 2.9), both of which are Maxwell's electromagnetic equations:

$$\nabla \times H = \sigma E + \frac{\partial D}{\partial t} \tag{2.8}$$

relates magnetic field H to the electric flux density D, while

$$\nabla \times E = -\frac{\partial B}{\partial t} \tag{2.9}$$

relates electric field E to the magnetic flux density B.

We can, however, choose a mode of the microwaves such that the magnetic component is eliminated. For a dielectric material we then get Eq. (2.10):

$$\nabla^2 E = -\mu \epsilon \frac{\partial^2 E}{\partial t^2} \tag{2.10}$$

Here μ is the permeability representing the interaction with the magnetic field, and ϵ is the dielectric constant representing the interaction of nonconducting material with the electric field. The power dissipated per unit volume is manifested as heat generation rate $\dot{Q}$. Hence we have Eq. (2.11) for the heat generated in the microwave.

$$\dot{Q} = \left(\sigma_e |E|^2\right)/2 \tag{2.11}$$

The conductivity term σ includes the electrical conductivity and an imaginary part of the permittivity ϵ or the loss factor. This is the starting point for any industrial microwave heating calculation, whether for food or nonfood applications. However, when we consider foods we need to include innumerable influencing factors and corrections for unknowns. It is therefore easier to model nonfood processing systems.

2.5 Microwave heating variables

The microwave heating rates and potential nonuniformity are functions of oven factors and load characteristics (size, shape, dielectric properties, etc.). Any change in those parameters significantly affects the microwave heating process (Peyre, Datta, & Seyler, 1997). The most interesting and applied influences on dielectric properties include dielectric polarization, dipolar polarization, interfacial polarization, conduction effects, and combination effects.

Hossan et al. (2010) studied and related the heat transfer coefficient, cylinder length, and microwave frequency with the temperature distribution inside a food sample. Their studies concluded that a change in length of the particulate samples led to temperature variations within the body whereas a change in radius (for cylindrical shapes) had a very minor effect. In addition, they concluded that the dielectric constant and microwave frequency are critical parameters for uniform temperature distribution, whereas high convective heat transfer can bring uniform temperature distribution within foodstuffs.

Two major effects are responsible for the heating which results from this interaction. If the charged particles are free to travel through the material (electrons in a sample of carbon, for example), a current will be induced which will travel in phase with the field. If, on the other hand, the charged particles are bound within regions of the material, the electric field component will cause them to move until opposing forces balance the electric force. The result is a dipolar polarization in the material. Conduction and dipolar polarization may both give rise to heating under microwave irradiation, and are discussed in more details below.

It is important to note that microwave heating is quite distinct from microwave spectroscopy. The latter is a quantum phenomenon in which photons of particular energies (and therefore frequencies) excite the rotation levels of molecules. Whilst the absorption of microwaves in solid and liquid samples is frequency dependent, it is by no means quantized. Rather, the material behaves as though reacting to a high frequency electric field, and so may be subjected to classical analysis. Details of this analysis are beyond the scope of this chapter, although some of its chemically significant aspects will be introduced in the following sections.

2.5.1 Dielectric polarization

The inability of partially bound charges to follow the rapid changes in a high frequency electric field gives rise to microwave heating. The total polarization of the material arising from the displacement of charges may be expressed as the sum of a number of components resulting from the displacement of electron charges in relation to the nuclei in a material, and the displacement of nuclei relative to one another in materials with unequal charge distributions. Polarization of both operates on timescales which are very much smaller than that required for microwave frequency field reversals, and therefore follow microwave frequency fields almost exactly. As such they do not contribute to the microwave heating effect.

2.5.2 Reorientation of polar molecules/other permanent dipoles in the material

This is the most important of the polarization phenomena in relation to microwave heating. The role of the interfacial polarization (Maxwell-Wagner) effect $\mathbf{a}_i$, which results from interfacial phenomena in inhomogeneous materials, is limited at microwave frequencies, and in general its contribution is limited. In those cases where it is thought to be important, theoretical studies are thus very complex due to the large number of variables involved.

A useful dielectric parameter, the loss angle δ, is commonly used in the literature, and is more usually given in the form of its tangent. It is related to the complex dielectric constant by $\tan\delta = \epsilon''/\epsilon'$. The angle is the phase difference between the electric field and the polarization of the material. Magnetic polarization may also contribute to the heating effect observed in materials where magnetic properties exist, and a similar expression for the complex permeability of such materials may be formulated.

2.5.3 Dipolar polarization

Dipolar polarization is the phenomenon responsible for the majority of microwave heating effects observed in solvent systems. In substances such as water, the different electronegativities of individual atoms result in the existence of a permanent electric dipole on the molecule. The dipole is sensitive to external electric fields, and will attempt to align with them by rotation, the energy for this rotation being provided by the field. This realignment is rapid for a free molecule, but in liquids the instantaneous alignment is prohibited by the presence of other molecules. A limit is therefore placed on the ability of the dipole to respond to a field, which affects the behavior of the molecule with different frequencies of the electric field.

Under low frequency irradiation, the dipole may react by aligning itself in phase with the electric field. Whilst the molecule gains some energy by this behavior, and some is also lost in collisions, the overall heating effect is small. Under the influence of a high frequency electric field, on the other hand, the dipoles do not have sufficient time to respond to the field, and so do not rotate. As no motion is induced in the molecules, no energy transfer takes place, and therefore, no heating.

Between these two extremes, at frequencies which are approximately those of the response times of the dipoles, is the microwave region. The microwave frequency is low enough that the dipoles have time to respond to the alternating field, and therefore to rotate, but high enough that the rotation does not precisely follow the field. As the dipole reorients to align itself with the field, the field is already changing, and a phase difference exists between the orientation of the field and that of the dipole. This phase difference causes energy to be lost from the dipole in random collisions, and to give rise to dielectric heating.

The range of frequencies over which the dielectric loss is nonzero, indicating that microwave absorption occurs, is relatively large (Fig. 2.1). This is in contrast to the line-widths of quantum spectroscopic absorption, which are typically some orders of magnitude smaller. There is a clear maximum in the dielectric loss for water at a frequency of approximately 20 GHz, the same point at which the dielectric constant ϵ' goes through a point of inflexion as it decreases with increasing frequency. The 2.45 GHz operating frequency of domestic ovens is selected to be some way from this maximum in order to limit the efficiency of the absorption.

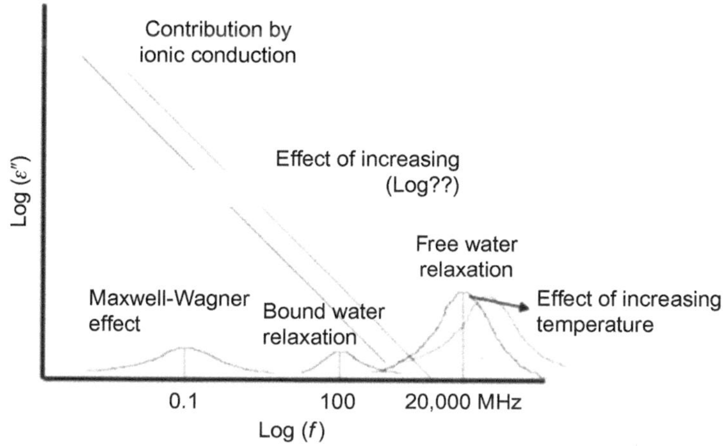

Figure 2.1 Contributions of various mechanisms to the loss factor of moist materials as a function of frequency (f). The critical frequencies are not accurate and show only the relative locations of the peaks.
Source: Adapted from Tang, J., Hao, F., & Lau, M. (2002). Microwave heating in food processing. *Advances in Bioprocessing Engineering*, 1–44, Chapter 1 (Tang, Hao & Lau, 2002) and Tang, J. (2005). Dielectric properties of foods. In H. Schubert & M. Regier (Eds.), *Microwave processing of food* (pp. 22–40, Chapter 2). Cambridge, UK: Woodhead Publishing Limited.

Too efficient an absorption by the outer layers would inevitably lead to poor heating of the internal volume within large samples.

In solids, the molecular dipoles are no longer free to rotate as they are in liquids, but are restricted to a number of equilibrium positions, separated by potential barriers. Theoretical treatments of this behavior have been formulated and are similar to those developed for liquids. The simplest model for this behavior assumes that there are two potential wells separated by a potential barrier of energy. This represents the two possible orientations of the dipole (Meredith, 1998).

2.5.4 Maxwell-Wagner (interfacial) polarization

Where a dielectric material is not homogeneous, but consists of inclusions of one dielectric in another, it is still possible to treat the material theoretically. If the dielectric properties and geometry of the inclusions are known, it is possible to arrive at expressions for the dielectric behavior of the bulk sample (see also Chapter 3: Determination of the dielectric properties of foods). The reverse problem—that of determining the dielectric properties of the components from that of the system—is generally insoluble except in the simplest of cases. The addition of dissolved salts in water markedly affects the dielectric properties as conduction increases, and may become important enough to swamp the dielectric losses. On the other hand, the

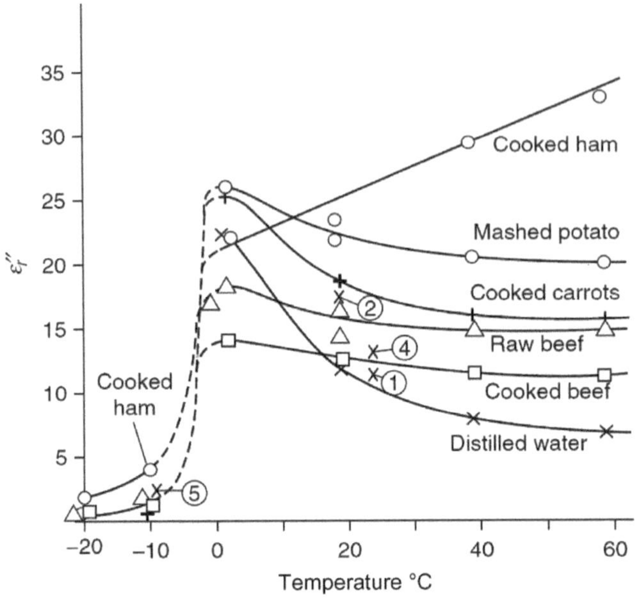

Figure 2.2 The dielectric loss factor (ϵ″) of selected foods at 3000 MHz as affected by temperature.
Source: From Bengtsson, N. B., & Risman, P. O. (1971). Dielectric properties of foods at three GHz as determined by a cavity perturbation technique. *Journal of Microwave Power and Electromagnetic Energy*, *6*, 107–123 (Bengtsson and Risman, 1971); and Tang, J. (2005). Dielectric properties of foods. In H. Schubert & M. Regier (Eds.), *Microwave processing of food* (pp. 22–40, Chapter 2). Cambridge, UK: Woodhead Publishing Limited.

dielectric losses of the majority of solids arise predominantly from these conduction terms, and may be strongly affected by temperature (Fig. 2.2).

2.6 Product formulation to optimize microwave heating

For food designed for microwavable applications, what consumers desire and what is currently available on the supermarket shelves can often be apart. Working families want healthy processed foods that are easy and rapid to prepare (microwavable is favored), and they want their foods to look, smell, and most of all taste as good as those which come out of a traditional oven. In order to accomplish this, more care is needed in the formulation of microwavable foods.

Companies are no longer content with adapting existing products for microwave cooking. Products are beginning to be designed specifically for microwave ovens, whether by means of changes in packaging or changes within the product formulation. Cylindrical-shaped products, e.g., reduce the number of sharp edges and therefore reduce the uneven heating pattern. Browning of the product may be addressed by changes in packaging (see Chapter 13: Packaging for microwave foods) and the use of susceptors that collect microwave energy and concentrate the heat generated at the surface. It is also important to understand how composition influences the heating profile of different food products. If these drawbacks are addressed, the opportunity to develop products that can be fully cooked by microwave technology is possible.

There is a considerable increase in availability and choice of convenience microwavable prepared foods in today's grocery stores. Product development has focused on both quality and safety aspects, targeting principally the microwave downfalls of lack of crisping and browning, uneven heat distribution, and compromised flavor development (Fraser & LeBlanc, 1989).

Foods cooked by conventional heating and microwave heating can produce significantly different flavors, hence the need for improved ingredients for use in microwave products formulation. The success of new applications of microwave heating in foods depends on the understanding of the interactions between the microwave energy field and the material components. Any change in material properties will directly impact on the heating pattern (Bows, 2000).

In microwavable food product development, ingredient interaction is key to successful product development. Each material component possesses a different molecular makeup which gives it a unique heating property in the microwave oven. Water, the major constituent of most food products, is the main source for microwave interactions (Oliveira & Franca, 2002). Along with water, a food item is comprised of various components that each contribute their dielectric property to make a mass that has a unique and complex microwave heating potential (Hegenbart, 1992). The principal factors governing the way microwaves heat food are the component's dielectric properties, the specific heat, mass, and shape.

2.6.1 Microwave browning

Color of processed food is a significant factor to be considered during microwave heating (Pathare, Opara, & Al-Said, 2013). The biochemical index can be used to quantify the level of enzymatic browning (Lunadei, Galleguillos, Diezma, Lleo, & Ruiz Garcia, 2011). For many foods, it is the browning during heating/cooking at temperatures above 140°C which imparts distinctive and sought-after flavors. Without this browning, foods appear tasteless and dull. The browning index is linked to the ratio of microwave output to the quantity of treated sample and as a result a significant change in color can be observed, e.g., at 9 W/g browning was experienced in okra (Dadali, Kilik Apar, & Ozbek, 2007). Two factors, time and temperature, which are typically involved in the browning process, are usually absent in microwave heating. To compensate for that, browning formulations can be considered as an added ingredient to food formulations for acceptable quality in microwave processed foods (Decareau, 1992).

Browning formulations are essentially compounds (such as glucose, sodium carbonate, lecithin, salts, yeast extract, etc.) which can help to give surface browning in a microwave oven or simply give a browned appearance (color). It is the concentration of sugars with amino acids, peptides, and proteins which results in the formation of key flavor profiles such as pyrazines, furans, and thiazoles (Liao, Raghavan, Meda, & Yaylayan, 2001).

Browning devices can also be considered to improve surface browning during microwave cooking of foods. Disposable susceptors are used to direct a part of the microwave energy to provide intense heat to the surface of the food to create a crisping and browning effect (Mehdizadeh, Chi, Riegert, Blankenbeckler, & Corcoran, 2015). The principle here lies in the conversion of microwave energy into heat in a supporting material in contact with the surface of the food to provide sufficient heat to brown the food (Decareau, 1992). The browning devices have been designed in a number of ways, dish, packaging material with susceptors, and ferromagnetic compounds (see Chapter 13: Packaging for microwave foods). The success of these devices depends on their capacity to heat faster than the food item itself.

A study was conducted to examine the effect of temperature and moisture content on color change and the result supported the fact that the change in color occurs with a significant increase in temperature (Rhim & Hong, 2011).

Recently, food ingredients manufacturers have developed browning reaction accelerators. Derived from cellulose and dextrose, they take their coloring action from the Maillard reaction and develop an attractive brown color when heated, including in microwave ovens. Caramelization of sugars occurs when a certain temperature is reached, which may be difficult to reach in a microwave oven (Davis, 1995). However, saturated sugar solutions can have two to three times higher microwave power absorption compared to pure water. Adding lactose to baked goods improves moisture retention, enhances crumb texture, increases volume, and intensifies the Maillard reaction, especially at low temperatures, a handy trick for promoting microwave browning (Knehr, 2001). Size of the added sugar crystals can also play an important role. Indeed, in bakery products, sugar is very important for the development of the dough structure through starch gelatinization. If the sugar cannot go into solution fast enough (small crystals), the process of starch gelatinization will be much too long for the short microwave cooking period (Hegenbart, 1992).

2.6.2 Microwave popcorn

In a study by Singh and Singh (1999), coating systems were tested to improve popcorn popping. Coating systems studied consisted of hydrogenated oil, butter, sodium chloride, and sodium bicarbonate. Effects on the bulk density, expansion index, and percentage of popped kernels were significant for all ingredients tested and it was demonstrated that a given coating formulation can significantly reduce the bulk density while maximizing the percentage of popped kernels and expansion index. On the other hand, sodium bicarbonate was found to have no positive (or negative) effects on microwave popping properties, while moderate levels of salt have a desirable effect on popping quality (Ceylan & Karababa, 2004).

2.6.3 Microwave baking (see also Chapter 6: Microwave-assisted baking)

In the case of microwave-baked breads, considerable dough reformulation must be considered to achieve a minimum quality. Indeed microwave-baked breads traditionally have unacceptable texture and color. Since the time of microwave heating is very fast, there is not enough time for proper starch gelatinization, enzymatic starch conversion, or dough expansion (Sumnu, 2001). Trials have shown that the gluten content is an important factor affecting microwave-baked breads. The lower the content of gluten in the flour (around 25%) the better the quality. Addition of a dough emulsifier (lecithin, soy protein, and gums) can also help to increase the dough volume and reduce the bread firmness (Ozmutlu, Sumnu, & Sahin, 2001). The addition of food-grade methylcellulose and hydroxypropyl methylcellulose in cake formulas has been demonstrated to increase final cake volume and improved crumb mouthfeel for microwave-baked cakes (Bell & Steinke, 1990). Clarke and Farrell (2000) obtained pizza dough improvements in microwave heating with the addition of a combined dough emulsifier and an oat fiber.

With the selection of proper ingredients to create the parameters needed to control moisture migration and by utilizing specially designed susceptors for browning and crisping, microwave bakery products have now reached unparalleled results (Mast, 2000; Ozmutlu et al., 2001). The selected ingredients are used to control moisture migration, as well as enhance browning and crisping. A blend of cellulose powder, food starch, gums, and vegetable oil can be designed to enhance moisture retention in dough products. A blend of alginate, cellulose powder, milk protein, gums, and enzymes can be designed to control moisture migration and enhance volume and strength of cell structure in dough products (Keskin, Sumnu, & Sahin, 2004). Whereas a blend of sugar powders, dextrin, shortening, browning agent, and vegetable oil procures good dielectric properties that enhance browning and crisping of dough products when blended with other carrier agents (Seihun, Sumnu, & Sahin, 2003; Torrealba-Meléndez et al., 2015).

Specialty starches are being formulated to perform a variety of functions in food matrices subjected to microwaves (National Starch and Chemical Co, Bridgewater, NJ, USA). These include achieving optimum crispiness, improving texture and mouthfeel, controlling oil absorption, and controlling expansion and product density. Starch can be physically modified with pregelatinization for quick viscosity development and swelling where process-tolerant products are needed such as in microwaves. Modified food starches such as amylose starch are less sensitive to water and can therefore limit sogginess to provide a crisper coating.

In a frozen ready-prepared microwave meal, the starch properties are required to have a smooth short texture to maintain viscosity in case of overheating in the microwave.

2.6.4 Microwave cooking

Cooking fresh meat in a microwave can represent a challenge since the cool air surrounding the meat results in lack of browning. While the meat heats up inside, there is mass transfer from the interior to the outside, resulting in a tough, dry and flavorless product (Taki, 1991). Ingredient selection can target these problems, such as a salt-based coating to attract microwave energy to the surface of the meat along with a coloring agent, a water-binding ingredient (starch) to reduce moisture loss, and an enzyme to retain tenderness. Flavors can be added for overall improvements.

In general, it is recommended to keep the salt concentration as low as possible to improve the penetration depth and slow the heating at the surface while providing more uniform heating and avoiding thermal runaway (Schiffmann, 1993). On the other hand, the addition of salt may be recommended in some cases as it can increase the heating rate at the surface for particular applications (Schiffmann, 1986).

2.6.5 Water migration

Water is probably the most important ingredient in any microwavable food and care has to be given to its proper management in food systems. Usually, the higher the water content, the higher the dielectric loss factor and the better the heating. Many products can benefit from having higher than normal water content to provide tolerance to rapid heating and reduce uneven heating. In microwave baking, products that contain additional water with the help of water-binding ingredients yield better results (Miller & Hoseney, 1997).

In conventional cooking, the high temperature at the surface of the product leads to dehydration and crisping at the surface. The low temperature of the product surface in microwave heating and the internal temperature gradient forces the moisture to transfer to the surface, leading to sogginess. Special product formulations have been studied to address this problem in microwave assisted (Fraser & LeBlanc, 1989). Food gums can be used to maintain consistent viscosities in microwavable foods, to emulate the functional and sensory quality of fats and to provide a film barrier between the food item and the breading to preserve its crispness while preventing the escape of moisture from the food (Dow Chemical Company, Midland, MI, USA).

Research has been conducted on nonmeat ingredients as fat/moisture binders on sensory, cooking, and compositional properties of microwave-cooked ground beef. With the addition of vegetable fiber (pea fiber), there is swelling of the fiber and fat absorption which interacts with the protein of the ground beef to form a complex matrix preventing fat and moisture release (Anderson & Berry, 2001).

Adjusting the meat composition in a microwave burger/bun combination can offer the food developer the flexibility to alter the temperature development during microwave heating (Lyng, Scully, Mckenna, Hunter, & Molloy, 2002). It is important to adjust the food with emulsifiers, gums, proteins, and sugars to manage the water content of the meat and the bun separately so that, when processed together in a microwave environment, the bun does not burn while the meat is cooking (Parker & Vollmer, 2004).

2.6.6 *Flavors*

The flavor industry has been working to meet the challenges of the microwave processing food developers with flavor systems specifically designed for microwavable food formulations.

Full-flavored taste and aromas are frequently absent in microwave-processed foods. To address this issue, food developers use ingredients that produce more complex and varied flavor precursors. Roasted flavor compounds have been formulated to ensure stability over wide ranges for processing conditions (Griffith Laboratories Worldwide Inc.).

Flavors added to microwave food formulations are very complex, since they must provide characteristic flavors, such as lemon, butter, or vanilla, along with typical roasted, toasted, and baked flavors not developed enough during microwave heating (Steinke, Frick, Strassburger, & Gallagher, 1989). The strong interaction of microwave energy with the water molecules of the food product can result in rapid loss of volatile fatty acids, leading to objectionable off-flavor developments. The presence of oil may help to limit the losses of volatile fatty acids, as was demonstrated by Steinke et al. (1989) for a variety of volatile acids in oil/water mixtures.

New encapsulated techniques are commercially being offered for controlling the release of active/functional food ingredients. One such product is hydrophobic nanospheres which are designed to be utilized to release active ingredients and flavors at a certain temperature upon heating in a microwave oven, thus providing retention of volatile flavors (Shefer & Shefer, 2003).

2.7 Future trends

As materials undergo microwave processing, they experience physical and structural transformations that, in turn, change their electrical properties and behavior. The ability of microwaves to convert electromagnetic energy into thermal energy thus varies during the process. Understanding the generation and interactions of microwaves with different materials is thus critical to the development of any agri-food microwave application. The chemical composition of the material being processed, its size and shape, and the physics of the microwave/material interactions govern their success.

Future developments for microwave heating appear very promising in chemical applications for the fractionation of biomass (Kurian, Gariepy, Orsat, & Raghavan, 2015), for material synthesis (Wang, Qu, Xu, Rena, and Qu, 2011), to breakdown emulsions in oil sands (Sawsan & Mortatha, 2013), and for biomolecule extraction (Khodaei, Karboune, & Orsat, 2016). Furthermore applications in microwave assisted drying should continue to grow as the process permits significant energy savings and the production of high quality dried commodities (Routray, Orsat, & Gariepy, 2014) (see Chapter 8: Microwave–assisted drying).

References

Anderson, E. T., & Berry, B. W. (2001). Effects of inner pea fiber on fat retention and cooking yield in high fat ground beef. *Food Research International*, *34*(8), 689–694.

Bell, D. A., & Steinke, L. W. (1990). Improved structure and texture in microwave cakes by the addition of methylcellulose and hydroxypropyl methylcellulose. *Cereal Foods World*, *35*(8), 843.

Bengtsson, N. B., & Risman, P. O. (1971). Dielectric properties of foods at three GHz as determined by a cavity perturbation technique. *Journal of Microwave Power and Electromagnetic Energy*, *6*, 107–123.

Bows, J. R. (2000). A classification system for microwave heating of food. *International Journal of Food Science and Technology*, *35*, 417–430.

Ceylan, M., & Karababa, E. (2004). The effects of ingredients on popcorn popping characteristics. *International Journal of Food Science and Technology*, *39*, 361–370.

Chan, T. V. C. T., & Reader, H. C. (2000). *Understanding microwave heating cavities*. Boston, MA: Artech House Publishers.

Chandrasekaran, S., Ramanathan, S., & Basak, T. (2013). Microwave food processing—A review. *Food Research International*, *52*, 243–261.

Clarke, C. I., & Farrell, G. (2000). The effects of recipe formulation on the textural characteristics of microwave-reheated pizza bases. *Journal of the Science of Food and Agriculture*, *80*, 1237–1244.

Curet, S., Rouaud, O., & Boillereaux, L. (2014). Estimation of dielectric properties of food materials during microwave tempering and heating. *Food Bioprocess & Technology*, *7*, 371–384.

Dadali, G., Kilik Apar, D., & Ozbek, B. (2007). Color change kinetics of okra undergoing microwave drying. *Drying Technology*, *25*(5), 925–936.

Datta, A. K., & Rakesh, V. (2013). Principles of microwave combination heating. *Comprehensive Reviews in Food Science and Food Safety*, *12*, 24–39.

Davis, E. A. (1995). Functionality of sugars: Physicochemical interactions in foods. *American Journal of Clinical Nutrition*, *62*(Suppl.), 170S–177S.

Decareau, R. V. (1992). *Microwave foods: New product development*. Connecticut, CT: Food and Nutrition Press, Inc.

Dow Chemical Company. Available from http://www.dow.com.

Fraser, J. D., & Leblanc, E. L. (1989). Microwaveability: The new wave considerations in food product formulation. *Canadian Home Economics Journal*, *39*(4), 139–143.

Gowen, A., Abu Ghanam, N., Frias, J., & Oliviera, J. (2006). Optimisation of dehydration and rehydration properties of cooked chickpeas (*Cicer arietinum* L.) undergoing microwave–hot air combination drying. *Trends in Food Science & Technology*, *17*(4), 177–183.

Griffith Laboratories Worldwide Inc. USA. http://www.griffithlabs.com.

Hegenbart, S. (1992). Microwave quality: Coming of age, *Food Product Design*, *June 1992*.

Hossan, M. R., Byun, D., & Dutta, P. (2010). Analysis of microwave heating for cylindrical shaped objects. *International Journal of Heat and Mass Transfer*, *53*(23), 5129–5138.

Keskin, S. O., Sumnu, G., & Sahin, S. (2004). Usage of enzymes in a novel baking process. *Nahrung/Food*, *48*(2), 156–160.

Khodaei, N., Karboune, S., & Orsat, V. (2016). Microwave-assisted alkaline extraction of galactan-rich rhamnogalacturonan I from potato cell wall by-product. *Food Chemistry*, *190*(1), 495–505.

Knehr, E. (2001). Dairy strikes a balance: Nutrition, flavour and functionality. *Food Product Design, June 2001*(Suppl.).

Kurian, J. K., Gariepy, Y., Orsat, V., & Raghavan, G. S. V. (2015). Comparison of steam-assisted versus microwave-assisted treatments for the fractionation of sweet sorghum bagasse. *Bioresources and Bioprocessing, 2*, 30. Available from: http://dx.doi.org/10.1186/s40643-015-0059-3.

Lew, A., Krutzik, P. O., Hart, M. E., & Chamberlin, A. R. (2002). Increasing rates of reaction: Microwave-assisted organic synthesis for combinatorial chemistry. *Journal of Combinatorial Chemistry, 4*(2), 95–105.

Liao, X., Raghavan, G. S. V., Meda, V., & Yaylayan, V. A. (2001). Dielectric properties of supersaturated α-D-glucose aqueous solutions at 2450 MHz. *Journal of Microwave Power & Electromagnetic Energy, 36*(3), 131–138.

Lunadei, L., Galleguillos, P., Diezma, B., Lleo, L., & Ruiz Garcia, L. (2011). A multispectral vision system to evaluate enzymatic browning in fresh-cut apple slices. *Postharvest Biology and Technology, 60*, 225–234.

Lyng, J. G., Scully, M., Mckenna, B. M., Hunter, A., & Molloy, G. (2002). The influence of compositional changes in beef burgers on their temperatures and their thermal and dielectric properties during microwave heating. *Journal of Muscle Foods, 13*(2), 123–142.

Mast, R. (2000). Application of microwave technology to bakery products, British Society of Baking, BSB Conference, October 2000, Paper 441. http://britishsocietyofbaking.org.uk/conference-2000/.

Mehdizadeh, M., Chi, H., Riegert, R., Blankenbeckler, N., & Corcoran, W. (2015). Microwave browning of foods: Recent advancements in improving uniformity with a novel susceptor system. In *Conference presentation, 49th annual international microwave power symposium (IMPI 49)*, San Diego, California, June 2015.

Meredith, R. J. (1998). *Engineer's handbook of industrial microwave heating*. London, UK: The Institution of Electrical Engineers IEEE Publications.

Miller, R. A., & Hoseney, R. C. (1997). Method to measure microwave induced toughness in bread. *Journal of Food Science, 62*, 1202–1204.

Miura, N., Yagihara, S., & Mashimo, S. (2003). Microwave dielectric properties of solid and liquid foods investigated by time-domain reflectometry. *Journal of Food Science, 68*(4), 1396–1403.

National Starch And Chemical Co. Available from: http://www.nationalstarch.com.

Ohlsson, T. (1983). Fundamentals of microwave cooking. *Microwave World, 4*(2), 4.

Oliveira, M. E. C., & Franca, A. A. (2002). Microwave heating of foodstuffs. *Journal of Food Engineering, 53*, 347–359.

Ozmutlu, O., Sumnu, G., & Sahin, S. (2001). Effects of different formulations on the quality of microwave baked breads. *European Food Research and Technology, 213*(1), 38–42.

Parker, K., & Vollmer, M. (2004). Bad food and good physics: The development of domestic microwave cookery. *Physics Education, 39*(1), 82–90.

Pathare, P. B., Opara, U. L., & Al-Said, F. A. J. (2013). Colour measurement and analysis in fresh and processed foods: A review. *Food and Bioprocess Technology, 6*(1), 36–60.

Peyre, F., Datta, A., & Seyler, C. (1997). Influence of the dielectric property on microwave heating patterns: Application to food materials. *Journal of Microwave Power and Electromagnetic Energy, 32*(1), 3–14.

Rhim, J. W., & Hong, S. I. (2011). Effect of water activity and temperature on the color change of red pepper (*Capsicum annuum* L.) powder. *Food Science and Biotechnology*, *20*(1), 215–222.

Routray, W., Orsat, V., & Gariepy, Y. (2014). Effect of different drying methods on the microwave extraction of phenolic components and antioxidant activity of highbush blueberry leaves. *Drying Technology*, *32*(16), 1888–1904.

Salazar-Gonzalez, C., San Martin-Gonzalez, M. F., Lopez Malo, A., & Sosa-Morales, M. E. (2012). Recent studies related to microwave processing of fluid foods. *Food and Bioprocess Technology*, *5*(1), 31–46.

Sawsan, M., & Mortatha, M. (2013). The application of microwave technology in demulsification of water-in-oil emulsion for missan oil fields. *Iraqi Journal of Chemical and Petroleum Engineering*, *14*(2), 21–27.

Schiffmann, R. F. (1986). Food product development for microwave processing. *Food Technology*, *June 1986*, 94–98.

Schiffmann, R. F. (1993). Understanding microwave reactions and interactions. *Food Product Design*, *April 1993*.

Seihun, N., Sumnu, G., & Sahin, S. (2003). Effects of different emulsifier types, fat contents, and gum types on retardation of staling of microwave-baked cakes. *Nahrung/Food*, *47* (4), 248–251.

Shefer, A., & Shefer, S. (2003). Novel encapsulation system provides controlled release of ingredients. *Food Technology*, *57*(11), 40–42.

Singh, J., & Singh, N. (1999). Effects of different ingredients and microwave power on popping characteristics of popcorn. *Journal of Food Engineering*, *42*, 461–465.

Singh, R. P., & Heldman, D. R. (2009). *Introduction to food engineering* (4th ed.). Burlington, San Diego, London: Academic Press.

Sosa-Morales, M. E., Valerio-Junco, L., Lopez-Malo, A., & Garcia, H. S. (2010). Dielectric properties of foods: Reported data in the 21st century and their potential applications. *LWT Food Science and Technology*, *43*, 1169–1179.

Steinke, J. A., Frick, C., Strassburger, K., & Gallagher, J. (1989). Interaction of flavor systems in the microwave environment. *Cereal Foods World*, *34*(4), 330–332.

Suarez, C., Viollaz, P. E., Rovedo, C. O., Tolaba, M. P., & Haros, M. (2000). Improved drying techniques and microwave food processing. *Minimally Processed Fruits and Vegetables*, 175–188.

Sumnu, G. (2001). A review on microwave baking of foods. *International Journal of Food Science and Technology*, *36*, 117–127.

Taki, G. H. (1991). Functional ingredient blend produces low-fat meat products to meet consumer expectations. *Food Technology*, *November 1991*, 71–74.

Tang, J. (2005). Dielectric properties of foods. In H. Schubert, & M. Regier (Eds.), *Microwave processing of food* (pp. 22–40). Cambridge, UK: Woodhead Publishing Limited, Chapter 2.

Tang, J., Hao, F., & Lau, M. (2002). Microwave heating in food processing. *Advances in Bioprocessing Engineering*, 1–44, Chapter 1.

Torrealba-Meléndez, R., Sosa-Morales, M. E., Olvera-Cervantes, J. L., & Corona-Chávez, A. (2015). Dielectric properties of cereals at frequencies useful for processes with microwave heating. *Journal of Food Science and Technology*, *52*(12), 8403–8409.

Vadivambal, R., & Jayas, D. S. (2010). Non-uniform temperature distribution during microwave heating of food materials—A review. *Food and Bioprocess Technology*, *3* (2), 161–171.

Venkatesh, M. S., & Raghavan, G. S. V. (2004). An overview of microwave processing and dielectric properties of agri-food materials. *Biosystems Engineering*, *88*(1), 1–18.

Wang, J., Tang, J., Wang, Y., & Swanson, B. (2009). Dielectric properties of egg whites and whole eggs as influenced by thermal treatments. *LWT – Food Science and Technology*, *42*, 1204–1212.

Wang, X., Qu, K., Xu, B., Rena, J., & Qu, X. (2011). Microwave assisted one-step green synthesis of cell-permeable multicolor photoluminescent carbon dots without surface passivation reagents. *Journal Materials Chemistry*, *21*, 2445–2450.

Wang, Y., & Wang, J. (2009). Computer simulation of radio frequency heating. In S. Jun, & J. M. Irudayaraj (Eds.), *Food processing operations modeling* (pp. 81–112). Boca Ratón, FL: CRC Press.

Yam, K. L., & Lai, C. C. (2006). Microwavable frozen food or meals. In Y. H. Hui (Ed.), *Handbook of food science, technology and engineering* (pp. 1131–1138). Boca Ratón, FL: CRC Press.

Zhang, M., Tang, J., Mujumdar, A. S., & Wang, S. (2006). Trends in microwave-related drying of fruits and vegetables. *Trends in Food Science & Technology*, *17*(10), 524–534.

3 Determination of the dielectric properties of foods

M. Regier[1], K. Knoerzer[2] and H. Schubert[3]
[1]Trier University of Applied Sciences, Trier, Germany, [2]CSIRO Agriculture and Food, Melbourne, VIC, Australia, [3]Karlsruhe Institute of Technology (KIT), Karlsruhe, Germany

3.1 Introduction

In the literature, many different methods for the measurement of dielectric properties, the properties that describe the ability to store and dissipate electric field energy (see Chapter 2: Microwave heating and the dielectric properties of foods), can be found, depending on the frequency range of interest and the values of the dielectric properties. In the microwave frequency range, most of the methods have been originally developed for materials that are used in high frequency and radio frequency telecommunication devices. Thus, the majority of these techniques aimed originally for materials with low losses in opposition to most of the (high moisture) food substances.

The principles of microwave dielectric properties determination have been reviewed previously, e.g., in Engelder and Buffler (1991), Kaatze and Giese (1980), and Rost (1978). In this chapter, an overview of the most common measurement techniques for dielectric properties of foods in the microwave frequency range will be given. After the classification and descriptions of the different methods on a phenomenological level, further analysis techniques, yielding more information on unknown dielectric properties are shown. Summarizing the main advantages and disadvantages should help to choose the best technique for the material (depending on its state, geometric form, and expected value of dielectric properties) and frequency range of interest. The chapter is completed by a literature reference section.

3.2 Classification

In general, the measurement techniques can be divided into time domain and frequency domain methods, whereby for food substances the frequency domain methods are more common and provide greater accuracy, also due to the availability of modern network analyzers. Therefore, in this chapter only the frequency domain methods are presented.

The classification continues by the differentiation between the measurement in open and closed structures.

The Microwave Processing of Foods. DOI: http://dx.doi.org/10.1016/B978-0-08-100528-6.00003-6

Closed systems are distinguished between transmission lines and resonators (which can be further divided into completely filled and partially filled resonators), the common distinction in open structure measurements is between open-ended probes and real free-space techniques.

3.2.1 Closed structures

Closed systems are setups in which the electromagnetic waves are restricted in a confined space, surrounded by metallic walls.

3.2.1.1 Transmission lines

As transmission lines, both coaxial lines as well as simple waveguides (commonly in rectangular or circular configuration) can be used. To measure dielectric properties within such transmission lines, the transmission line first has to be characterized without the material to be measured. After introducing the material, the measurement of the standing wave ratio (SWR) or of the reflection or transmission coefficients by network analyzers (with wide frequency ranges) can yield the unknown dielectric properties of the material. The detection of the reflected waves is more common than transmission measurements.

In the case of SWR measurements, the shift of nodal positions of the developed standing wave and the node width related to SWRs, the sample and waveguide dimensions, as well as the incident field mode are used (Nelson, 2010; Nelson, Stetson, & Schlaphoff, 1974) to determine the dielectric properties; analysis is often performed through commercially available software, as well as user-developed programs.

For these reflection measurements, two end types of transmission lines have been used: either a fixed shortening end, on which the sample is directly placed, or a sliding short circuit-plunge. Examples of both types are shown in Fig. 3.1. The basics of the measurement are illustrated by Roberts and von Hippel (1946) and have been generalized and ameliorated by Nelson et al. (1974) and Roussy, Ghanem, Thiebaut, and Dichtel (1990).

In all cases, the material to be measured has to be geometrically well defined and provide a smooth fit into the prepared sample holder to minimize errors that may occur due to air gaps or twists. Additionally, there is a limit on the maximum permittivity of the material to be determined, which is important to prevent the waveguide from containing field modes that are different from the desired mode. Thus, in an application note of an instrument manufacturer (Agilent, 2014), the method is claimed best for lossy to low loss machineable solids. Although special calibration methods to eliminate errors caused by unknown reflections in the waveguide exist, the above prerequisites often prevent these techniques from the use for food materials (Engelder & Buffler, 1991). Nonetheless, various successful measurements have been done for bulks of cereals, seed and powdered or pulverized materials by, for example, Nelson (1972, 1983), Nelson and You (1989), and You and Nelson (1988).

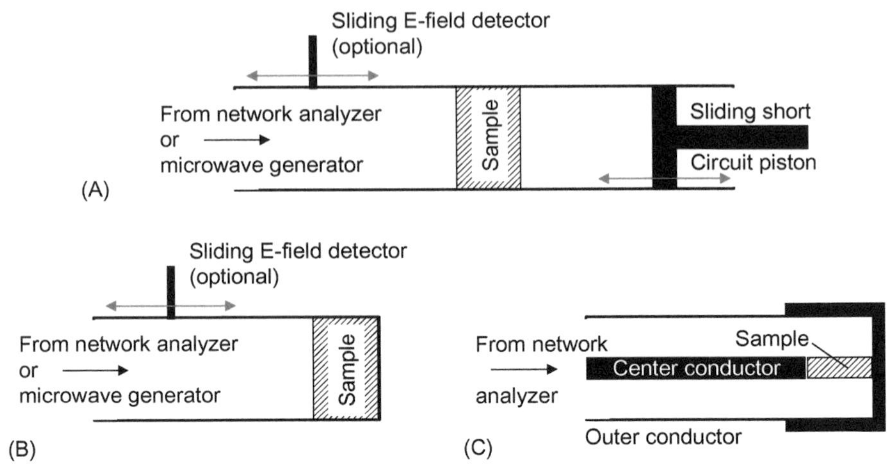

Figure 3.1 Various forms of transmissions lines for the determination of dielectric (A) waveguide with sliding shortening; (B) waveguide (sample at shortened end); (C) coaxial line shortened.
Source: Adapted from Roussy, G., & Pearce, J. A. (1995). *Foundations and industrial applications of microwaves and radio frequency fields*. New York: Wiley (Roussy & Pearce, 1995).

3.2.1.2 Cavity resonators

Whereas in transmission lines it is advantageous to measure materials with high losses, materials with low losses require a different measuring method since the attenuation by these materials is often too weak for exact permittivity calculations. Cavity resonators are used instead, which are able to amplify the electric field (in certain regions) by the quality factor Q and, thus, the interaction between electric field and material characterized by the dielectric properties.

A resonator is a cavity enclosed by conducting (metallic) walls, in which electromagnetic standing waves can be excited by an oscillating coupling device. Common resonators for dielectric measurements are cuboid or circular cylinders, since for these geometries analytical solutions for the standing waves based on Maxwell's equations with boundary conditions (e.g., no electric field component parallel to conducting walls may exist) can be found. As in the case of measurements with transmission lines, in the case of resonators only measurements at discrete, fixed (resonant) frequencies can be performed. If measurements at different frequencies should be performed, either another resonant mode or even another resonator with changed geometrical dimensions has to be chosen. Since the dielectric properties are determined by a change of the resonance (resonant frequency and quality factor), the measurement frequency is furthermore dependent on the material (see Fig. 3.2).

Thus, to determine the dielectric properties, the resonant curve, which means the reflection or transmission coefficient of the resonator over the frequency, has to be

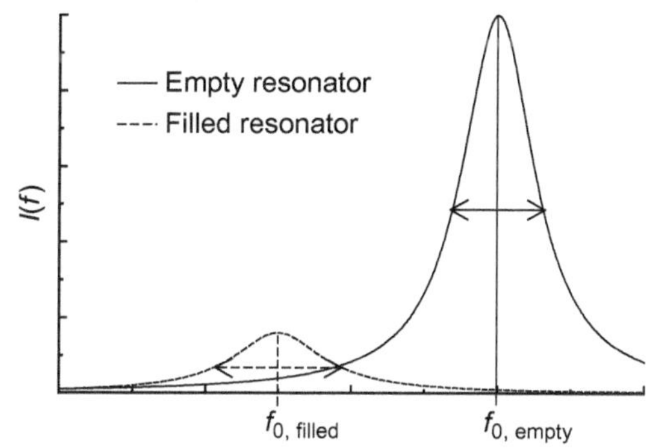

Figure 3.2 Resonant frequency f_0 shift and broadening of the resonant curve of a resonator due to a dielectric filling. The arrows show the free width at half maximum $\Delta f_{1/2}$.

determined. For this task, network analyzers can be effectively used as the source of electromagnetic waves of defined frequency and the detector of the reflections.

One can distinguish between completely filled and partially filled resonators, depending on the part of the space that is occupied by the dielectric material. Whereas for completely filled resonators cuboid and circular cylindrical resonators are common for the dielectric measurements of foods, in the case of partially filled resonators the circular cylindrical form is preferred (Bengtsson & Risman, 1971; Gallone, Lucardesi, Martinelli, & Rolla, 1996; Janezic, 2003; Persch, 1997; Risman & Bengtsson, 1971).

For both types of cavities, two types of measurements are possible:

An absolute measurement, which means, that the dielectric properties are derived from a theoretical solution of the resonance in the cavity, by (an approximate) solution of Maxwell's equations or a measurement by calibration, where first the response (of frequency and quality factor) of the resonance on insertion of materials with known dielectrics are studied, yielding calibrations curves, before inserting the material with the unknown properties.

As will be shown later, for completely filled cavities, it is relatively simple to derive the describing theoretical equations, hence, the absolute measurement method is preferred. In partially filled resonators, this derivation is only possible for simple geometries where symmetries can be utilized (see corresponding Section Partially filled resonators or Regier & Schubert (2000)). This is significantly more complex. Therefore, in practice the calibration method is used more commonly.

Completely filled resonators

If both the real and the imaginary part of the dielectric properties to be measured are rather small, completely filled resonators are better suited, since then the interaction between the electric field and the material and, therefore, the sensitivity is maximized. Another strength of this form of resonators is the ease of construction and the theoretical description of the resonance and, thus, its analysis yielding the dielectric properties.

The derivation of the resonance formulas should be given in detail for the simplest case of a cuboid cavity resonator, in order to show the principle proceeding.

As shown in Chapter 1, Introducing microwave-assisted processing of food: fundamentals of the technology, the wave Eqs. (3.1 and 3.2) for the electric and the magnetic field can be derived from the Maxwell's equations.

$$\left(\Delta - \frac{1}{\mu\mu_0\varepsilon\varepsilon_0}\cdot\frac{\partial^2}{\partial t^2}\right)\vec{E}(\vec{r},t) = 0 \tag{3.1}$$

$$\left(\Delta - \frac{1}{\mu\mu_0\varepsilon\varepsilon_0}\cdot\frac{\partial^2}{\partial t^2}\right)\vec{B}(\vec{r},t) = 0 \tag{3.2}$$

Additionally, a dependence of the magnetic on the electric field and vice versa is described by Maxwell's equations (see Chapter 1: Introducing microwave-assisted processing of food: fundamentals of the technology, Eq. 1.3, or Table 1.1). In the case of an enclosed metallic cavity, the boundary conditions of Table 1.2 are valid. That means that at the metallic boundaries no parallel component of the electric field and no orthogonal component of the magnetic flux may exist (Eqs. 3.3 and 3.4).

$$\vec{E}_{||} = 0 \text{ at metallic walls} \tag{3.3}$$

$$\vec{B}_{\perp} = 0 \text{ at metallic walls} \tag{3.4}$$

For the solution, we take the x-component of the electric field E_x and use a separation approach:

$$E_x = f(x)\cdot g(y)\cdot h(z)\cdot e^{-i\omega t} \tag{3.5}$$

where a time harmonic behavior is suggested (and only the real part of the solution is physically relevant).

With this approach, the wave Eq. (3.1) yields:

$$\begin{aligned} &f''(x)\cdot g(y)\cdot h(z)\cdot e^{-i\omega t} + f(x)\cdot g''(y)\cdot h(z)\cdot e^{-i\omega t} + f(x)\cdot g(y)\cdot h''(z)\cdot e^{-i\omega t} \\ &+ \frac{\omega^2}{\mu\mu_0\varepsilon\varepsilon_0} f(x)\cdot g(y)\cdot h(z)\cdot e^{-i\omega t} = 0 \\ &\Rightarrow \frac{f''(x)}{f(x)} + \frac{g''(y)}{g(y)} + \frac{h''(z)}{h(z)} + \frac{\omega^2}{\mu\mu_0\varepsilon\varepsilon_0} = 0 \end{aligned} \tag{3.6}$$

Since every part of the sum is only dependent on one variable and the equation is valid for all combinations, these parts have to be equal to constants $\frac{f''}{f} = -k_x{}^2$, $\frac{g''}{g} = -k_y{}^2$, $\frac{h''}{h} = -k_z{}^2$, whereby:

$$-k_x{}^2 - k_y{}^2 - k_z{}^2 + \frac{\omega^2}{\mu\mu_0\varepsilon\varepsilon_0} = 0 \tag{3.7}$$

Again, a harmonic approach can be chosen:

$$
\begin{aligned}
f(x) &= \sin(k_x x + \phi_x) \\
g(y) &= \sin(k_y y + \phi_y) \\
h(z) &= \sin(k_z z + \phi_z)
\end{aligned}
\tag{3.8}
$$

so that the general solution can be written as:

$$
E_x(x, y, z, t) = A_x \cdot \sin(k_x x + \phi_x) \cdot \sin(k_y y + \phi_y) \cdot \sin(k_z z + \phi_z) \cdot e^{-i\omega t} \tag{3.9}
$$

This solution can be easily generalized to E_y and E_z and has to fulfil the boundary conditions Eqs. (3.3 and 3.4). In the case of a cuboid cavity bordered by metallic walls at $x = 0$, as well as $x = L_x$, $y = 0$, as well as $y = L_y$ and $z = 0$, as well as $z = L_z$, E_x is parallel to the walls at $y = 0$, $y = L_y$, $z = 0$ and $z = L_z$, where E_x has to vanish, i.e., $E_x = 0$.

This means, that the corresponding phase factors disappear diminish: $\phi_y = 0 = \phi_z$ and the corresponding wave numbers have to fulfil Eq. (3.10), with m and n natural numbers.

$$
\begin{aligned}
k_y &= \frac{m \cdot \pi}{L_y} \\
k_z &= \frac{n \cdot \pi}{L_z}
\end{aligned}
\tag{3.10}
$$

For the y- and z-component of the electric field, very similar equations, but with generally different numbers l, l', m', and n' (all natural numbers) and frequencies (ω', ω'') are valid. Only the coupling by Maxwell's and the constitutive equations Eqs. (1.2 and 1.6) yields the equality of $l = l'$, $m = m'$, $n = n'$, and $\omega = \omega' = \omega''$:

$$
\begin{aligned}
E_x &= A_x \cdot \cos\left(\frac{l \cdot \pi}{L_x} \cdot x\right) \cdot \sin\left(\frac{m \cdot \pi}{L_y} \cdot y\right) \cdot \sin\left(\frac{n \cdot \pi}{L_z} \cdot z\right) \cdot e^{-i\omega t} \\
E_y &= A_y \cdot \sin\left(\frac{l \cdot \pi}{L_x} \cdot x\right) \cdot \cos\left(\frac{m \cdot \pi}{L_y} \cdot y\right) \cdot \sin\left(\frac{n \cdot \pi}{L_z} \cdot z\right) \cdot e^{-i\omega t} \\
E_z &= A_z \cdot \sin\left(\frac{l \cdot \pi}{L_x} \cdot x\right) \cdot \sin\left(\frac{m \cdot \pi}{L_y} \cdot y\right) \cdot \cos\left(\frac{n \cdot \pi}{L_z} \cdot z\right) \cdot e^{-i\omega t}, \qquad l, m, n \in N_0
\end{aligned}
\tag{3.11}
$$

and the coupling of the amplitudes A_x, A_y, A_z, so that only two of them are independent. The corresponding magnetic fields, which also fulfil the boundary conditions, can be determined by Maxwell's equations of Table 1.1. By adaption of Eq. (3.7) the so-called "resonant frequencies" ω or f can be calculated by Eq. (3.12a):

$$f_{lmn} = \frac{\omega_{lmn}}{2\pi} = \frac{\sqrt{\mu\mu_0\varepsilon\varepsilon_0}}{2}\sqrt{\frac{l^2}{L_x^2} + \frac{m^2}{L_y^2} + \frac{n^2}{L_z^2}} \tag{3.12a}$$

The numbers l, m, and n describe the resonant or eigen-mode, in which the resonator is applied. Thus, l, m, and n are the number of local maxima of the electric field observed along the x-, y-, and z-axis, respectively. Conversely, Eq. (3.12a) shows that the dielectric filling of a resonator with known dimensions and applied mode changes the resonant frequency and thus the dielectric constant can be determined.

$$\varepsilon' = \frac{4 \cdot f_{lmn}^2}{\mu\mu_0\varepsilon_0\sqrt{\frac{l^2}{L_x^2} + \frac{m^2}{L_y^2} + \frac{n^2}{L_z^2}}} \tag{3.13}$$

Whereas in cuboid resonators no special axis can be generally marked, in the case of circular cylindrical cavities, this role is occupied by the cylinder axis. Thus, transversal electric (TE) and transversal magnetic modes (TM), which mean that no electric or no magnetic field component parallel to the cylinder axis exists, can be distinguished.

Similar derivations for the circular cylindrical cavities with radius R and height L_z yield:

$$f_{lmn}^{\text{TE}} = \frac{\omega_{lmn}}{2\pi} = \frac{\sqrt{\mu\mu_0\varepsilon\varepsilon_0}}{2}\sqrt{\left(\frac{\alpha'_{lm}}{\pi R}\right)^2 + \frac{n^2}{L_z^2}}, \quad \text{for TE modes} \tag{3.12b}$$

$$f_{lmn}^{\text{TM}} = \frac{\omega_{lmn}}{2\pi} = \frac{\sqrt{\mu\mu_0\varepsilon\varepsilon_0}}{2}\sqrt{\left(\frac{\alpha_{lm}}{\pi R}\right)^2 + \frac{n^2}{L_z^2}}, \quad \text{for TM modes} \tag{3.12c}$$

Due to the cylindrical symmetry, Bessel's functions are needed as field solutions and the mth zeros of the Bessel's function of order l, α_{lm}, and of its first derivative, α'_{lm}, play a role in the resonant frequency.

For practical purposes, only modes with very low mode-numbers l, m, and n are chosen, since the mode density (the number of modes per frequency interval) strongly increases with these numbers and thus ambiguities and even degenerations are more prominent.

Considering the dielectric losses, the resonant curve is not only one peak at one discrete resonant frequency, but also a resonant curve with a finite line width (or a

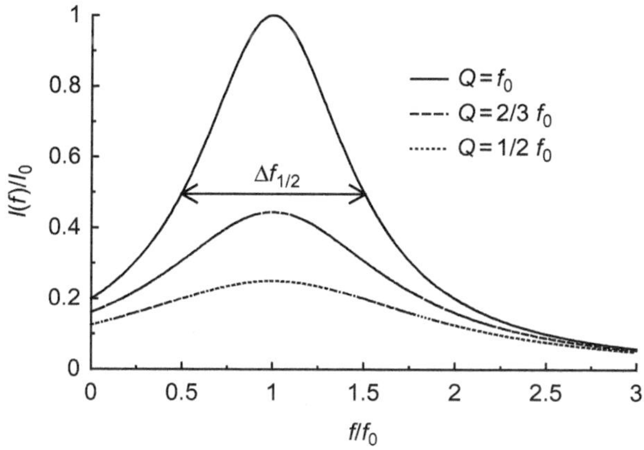

Figure 3.3 Quadratic Lorentzian curves with quality factor as parameter.

finite quality factor). This resonant curve can be described by a quadratic Lorentzian line Eq. (3.14) (see Fig. 3.3):

$$I(f) \propto \frac{f_0{}^2}{(f-f_0)^2 + \left(\frac{f_0}{2Q}\right)^2} \tag{3.14}$$

with f_0 the resonant frequency and Q the quality factor. This means that the intensity at the resonant frequency is scaled by a factor Q^2.

If $\Delta f_{1/2}$ means the full width at half maximum, the quality factor can be derived to:

$$Q = \frac{f_0}{\Delta f_{1/2}} \tag{3.15}$$

Nonetheless, experimentally, a fit of a (modified) Lorentzian line to the measured resonant curve will yield more reliable results than only taking the two parameters resonant frequency and full width at half maximum.

By introducing dielectric losses, the quality factor of the resonator can be calculated by Metaxas and Meredith (1983):

$$Q_{\text{loss}} = \frac{\varepsilon'}{\varepsilon''} + \frac{\iiint\limits_{V_{\text{empty}}} \left|\vec{E}\right|^2 dV}{\iiint\limits_{V_{\text{filled}}} \left|\vec{E}\right|^2 dV} \quad \underset{\text{filled cavity}}{\overset{\text{completely}}{=}} \quad \frac{\varepsilon'}{\varepsilon''} \tag{3.16}$$

In the case of a completely filled cavity, the integral in the nominator vanishes and the equation becomes simple. Additional to the losses in the dielectric material, broadening the resonance, losses in the nonideally conducting cavity walls, which are also present in the empty cavity (and which are assumed to stay nearly unchanged for one mode), have to be taken into account by an empty quality factor Q_{empty}. Thus, the measured quality factor Q is determined by:

$$\frac{1}{Q} = \frac{1}{Q_{\text{empty}}} + \frac{1}{Q_{\text{loss}}} = \frac{1}{Q_{\text{empty}}} + \frac{\varepsilon''}{\varepsilon'} \tag{3.17}$$

Accordingly, the dielectric loss factor can be calculated from Eq. (3.18):

$$\varepsilon'' = \varepsilon' \cdot \left(\frac{1}{Q} - \frac{1}{Q_{\text{empty}}}\right) \tag{3.18}$$

Thus, measuring the quality factor of the empty and the filled resonator yields the dielectric loss factor, assuming the real part has been determined earlier by the line shift Eq. (3.13).

Although the governing Eqs. (3.12a,b,c and 3.18) are relatively simple for this kind of completely filled resonators, also in this case calibration methods have often been used (Risman & Bengtsson, 1971).

In the following section, some governing equations for special partially filled cavities with simple geometry are presented, as well as other geometries reported in literature, where exclusively calibration methods are common.

Partially filled resonators

The solution of Maxwell's equations or the following wave equations for partially filled cavities starts very similar to the completely filled case. The only difference is to include the boundary conditions at the interfaces between the dielectric filling and the empty space. Analytically, this can only be covered for easy geometries or symmetries, still yielding complicated equations. For cuboid cavities, this means that the dielectric material has to be located directly in front of one metallic wall, whereas for the cylindrical geometry concentric dielectric fillings are appropriate (see Fig. 3.4). The following section is limited to the latter ones, which are more common in practice.

For the concentric circular cylinder geometry, the governing equations can be found in literature for TM010 (Metaxas, 1974, 1996) and TE011 modes (Regier, 2003; Regier & Schubert, 2000), respectively. In both cases, TM010 and TE011, a completely closed analytical expression cannot be derived for ε', but the numerical solution is a simple task for numerical solvers.

The disadvantage of TM010 mode is its maximum electric field value at the cylinder axis where the dielectric sample has to be placed. On one hand, this yields a strong interaction also with low loss and low dielectric constant materials, allowing accurate measurements on these materials. On the other hand, the strong decrease of the resonant frequency may shift it away from the frequency range of interest. Besides, already for medium lossy materials, the strong damping makes very small

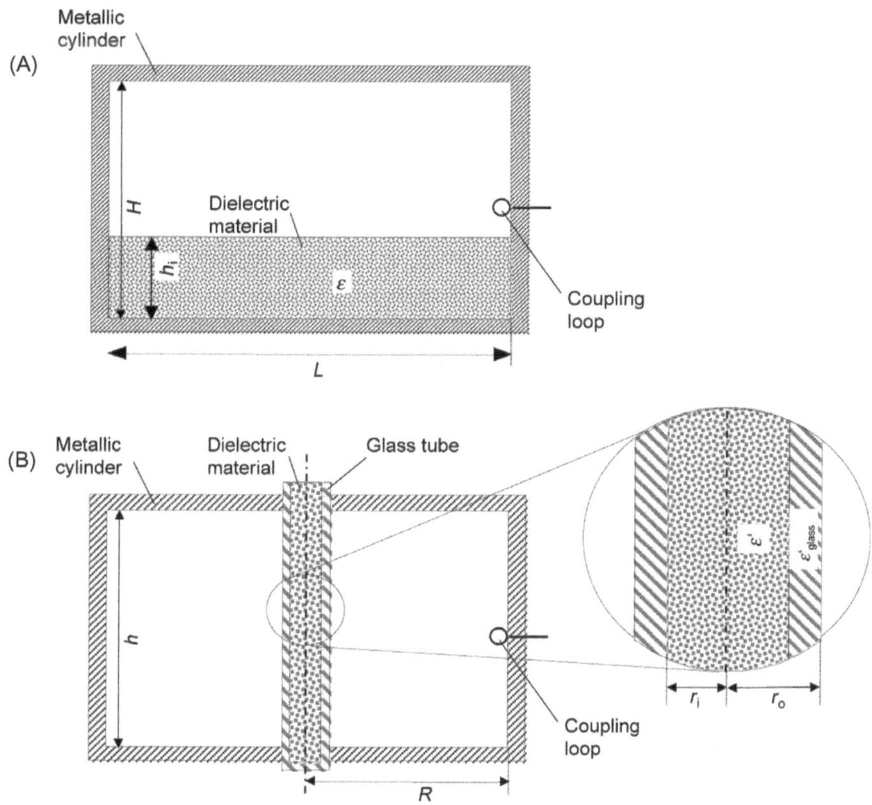

Figure 3.4 Typical geometries for the determination of dielectric properties in partially filled cavities. (A) Cuboid geometry; (B) Concentric cylinder geometry.

sample tubes (down to a diameter of 0.6 mm; Metaxas & Meredith, 1983) necessary, which is not well suited for often heterogeneous food materials, such as emulsions or granules.

Thus, for materials with higher dielectric constants and losses, the cylindrical cavity in TE011 mode with central dielectric filling is better suited. For TE011, the electric field minimum is at the center axis (Regier & Schubert, 2000), hence, the dielectric field interaction is strongly reduced. Therefore, the resonant frequency (the frequency where the measurement takes place) of the filled resonator is changed only marginally, but strongly enough for the accurate dielectric constant determination, and the damping by dielectric losses can also be analyzed. This makes it possible to use glass tubes of larger radii than required in TM010 mode, and hence, also more heterogeneous materials, such as powders or granules can be characterized.

3.2.2 Open structures

As opposed to the closed structures explained above, the open structure techniques may have the advantage to cover not only discrete frequencies but also to characterize the material over a wide frequency range. This is due to the fact that in open structures, with negligible multiple reflections, no characteristic modes for wave propagation exist and, therefore, wave propagation, transmission, and reflection is possible at (more or less) every frequency.

3.2.2.1 Free-space techniques

Although free-space systems with open structures have some advantages, e.g., concerning the possible geometry, the possible heterogeneity of samples, and the contactless measurements, free-space techniques are not commonly preferred for food substances. However, it is a well-established technique for the characterization of ceramic composite materials (Maurens et al., 1992; Musil & Zacek, 1986). This is due to the fact that for high loss materials, as is common for food materials, high microwave power would be needed for measurements. Nonetheless, in some cases, especially for larger (and rather dry) grain samples, free-space techniques have been used, particularly at high frequencies above 4 GHz (Kraszewski & Nelson, 1990; Nelson, 2010; Nelson & Trabelsi, 2004; Trabelsi & Nelson, 2003).

Common to all free-space techniques is the position of the unknown sample between the transmitting and receiving antennas. Similar to the waveguide techniques, the complex permittivity is determined from amplitude and phase variations of (mostly) plane waves transmitted through or reflected by the sample.

The most simple measurement setup geometry is a slab of the dielectric material with parallel plane surfaces. It is important that the sample dimension is chosen large enough, so that edge diffraction effects are negligible and the governing equations remain simple. In order to reduce the necessary sample dimensions, the transmitted or reflected wave is focused by dielectric lenses (Musil & Zacek, 1986) or specially formed metallic reflectors (Maurens et al., 1992). Furthermore, these setups help to generate a plane wave (defined by an identical phase in a plane perpendicular to the wave propagation) from an incident spherical wave as supplied by a microwave horn antenna (Maurens et al., 1992 and references therein). Through calibration routines, errors due to multiple reflections between the horns and the sample's surface can be eliminated.

Whereas dielectric lenses have the disadvantage of a limited bandwidth due to the frequency dependence of diffraction index or of the dielectric properties (the so-called dispersion), for metallic reflectors shadow effects of the feed horn may be problematic.

The two typical setups are shown in Fig. 3.5. They consist of a PC controlling a network analyzer (or a generator and detector), that generates and detects the electromagnetic waves at a defined frequency. In both cases, the waves are emitted by one of the horn antennas. Either the dielectric lens or the ellipsoidal reflector (one of the foci of the reflector is at the position of the horn, the other one is located at the sample's position) focuses the wave to the position of the sample. The interaction of the electromagnetic wave with the sample changes characteristically for the amplitude and the phase of the wave, which is received by the receiver horn and detected and analyzed by the network analyzer and the PC. An application

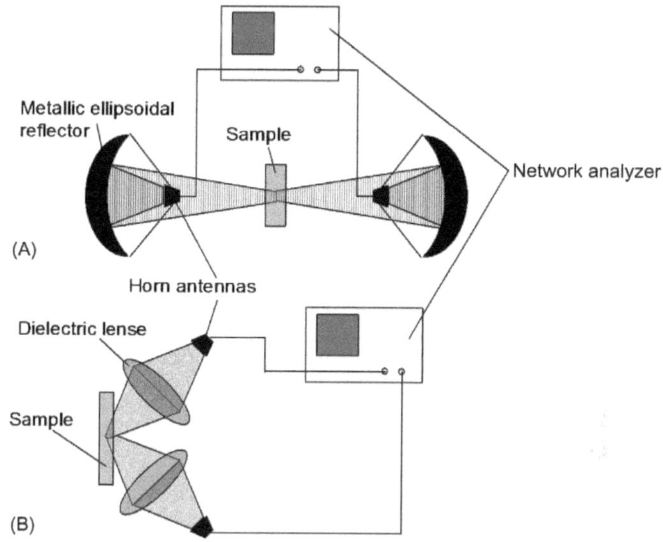

Figure 3.5 Typical setups for free-space measurement techniques. (A) Transmission measurements; (B) Reflection measurements. Although the transmission measurement is shown here with metallic ellipsoidal reflectors and the reflection measurement with dielectric lenses, each type can be operated in the same way with the other method, respectively.

note of an instrument manufacturer (Agilent, 2014) mentions the possibility to place the sample into a furnace with microwave transparent windows to allow for temperature-dependent measurements even at high temperatures.

3.2.2.2 *Open-ended coaxial probe*

The open-ended coaxial probe is generally also an open structure method, or even a "free-space" method, but is normally not referred to in that way. Since this method is commonly used for the dielectric measurement of foods (with high dielectric losses), it will be described here in more detail.

The source (and detector) of electromagnetic waves of defined frequency is again a network analyzer. The generated waves are coupled into a coaxial line, which has an open end, formed as a tip being in direct contact with the material under test (see Fig. 3.6).

Thus, in first approximation (Stuchly & Stuchly, 1980), the impedance Z of the probe is assumed to be purely capacitive and consists of two parallel capacitances, one within the probe C_0 (which is unchanged by a sample) and the second one the fringing field capacitance $C_f = \varepsilon \cdot C_{f,0}$, which is linearly correlated with the permittivity of the sample.

$$C = C_0 + C_f = C_0 + \varepsilon \cdot C_{f,0} \tag{3.19}$$

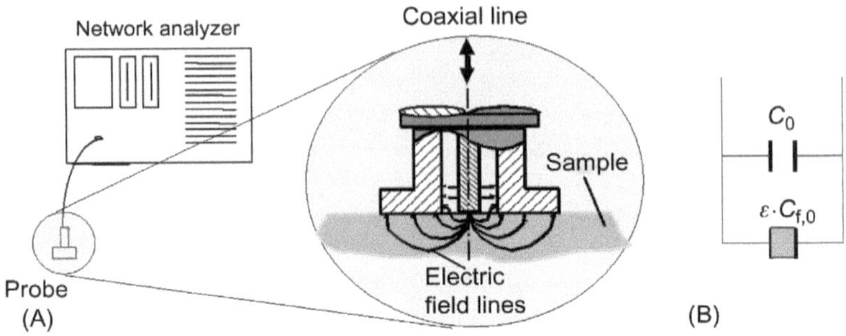

Figure 3.6 (A) Schematic view of an open-ended coaxial line measurement setup and (B) the simplified equivalent circuit of the probe.

On one hand, the impedance of the system Z can be expressed by its capacitance:

$$Z = \frac{1}{i\omega C} = \frac{1}{i\omega(C_0 + \varepsilon \cdot C_{f,0})} \tag{3.20}$$

On the other hand, the impedance can be calculated from the line's characteristic impedance Z_0 and the complex voltage reflection coefficient $\Gamma = \Gamma_0 \cdot e^{i\phi}$:

$$Z = Z_0 \cdot \frac{1 + \Gamma}{1 - \Gamma} \tag{3.21}$$

Taking the Eqs. (3.20 and 3.21), separating for real and imaginary part and solving for ε' and ε'' yields:

$$\begin{aligned} \varepsilon' &= \frac{-2\Gamma_0 \cdot \sin\phi}{\omega \cdot C_{f,0} \cdot Z_0 \cdot (1 + 2\Gamma_0 \cdot \cos\phi + {\Gamma_0}^2)} - \frac{C_0}{C_{f,0}} \\ \varepsilon'' &= \frac{1 - {\Gamma_0}^2}{\omega \cdot C_{f,0} \cdot Z_0 \cdot (1 + 2\Gamma_0 \cdot \cos\phi + {\Gamma_0}^2)} \end{aligned} \tag{3.22}$$

These equations are relatively simple, but are only valid for certain assumptions (Grant, Clarke, Symm, & Spyrou, 1989): The capacitances C_0 and $C_{f,0}$ have to be frequency-independent and also independent from the permittivity of the sample, and the probe does not launch traveling waves. Since these assumptions are not strictly fulfilled in most cases, alternative concepts are also proposed (Grant et al., 1989), but they are much more complicated and thus not as popular. The unknowns C_0, $C_{f,0}$, and the line's characteristic impedance are calculated by calibration measurements of three known conditions. In most cases, an open end, a metallic short cut, and a test substance with known dielectric properties (mostly water) are taken.

In practice, these systems cover a frequency range between 200 MHz and 20 GHz. At lower frequencies, the accuracy is strongly decreased, but adapted probe geometries (Stuchly, Gregory, Anderson, & Kraszewski, 1986) may overcome this disadvantage. However, the probe has a limited accuracy especially at low values of ε' and ε'' (compared to resonator systems), but such precision is usually adequate for microwave heating studies (Engelder & Buffler, 1991).

Its main advantage is the simplicity of use (especially for liquids), due to the commercial availability (Agilent, 2014; Hewlett Packard, 1993; Speag, 2016), and in the small size of the probes. Whereas some years ago the only commercially available probe had dimensions of some centimeters, which could sometimes be problematic, together with the minimum sample thickness $h_{\min}$, Hewlett Packard (1993) mentioned the empirical formula $h_{\min} = \frac{20\ \text{mm}}{\sqrt{|\varepsilon|}}$. Today, the open-ended coaxial probe is available in different configurations with different diameters (down to diameters of some millimeters), for wide temperature ranges and even for very thin (sub-mm-range) layers (Speag, 2016) and demanding applications. However, intrinsically for solid samples there still remains the need for a flat surface, since air gaps can lead to significant errors.

3.3 Further analysis of dielectric properties

Although there are numerous methods to characterize the material dielectrically, as shown above, sometimes some more information can be gained from already existing data. Two examples will be covered here, the inter- or extrapolation of dielectric properties of mixtures and the relations between the real and the imaginary part of dielectric properties of the material. The first method helps to estimate either the dielectric properties of mixtures starting from the pure material values or to infer the pure material values from measured data on mixtures with varied composition.

The second method shows how the real and imaginary part of the dielectric properties are coupled by Kramers–Kronig relations, so that the real part can be determined when the frequency dependent imaginary part is known or vice versa.

3.3.1 *Dielectric properties of mixtures*

Most food materials are often complex mixtures of several components of varying concentrations. To avoid the necessity of excessive experiments, one may try to predict the dielectric properties of a mixture from the values of the pure components. On the other hand, sometimes the pure component values are needed, but cannot be measured easily, as in the case of pulverized or granular sample, so that each measurement of dielectric properties gives the values for the mixture of air and sample, but not the pure solid value.

For both cases, models and formulas have been developed and can be found in literature. Besides pure empirical formulas, which have a very limited area of validity (only for one certain material, and for the specific frequency range investigated), some more theoretical equations exist, and will be presented in this section.

Since all of these models do not go into the details of molecular dynamics, they can only cover practically noninteracting mixtures, like dispersions; on the other hand for interacting mixtures, like, for example, water–alcohol mixtures, or solutions of salts, at the moment no simple general approaches can be found; the reader has to be referred to molecular dynamic modeling.

Indeed, there is one work (Bergman, 1978), that is unfortunately not well known in the area of food technology, that solves the problem of (noninteracting) dielectric mixtures on a theoretical basis, when the "microgeometry" of the distribution of one dielectric within another is known. Obviously, due to its mathematical complexity other simpler models have been preferred often, although the general work of Bergman covers all of them, assuming the wavelength to be much larger than the particle size.

Wiener (1913) based his equations on the simple model of a plate capacitor, in which layers of different dielectric media with corresponding volume fractions exist. Depending on the layer arrangement, parallel or orthogonal to the capacitor plates, the equivalent circuit of serial or parallel capacitors yield Eqs. (3.23 and 3.24), respectively, for the dielectric property of the mixture:

$$\frac{1}{\varepsilon_m} = \frac{\Phi_2}{\varepsilon_2} + \frac{1-\Phi_2}{\varepsilon_1} \tag{3.23}$$

$$\varepsilon_m = \Phi_2 \cdot \varepsilon_2 + (1-\Phi_2) \cdot \varepsilon_1 \tag{3.24}$$

Since real mixtures are usually not of a layered kind, by combining these equations with a shape coefficient u, the universal Wiener equation is formed:

$$\frac{\varepsilon_m - \varepsilon_1}{\varepsilon_m + u \cdot \varepsilon_1} = \Phi_2 \cdot \frac{\varepsilon_2 - \varepsilon_1}{\varepsilon_2 + u \cdot \varepsilon_1} \tag{3.25}$$

Fricke (1955) created a model for the calculation of the conductivity, which was used by Mudgett, Wang, and Goldblith (1974) for the calculation of the dielectric loss of a mixture, due to the strong relationship (see Chapter 1: Introducing microwave-assisted processing of food: fundamentals of the technology) between loss factor and conductivity.

Valid for mixtures with a high dielectric continuous phase (1) and a low dielectric dispersed phase (2), this Eq. (3.26) has also a shape coefficient called a, that should be $a = 2$ for spheres (like in an emulsion) and $a = 1$ for needle-shaped objects.

$$\varepsilon_m = \varepsilon_1 \cdot \frac{\varepsilon_2 \cdot (1 + a \cdot \Phi_2) + \varepsilon_1 \cdot a \cdot (1-\Phi_2)}{\varepsilon_1 \cdot (a + \Phi_2) + \varepsilon_2 \cdot (1-\Phi_2)} \tag{3.26}$$

From a mathematical point of view, a suitable function for the dielectric property of the mixture from the values of the pure substances should be continuous, monotonic, and differentiable for all variables ε_1, ε_2, and Φ_2. Moreover, it should be applicable to mixtures of mixtures.

Besides Eq. (3.27) of Lichtenecker and Rother (1931),

$$\ln\varepsilon_m = \Phi_2 \cdot \ln\varepsilon_2 + (1 - \Phi_2) \cdot \ln\varepsilon_1 \tag{3.27}$$

the equations of type Eq. (3.28) may be possible:

$$\varepsilon_m{}^y = \Phi_2{}^x \cdot \varepsilon_2{}^y + (1 - \Phi_2)^x \cdot \varepsilon_1{}^y \tag{3.28}$$

An example where that kind of equation is commonly applied, is the case of the mixture of granular or pulverized material with air (Nelson, 1996). When the air mass is neglected, the approximate Eq. (3.28a) with bulk density ρ and solid density ρ_{solid} may be derived, using the exponents $x = 3$ and $y = 1$:

$$\varepsilon_m \propto \left(\frac{\rho}{\rho_{\text{solid}}}\right)^3 \tag{3.28a}$$

This equation may be helpful when the dielectric properties of a solid material are of interest, but only pulverized formulations are available. Taking measurements of the material with varied bulk density may yield the solid dielectric property through solid density extrapolation using Eq. (3.28a). In Fig. 3.7, this is shown using the example of limestone powders (Regier, 2003).

A comparison of the validity in the case of emulsions and suspensions can be found, for example, in Erle, Regier, Persch, and Schubert (2000) or for granular material in Nelson (1996). It can be concluded that although a general formula for the exact description of the dielectric property of mixtures (of noninteracting dielectrics) exists, in many cases it seems to be simpler (mathematically or due to unknown "microgeometries") to find and adapt the most suitable expression for

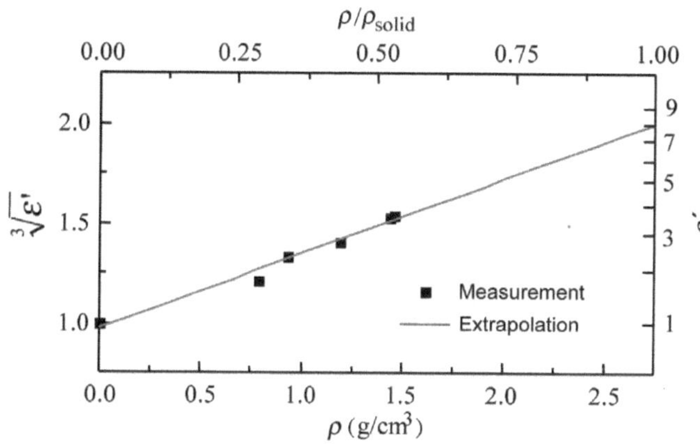

Figure 3.7 Extrapolation of density dependent measurement of dielectric constant of limestone to determine the solid value. The extrapolation yield a dielectric constant of $\varepsilon' = 8.0$, which agrees well with the literature value of 7.6 (Nelson, 1996).

each product. Nevertheless, all of the equations above may give an approximate first estimation of the dielectric properties of mixtures, which may already be helpful for further calculations.

3.3.2 Relations between real and imaginary part of dielectric properties (Kramers-Kronig-relation)

As shown in Chapter 1, Introducing microwave-assisted processing of food: fundamentals of the technology, the dielectric property ε or more exactly the susceptibility $\chi = \varepsilon - 1$ is responsible for a linear response of the material, the so-called polarization $\vec{P}$, to an outer electric field $\vec{E}$. In the following, only homogeneous behavior is suggested, so that only the absolute values E and P are important.

Although, in most cases a history dependence is neglected, which means:

$$P(t) = \chi(t) \cdot E(t) \tag{3.29}$$

And a more general approach is the convolution:

$$P(t) = \int_{-\infty}^{+\infty} \chi(t - t') \cdot E(t')dt' \tag{3.30}$$

This means that the response of the material at time t, the polarization, consists of reactions to electric fields at times t' (this is called a causal behavior).

If the variables E, P, and χ are Fourier-transformed to the variables $\tilde{E}$, $\tilde{P}$, and $\tilde{\chi}$ (shown for the example of E):

$$\begin{aligned} \tilde{E}(\omega) &= \int_{-\infty}^{+\infty} E(t) \cdot e^{i\omega t} dt \\ E(t) &= \frac{1}{2\pi} \int_{-\infty}^{+\infty} \tilde{E}(\omega) \cdot e^{-i\omega t} d\omega \end{aligned} \tag{3.31}$$

for the transformed variables the convolution of Eq. (3.29) becomes a simple product:

$$\tilde{P}(\omega) = \tilde{\chi}(\omega) \cdot \tilde{E}(\omega) \tag{3.32}$$

Another physical assumption is the causality of $\chi(t)$, which means that for the effect at time t only reasons ($E(t')$) before the effect ($t' \leq t$) are considered. This yields:

$$\chi(t - t') \equiv 0, \text{if } t' > t \tag{3.33}$$

Thus, the Fourier transformation for $\tilde{\chi}(\omega)$ becomes:

$$\tilde{\chi}(\omega) = \int_{0}^{\infty} \chi(t) \cdot e^{i\omega t} dt = \tilde{\chi}' + i\tilde{\chi}'' \tag{3.34}$$

Since $\chi(t)$ is defined as a real function, there has to be a dependence between the real $\tilde{\chi}'$ and the imaginary part $\tilde{\chi}''$ of the susceptibility. Additionally, $\tilde{\chi}(\omega)$ has to be analytic for $\Im(\omega) > 0$, bounded for $\omega \in R$, and should diminish asymptotically ($\lim_{\omega \to \infty} \tilde{\chi}(\omega) = 0$), which is physically satisfied.

By functional theory, the correlation between real $\tilde{\chi}'$ and the imaginary part $\tilde{\chi}''$ of the susceptibility or in terms of dielectric properties may be expressed as:

$$\begin{aligned} \varepsilon'(\omega) &= \tilde{\chi}'(\omega) + 1 = \frac{1}{\pi} \int_{-\infty}^{+\infty} \frac{\varepsilon''(\omega)}{\omega' - \omega} d\omega' + 1 \\ \varepsilon''(\omega) &= -\tilde{\chi}''(\omega) = \frac{1}{\pi} \int_{-\infty}^{+\infty} \frac{\varepsilon'(\omega) - 1}{\omega' - \omega} d\omega' \end{aligned} \tag{3.35}$$

In both cases the integral is defined as Cauchy's main value:

$$\int_{-\infty}^{+\infty} \frac{X}{\omega' - \omega} d\omega' \equiv \lim_{\delta \to 0^+} \left[\int_{-\infty}^{\omega - \delta} \frac{X}{\omega' - \omega} d\omega' + \int_{\omega + \delta}^{+\infty} \frac{X}{\omega' - \omega} d\omega' \right] \tag{3.36}$$

Thus, if the frequency dependence of either the real or the imaginary part is known, e.g., by measurements, the other part can be calculated. Although this still seems to be not very common, some special examples can be found, e.g., to distinguish between various parts of dielectric losses (ionic, dipolar, electronic, etc.) (see e.g., Steeman & van Turnhout, 1997), or for the case of effective dielectric functions of mixtures as in Section 3.3.1 (Gorges, Grosse, & Theiß, 1995).

3.4 Summary

This chapter covered several common methods for the measurement of dielectric properties at microwave frequencies. After the classification of the different types to closed and open structures, the most prominent examples were shown. The application areas, as well as the advantages and disadvantages of the different methods, were given, so that the most suitable technique for the particular measurement problem can be chosen.

Additionally, some information was given for the further analysis of dielectric property measurements. On one hand, the possibilities of mixture equations to predict dielectric properties of mixtures, as well as of the pure substances, were shown.

On the other hand, the Kramers–Kronig relation has been derived to extract information of the real part of the permittivity, when the frequency dependent imaginary part is known and vice versa.

3.5 Nomenclature

a	shape coefficient
A_x, A_y, A_z	electric field amplitude coefficients
$\vec{B}$	magnetic flux density
$C, C_0, C_f, C_{f,0}$	capacity, inner capacity, fringing field capacity, fringing field capacity without dielectric sample
$\vec{E}$	electric field
f_0, f_{lmn}	resonant frequency
$f_{0,\text{ empty}}$	resonant frequency of empty cavity
$f_{0,\text{ filled}}$	resonant frequency of filled cavity
f, g, h	functions
$h_{\min}$	minimal sample thickness
i	imaginary unit
I	intensity
$\Im(x)$	imaginary part of x
k_x, k_y, k_z	wave number in x-, y-, z-direction
L_x, L_y, L_z	length in x-, y-, z-direction
l, m, n	natural "quantum" numbers
$\vec{P}$	electric polarization
Q	quality factor
Q_{empty}	quality factor of empty cavity
Q_{loss}	quality factor due to dielectric losses
$\vec{r} = (x, y, z)$	local vector
R	cylinder radius
$\Re(x)$	real part of x
t	time
u	shape coefficient
x, y	constant exponents
α_{lm}	m'th zero of Bessel's function of order l
α'_{lm}	m'th zero of the first derivative of Bessel's function of order l
$\Delta f_{1/2}$	free width at half maximum
Γ	voltage reflection coefficient
ε_0	dielectric constant of vacuum
$\varepsilon = \varepsilon' - i\varepsilon''$	relative permittivity
$\varphi, \varphi_x, \varphi_y, \varphi_z$	phase constants
Φ	volume fraction
μ	relative permeability
μ_0	permeability of vacuum
ρ	bulk density
ρ_{solid}	solid density
$\chi = \varepsilon - 1$	electric susceptibility
ω	circular frequency

References

Agilent. (2014). *Basics of measuring dielectric properties of materials – Application note.*

Bengtsson, N. E., & Risman, P. O. (1971). Dielectric properties of foods at 3 GHz as determined by cavity perturbation technique, II. *Journal of Microwave Power, 6*(2), 107–123.

Bergman, D. J. (1978). The dielectric constant of a composite material—A problem in classical physics. *Physics Reports, 43*(9), 377–407.

Engelder, D. S., & Buffler, C. R. (1991). Measuring dielectric properties of food products at microwave frequencies. *Microwave World, 12*(2), 6–15.

Erle, U., Regier, M., Persch, C., & Schubert, H. (2000). Dielectric properties of emulsions and suspensions: Mixture equations and measurement comparisons. *Journal of Microwave Power and Electromagnetic Energy, 35*(3), 185–190.

Fricke, H. (1955). The complex conductivity of a suspension of stratified particles of spherical or cylindrical form. *Journal of Physical Chemistry, 59*(2), 168–170.

Gallone, G., Lucardesi, P., Martinelli, M., & Rolla, P. A. (1996). A fast and precise method for the measurement of dielectric permittivity at microwave frequencies. *Journal of Microwave Power and Electromagnetic Energy, 31*, 158–164.

Gorges, E., Grosse, P., & Theiß, W. (1995). The Kramers-Kronig-Relation of effective dielectric functions. *Zeitschrift für Physik B, 97*, 49–54.

Grant, J. P., Clarke, R. N., Symm, G. T., & Spyrou, N. M. (1989). A critical study ofthe open-ended coaxial line sensor technique for RF and microwave complex permittivity measurements. *Journal of Physics E, 22*, 757–770.

Hewlett Packard. (1993). *User's Manual – HP 85070B Dielectric Probe Kit.*

Janezic, M. D. (2003). *Nondestructive relative permittivity and loss tangent measurements using a split-cylinder resonator. PhD thesis.* University of Colorado.

Kaatze, U., & Giese, K. (1980). Dielectric relaxation spectroscopy of liquids: Frequency domain and time domain experimental methods. *Journal of Physics E, 13*, 133–141.

Kraszewski, A., & Nelson, S. O. (1990). Study on grain permittivity measurements in free space. *Journal of Microwave Power and Electromagnetic Energy, 25*(4), 202–210.

Lichtenecker, K., & Rother, K. (1931). Die Herleitung des logarithmischen Mischungsgesetzes aus allgemeinen Prinzipien der stationären Strömung. *Physikalische Zeitschrift, 32*, 255–260.

Maurens, M., Priou, A., Brunier, P., Aussudre, S., Lopez, M., & Combes, P. (1992). Free-space microwave measurement technique for composite minerals. In A. Priou (Ed.), *Dielectric Properties of Heterogeneous Materials.* Elsevier.

Metaxas, A. C. (1974). Design of a resonant cavity as a heating device at 2.45 GHz. *Journal of Microwave Power, 9*(2), 123–128.

Metaxas, A. C. (1996). *Foundations of electroheat.* Wiley.

Metaxas, A. C., & Meredith, R. (1983). *Industrial microwave heating.* Peter Peregrinus.

Mudgett, R. E., Wang, D. I. C., & Goldblith, S. A. (1974). Prediction of dielectric properties in oil-water and alcohol-water mixtures at 3000 MHz, 25°C based on pure component properties. *Journal of Food Science, 39*(4), 632–635.

Musil, J., & Zacek, F. (1986). *Microwave measurements of complex permittivity by free-space method and their applications.* Elsevier.

Nelson, S. O. (1972). A system for measuring dielectric properties at frequencies from 8.2 to 12.4 GHz. *Transactions of the ASAE, 15*(6), 1094–1098.

Nelson, S. O. (1983). Dielectric properties of some fresh fruits and vegetables at frequencies of 2.45 to 22 GHz. *Transactions of the ASAE, 26*(2), 613–616.

Nelson, S. O. (1996). Determining dielectric properties of coal and limestone by measurements on pulverized samples. *Journal of Microwave Power and Electromagnetic Energy*, *31*(4), 215–220.
Nelson, S. O. (2010). Fundamentals of dielectric properties measurements and agricultural applications. *Journal of Microwave Power and Electromagnetic Energy*, *44*(2), 98–113.
Nelson, S. O., Stetson, L. E., & Schlaphoff, C. W. (1974). A general computer program for precise calculation of dielectric properties from short-circuited waveguide measurements. *IEEE Transactions on Instrumentation and Measurement*, *IM-23*(4), 455–460.
Nelson, S. O., & Trabelsi, S. (2004). Principles for microwave moisture and density measurement in grain and seed. *Journal of Microwave Power and Electromagnetic Energy*, *39*(2), 107–114.
Nelson, S. O., & You, T. S. (1989). Microwave dielectric properties of corn and wheat kernels and soybeans. *Transactions of the ASAE*, *32*(1), 242–249.
Persch, C. (1997). *Messung von Dielektrizitätskonstanten im Bereich von 0.2 bis 6 GHz und deren Bedeutung für die Mikrowellenerwärmung von Lebensmitteln, Dissertation.* Universität Karlsruhe.
Regier, M. (2003). *Über dielektrische und Magnetresonanz-Methoden zur Charakterisierung disperser Systeme. Dissertation.* Logos Verlag: Universität Karlsruhe.
Regier, M., & Schubert, H. (2000). Dielectric properties at microwave frequencies studied in partially filled cylindrical cavities. *Journal of Microwave Power and Electromagnetic Energy*, *35*(1), 25–33.
Risman, P. O., & Bengtsson, N. E. (1971). Dielectric properties of foods at 3 GHz as determined by cavity perturbation technique, I. *Journal of Microwave Power*, *6*(2), 101–106.
Roberts, S., & van Hippel, A. (1946). A new method for measuring dielectric constant and loss in the range of centimeter waves. *Journal of Applied Physics*, *17*(7), 610–616.
Rost, A. (1978). *Messung dielektrischer Stoffeigenschaften*. Vieweg.
Roussy, G., Ghanem, H., Thiebaut, J. M., & Dichtel, B. (1990). Six-port waveguide used for simultaneous measuring permittivity and permeability of solid materials in microwave region. *Journal of Microwave Power and Electromagnetic Energy*, *25*, 67–75.
Roussy, G., & Pearce, J. A. (1995). *Foundations and industrial applications of microwaves and radio frequency fields*. New York: Wiley.
Speag. (2016). *Schmid & Partner Engineering AG*. Available from: <http://www.speag.com/products/dak/dielectric-measurements/> Accessed 2016.
Steeman, P. A. M., & van Turnhout, J. (1997). A numerical Kramers-Kronig transform for the calculation of dielectric relaxation losses free from Ohmic conduction losses. *Colloid and Polymer Science*, *275*(2), 106–115.
Stuchly, M. A., & Stuchly, S. S. (1980). Coaxial line reflection methods for measuring dielectric properties of biological substances at radio and microwave frequencies – A review. *IEEE Transactions on Instrumentation and Measurement*, *IM-29*(3), 176–183.
Stuchly, S. S., Gregory, G., Anderson, L., & Kraszewski, A. (1986). A new sensor for dielectric measurements. *IEEE Transactions on Instrumentation and Measurement*, *IM-35*(2), 138–141.
Trabelsi, S., & Nelson, S. O. (2003). Free-space measurement of dielectric properties of cereal grain and oilseed at microwave frequencies. *Measurement Science and Technology*, *14*, 589–600.
Wiener, O. (1913). Die Theorie des Mischkörpers für das Feld der stationären Strömung. *Abhandlungen der mathematisch-physikalischen Klasse der königlich sächsischen Gesellschaft der Wissenschaft*, *32*, 507–604.
You, T. S., & Nelson, S. O. (1988). Microwave dielectric properties of rice kernels. *Journal of Microwave Power*, *23*(3), 150–159.

Impact of microwave processing on nutritional, sensory, and other quality attributes

4

C. Contreras[1], *M. Benlloch-Tinoco*[2], *D. Rodrigo*[3] *and*
N. Martínez-Navarrete[1]
[1]Universitat Politècnica de València, Valencia, Spain, [2]University of Nottingham, Loughborough, United Kingdom, [3]Instituto de Agroquímica y Tecnología de Alimentos (IATA-CSIC), Valencia, Spain

4.1 Introduction

In recent years, consumers in developed countries have become aware of the influence of diet on health and well-being (Lado & Yousef, 2002). Consequently, there has been a growth in demand for high quality, minimally processed food, which meets the nutritional needs, brings health benefits, and allows the risk of certain diseases to be reduced; further benefits can include an extended shelf life, economical processing, improved palatability, and simplicity to prepare (Costa, Dekker, Beumer, Rombouts, & Jongen, 2001). For the industry, obtaining minimally processed foods that are stable and safe for extended periods of time is challenging, and it generally involves the development and implementation of new preservation strategies capable of preventing the physical and/or chemical deterioration, inhibiting enzymatic activity, and, above all, reducing the concentration of, or controlling, outgrowth of pathogenic and spoilage organisms. The process should affect the natural properties of the product as little as possible, and a substantial price increase should be avoided.

One way to address this challenge is to utilize alternatives to conventional thermal preservation technologies, which can be applied either to replace conventional heating techniques for the purposes of improving the quality of products that are available on today's market, or to develop new products. Microwave processing is one of these alternative technologies that have successfully been used for different chemical industry applications and, more recently, in diverse unit processes in the food industry, like thawing/tempering, dehydration/drying, cooking, blanching, baking, sanitization, or pasteurization of different food matrices. Microwaves are non-ionizing electromagnetic wave radiation, whose frequency between 300 MHz and 300 GHz, with a more or less stable ISM frequency near 2.45 GHz (see Chapter 1: Introducing microwave-assisted processing of food: fundamentals of the technology) lies between that of infrared rays and radio and television waves. When applied, an internal heat generation or volumetric heating takes place in parallel, although at very different time scales, with convection and conduction phenomena. That is the reason why the use of microwave heating generally offers many advantages when

The Microwave Processing of Foods. DOI: http://dx.doi.org/10.1016/B978-0-08-100528-6.00004-8

compared to conventional thermal methods, in terms of a shorter start-up time, faster heating, greater energy efficiency, smaller footprint, precise equipment control, selective heating, and final products with an improved nutritional quality (Chandrasekaran, Ramanathan, & Basak, 2013; Salazar-González, San Martín-González, López-Malo, & Sosa-Morales, 2012; Sumnu, 2001). Since microwaves penetrate into a food product and do not act at the surface level only, the conversion of energy into heat throughout the product is more efficient. In most cases, combining microwave energy with conventional operations may lead to fast internal heating with desirable surface-heating effects (Martín, Albors, Martínez-Navarrete, Chiralt, & Fito, 2000). This occurs to differing extents as a function of the treatment conditions and the food matrix characteristics.

The progress made in the field of microwave technology has already been the subject of many discussions; and several studies have focused on the interest which exists in applying microwave heating for industrial purposes. Nevertheless, its practical application is still limited. What is presented in this chapter is an overview of microwave heating as a replacement or complementary method for food processing. Every processing technique causes important changes to the sensory, biochemical, and nutritional characteristics of foods which must be evaluated; therefore, diverse food products are employed in order to comment on some relevant quality aspects and, in the cited references, more detailed information can be found about the special items discussed in this section. At the same time, general comments are made in the following paragraphs on unit operations applied to many different foods and, more specifically, for each group of food, in the corresponding section.

4.2 Microwave processing of different food matrices

Some examples of microwave application on the processing of different fruits, vegetables, meat, fish, and other products are shown in Table 4.1. What is of particular interest is the application of microwaves in the drying (see Chapter 8: Microwave-assisted drying) of a wide range of products due to the fact that the process is accelerated, something which is limited when no microwaves are used because of the low thermal conductivity of foods, especially of products with low water content, or of course products that progressively lose water during processing. There are two possible forms of microwave drying; drying at atmospheric pressure and under applied vacuum conditions. Combined microwave–air dryers are especially widespread in the food industry, and can be found in a serial or a parallel combination of both methods (Shaheen, El-Massry, El-Ghorab, & Anjum, 2012). Microwave–vacuum drying in particular, substantially reduces both the process time and the product temperature, as compared to heated air drying or microwave drying, and has great potential as a means of improving the quality of the dried product, since the levels of both the thermal and oxidative stress are reduced (Lin, Durance, & Scaman, 1998; Orsat, Changrue, & Raghavan, 2006; Torringa, Van Dijk, & Bartels, 1996). Microwave-assisted freeze-drying has also been studied, but to the best of the authors' knowledge, no commercial systems are currently installed and used in the

Table 4.1 Practical microwave application in domestic and industrial food processing

Food commodities	Food processing techniques with microwave application	Examples of practical uses on
Fruit & vegetable	Drying	Apples, strawberries, bananas, raisins, potatoes, onions, carrots, garlic
	Blanching	Potatoes, mushrooms
	Cooking	Legume seeds
	Roasting	Cumin seeds
	Thawing	Fruit juice
	Pasteurization	Fruit juice
	Sanitization	Herbs and spices, such as pepper, origanum
Meat & fish	Tempering/thawing	Frozen meats, such as pork, beef
	Cooking/roasting	Meats, such as pork, chicken, lamb chops. Bacon. Fish
	Drying	Fillets of tilapia fish
Other foodstuffs	Heating	Precooked products
	Blanching	Peanuts
	Cooking	Spaghetti, rice, chickpeas
	Baking	Bread, cakes, doughnuts, cookies
	Drying	Macadamia nuts
	Pasteurization/sterilization	Fluids, such as milk, peanut-based beverages. Eggs

food industry. The main reason for this is the high cost involved and the small market size for freeze-dried food products. In general, according to Shaheen et al. (2012), the quality of microwave-dried food products lies somewhere between air-dried and freeze-dried ones. In this way, the high quality of the products could also make this relatively expensive process economical in some cases.

Another frequent industrial and domestic practice in which microwaves are applied is for product tempering or thawing (see Chapter 12: Microwave-assisted thawing and tempering). Frozen meat, fish, vegetables, fruit, butter, and juice concentrate are common raw materials for many food manufacturing operations. Few processes can handle the frozen material and it is usually either thawed or tempered before further processing (Shaheen et al., 2012). Thawing is regarded as complete when all the material has reached 0°C and no free ice is present. Conventional thawing systems supply heat to the product surface and then rely on conduction to transfer that heat into the center of the product. Microwave systems, however, use electromagnetic radiation to generate heat within the food (Yarmand & Homayouni, 2009). Water has a lower thermal conductivity than ice. Therefore, as the ice melts during thawing, the rate at which heat is conducted to the resting frozen food decreases. In this sense, the internal heat generation caused by microwaves is desirable.

With respect to safety aspects, the problem with microwave heating is the elevated number of factors that affect the absorption of microwave energy and, thus, the heating behavior, resulting in a nonuniform temperature distribution (see Chapter 17: Improving the heating uniformity in microwave processing) within the heated products. This is one of the major drawbacks associated with microwave heating and can lead to the survival of pathogens in the less heated zones (Hamoud-Aghaa, Cureta, Simoninb, & Boillereauxa, 2014; Salazar-González et al., 2012). To avoid an undesirable nonuniform temperature distribution during microwave heating, several methods have been proposed; e.g., combining conventional and microwave heating (Hamoud-Aghaa et al., 2014). Nevertheless, as will be commented on more extensively below, in recent years, the suitability of microwave heating for the purposes of enhancing food microbial safety (pasteurization and sterilization processes) has been successfully tested in various animal and vegetable food products (Benlloch-Tinoco, Martínez-Navarrete, et al., 2014; Benlloch-Tinoco, Pina-Pérez, et al., 2014; Cañumir, Celis, De Brujin, & Vidal, 2002; Huang, Sheng, Yang, & Hu, 2007; O'donnell, Tiwari, Bourke, & Cullen, 2010; see Chapter 10: Microwave-assisted pasteurization and sterilization—commercial perspective in this book).

Food quality often deals only with the influence of primary production and industrial processing. Nevertheless, food preparation at home as the final step of the chain also has a great influence on quality-determining parameters like the sensory and functional attributes. The heating of precooked products is the principal practical application of microwave ovens, both domestically and in the catering industry, since it is a rapid, safe, and hygienic process.

The main observations related to the application of microwave energy in the processing of different groups of foods are discussed below.

4.2.1 Fruit and vegetable products

It is well known that fruits and vegetables are important sources of essential dietary nutrients, such as vitamins, minerals, and fiber. Keeping the product fresh is the best way to maintain its nutritional value, but fresh consumption is not always possible. In this sense, preservation methods must be applied in order to maintain the health benefits derived from these products. Dehydration is one of the oldest methods of food preservation and it covers a very important field in fruit and vegetable processing (Lin et al., 1998). This process leads to several physical effects due to the complex morphology of plant tissues. Disturbances in the cellular structure not only cause some macroscopic changes, such as shrinking, case hardening, textural changes, among others, but also bring about alterations in the mass transfer process (Martín et al., 2000).

In order to ensure that the dried fruits and vegetables look good while retaining their nutrient content, combined drying techniques are recommended. In most cases, the pretreatment prior to microwave fruit dehydration affords several advantages both from a technological point of view and from that of the consumer, resulting, e.g., in the improved reconstitution of the dried material (Contreras, Martín-Esparza, & Martínez-Navarrete, 2012; Torringa et al., 1996). Internal heat generated in food materials when microwaves are applied may be improved by vacuum impregnation or osmotic pretreatment (Contreras, Martín, Martínez-Navarrete, & Chiralt, 2005; Orsat et al., 2006). Vacuum impregnation causes the substitution of the gas fraction occluded in the pores of the fruit by a liquid phase, which may increase the proportion of dipolar molecules which interact with the applied electromagnetic field, thus causing the molecular friction that generates the heat (Martín et al., 2000). On the other hand, water removal during osmotic pretreatment is accompanied by solute uptake from the osmotic solution; when ions are present in the osmotic solution, they will normally enter the tissue, thereby increasing its dielectric loss factor, which makes microwave drying more efficient (Erle & Schubert, 2001). The resulting changes in composition yield better properties of the final product, e.g., in terms of structure and volume (Contreras, Martín-Esparza, Martíinez-Navarrete, & Chiralt, 2007; Martín et al., 2000). In the last 25 years, a number of studies have shown the advantages of drying fruits and vegetables by means of a microwave-assisted processes, some of which are presented in the following paragraphs.

Blanching is an important step in the industrial processing of fruits and vegetables. Its main purpose is the inactivation of enzymes that may cause a deterioration of the quality attributes of vegetables and fruits, such as color, texture or flavor, during the subsequent processes or storage. It consists of a thermal process that can be performed by immersing vegetables in hot water, hot and boiling solutions containing acids and/or salts or steam. In order to achieve blanching, microwave heating can also be applied (Regier, 2014; Schirack, Drake, Sanders, & Sandeep, 2006; see Chapter 9: Microwave-assisted blanching in this book). In this case, some of the advantages, when compared with conventional heating methods, include the speed of the operation, energy savings, the efficiency of the process control, and, in some cases, the fact that no additional water is required. In these cases, there is less leaching of vitamins and other soluble nutrients, and the generation of wastewater is eliminated or greatly reduced. The use of microwaves enables the central sections to be blanched more quickly, unaffected by any thick or nonuniform sections, providing a fresher vegetable flavor. Nevertheless, the combination of initial hot water or steam blanching is economically advantageous. This is because low-cost hot water or steam is used to first partially raise the temperature, while microwave power, which is more expensive, does the more difficult task of internally blanching the food product. Optimal conditions of time, temperature, vapor pressure, and microwave power depend on the particular vegetable that is being processed and must be empirically determined (Shaheen et al., 2012). According to Dorantes-Alvarez, Barbosa-Caanovas, and Gutiearrez-Loa Pez (2000) and Carranza-Concha, Camacho, and Martínez-Navarrete (2012), the quality of fruits and vegetables is better preserved when heating rapidly. In the near future, it is expected that researchers interested in this matter will discover more specific effects that may be advantageous in the processing of food by microwave blanching. This would give an additional value to food products and would overcome the cost of microwave energy for this particular application (Shaheen et al., 2012).

On the other hand, the use of microwaves for the sanitization of food has acquired an ever-increasing importance in recent years. In particular, this treatment is very efficient for the sanitization of herbs and spices. The effectiveness of the microwave treatment for spices depends on the high penetration of the microwaves into the products (Plessi, Bertelli, & Miglietta, 2002). Herbs and spices are particularly characterized by their flavors; therefore, it is fundamental to understand the influence of each technological treatment they are submitted to on their volatile compounds. This is particularly true for treatments that involve the heating of material, such as in the case of microwave treatment (Bertelli, Plessi, & Miglietta, 2003). Microwave pasteurization has also been proven to be much more effective at destroying *Listeria monocitogenes* in kiwifruit puree than the conventional thermal treatment (Benlloch-Tinoco, Martínez-Navarrete, et al., 2014; Benlloch-Tinoco, Pina-Pérez, et al., 2014; see Chapter 10: Microwave-assisted pasteurization and sterilization—commercial perspective in this book).

The abovementioned processes affect the overall quality of fruits and vegetables to a greater or lesser extent, as a function of the raw materials, the type of process, and the process conditions. Some effects of applying microwave energy on the sensory properties, functional value, and macro- and microcomponents of fruits and vegetables, are discussed below.

4.2.1.1 Impact on sensory properties

Sensory aspects are essential in food quality evaluation. Several factors, like the raw materials or technological food processes, can exert either a positive or a negative sensorial influence. Drying of fruits and vegetables is a widely used unit operation which can have a major influence on the physicochemical properties of products and, unfortunately, the sensorial quality of dehydrated products is very often poor. There are many quality properties which are important for the consumer, such as the rehydration capacity, color, texture, bulk density, and shrinkage.

As previously commented on, the most important characteristic of microwaves is that they allow processing times to be shortened. This seems to result in a better preservation of the organoleptic, nutritional, and functional properties, with a particular effect on the color and textural characteristics (De Ancos, Cano, Hernández, & Monreal, 1999; Igual, García-Martínez, Camacho, & Martínez-Navarrete, 2010). As described in the study performed by Benlloch-Tinoco, Varela, Salvador, and Martínez-Navarrete (2012), microwave application may be considered an interesting technique with which to pasteurize kiwifruit puree samples, although much care has to be taken with the intense treatments, which mainly affect the color and taste. Microwaved kiwifruit puree was preferred by the consumers when compared with a conventionally treated one (Benlloch-Tinoco, Igual, et al., 2014). Table 4.2 shows some other studies on microwave application in fruit/vegetable processing and its influence on the sensorial characteristics.

4.2.1.2 Impact on nutritional and functional value

Fruits and vegetables can be considered functional foods, since they provide health benefits beyond basic nutrition. They are good sources of natural antioxidants,

Table 4.2 **General review of microwave energy application on different fruits and vegetables. Effect on sensory properties**

Thermal method	Practical application on (and reference)	Microwave overview
Microwave drying as opposed to air drying	Raisin (Tulasidas, Raghavan, & Mujumdar, 1995)	The quality of raisins dried by microwave was found to be superior to air-dried samples in terms of color, crystallized sugar, stickiness, and product damage and uniformity. Increasing the air velocity resulted in better quality raisins whereas microwave power and air temperature had the opposite effect. An optimum selection of the three process conditions was found to be critical in order to achieve an energy efficient process for a quality product.
	Apple and strawberry (Contreras, Martín-Esparza, Chiralt, & Martínez-Navarrete, 2008)	A dehydrated product with less color change and a more rigid and firmer structure was obtained at higher air temperature or when applying microwave. However, the higher temperatures during microwave treatment increased pigment degradation. In the case of strawberry, anthocyanin loss and an increase in luminosity were observed when microwave was applied.
	Garlic (Sharma & Prasad, 2001)	The color of the dried garlic cloves was measured considering the CIE L*a*b* color coordinates. The volatile oil, comprising sulphur compounds that are responsible for the pungency of garlic, was also compared. The dried garlic cloves were lighter in color when dried by a combined microwave–air drying process as compared to air drying. A greater amount of the volatile components responsible for the strength of the flavor was also retained in microwave–air drying. Thus, the microwave–air drying gave superior quality dehydrated garlic cloves.
Microwave–vacuum drying as opposed to air drying	Banana (Drouzaf & Schuber, 1996)	Microwave–vacuum drying was used for the preservation of high quality products. It is preferable to conventional dying techniques in order to avoid product degradation caused by the high temperatures of convective dying. The structure of dried products before and after processing plays an important role in the final product quality. This process produced dehydrated banana slices of excellent quality, as validated by taste, aroma, smell and rehydration tests.
	Cranberry (Yongsawatdigul & Gunasekaran, 1996)	Microwave–vacuum drying improved the color quality of the dried cranberries (redder samples) when compared to the conventional method. At the same moisture level, the cranberries that were microwave–vacuum dried required less cutting force (softer in texture).

(Continued)

Table 4.2 (Continued)

Thermal method	Practical application on (and reference)	Microwave overview
Microwave–vacuum drying as opposed to air drying or freeze-drying	Carrot (Lin et al., 1998)	Microwave–vacuum dried carrot slices had greater rehydration potential, lower density, less color deterioration, and softer texture than those prepared by air drying. Although the freeze-drying of carrot slices yielded a product with improved rehydration potential and appearance, the microwave–vacuum dried samples were rated as equal to or better than freeze dried samples by a sensory panel for color, texture, flavor and overall preference, in both the dry and rehydrated state. Furthermore, in the dry state, microwave–vacuum dried carrot slices possess a unique puffed texture which was preferred by sensory panelists and could be a desirable property for developing a snack-type product.
Osmodehydration & microwave–vacuum drying	Strawberry and apple (Erle & Schubert, 2001)	The microwave–vacuum drying of osmotically pretreated fruits combines the benefits of both unit operations. Selecting the conditions during the osmotic treatment offers the possibility of influencing both the efficiency of microwave dehydration and the properties of the final product. High quality dried fruit products, in terms of their color, taste, structure and volume, can be obtained.
Microwave–vacuum drying	Saskatoon berries (Meda, Gupta, & Opoku, 2008)	The influence which microwave–vacuum drying had on color changes was investigated. Different microwave power levels and vacuum pressure levels were used and the CIE $L^*a^*b^*$ color coordinates were considered. The yellowness of the samples was not affected. However, the drying process produced significant color changes in the L^* and a^* values. The total color difference was dependent on the microwave power and vacuum pressure levels.
Air drying & microwave–vacuum drying as opposed to air drying or freeze-drying	Apple (Ferenczi, Czukor, & Cserhalmi, 2014)	A combination of microwave–vacuum with hot air drying is capable of creating a puffed crunchy product with preferable appearance, color, consistency and taste than the dried products made by means of hot air or freeze-drying.
Microwave drying or halogen lamp–microwave drying as opposed to air drying.	Carrot (Sumnu, Turabi, & Oztop, 2005)	Microwave and halogen lamp–microwave combination methods can be recommended for the drying of carrots which can be used in the instant soup and snack food industry. They lead to a significant improvement in the rehydration ratio and less color deterioration. In addition, the use of a halogen lamp–microwave combination oven can be recommended if the moisture content of the product needs to be reduced to very low values.

Microwave drying & spouted bed as opposed to air drying or spouted bed drying	Apple (Feng & Tang 1998)	The color of the dehydrated diced apple was observed as an indicator of heating uniformity, because uneven heating would cause partial scorching and no obviously uniform color. Using the combination method, the dried products exhibited less discoloration and better reconstitution characteristics as compared to conventional hot air drying or spouted bed drying. The lower degree of color degradation may be due to the substantial reduction in drying time.
Microwave blanching & microwave drying as opposed to water/sugary-saline solution blanching & air drying or belt dryer	Potato (Severini, Baiano, De Pilli, Carbone, & Derossi, 2005)	Potato was microwave blanched in distilled water, in saline solution, in hot distilled water, or in hot sugary saline solution. The drying was carried out in a microwave oven, in an air cabinet, or a belt dryer. Dehydration by microwaves gave an attractive product with good color retention and water holding capacity. Nevertheless, an entire blanching-drying process carried out by microwaves would be too expensive. In terms of product quality (color and rehydration capacity) the best combinations were the microwave blanching coupled to dehydration on the belt dryer.
Microwave heating & air drying as opposed to air drying	Apple (Funebo, Ahrné, Kidman, Langton, & Skjöldebrand, 2000)	A pretreatment with high output power microwave energy has significant effects on the final product quality. Microwave heating for a few minutes initially resulted in hard and shrunken apple cubes, which can be very good for the transportation and storage of dehydrated apples. In short, an initial microwave heat treatment results in higher stress needed to puncture dehydrated apples, lower stress needed to puncture rehydrated apples, greater shrinkage and higher rehydration capacity of dehydrated apples.
Microwave heating with or without osmotic dehydration pretreatment as opposed to conventional heating	Grapefruit jam (Igual, Contreras, & Martínez-Navarrete, 2014)	Microwave heating and osmotic dehydration were considered as nonconventional grapefruit jams manufacturing techniques. When compared to the fresh fruit, the greatest color changes took place in those jams processed by microwave and conventional heating. The higher yield stress, greater consistency, and more viscoelastic behavior was displayed by the jams obtained by combining osmotic dehydration and microwave processes. In the sensory analysis, the judges awarded this sample a better score.

(*Continued*)

Table 4.2 **(Continued)**

Thermal method	Practical application on (and reference)	Microwave overview
Microwave roasting as opposed to conventionally roaster	Cumin seeds (Behera, Nagarajan, & Jagan Mohan Rao, 2004)	The processing of spices using microwaves is considered as an alternative methodology thanks to the ease of handling. The optimum thermal conditions were standardized. Microwave heated samples showed a better retention of the characteristic flavor compounds of cumin (total aldehydes), than did conventionally roasted samples. Thus, the microwave treatment retained aldehydes in the volatile oil, making microwaves the best choice as an alternative heating medium for cumin processing.
Microwave sanitization	Pepper (Plessi et al., 2002) Origanum (Bertelli, et al., 2003)	Microwave treatment seems to be a sanitization method that guarantees the preservation of the principal aroma compounds of this kind of food. The treatment appears to affect these two fundamental volatile compounds only marginally, unable to produce a real differentiation between treated and nontreated samples.
Microwave pasteurization as opposed to conventional pasteurization	Grapefruit juice (Igual, Contreras, Camacho, & Martínez-Navarrete, 2014)	Freshly squeezed grapefruit juice was compared with a juice pasteurized by microwave or by following a conventional heating method. In general, the flow behavior, turbidity, and color of grapefruit juice were significantly affected by the heat treatment, especially in the case of the conventional process. However, from a sensory point of view, pasteurized samples were similar to fresh ones. The smallest color change was experienced by microwave pasteurized juice. For these reasons, and also due to the reduction in the process time, microwave treatment can be recommended as a method for the pasteurization of grapefruit juice.
	Kiwifruit puree (Benlloch-Tinoco et al., 2013; Benlloch-Tinoco, Igual, et al., 2014)	The processing conditions were chosen in order to simulate pasteurization treatments in terms of the enzyme and microbial inactivation achieved. For the sensory assessment, the consumers had to score their liking for the appearance, color, odor, taste, sweetness, acidity, consistency and overall acceptance. Although pasteurization significantly affected the consumer acceptance of kiwifruit puree, based on all the sensory attributes evaluated, microwaved kiwifruit puree was clearly preferred. Microwave heating did not have an effect on sweetness and acidity and those sensory attributes of kiwifruit puree that most influence consumer acceptance. Moreover, microwave technology produced smaller color changes.

mainly responsible for the benefits to human health commonly attributed to the regular intake of products of vegetable origin (Antunes, Dandlen, Cavaco, & Graça, 2010; Benlloch-Tinoco, Igual, Rodrigo, & Martínez-Navarrete, 2013; Du, Li, Ma, & Liang, 2009). In this sense, much care has to be taken to preserve antioxidant compounds during processing. The major antioxidants present in vegetables are vitamins C and E, carotenoids, and phenolic compounds, especially flavonoids (Podsędek, 2007). The industrial processing of vegetables, as well as domestic cooking, is expected to affect the functional characteristics (e.g., antioxidant content or composition, antioxidant activity, and bioavailability of antioxidants). Generally, there are fewer antioxidants present in thermally processed vegetables, and they are less active, than in the corresponding raw material, which is caused by their degradation during the heating step (Podsędek, 2007). Cooking may affect the antioxidant content due to antioxidant release and to the destruction or creation of redox-active metabolites (Wachtel-Galor, Wong, & Benzie, 2008). The effect of thermal procedures on nutrient availability or loss, especially the antioxidant capacity, needs to be further investigated in order to better guide food processing methods so as not to deplete the rich antioxidant capacity of foods, such as fruits or vegetables. According to Drouzas and Schubert (1996), a major disadvantage of conventional drying operations is the thermal degradation of important flavor and nutritional substances. In order to prevent the significant loss of these substances, microwave–vacuum drying has been successfully used instead. Table 4.3 summarizes some of the extensive information that exists on the application of microwaves in fruit/vegetable processing and their influence on the nutritive and functional aspects.

4.2.2 Meat and fish products

In meat and fish processing, one of the main applications of microwave energy is that of thawing. Thawing is often considered as simply the reversal of the freezing process. The most commonly used methods for the purposes of thawing frozen meat products include room temperature thawing, cold water thawing, steam thawing, and contact thawing. The main detrimental effect of freezing and thawing of meat and fish tissues is the great increase in the amount of proteinaceous fluid (drip) released on final cutting. Recently, various alternative thawing methods have been actively researched; these include high-pressure, microwave, ohmic, and ultrasound processing, with the aim to minimize the amount of drip that takes place and the consequent deterioration in the meat quality (Kim et al., 2013).

Microwaves can rapidly thaw small pieces of meat, but difficulties arise with large masses of frozen meat used in the industrial processes (see Chapter 12: Microwave-assisted thawing and tempering). Thawing does not occur uniformly, and some parts of the meat may cook while others remain frozen. Applying cyclic heating may minimize the runaway heating. Thawing rates of frozen samples depend on the sample's material properties, dimensions, and on the power and frequency of electromagnetic radiation (Taher & Farid, 2001). Most studies have compared traditional thawing methods with microwave thawing. The experiments carried out by Taher and Farid (2001) on the cyclic microwave thawing of frozen

Table 4.3 General review of microwave energy application on different fruits and vegetables. Effect on nutritional and functional value

Thermal method	Practical application on (and reference)	Microwave overview
Microwave–air drying as opposed to air drying	Potato (Khraisheh, Mcminn, & Magee, 2004)	An oven operated in either the microwave or convective drying mode, was used to dry cylindrical potato samples. The system was operated in convective mode and in the microwave mode at various output power levels. Samples were removed at predetermined time intervals throughout the experimental run for vitamin C content measurement. Microwave dried samples retained at least twice the vitamin C content of convective dried samples (for comparable moisture contents).
Microwave–vacuum drying as opposed to air drying or freeze-drying	Carrot (Lin et al., 1998)	Air drying caused a decrease in both α and β-carotene content, while a smaller α-carotene depletion occurred with microwave–vacuum drying. The rapid heating rate and the depletion of oxygen produced by microwave–vacuum reduced the loss of α-carotene during drying. No significant change in α and β-carotene content was observed upon freeze-drying. Important losses in vitamin C occurred during air drying, due to the oxidation of ascorbic acid in the presence of hot air and by the relatively long drying time. Vitamin C was better retained in microwave–vacuum-dried carrots. No significant loss in vitamin C occurred during freeze-drying because the temperature was very low during the drying process. Microwave–vacuum drying is better at retaining vitamins, especially those sensitive to thermal damage and oxidation, than hot air drying.
Microwave–vacuum drying & air pre- and post-drying as opposed to air drying or freeze-drying	Strawberry (Böhm, Kühnert, Rohm, & Scholze, 2006)	Both microwave–vacuum drying and convective drying significantly reduce ascorbic acid, the content of phenolic compounds, and the antioxidative capacity of dried strawberries. No reduction was observed in freeze dried strawberries. However, after improving the processing conditions, it was possible to increase the recovery of ascorbic acid, the stability of phenolic compounds and the minimal reduction of the antioxidative capacity. The residual concentration of nutritionally relevant compounds after improving the processing conditions was close to the values observed for freeze-dried products.

	Apple (Ferenczi et al., 2014)	Hot air predrying was applied prior to microwave–vacuum drying. The total phenolic compounds, free radical-scavenging activity and aromatic content were measured. The dried apple obtained by combined microwave–vacuum and air drying contains notable aromatic compounds, significantly more phenolic compounds, although not significantly higher antioxidant capacity, than the other two products obtained by air drying or freeze-drying. These properties make the combined microwave–vacuum drying a strong competitor to hot air-dried or freeze-dried products in many areas of food manufacturing.
Microwave cooking as opposed to conventional boiling	Legume seedes (Hernández-Infante, Sousa, Montalvo, & Tena, 1998)	Microwave heating constitutes an adequate method for destroying hemagglutinins and trypsin inhibitors (antinutritional substances) in a similar way to conventional boiling, without affecting the protein quality of most legume seeds (faba beans, peas, chickpeas, soybeans, lentils); with the exception of common beans which, despite this process, retained the toxic substances.
Microwave cooking as opposed to stir-frying and boiling in water	Pepper (Chuah et al., 2008)	Microwave heating and stir-frying did not affect the radical-scavenging activity, total polyphenol content, ascorbic acid content or total carotenoid content, whereas their levels were partly degraded during boiling. The loss in antioxidant activity was greater in the tissues boiled over a prolonged time. This finding indicated that cooking peppers by microwave heating and stir-frying is better as a means of ensuring a greater retention of the bioactive components in peppers.
Microwave cooking as opposed to pressure or usual cooking	Soybean (Kaushik, Satya, & Naik, 2010)	Presoaked seeds were cooked by means of a microwave system (in a domestic microwave oven), following the usual procedure (on a hot plate) and the pressure procedure (in autoclave), until the seeds became tender. Microwave cooking was suggested for soybean preparation due to the fact that it resulted in a greater retention of minerals and vitamins as compared to pressure cooking and the usual methods.
Microwave heating as pretreatment before juice yield as opposed to unheated samples	Apple mashes (Gerard & Roberts, 2004)	The apples were mashed and then pressed at different temperatures. The apple mashes were heated using a microwave oven (different process times in order to achieve optimal bulk temperatures); the control mash was not heated.

(*Continued*)

Table 4.3 **(Continued)**

Thermal method	Practical application on (and reference)	Microwave overview
		The resulting cider was evaluated based on the phenolic and flavonoid content. The microwave heat treatment of the mash increased the extraction of phenolics and flavonoids from the apple mash and resulted in juice with increased concentrations of total phenolics and flavonoids
Microwave pasteurization as opposed to conventional pasteurization	Grapefruit juice (Igual et al., 2010; Igual, García-Martínez, Camacho, & Martínez-Navarrete, 2011)	Microwave energy may be proposed as an alternative to traditional heat pasteurization in order to preserve the essential thermolabile nutrients of grapefruit juice. Contrary to conventional treatment, which leads to a significant decrease in citric acid and ascorbic acid in grapefruit juice, the microwave treatment preserved these compounds and led to a greater retention of flavonoids. Moreover, frozen microwave pasteurized juices better preserved the total phenols and antioxidant capacity when compared with the fresh or conventionally pasteurized ones and maintained the amount of ascorbic acid and vitamin C, especially in the pasteurized samples.
	Kiwifruit puree (Benlloch-Tinoco et al., 2013; Benlloch-Tinoco, Igual, et al., 2014)	The effect of microwave and conventional thermal procedures were compared, in terms of both the inactivation of kiwifruit puree enzymes. Both methods were designed to cause the same level of peroxidase inactivation. The microwaved puree exhibited a superior preservation of the nutritive (vitamins C, A, and E and the total flavonoid, phenol, and tannin) and functional (antioxidant capacity) properties of the fruit, and a content of inactivated enzymes and microorganisms equal to or greater than the conventionally heated sample. Accordingly, microwave heating may be used to obtain pasteurized kiwifruit puree with enhanced safety, stability, nutritional and functional value, and consumer acceptance.

minced beef (1.50 kg preformed to 235 mm diameter and 40 mm height) showed that the thawing process starts from the surface of the sample and progresses slowly to the core, which is due to the strong absorption of microwave at positions close to the surface. This should respond to the higher loss factor of water compared to that of ice. They also noted that microwave thawing time is found to be less than one-fifth of that required in the conventional thawing of frozen minced beef. Furthermore, it was demonstrated that microwave thawing at a controlled surface temperature might provide a means of heating efficiently without causing an excessive rise in the surface temperature of the meat. It is possible to thaw the meat under controlled conditions in such a way that its surface temperature never exceeds 10°C, which maintains the quality of meat. Microwave thawing supplies a better quality product than conventional thawing at 4°C or 21°C. Moreover, Lee and Park (1999) compared refrigeration thawing (at 4°C) and oven thawing (at 28°C and 50°C) with microwave thawing. They found the latter to be a rapid thawing method which prevents excessive drip loss and the local overheating of frozen pork. Microwave energy accelerated the thawing rate, and had a significant effect on drip loss, water holding capacity and total aerobes. It was concluded that the cotreatment of microwave and convection heating appeared to be a suitable thawing process for the food industry.

On the other hand, microwave cooking is a method that has only recently been introduced into the meat industry. However, earlier reports on the commercial use of microwave cooking of bacon date back as early as 1986 (Edgar, 1986). More recently, microwave-cooked roasts are commercialized, which take less time to reach endpoint temperature as compared to conventional methods. However, meat cooked using microwaves does not have the typical browned surface associated with other methods of cooking because of the short cooking time and the surface temperature being too low to allow for the browning to develop. A study carried out by James, Barlow, James, and Swain (2006) established the optimum processing and product parameters that would give an adequate quality of cooked bacon by using a domestic microwave oven and a pilot-scale industrial microwave-cooking tunnel.

According to Duan, Jiang, Wang, Yu, and Wang (2011), hot air–microwave drying technology can be used for the dehydration of fresh tilapia fillets due to the fact that it improves the rehydration rate and shortens the drying time. The authors recommended low air temperature and microwave power in order to preserve the quality of fish fillets. On the other hand, Duan, Zhang, Mujumdar, and Wang (2010) proposed that microwave freeze-drying could replace conventional freeze-drying of sea cucumber because it greatly reduces both the process time and energy consumption, while producing the same product quality as that of conventional freeze-drying.

4.2.2.1 Impact on sensory properties

In this section, different observations are summarized and discussed on the effect of microwave application on the thawing and cooking of meat and fish and its influence on sensory properties.

Comparative studies were conducted on the quality characteristics of frozen beef, thawed by conventional methods and microwaves (Kim et al., 2013).

The sensory evaluation of meat was performed by a trained panel, and based on the appearance, flavor, texture, taste, and overall acceptability. The samples thawed by microwave yielded better sensory attributes and a lower drip loss than those of the conventional thawing methods (refrigeration thawing at 4°C, room temperature thawing at 25°C, and cold water thawing at 15°C). Therefore, the authors suggest that microwave thawing is an appropriate means of reducing the deterioration of quality caused by freezing. Complementary, the effect of the microwave power level on the physicochemical properties of thawed chicken breast becoming from fresh boilers 1.5–2.0 kg suggests that, as the microwave power level increased, the thawing rate increased and the differences in temperature between the edge position and the center position of the samples also grew. The treatment at an intermediate power level of 250 W for 144 s was the most suitable with which to minimize changes in the meat quality, resulting in a better color stability and textural properties and less pronounced thawing loss compared to that of other thawing methods, including the use of higher or lower microwave powers or conventional refrigeration thawing (Kim et al., 2011).

The sensory characteristics of buffalo meat processed by air-forced oven cooking and microwave heating were evaluated by Hoda, Ahmad, and Srivastava (2002). Better sensory scores for color (on a hedonic scale) were obtained for microwave-cooked buffalo meat and, in addition, the tenderness was improved. However, the flavor was superior in the case of the hot air oven product. As would be expected, using the microwave oven reduced the time needed to cook meat as compared to air-forced oven cooking and the meat was heated uniformly.

4.2.2.2 Impact on nutritional and functional value

Meat and fish are highly nutritive foods, hence the interest shown by different researchers in investigating the effects caused by microwaves on their nutritive and functional components. The effects of microwave heating on the retention of thermolabile vitamin B_6 and thiamine after the thermal treatment of pork and chicken were studied by Uherová, Hozová, and Smirnov (1993). In a microwave oven, the samples exhibited a higher degree of vitamin B_6 and thiamine retention as compared to a conventional oven. The measured values of vitamin retention demonstrated clearly that heating the muscle tissue using microwaves is less destructive to heat-sensitive vitamins than conventional roasting. Therefore, from this point of view, the authors recommended microwave cooking and roasting for the purposes of meat preparation in the food industry, households, restaurants, and hospitals. In another study, Kumar and Aalbersberg (2006) evaluated the earth-oven, a traditional Pacific style of cooking, and compared the nutrient retention with those of microwaved and oven-roasted chicken and lamb chops and also compared it with microwave-cooked fish. They commented that microwave cooking, overall, proved to be the most effective method with an enhanced retention of nutrients in meat samples. They observed the greatest retention of thiamine, retinol, and riboflavin in microwave-cooked meat. The difference in vitamin retention could be due to the higher temperature when cooking in an earth-oven compared with that in a microwave. The water content was found to be dependent on the cooking method and also on the food sample. Generally, the highest loss in the water content of all of the samples was associated with earth-oven cooking. This could mean that the greatest dripping losses occurred in earth-oven cooking, compared to microwave cooking or oven roasting. Microwave cooking or oven roasting only had a very slight deterioration effect on the protein in the cooked samples.

Meanwhile, Maranesi et al. (2005) studied the effect of microwave cooking or broiling on selected nutrient contents in separable lean from lamb rib loins. The traditional cooking method for this meat is by broiling the steaks. Microwaving, followed by a final grilling of the roasts, was proposed as an alternative cooking method. Significant differences between the two methods were also found in the true retention of moisture, protein, and several fatty acids, again to the advantage of microwave processing. Microwave cooking produced compositional changes that were less pronounced than those produced by broiling fat on lamb steaks. Some studies explained the significant role played by fat in the palatability of meat, and the cooking process affects the fat content. The effect of microwave cooking on goat fat reduction was compared with samples roasted in a conventional oven. The overall migration of fat globules during microwave cooking was higher (i.e., a more pronounced fat reduction) compared to conventional cooking. In this regard, an adverse effect on the quality of the meat can occur.

4.2.3 Other food products

Besides the previously mentioned foods, microwaves have also been used for processing of rice, pasta, or baked goods. The most common processes utilizing microwaves are cooking, drying, or baking. In many cases, a comparison between microwave and traditional cooking has been carried out. For example, Lakshmi, Chakkaravarthi, Subramanian, and Singh (2007) compared the energy consumption in the microwave cooking of rice with an electric cooker and a pressure cooker. They concluded that microwave cooking offered the advantage of the shortest cooking time of all of the various methods and also substantially reduced energy consumption. The most frequently employed process for the drying of pasta is that using hot air, although De Pilli, Giuliani, Derossi, and Severini (2009) studied the effects of microwave drying on the characteristics of cooked spaghetti. They suggest that the quality of cooked samples, dried by different methods, was comparable, with an important drying time reduction. Several researchers have studied a combined microwave–convective method of drying nuts, such as macadamia nuts (Silva, Marsaioli, Maximo, Silva, & Gonçalves, 2006) or pistachios (Kouchakzadeh & Shafeei, 2010). As far as baking is concerned, detailed references can be found on the microwave-baking process for bread, cakes, doughnuts, cookies, etc. The reader is referred to Sumnu (2001) and Chapter 6, Microwave-assisted baking for a comprehensive report on microwave-baked products.

On the other hand, as mentioned before, studies on microwave-assisted pasteurization and sterilization (see Chapter 10: Microwave-assisted pasteurization and sterilization—commercial perspective) have been motivated by the fast and effective microwave heating of many foods containing water or salts. The applications of microwave pasteurization or sterilization studied cover prepacked food, like yoghurt, or pouch-packed meals, as well as the continuous pasteurization of fluids, such as milk (Schubert & Regier, 2007). Meanwhile, Sabliov, Boldor, Coronel, and Sanders (2008) suggested that microwave heating could be used as a sterilization step in a UHT process for peanut-based beverages due to the short time required and the uniformity of the temperature distribution. An interesting study was conducted by Dev, Raghavan, and Gariepy (2008) who considered microwave heating for the in-shell pasteurization of eggs due to the very good transparency of the eggshell and shell membrane to microwaves. They noted that microwaves will have the potential for in-shell egg pasteurization and concluded that microwaves appeared to perform better in terms of the process time and product quality when compared to the current water bath pasteurization technique. On the other hand, the concept of pasteurizing prepared meals using microwaves to extend their shelf life has been considered for many years. In recent studies Tang and coworkers (Tang, 2015; Tang et al., 2008) indicated that microwave sterilization technology was effective for the processing of inhomogeneous food prepackaged in polymeric trays.

4.2.3.1 Impact on sensory properties

In baking processes, it is essential to obtain both browning and a good texture at a fixed moisture level (see Chapter 6: Microwave-assisted baking). In this regard, conventional baking using hot air provides both suitable color and texture. In microwave baking, as has been previously commented on for other processes, it is not possible to achieve either enough browning on the surface of products or crust formation (Chandrasekaran et al., 2013). For this reason, microwaves in baking are used in combination with conventional or infrared surface baking (Shaheen, et al., 2012). As explained by Sumnu (2001) and in Chapter 6, Microwave-assisted baking in this book, due to heating in the microwave oven being a very fast process, there is not enough time for either starch gelatinization, starch conversion by enzymes or for the sufficient expansion of the dough/batter and the final setting of the dough/batter into a rigid crumb structure, or for flavor development. The result is that microwave-baked products are often of poor quality, mainly with respect to the sensory properties. The volume of microwave-baked products is low, they are tough or firm in texture, and they lack both browning and crust formation. In this respect, as reported by Sumnu (2001) and in Chapter 6, Microwave-assisted baking in this book, some alternative means and combinations to improve the quality of microwave-baked products were suggested. The problem of low volume as a result of the incomplete gelatinization of starch can be overcome by, e.g., using pregelatinized starches or by starches having high dielectric properties and low gelatinization enthalpy. The great moisture loss in microwave-baked cakes can be made up for by using hydrocolloids. Developing a new formulation or improving the conventional formulation by using specific additives can solve the problem of the firmness or toughness of baked products. Susceptors (material used for its ability to absorb electromagnetic energy and convert it into heat) (see Chapter 13: Packaging for microwave foods) can be used to increase the surface temperature of the dough which is needed for the development of the browning reaction. The browning or crisping of products baked in a microwave oven can also be achieved by using special coatings or by heating the baked product in a conventional oven for a short time.

In the experimental study on the microwave baking of Madeira cake batter performed by Megahey, Mcminn, and Magee (2005), different microwave powers were tested. At a specific moderate microwave power (250 W, 70 s), the cake (30 g) exhibited improved textural properties, such as better springiness and less firmness, as compared to cake baked in a convective oven. Some studies investigated the effects of emulsifiers and gums (Seyhun, Sumnu, & Sahin, 2003) and added starch (Seyhun, Sumnu, & Sahin, 2005) to obtain microwaved products with similar sensory characteristics to conventionally baked products. Sumnu, Sahin, and Sevimli (2005) compared a nonconventional baking method, such as infrared, with and without the combination with microwave. As control, conventionally baked cakes were used. An infrared–microwave combination oven produced cakes which had similar firmness and color values to conventionally baked ones. Comparable results were obtained previously in bread baked in an infrared–microwave combination oven (Keskin, Sumnu, & Sahin, 2004).

In the cookie and biscuit industry, breakage can occur as a result of the cracking phenomenon known as "checking." Checking usually manifests itself as cracks that extend partially across the surface of the product, rendering it prone to complete breakage upon subsequent handling. The influence of microwave energy on quality in terms of checking and the mechanical strength of low-moisture-baked products were investigated by Ahmad, Morgan, and Okos (2001). Conventional biscuits were baked in a reel oven and microwave-baked biscuits were first prebaked in the reel oven, and the baking process was then finished in a microwave oven. Microwave baking was found to significantly reduce checking compared to conventionally-baked biscuits. The most important reason for the reduction in checking and the increase in the mechanical strength was found to be the development of more uniform moisture profiles while the biscuit temperature during microwaving was maintained at above 110°C. Microwaved biscuits were also found to be less susceptible to checking than conventional biscuits upon exposure to high atmospheric humidity.

For the purposes of optimizing a traditional cooking process, Cocci, Sacchetti, Vallicelli, Angioloni, and Dalla Rosa (2008) investigated the quality of spaghetti cooked by the traditional water boiling method and by microwave, using an innovative oven suitable for pasta cooking. In particular, microwave cooking led to a better color retention, a higher gel degree, and a more compact gluten network in the outer layer of the spaghetti which, in turn, resulted in less pronounced losses during cooking. The higher gel degree of microwave-cooked spaghetti gave firmness values that were lower than those of traditionally cooked spaghetti, but this textural difference could be minimized by choosing an adequate cooking time. Considering that, apart from color, the sensory profile of microwave-cooked spaghetti was not different from that of traditionally cooked spaghetti, and that microwave cooking could shorten the preparation time, microwave heating in water can be regarded as a convenient cooking alternative for this product. Along the same lines, Zhao, Xiong, Qiu, and Xu (2007) studied the effect of microwaves on the quality of cooked rice. Rice was cooked in an electric cooker and in an industrial continuous microwave oven at different specific energies. The specific energy consumption of the microwave oven was defined as the quotient of microwave power divided by the flow rate of the rice samples. The sensory evaluation of cooked rice was tested immediately after cooking. The odor, appearance, taste, and texture of the cooked rice were scored individually. It was observed that the sensory quality of cooked rice improved as the microwave-energy consumption increased. The authors emphasized that the taste score of the microwave-treated rice during storage was higher than that of the rice cooked in an electric oven.

Previously, the use of microwaves as an alternative process in fluid pasteurization has been mentioned. To this end, the sensory changes in conventionally pasteurized milk and in milk pasteurized with microwave application were studied by Valero, Villamiel, Sanz, and Martínez-Castro (2000). Raw milk was submitted to a continuous-flow microwave treatment or placed in a conventional heat exchanger. A trained sensory panel was utilized for sample differentiation. The acceptance scores of the milk were recorded from extremely unacceptable to extremely acceptable. The milk was also evaluated for overall flavor. The differences between microwaved- and conventionally treated milks could not be distinguished by the sensory panel directly after processing and during the storage period. Since the taste and odor of microwave-treated milk were not affected either during the treatment or the subsequent storage, microwave heating can be considered a good alternative to conventional pasteurization methods. Continuing with the same topic of research, Clare et al. (2005) compared the sensory parameters of skim milk (white and chocolate flavored) heated using a UHT-processing unit and a microwave processing unit under equivalent conditions (temperature and residence time) to those used in the UHT system. Every 3 months, samples were taken for descriptive sensory evaluation (up to the 12 months of the control period). The sensory analyses established that UHT-treated milks were visibly darker, and exhibited more pronounced caramelized and stale/fatty flavors with increased astringency when compared with the microwave samples. The sweet aromatic flavor and sweet taste of both UHT and microwaved milk products decreased during storage, whereas the stale/fatty flavors increased over time. The sensory effects were more apparent in white milks than in the chocolate varieties. These studies suggest that microwave technology may provide a useful alternative processing method for supplying aseptic milk products that retain a long shelf life.

With respect to the potential use of microwaves for blanching purposes as an alternative to traditional blanching methods, off-flavors have been observed (Schirack et al., 2006; Schirack, Sanders, & Sandeep, 2007). In this sense, the resulting off-flavors detected in microwave blanched peanuts may be related to the increased concentrations in phenylacetaldehyde, guaiacol, and 2,6-dimethylpyrazine, which may have developed as a consequence of Maillard reactions and thermal degradation when high temperatures are reached (Schirack et al., 2006).

In order to avoid this negative sensory characteristic, Schirack et al. (2007) described effective processing parameters for microwave blanching and provided guidelines on the appropriate process parameters that will ensure rapid blanching and an adequate quality (absence of off-flavors) of the peanuts.

As for nuts, Uysal, Sumnu, and Sahin (2009) concluded that microwave-roasted hazelnuts had a comparable quality to conventionally roasted ones with respect to their color (based on the L* and a* values of CIE measurement) and texture (based on the fracture force). The results obtained from the study carried out by Silva et al. (2006) led to the conclusion that macadamia kernels dried by microwaves and hot air did not differ significantly from the conventionally dried ones in terms of the appearance, crispness, and overall acceptance.

4.2.3.2 Impact on nutritional and functional value

The effects of microwave cooking and other traditional cooking methods, such as boiling and autoclaving, on the nutritional composition and antinutritional factors of chickpeas (*Cicer arietinum* L.) were studied by Alajaji and El-Adawy (2006). Each of the cooking methods decreased the minerals, fat, total ash, and carbohydrate fractions of this legume. The cooking treatments also decreased the concentrations of lysine, tryptophan, and sulfur-containing amino acids. However, the in vitro protein digestibility, protein efficiency ratio, and essential amino acid index were improved by the cooking treatments. Nevertheless, what characterizes the microwave cooking of chickpea is the fact that it saves time and also retains the highest nutritional value, increasing the protein digestibility and retention rates of both B vitamins and minerals. In microwave cooking, fewer B vitamins (riboflavin, thiamin, niacin, and pyridoxine) were lost than in traditional cooking. Based on this study, microwave cooking was recommended for chickpea preparation as a means of improving the nutritional quality by reducing the level of flatulence and antinutritional factors (trypsin inhibitor, haemagglutinin activity, tannins, saponins, and phytic acid). With respect to the cooking of rice, microwave application caused significant compositional differences when compared with rice cooked in an electric oven at the beginning of storage but not after 120 days (Zhao et al., 2007). The greater the microwave-energy consumption, the more pronounced the initial reduction of the water, free fatty acid, and protein content. Considering the overall results of this study, the application of microwaves in a controlled process can be considered as a possible method to be used for cooking rice.

The influence of microwave heating on several nutrients of milk has also been studied previously. The conclusions on the effect of microwave on vitamin losses in the public domain are diverse. Some researchers say that hardly any vitamins are lost in microwave-treated milk, or only in very small amounts, and also that fat and protein concentration remain unchanged. Sieber, Eberhard, and Gallmann (1996) emphasize diverse studies about nutrient alterations in microwave treated cows' milk, and concluded that microwave-treated milk does not differ significantly from conventionally heated milk. An investigation conducted by Villamiel, Corzo, Martínez-Castro, and Olano (1996) exposed the main chemical changes taking place during milk heating processes. For this purpose, 25 mL raw milk was heated to controlled different temperatures (70–120°C) in a microwave oven or treated by conventional heating. The microwaved milk samples exhibited a more pronounced isomerization of lactose, Maillard reaction, and protein denaturation than what was observed in conventionally heated milks. This can likely be attributed to the previously reported nonuniform temperature distribution during microwave treatment. The researchers suggest that improved equipment and process control could improve the problem of nonuniform temperature distribution, and, consequently, the chemical changes in microwave-processed foods would be reduced.

El-Badrawy, El-Zainy, Shalaby, and El-Sayed (2007) investigated the application of microwaves for roasting peanut seeds and compared it with a conventional roasting method. The chemical composition, mineral content, individual amino acids, and oligosaccharides, as well as the fatty acids were investigated in raw, microwave-roasted and ordinary-roasted peanut oils. The results revealed that the microwave roasting of peanut is better than conventional roasting in terms of the preservation of chemical composition and mineral contents. In addition, it raised the protein efficiency ratio (PER) and biological value (BV) of peanut protein more than when using ordinary roasting. PER evaluates protein quality, based on the amino acids composition, and BV is a measure of the proportion of the absorbed protein from a food. Whereas microwave roasting did not cause a significant chemical change in peanut oil, conventional roasting did when compared with raw peanut oil. Microwave roasting did not increase the malondialdehyde content, whereas conventional roasting did when compared with raw peanut oil. Microwave roasting showed to be highly efficient at preserving the characteristics of raw peanut oil. The degree of unsaturation of microwave-roasted peanut oil was higher than that of the conventionally roasted, whereas the trans-fatty acid content of microwave-roasted peanut oil was lower than that of the conventionally roasted. Meanwhile, in their previously cited study, Uysal et al. (2009) recommended the microwave–infrared combination oven for roasting hazelnuts due to the fact that a similar fatty acid composition was obtained when compared with the conventionally roasted ones. Nevertheless, in his study on the microwave roasting of peanuts, Megahed (2001) rejects its use due to the fact that microwave accelerates the formation of some undesirable and harmful compounds (e.g., oxidation products, transformed pigments, etc.), together with oil darkening. Other studies related to the use of a microwave oven in the roasting of nuts (Zook, Macku, & Deming, 1995), peanut seeds (Yoshida, Hirakawa, Tomiyama, Nagamizu, & Mizushina, 2005), and sunflower seeds (Anjum, Anwar, Jamil, & Iqbal, 2006) can be consulted in order to further knowledge on the nutritional changes in these kinds of products.

Coffee beans have usually been roasted by using the convection method. However, studies on coffee microwaving have also been carried out. Nebesny and Budryn (2003) estimated the antioxidative activity of coffee beans roasted by using either convection or microwave methods. Their results indicate that coffee beans roasted by using either the traditional (convection) or the microwave methods are highly active DPPH (1,1-diphenyl-2-picryl hydrazyl) scavengers and protect linoleic acid from oxidation. However, the process of roasting resulted in a decline in the initial antioxidative capacity. The antioxidative effectiveness is very closely correlated with the polyphenol concentration. Microwaving caused a less pronounced decrease in their content and, therefore, coffee roasted in this way exhibited greater antioxidative activity. The test based on the linoleic acid oxidation did not show such a distinct correlation, but it revealed that microwaved samples were more active as antioxidants.

Honey, in its liquid state, presents significant handling problems for a few processes due to its viscosity and stickiness. There is a demand in the food manufacturing industry for dried honey as a more convenient ingredient for processing. Honey powder can be directly added into dry mixes, seasonings, or dry coatings, and may be easily blended with other dry ingredients preserving the full honey flavor. Drying processes, such as spray drying, tunnel drying, and drum drying, are commonly used for the purposes of preparing the honey powder. Cui, Sun, Chen, and Sun (2008) investigated the optimization of operating parameters of the microwave–vacuum drying of honey and evaluated the quality of the dry product. Different microwave powers, vacuum pressure levels, and sample thicknesses were considered and the best parameters for the microwave–vacuum drying were identified. Under optimal process conditions, the contents of the main sugars (fructose, glucose, maltose, and sucrose) were not significantly changed. Volatile acids, alcohols, aldehydes and esters make up the bulk of the identified aroma compounds present in the liquid honey used today. While the content of the alcohols and esters changed slightly during microwave–vacuum drying, that of the acids decreased markedly. On the other hand, the aldehyde and ketone content of the honey which was dehydrated by microwave–vacuum drying increased remarkably. Cui et al. (2008) do not discuss the potential advantages or disadvantages of increasing aldehyde and ketone content, although it may be expected that these changes due to processing do not significantly alter the overall volatile fraction and consequently do not alter significantly the intrinsic flavor and aroma of honey (Escriche, Visquert, Juan-Borrás, & Fito, 2009). Additionally, no darkening took place during drying.

4.3 Superiority of microwaves over conventional heating techniques to preserve food quality

Ever since the use of microwaves for food processing purposes was initiated in the 1930s (Schubert & Regier, 2006), there was clear evidence of their potential for the preservation and/or transformation of different food matrices, with the consequent rise of the expectations of this technology representing a likely alternative to conventional thermal processes (Vadivambal & Jayas, 2007, 2010). As it has been previously reported by many authors (Salazar-González et al., 2012; Vadivambal & Jayas, 2007, 2010), the attractiveness of microwaves mainly relies on its apparent ability to maintain the quality of food products to a greater extent, when compared with conventional heating, while still securing safety and a suitable extension of their shelf life (De Ancos et al., 1999; Salazar-Gonzalez et al., 2012).

It should be noted, however, that several decades later, the applicability of microwave heating in industrial processes continues to be limited. Despite numerous scientific studies having been focused on evaluating the use of microwaves in food processing, with the suitability of this technology for pasteurization, sterilization, and dehydration steps having been widely demonstrated (Cañumir et al., 2002; Huang et al., 2007; O'Donnell et al., 2010), there are still serious doubts on the real benefits of its application and there seems to be some controversy around its superiority over conventional heating techniques, when these are compared (Banik, Bandyopadhyay, & Ganguly, 2003).

Properly comparing thermal processes based on the application of alternative and conventional heating techniques represents a challenge, given the different principles of heat generation. Unfortunately, the lack of common criteria on how to establish an effective comparison on the impact of microwave and conventional heating on several food safety and quality issues does not positively contribute to get a clear overview of the real benefits of microwaves, or, going a step further, in the understanding of the main differences between the interaction of each one of these two technologies and the food matrix. In other words, this lack of homogeneity in comparison procedures may result in erroneous interpretations and restrict the feasibility of comparing different studies in the public domain. An example of this can be found in how different approaches on selecting the processing conditions are studied, such as temperature exposure or treatment times, as reviewed and discussed by Banik et al. (2003), or simply, poor kinetic data acquisition, where applicable, frequently results in different authors reaching contradictory conclusions.

Despite the desperate need for the standardization of criteria when it comes to the assessment of the benefits of microwaves over conventional thermal processes, the big majority of authors agree on the superiority of microwaves, when applied to several food processing operations, to preserve the quality of food products to a greater extent. The studies published by Lau and Tang (2002) and Igual et al. (2010) are highlighted below as two examples of the numerous research studies conducted on this topic over the last decades (Salazar-González et al., 2012; Vadivambal & Jayas, 2007, 2010). Lau and Tang (2002) pointed out that greater textural changes were found in pickled asparagus when conventional heating, instead of microwaves, was used to pasteurize the product. Similarly, Igual et al. (2010) proved how microwave pasteurization of grapefruit juice secured better preservation of citric acid and ascorbic acid in the product than the conventional treatment. Interestingly, this lower thermal degradation of the quality attributes of several food products that appears to be associated with the application of microwaves tends to be attributed to the higher effectiveness of this technology to inactivate enzymes and microorganisms, allowing the processing time to be reduced, as it has been pointed out by various authors before. In this respect, e.g., Latorre, Bonelli, Rojas, and Gerschenson (2012) observed that microwave blanching of red beet gave rise to a faster inactivation of peroxidase enzyme than the conventional treatment. Similarly, Matsui, Gut, De Oliveira, and Tadini (2008) reported microwave inactivation of peroxidase and polyphenoloxidase enzymes to be significantly faster in coconut water when compared to conventional heat processes.

Furthermore, Yaghmaee and Durance (2005) observed *Escherichia coli* to be more sensitive to temperature changes under microwave than conventional heating.

It should be highlighted, however, that the reasons behind this apparent superiority are not clearly understood yet, with the different authors not being able to reach an agreement when analyzing the results and providing an explanation for the outcomes of their research. In this respect, it must be kept in mind that deeper understanding of how the microwave energy interacts with the food matrix during processing is critical if this technology is intended to be exploited to its fullest potential at an industrial level. This knowledge will then contribute to safe and secure control of the heating process, giving rise to safer and effective, and economically more viable microwave treatments.

When focusing on the preservation area (pasteurization and sterilization), three different hypotheses have, up to now, been considered to explain the greater effectiveness of microwaves over conventional heating techniques to process and preserve foods: (1) faster heating rates; (2) enhanced effects; and (3) nonthermal effects. In this regard, as reviewed and discussed by Chandrasekaran et al. (2013), despite several scientific studies arguing on the nonthermal effects of microwaves leading to faster microbial and enzyme inactivation and better preservation of the nutritional, functional, and sensory properties of several food products (Banik et al., 2003), no irrefutable proof of the lethal effects of microwave radiation in the absence of temperature or pH stress conditions have been found yet. From a chemical point of view, a recent work (Kappe, Pieber, & Dallinger, 2013) showed for organic synthesis, that nonthermal microwave effects do not exist. Therefore, faster heating rates and enhanced effects are going to be exclusively considered and further discussed in this chapter.

4.3.1 Faster heating rates

As it has been extensively discussed in the literature (Salazar-González et al., 2012; Vadivambal & Jayas 2007; Zhang, Tang, Mujumdar, & Wang, 2006) and in various chapters in this book, the particular and distinctive way in which heat is generated during microwave processes (volumetric heating) gives rise to a number of advantages over traditional heating systems, which basically rely on a heating medium (e.g., water, air, or steam) to convey heat from the source to the product by means of conduction and convection. These include faster heating rates, reduced processing times, selective heating, high thermal efficiency, more efficient water removal, among others (Chandrasekaran et al., 2013; De Ancos et al., 1999; Huang et al., 2007; Salazar-González et al., 2012; Vadivambal & Jayas, 2010). It should be highlighted that, from the perspective of securing the maximum preservation of the quality of the product, faster heating rates seem to be one of the most acknowledged features of microwaves (Gowen, Abu-Ghannam, Frias, & Oliveira, 2006; Huang et al., 2007; Queiroz et al., 2008). Examples of this can be found in the studies published, e.g., by Hebbar, Nandini, Lakshmi, and Subramanian (2003), Salazar-González et al. (2012), and Vadivambal and Jayas (2010). Hebbar et al. (2003) reported that microwave heating provided a rapid means of achieving the desired level of microbial reduction with reduced thermal damage. The reason behind faster heating rates leading to a better preservation of the quality of food products are evidently a shorter treatment time and less thermal exposure. During the last decades, the use of low temperature, long time (LTLT) heat treatments to preserve foods has been progressively replaced by high temperature, short time (HTST) ones, because of HTST processes allowing a better retention of the product quality, due to the higher thermosensitivity of microorganisms than that of most quality attributes of foods (Awuah, Ramaswamy, & Economides, 2007). Accordingly, and taking into account the ability of microwaves to heat foods directly and internally, processes based on the application of this technology can be seen as an opportunity to take the HTST approach to the next level, with all the implicit advantages of it.

4.3.2 Enhanced effects

Despite faster heating rates being a proven and widely accepted benefit of microwave technology, in some cases, this premise cannot be accepted to explain and support the differences observed between some microwave and conventional heat processes. This is the case, for instance, of those studies, such as the ones carried out by Tajchakavit and Ramaswamy (1997), Tajchakavit, Ramaswamy, and Fustier (1998), or Soysal and Söylemez (2005), among others, in which kinetic parameters were calculated and compared, with the outcome of that comparison proving higher effectiveness of microwaves to inactivate enzymes and/or microorganisms, or simply, when intensity of equivalent microwave and conventional treatments was compared in terms of thermal load, with a lower thermal load being required for microwave heating to achieve an equal level of enzyme and microbial inactivation in the food product, as it was published in one of our previous studies (Benlloch-Tinoco, Martínez-Navarrete, et al., 2014). It should be taken into account that, in all these cases, the differences observed between both technologies should not be attributed to faster heating rates alone, given that the experimental data were calculated as if the treatments had taken place under isothermal conditions. Accordingly, authors discussed the possibility of their results indicating a magnification or enhancement of the lethal effects associated with the temperature rise in the product due to the exposure to microwave radiation. However, to date no definitive proof has been provided that there are indeed nonthermal effects associated with microwave processing. Some authors have proposed that contributory effects of microwaves may be caused in part by the alignment of molecules in the electromagnetic field, which takes place during exposure, with this affecting the protein structure of the microorganisms (Banik et al., 2003; Rosen, 1972). Another hypothesis considered is that the microwave field, which foods are exposed to, can affect the polar and/or charged moieties of proteins enhancing both enzyme and microbial inactivation during the treatment (Kermasha, Bisakwski, Ramaswamy, & Van De Voort, 1993). Therefore, more research is needed on this subject to gain a deeper understanding of what the real impact of microwave energy is on the food matrix and how both of them interact with each other, in order to achieve a complete comprehension of the real potential and benefits of the application of this technology in the food processing area.

4.4 Future trends

Microwave energy offers numerous advantages for the food processing industry that can contribute to improving traditional treatments; if applied under purpose-designed operating conditions, this technology can result in high throughput processes, enabling foods to be transformed and/or preserved in considerably less time and at lower energy consumption than that required by conventional thermal processes, and at the same time reducing quality losses.

Consequently, the future of the marketing and applications of microwave processing is a promising one, but it will rely on different factors: First, it is critical to develop a deeper understanding of the physical phenomena behind microwave heating (electromagnetic effects as well as heat and mass transfer phenomena) that implies the interaction of the fields and the physical properties of the food to be heated.

The possibility of predicting the microwave heating distribution in terms of process modeling will also provide a powerful tool, which will enable optimization of different microwave applications. In this sense, new methods, such as magnetic resonance thermal imaging, intrinsically produced compounds (chemical markers), and time–temperature indicator methods, will assist in the determination of the temperature distribution within the food material, the validation of the process simulations, and will eventually foster the widespread application of microwave pasteurization and sterilization systems in the food industry.

At the same time, there is a marked trend in today's food industry toward continuous processing lines. The faster the development and implementation of continuous processing microwave ovens, the greater the application of microwave heating as a unit operation for rapid and in-depth heating. The lack of uniformity is another important problem. Techniques, such as rotating and oscillating the food package, providing an absorbing medium (hot water) surrounding the product, equilibrating after heating, and cycling the power have been implemented to improve product homogeneity (Chandrasekaran et al., 2013). However, variable frequency microwave processing and phase control microwave processing, which have already succeeded in commercial applications to heat nonfood materials, are novel methods that are being studied as a means of minimizing the problem (FDA, 2012). More research is needed to apply such systems in the food industry.

4.5 Conclusions

Microwaves have been successfully used for many food processes, such as in the cooking, drying, and pasteurization of food materials. In this chapter, experimental investigations into various microwave-assisted food-processing techniques and their effects on the nutritional, sensory, and other quality attributes of different food products were reviewed.

Microwave-cooked products have the advantages of better retaining taste, color, quality, and nutritional value compared to those cooked by conventional methods. Faster heating and high energy efficiency are also advantages of microwave processing. At the same time, this technology has been found to be more effective in the destruction of pathogens or in the inactivation of enzymes, largely attributed to the enhanced heating rates, leading to significant enhancement or magnification of thermal effects. However, also nonthermal effects have been reported, where it has been postulated that nonthermal microwave effects may result from a direct stabilizing interaction of the electric field with specific polar molecules present in the food, which is not related with a macroscopic temperature increase. Although it has received considerable attention in recent years, it is still the subject of intense debate in the scientific community.

References

Ahmad, S. S., Morgan, M. T., & Okos, M. R. (2001). Effects of microwave on the drying, checking and mechanical strength of baked biscuits. *Journal of Food Engineering*, *50*(2), 63–75.

Alajaji, S. A., & El-Adawy, T. A. (2006). Nutritional composition of chickpea (*Cicer arietinum* L.) as affected by microwave cooking and other traditional cooking methods. *Journal of Food Composition and Analysis*, *19*(8), 806–812.

Anjum, F., Anwar, F., Jamil, A., & Iqbal, M. (2006). Microwave roasting effects on the physico-chemical composition and oxidative stability of sunflower seed oil. *Journal of the American Oil Chemists' Society*, *83*(9), 777–784.

Antunes, M. D. C., Dandlen, S., Cavaco, A. M., & Graça, M. (2010). Effects of postharvest application of 1-mcp and postcutting dip treatment on the quality and nutritional properties of fresh-cut kiwifruit. *Journal of Agricultural and Food Chemistry*, *58*(10), 6173–6181.

Awuah, G. B., Ramaswamy, H. S., & Economides, A. (2007). Thermal processing and quality: Principles and overview. *Chemical Engineering and Processing: Process Intensification*, *46*(6), 584–602.

Banik, S., Bandyopadhyay, S., & Ganguly, S. (2003). Bioeffects of microwave—A brief review. *Bioresource Technology*, *87*(2), 155–159.

Behera, S., Nagarajan, S., & Jagan Mohan Rao, L. (2004). Microwave heating and conventional roasting of cumin seeds (*Cuminum cyminum* L.) and effect on chemical composition of volatiles. *Food Chemistry*, *87*(1), 25–29.

Benlloch-Tinoco, M., Igual, M., Rodrigo, D., & Martínez-Navarrete, N. (2013). Comparison of microwaves and conventional thermal treatment on enzymes activity and antioxidant capacity of kiwifruit puree. *Innovative Food Science and Emerging Technologies*, *19*, 166–172.

Benlloch-Tinoco, M., Igual, M., Salvador, A., Rodrigo, D., & Martínez-Navarrete, N. (2014). Quality and acceptability of microwave and conventionally pasteurised kiwifruit puree. *Food Bioprocess Technology*, *7*, 3282–3292.

Benlloch-Tinoco, M., Martínez-Navarrete, N., & Rodrigo, D. (2014). Impact of temperature on lethality of kiwifruit puree pasteurization by thermal and microwave processing. *Food Control*, *35*, 22–25.

Benlloch-Tinoco, M., Pina-Pérez, M. C., Martínez-Navarrete, N., & Rodrigo, D. (2014). *Listeria monocytogenes* inactivation kinetics under microwave and conventional thermal processing in a kiwifruit puree. *Innovative Food Science and Emerging Technologies*, *22*, 131–136.

Benlloch-Tinoco, M., Varela, P., Salvador, A., & Martínez-Navarrete, N. (2012). Effects of microwave heating on sensory characteristics of kiwifruit puree. *Food Bioprocess Technology*, *5*, 3021–3031.

Bertelli, D., Plessi, M., & Miglietta, F. (2003). Effect of microwaves on volatile compounds in origanum. *LWT-Food Science and Technology*, *36*(6), 555–560.

Böhm, V., Kühnert, S., Rohm, H., & Scholze, G. (2006). Improving the nutritional quality of microwave-vacuum dried strawberries a preliminary study. *Food Science and Technology International*, *12*(1), 67–75.

Cañumir, J. A., Celis, J. E., De Brujin, J., & Vidal, L. V. (2002). Pasteurization of apple juice by using microwaves. *Lebensmittel-Wissenschaft und Technologie*, *35*, 389–392.

Carranza-Concha, J., Camacho, M. M., & Martínez-Navarrete, N. (2012). Effects of blanching on grapes (*Vitis vinifera*) and changes during storage in syrup. *Journal of Food Processing and Preservation*, *36*, 11–20.

Chandrasekaran, S., Ramanathan, S., & Basak, T. (2013). Microwave food processing—A review. *Food Research International*, *52*(1), 243–261.

Chuah, A. M., Lee, Y.-C., Yamaguchi, T., Takamura, H., Yin, L.-J., & Matoba, T. (2008). Effect of cooking on the antioxidant properties of coloured peppers. *Food Chemistry Volume*, *111*(1), 20–28.

Clare, D. A., Bang, W. S., Cartwright, G., Drake, M. A., Coronel, P., & Simunovic, J. (2005). Comparison of sensory, microbiological, and biochemical parameters of microwave versus indirect UHT fluid skim milk during storage. *Journal of Dairy Science*, *88*(12), 4172–4182.

Cocci, E., Sacchetti, G., Vallicelli, M., Angioloni, A., & Dalla Rosa, M. (2008). Spaghetti cooking by microwave oven: Cooking kinetics and product quality. *Journal of Food Engineering*, *85*, 537–546.

Contreras, C., Martín, M. E., Martínez-Navarrete, N., & Chiralt, A. (2005). Effect of vacuum impregnation and microwave application on structural changes which occurred during air-drying of apple. *Lebensmittel-Wissenschaft und-Technologie*, *38*, 471–477.

Contreras, C., Martín-Esparza, E., & Martínez-Navarrete, N. (2012). Influence of drying method on the rehydration properties of apricot and apple. *Journal of Food Process Engineering*, *35*, 178–190.

Contreras, C., Martín-Esparza, M. E., Chiralt, A., & Martínez-Navarrete, N. (2008). Influence of microwave application on convective drying: Effects on drying kinetics, and optical and mechanical properties of apple and strawberry. *Journal of Food Engineering*, *88*, 55–64.

Contreras, C., Martín-Esparza, M. E., Martíinez-Navarrete, N., & Chiralt, A. (2007). Influence of osmotic pre-treatment and microwave application on properties of air dried strawberry related to structural changes. *European Food Research and Technology*, *224*, 499–504.

Costa, A. I. A., Dekker, M., Beumer, R. R., Rombouts, F. M., & Jongen, W. M. F. (2001). A consumer-oriented classification system for home meal replacements. *Food Quality and Preference*, *12*, 229–242.

Cui, Z.-W., Sun, L.-J., Chen, W., & Sun, D.-W. (2008). Preparation of dry honey by microwave–vacuum drying. *Journal of Food Engineering*, *84*, 582–590.

De Ancos, B., Cano, M. P., Hernández, A., & Monreal, M. (1999). Effects of microwave eating on pigment composition and color of fruit purees. *Journal of the Science of Food and Agriculture*, *79*, 663–670.

De Pilli, T., Giuliani, R., Derossi, A., & Severini, C. (2009). Study of cooking quality of spaghetti dried through microwaves and comparison with hot air dried pasta. *Journal of Food Engineering*, *95*(3), 453–459.

Dev, S. R. S., Raghavan, G. S. V., & Gariepy, Y. (2008). Dielectric properties of egg components and microwave heating for in-shell pasteurization of eggs. *Journal of Food Engineering*, *86*, 207–214.

Dorantes-Alvarez, L., Barbosa-Caanovas, G., & Gutiearrez-Loa Pez, G. (2000). Blanching of fruits and vegetables using microwaves. In G. Barbosa-Cánovas & G. Gould (Eds.), *Innovations of Food Processing*. CRC Press, ISBN:1-56676-782-2.

Drouzaf, A. E., & Schuber, H. (1996). Microwave application in vacuum drying of fruits. *Journal of Food Engineering*, *28*, 203–209.

Du, G., Li, M., Ma, F., & Liang, D. (2009). Antioxidant capacity and the relationship with polyphenol and Vitamin C in *Actinidia* fruits. *Food Chemistry*, *113*(2), 557–562.

Duan, X., Zhang, M., Mujumdar, A. S., & Wang, S. (2010). Microwave freeze drying of sea cucumber (*Stichopus japonicus*). *Journal of Food Engineering*, *96*(4), 491–497.

Duan, Z.-H., Jiang, L.-N., Wang, J.-L., Yu, X.-Y., & Wang, T. (2011). Drying and quality characteristics of tilapia fish fillets dried with hot air-microwave heating. *Food and Bioproducts Processing*, *89*(4), 472–476.

Edgar, R. (1986). The economics of microwave processing in the food industry. *Food Technology*, *40*(6), 106–112.

El-Badrawy, E. E. Y., El-Zainy, A. R. M., Shalaby, A. O., & El-Sayed, N. Y. (2007). Effect of microwave roasting on chemical composition of peanut seeds and comparing it with the ordinary roasting process. In: *2th conference of faculty of specific education* (pp. 11–12).

Erle, U., & Schubert, H. (2001). Combined osmotic and microwave-vacuum dehydration of apples and strawberries. *Journal of Food Engineering*, *49*(2–3), 193–199.

Escriche, I., Visquert, M., Juan-Borrás, M., & Fito, P. (2009). Influence of simulated industrial thermal treatments on the volatile fractions of different varieties of honey. *Food Chemistry*, *112*(2), 329–338.

FDA Food and Drug Administration. (2012). *Kinetics of microbial inactivation for alternative food processing technologies – Microwave and radio frequency processing*. Available from http://www.fda.gov/Food/FoodScienceResearch/SafePracticesforFood Processes/ucm100250.htm Last accessed 05.02.16.

Feng, H., & Tang, J. (1998). Microwave finish drying of diced apples in a spouted bed. *Journal of Food Science*, *63*(4), 679–683.

Ferenczi, S., Czukor, B., & Cserhalmi, Z. (2014). Evaluation of microwave vacuum drying combined with hot-air drying and compared with freeze- and hot-air drying by the quality of the dried apple product. *Periodica Polytechnica Chemical Engineering*, *58*(2), 111–116.

Funebo, T., Ahrné, L., Kidman, S., Langton, M., & Skjöldebrand, C. (2000). Microwave heat treatment of apple before air dehydration – Effects on physical properties and microstructure. *Journal of Food Engineering*, *46*(3), 173–182.

Gerard, K. A., & Roberts, J. S. (2004). Microwave heating of apple mash to improve juice yield and quality. *LWT-Food Science and Technology*, *37*(5), 551–557.

Gowen, A., Abu-Ghannam, N., Frias, J., & Oliveira, J. (2006). Optimisation of dehydration and rehydration properties of cooked chickpeas (*Cicer arietinum* L.) undergoing microwave–hot air combination drying. *Trends in Food Science & Technology*, *17*(4), 177–183.

Hamoud-Aghaa, M. M., Cureta, S., Simoninb, H., & Boillereauxa, L. (2014). Holding time effect on microwave inactivation of *Escherichia coli* K12: Experimental and numerical investigations. *Journal of Food Engineering*, *143*, 102–113.

Hebbar, H. U., Nandini, K. E., Lakshmi, M. C., & Subramanian, R. (2003). Microwave and infrared heat processing of honey and its quality. *Food Science and Technology Research*, *9*(1), 49–53.

Hernández-Infante, M., Sousa, V., Montalvo, I., & Tena, E. (1998). Impact of microwave heating on hemagglutinins, trypsin inhibitors and protein quality of selected legume seeds. *Plant Foods for Human Nutrition*, *52*(3), 199–208.

Hoda, I., Ahmad, S., & Srivastava, P. K. (2002). Effect of microwave oven processing, hot air oven cooking, curing and polyphosphate treatment on physico-chemical, sensory and textural characteristics of buffalo meat products. *Journal of Food Science and Technology*, *39*(3), 240–245.

Huang, Y., Sheng, J., Yang, F., & Hu, Q. (2007). Effect of enzyme inactivation by microwave and oven heating on preservation quality of green tea. *Journal of Food Engineering*, *78*, 687–692.

Igual, M., Contreras, C., Camacho, M. M., & Martínez-Navarrete, N. (2014). Effect of thermal treatment and storage conditions on the physical and sensory properties of grapefruit juice. *Food Bioprocess Technology*, *7*, 191–203.

Igual, M., Contreras, C., & Martínez-Navarrete, N. (2014). Colour and rheological properties of non-conventional grapefruit jams: Instrumental and sensory measurement. *LWT-Food Science and Technology*, *56*, 200–206.

Igual, M., García-Martínez, E., Camacho, M. M., & Martínez-Navarrete, N. (2010). Effect of thermal treatment and storage on the stability of organic acids and the functional value of grapefruit juice. *Food Chemistry*, *118*(2), 291–299.

Igual, M., García-Martínez, E., Camacho, M. M., & Martínez-Navarrete, N. (2011). Changes in flavonoid content of grapefruit juice caused by thermal treatment and storage. *Innovative Food Science and Emerging Technologies*, *12*(2), 153–162.

James, C. H., Barlow, K. E., James, S. J., & Swain, M. J. (2006). The influence of processing and product factors on the quality of microwave pre-cooked bacon. *Journal of Food Engineering*, *77*(4), 835–843.

Kappe, C. O., Pieber, B., & Dallinger, D. (2013). Microwave effects in organic synthesis: Myth or reality? *Angewandte Chemie International Edition*, *52*, 1088–1094.

Kaushik, G., Satya, S., & Naik, S. N. (2010). Effect of domestic processing techniques on the nutritional quality of the soybean. *Mediterranean Journal of Nutrition and Metabolism*, *3*(1), 39–46.

Kermasha, S., Bisakwski, B., Ramaswamy, H., & Van De Voort, F. R. (1993). Thermal and microwave inactivation of soybean lipoxigenase. *LWT-Food Science and Technology*, *26*, 215–219.

Keskin, S. O., Sumnu, G., & Sahin, S. (2004). Bread baking in halogen lamp-microwave combination oven. *Food Research International*, *37*, 489–495.

Khraisheh, M. A. M., Mcminn, W. A. M., & Magee, T. R. A. (2004). Quality and structural changes in starchy foods during microwave and convective drying. *Food Research International*, *37*(5), 497–503.

Kim, T.-H., Choi, J.-H., Choi, Y.-S., Kim, H.-Y., Kim, S.-Y., Kim, H.-W., & Kim, C.-J. (2011). Physicochemical properties of thawed chicken breast as affected by microwave power levels. *Food Science and Biotechnology*, *20*(4), 971–977.

Kim, Y. B., Jeong, J. Y., Ku, S. K., Kim, E. M., Park, K. J., & Jang, A. (2013). Effects of various thawing methods on the quality characteristics of frozen beef. *Korean Journal for Food Science of Animal Resources*, *33*(6), 723–729.

Kouchakzadeh, A., & Shafeei, S. (2010). Modeling of microwave-convective drying of pistachios. *Energy Conversion and Management*, *51*(10), 2012–2015.

Kumar, S., & Aalbersberg, B. (2006). Nutrient retention in foods after earth-oven cooking compared to other forms of domestic cooking 2. Vitamins. *Journal of Food Composition and Analysis*, *19*, 311–320.

Lado, B. H., & Yousef, A. E. (2002). Alternative food-preservation technologies: Efficacy and mechanisms. *Microbes and Infection*, *4*, 433–440.

Lakshmi, S., Chakkaravarthi, A., Subramanian, R., & Singh, V. (2007). Energy consumption in microwave cooking of rice and its comparison with other domestic appliances. *Journal of Food Engineering*, *78*(2), 715–722.

Latorre, M. E., Bonelli, P. R., Rojas, A. M., & Gerschenson, L. N. (2012). Microwave inactivation of red beet (*Beta vulgaris* L. var. conditiva) peroxidase and polyphenoloxidase and the effect of radiation on vegetable tissue quality. *Journal of food Engineering*, *109*(4), 676–684.

Lau, M. H., & Tang, J. (2002). Pasteurization of pickled asparagus using 915 MHz microwaves. *Journal of Food Engineering*, *51*(4), 283–290.

Lee, J.-K., & Park, J.-Y. (1999). Rapid thawing of frozen pork by 915 MHz microwave. *Korean Journal of Food Science and Technology*, *31*(1), 54–61.

Lin, T. M., Durance, D. T., & Scaman, C. H. (1998). Characterization of vacuum microwave, air and freeze dried carrot slices. *Food Research International*, *31*(2), 111–117.

Maranesi, M., Bochicchio, D., Montellato, L., Zaghini, A., Pagliuca, G., & Badiani, A. (2005). Effect of microwave cooking or broiling on selected nutrient contents, fatty acid patterns and true retention values in separable lean from lamb rib-loins, with emphasis on conjugated linoleic acid. *Food Chemistry*, *90*, 207–218.

Martín, M. E., Albors, A., Martínez-Navarrete, N., Chiralt, A., & Fito, P. (2000). *Micro-structural changes in apple tissue subjected to combined air-microwave drying. Proceedings of the 12th international drying symposium IDS2000. Paper no. 419.* Elsevier Science, ISBN 0-444-50422-2 (CD-ROM).

Matsui, K. N., Gut, J. A. W., De Oliveira, P. V., & Tadini, C. C. (2008). Inactivation kinetics of polyphenol oxidase and peroxidase in green coconut water by microwave processing. *Journal of Food Engineering*, *88*(2), 169–176.

Meda, V., Gupta, M., & Opoku, A. (2008). Drying kinetics and quality characteristics of microwave-vacuum dried Saskatoon berries. *Journal of Microwave Power and Electromagnetic Energy*, *42*(4), 4–12.

Megahed, M. G. (2001). Microwave roasting of peanuts: Effects on oil characteristics and composition. *Nahrung/Food*, *45*(4), 255–257.

Megahey, E. K., Mcminn, W. A. M., & Magee, T. R. A. (2005). Experimental Study of Microwave baking of madeira cake batter. *Food and Bioproducts Processing*, *83*(4), 277–287.

Nebesny, E., & Budryn, G. (2003). Antioxidative activity of green and roasted coffee beans as influenced by convection and microwave roasting methods and content of certain compounds. *European Food Research and Technology*, *217*, 157–163.

O'donnell, C. P., Tiwari, B. K., Bourke, P., & Cullen, P. J. (2010). Effect of ultrasonic processing on food enzymes of industrial importance. *Trends in Food Science & Technology*, *21*, 358–367.

Orsat, V., Changrue, V., & Raghavan, G. S. V. (2006). Microwave drying of fruits and vegetables. *Stewart Postharvest Review*, *6*, 4–9.

Plessi, M., Bertelli, D., & Miglietta, F. (2002). Effect of microwaves on volatile compounds in white and black pepper. *Lebensmittel-Wissenschaft und-Technologie*, *35*, 260–264.

Podsędek, A. (2007). Natural antioxidants and antioxidant capacity of Brassica vegetables: A review. *LWT-Food Science and Technology*, *40*(1), 1–11.

Queiroz, C., Mendes, M. L., Fialho, E., & Valente-Mesquita, V. L. (2008). Polyphenol oxidase: Characteristics and mechanisms of browning control. *Food Reviews International*, *24*, 361–375.

Regier, M. (2014). Microwave processing of foods and its combination with electron beam processing. In S. Pillai, & S. Shayanfa (Eds.), *Electron beam pasteurization and complementary food processing technologies.* , ISBN:978-1-78242-100-9.

Rosen, C. G. (1972). Effects of microwaves on food and related material. *Food Technology*, *26*(7), 36–40.

Sabliov, C. M., Boldor, D., Coronel, P., & Sanders, T. H. (2008). Continuous microwave processing of peanut beverages. *Journal of Food Processing and Preservation*, *32*(6), 935–945.

Salazar-González, C., San Martín-González, M. F., López-Malo, A., & Sosa-Morales, M. E. (2012). Recent studies related to microwave processing of fluid foods. *Food Bioprocess and Technology*, *5*, 31–46.

Schirack, A. V., Drake, M. A., Sanders, T. H., & Sandeep, K. P. (2006). Characterization of aroma-active compounds in microwave blanched peanuts. *Journal of Food Science*, *71*, 513–520.

Schirack, A. V., Sanders, T. H., & Sandeep, K. P. (2007). Effect of processing parameters on the temperature and moisture content of microwave-blanched peanuts. *Journal of Food Process Engineering*, *30*(2), 225–240.

Schubert, H., & Regier, M. (2006). Novel and traditional microwave applications in the food industry. advances in microwave and radio frequency processing. In M. Willert-Porada (Ed.), *Report from the 8th international conference on microwave and high – frequency heating* (773 pp.). Editorial: Springer.

Schubert, H., & Regier, M. (2007). *The microwave processing of foods*. USA: Published by Woodhead Publishing Limited, England and CRC Press.

Severini, C., Baiano, A., De Pilli, T., Carbone, B. F., & Derossi, A. (2005). Combined treatments of blanching and dehydration: Study on potato cubes. *Journal of Food Engineering*, *68*(3), 289–296.

Seyhun, N., Sumnu, G., & Sahin, S. (2003). Effects of different emulsifier types, fat content and gum types on retardation of staling of microwave baked cakes. *Nahrung Food*, *47*(4), 248–251.

Seyhun, N., Sumnu, G., & Sahin, S. (2005). Effects of different starch types on retardation of staling of microwave-baked cakes. *Food and Bioproducts Processing*, *83*, 1–5.

Shaheen, M. S., El-Massry, K. F., El-Ghorab, A. H., & Anjum, F. M. (2012). Microwave applications in thermal food processing. In W. Cao (Ed.), *The development and application of microwave heating*. ISBN:978-953-51-0835-1.

Sharma, G. P., & Prasad, S. (2001). Drying of garlic (*Allium sativum*) cloves by microwave–hot air combination. *Journal of Food Engineering*, *50*(2), 99–105.

Sieber, R., Eberhard, P., & Gallmann, P. U. (1996). Heat treatment of milk in domestic microwave ovens. *International Dairy Journal*, *6*(3), 231–246.

Silva, F. A., Marsaioli, J. A., Maximo, G. J., Silva, M. A. A. P., & Gonçalves, L. A. G. (2006). Microwave assisted drying of macadamia nuts. *Journal of Food Engineering*, *77*, 550–558.

Soysal, Ç., & Söylemez, Z. (2005). Kinetics and inactivation of carrot peroxidase by heat treatment. *Journal of Food Engineering*, *68*(3), 349–356.

Sumnu, G. (2001). A review on microwave baking of foods. *International Journal of Food Science & Technology*, *36*(2), 117–127.

Sumnu, G., Sahin, S., & Sevimli, M. (2005). Microwave, infrared and infrared-microwave combination baking of cakes. *Journal of Food Engineering*, *71*(2), 150–155.

Sumnu, G., Turabi, E., & Oztop, M. (2005). Drying of carrots in microwave and halogen lamp–microwave combination ovens. *LWT-Food Science and Technology*, *38*(5), 549–553.

Taher, B. J., & Farid, M. M. (2001). Cyclic microwave thawing of frozen meat: experimental and theoretical investigation. *Chemical Engineering and Processing: Process Intensification*, *40*(4), 379–389.

Tajchakavit, S., & Ramaswamy, H. S. (1997). Thermal vs. microwave inactivation kinetics of pectin methylesterase in orange juice under batch mode heating conditions. *LWT-Food Science and Technology*, *30*(1), 85–93.

Tajchakavit, S., Ramaswamy, H. S., & Fustier, P. (1998). Enhanced destruction of spoilage microorganisms in apple juice during continuous flow microwave heating. *Food Research International*, *31*(10), 713–722.

Tang, J. (2015). Unlocking potentials of microwaves for food safety and quality. *Journal of Food Science*, *80*, 1776–1793.

Tang, Z., Mikhaylenko, G., Liu, F., Mah, J.-H., Pandit, R., Younce, F., & Tang, J. (2008). Microwave sterilization of sliced beef in gravy in 7-oz trays. *Journal of Food Engineering*, *89*(4), 375–383.

Torringa, E. M., Van Dijk, E. J., & Bartels, P. V. (1996). *Microwave puffing of vegetables: Modeling and measurements. Proceedings of the 31st microwave power symposium* (pp. 16–19). Manassas: International Microwave Power Institute.

Tulasidas, T. N., Raghavan, G. S. V., & Mujumdar, A. S. (1995). Microwave drying of grapes in a single mode cavity at 2450 MHz–II: Quality and energy aspects. *Drying Technology*, *13*(8/9), 1973–1992.

Uherová, R., Hozová, B., & Smirnov, V. (1993). The effect of microwave heating on retention of some B vitamins. *Food Chemistry*, *46*(3), 293–295.

Uysal, N., Sumnu, G., & Sahin, S. (2009). Optimization of microwave–infrared roasting of hazelnut. *Journal of Food Engineering*, *90*(2), 255–261.

Vadivambal, R., & Jayas, D. S. (2007). Changes in quality of microwave-treated agricultural products—A review. *Biosystems Engineering*, *98*(1), 1–16.

Vadivambal, R., & Jayas, D. S. (2010). Non-uniform temperature distribution during microwave heating of food materials – A review. *Food Bioprocess Technology*, *3*, 161–171.

Valero, E., Villamiel, M., Sanz, J., & Martínez-Castro, I. (2000). Chemical and sensorial changes in milk pasteurised by microwave and conventional systems during cold storage. *Food Chemistry*, *70*, 77–81.

Villamiel, M., Corzo, N., Martínez-Castro, I., & Olano, A. (1996). Chemical changes during microwave treatment of milk. *Food Chemistry*, *56*(4), 385–388.

Wachtel-Galor, S., Wong, K. W., & Benzie, I. F. F. (2008). The effect of cooking on *Brassica* vegetables. *Food Chemistry*, *110*(3), 706–710.

Yaghmaee, P., & Durance, T. D. (2005). Destruction and injury of *Escherichia coli* during microwave heating under vacuum. *Journal of applied Microbiology*, *98*(2), 498–506.

Yarmand, M. S., & Homayouni, A. (2009). Effect of microwave cooking on the microstructure and quality of meat in goat and lamb. *Food Chemistry*, *112*(4), 782–785.

Yongsawatdigul, J., & Gunasekaran, S. (1996). Microwave-vacuum-drying of cranberries: Part I: Energy use and efficiency. Part II: Quality evaluation. *Journal of Food Processing and Preservation*, *20*(2), 121–156.

Yoshida, H., Hirakawa, Y., Tomiyama, Y., Nagamizu, T., & Mizushina, Y. (2005). Fatty acid distributions of triacylglycerols and phospholipids in peanut seeds (*Arachis hypogaea* L.) following microwave treatment. *Journal of Food Composition and Analysis*, *18*(1), 3–14.

Zhang, M., Tang, J., Mujumdar, A. S., & Wang, S. (2006). Trends in microwave-related drying of fruits and vegetables. *Trends in Food Science & Technology*, *17*(10), 524–534.

Zhao, S., Xiong, S., Qiu, C., & Xu, Y. (2007). Effect of microwaves on rice quality. *Journal of Stored Products Research*, *43*(4), 496–502.

Zook, D. E., Macku, C., & Deming, D. (1995). Effect of microwave heating on roasted nut flavor. *Developments in Food Science*, *37*, 1493–1518.

5 Microwave technology for food processing: an overview of current and future applications

V. Orsat, G.S.V. Raghavan and K. Krishnaswamy
McGill University, Montreal, QC, Canada

5.1 Introduction

Microwave ovens have become an indispensable appliance in the 21st century household in countries like the United States and Canada, as well as many European countries. The increasing urbanization in developing countries and accessibility to power has also led to an increase in the usage of microwave ovens at the household level worldwide. Even today, microwave ovens are mainly used for heating and reheating of foods; however, trends show that they are used increasingly also for cooking and defrosting (WHO, 2015). The industrial applications using microwaves have seen an expansion in recent years and industrial scale microwave processing units have been developed for drying, precooking of meat, pasteurization of ready meals, and tempering of meat and fish. Modern food consumers are demanding high quality, minimally processed products (Hong & Wang, 2015), which has led to the development of novel microwave processing technologies for thawing (Shaheen, El-Ghorab, Anjum, & El-Massry, 2012), blanching (Adedeji, Gachovska, Ngadi, & Raghavan, 2008; Kowalska, Lenart, & Leszczyk, 2008), baking (Ahmad, Morgan, & Okos, 2001; Sumnu, 2001), pasteurization (Cañumir, Celis, de Bruijn, & Vidal, 2002; Lau & Tang, 2002), and microwave-assisted extraction of bioactive compounds (Krishnaswamy, Orsat, Gariépy, & Thangavel, 2013; Wang, Xi, Zheng, & Miao, 2010).

This chapter deals with the current and potential future applications of microwave technology in food processing, including a brief history of industrial applications of microwave systems and an overview of current food applications.

5.1.1 History: industrial application of microwave systems

The development of the cavity magnetron by two British physicists Henry A. Boot and John T. Randall (Brittain, 1985; Osepchuk, 1984) during World War II led to the discovery of microwave heating. Individuals working at Raytheon on microwave waveguides observed the popping of corn kernels near the radiating tubes due to the localized heat generation which initiated the concept of microwave heating. The potential development and appreciation of microwave heating is attributed to Percy Spencer at Raytheon, who patented many microwave heating applications.

The Microwave Processing of Foods. DOI: http://dx.doi.org/10.1016/B978-0-08-100528-6.00005-X

Scientists at Raytheon, GE Westinghouse, among many others, worked on developing equipment and applications for industrial microwave oven heating. Raytheon first launched the Radarange, a free-standing water-cooled unit which was upgraded during the early 1950s to air-cooled floor and countertop models. Most units were adopted by restaurants and food service kitchens. In the 1960s, Raytheon faced the growing competition from other manufacturers such as Litton, Tappan, and GE. At that time, Raytheon decided not to pursue the consumer microwave oven market, and instead licensed the patent to other companies such as Westinghouse and Whirlpool (Osepchuk, 1984). During the mid-1960s, Raytheon purchased the Amana company to acquire and expand their expertise in manufacturing and marketing to the consumer microwave oven market. The early 1970s saw considerable competition from actively developing countertop microwave oven companies such as Tappan, GE, Amana, Toshiba, Hitachi, and Sharp. Technical improvements have been made over the years with better door seals (Rocha, Facão, Sousa, & Viegas, 2008), solid-state power control, microprocessors, and indicators of microwave oven leakage (Lopez-Iturri et al., 2015; Narayan, Arun, Jha, & Bommer, 2012). The historical developments of industrial microwave systems are listed in Table 5.1.

5.2 Industrially available microwave applicators

Microwaves are part of the electromagnetic spectrum between the radio frequency and infrared regions, in the frequency range between 300 MHz to 300 GHz. The

Table 5.1 Historical development of industrial microwave systems (Ferrite, 2015)

Year	Development of industrial microwave systems
1948	Raytheon develops first *Radarange*
1965	Raytheon develops first *Single Mode dryer*
1970	Raytheon develops first *Continuous Tempering System at 25 kW*
1972	Raytheon develops first *25 kW microwave transmitter*
1973	*Industrial Microwave Equipment* operation is started by Raytheon
1982	Raytheon develops first *Microwave cooking system at 40 kW*
1984	Raytheon develops first *50 kW high power MW transmitter*
1992	Raytheon sells Industrial Microwave group *to Amana*
1996	*EnWave's* first prototype of Radiant Energy Vacuum *REV machine* was developed for dehydrating food and nutraceuticals
1997	Amana releases medium power *40 kW batch tempering line*
1999	*Ferrite Components Inc.* introduces circular waveguide technology for uniform heating and cooking
	Amana releases high power *75 kW Continuous Tempering line*
2003	Ferrite acquires Amana's *Industrial MW Division*
2010	Ferrite develops L, S, and X band above resonance 4-port ferrite devices for use in scientific, medical, and security applications.

2,450 MHz frequency is the most commonly used in consumer-based microwave ovens, and 915 MHz is used primarily for industrial-scale applications in the Americas. The list of various frequencies allowed for use in different countries is listed in Table 5.2.

A microwave system is composed of the following components: magnetron, step-up transformer, oven cavity, waveguide, applicator, and the sample to be processed. The underlying principle of all microwave systems is very much the same. The transformer inside the microwave system is used to step-up the alternating current from the domestic power line operating from a frequency of 50–60 Hz to a frequency of 2,450 MHz. The step-up transformer accounts for more than half the weight of the domestic microwave oven and powers the magnetron. The magnetron operates at 4,000 Volts and converts electric energy to microwave energy. Microwaves produced in the magnetron are channeled through a waveguide to reach the microwave cavity. The microwave cavity holds the product to be heated and is made up of reflecting walls to maximize the efficiency of microwave heating. A rectangular cavity, representing a multimode resonator, forms the microwave applicator in the common household microwave oven for batch processing. The most common applicators used in microwave systems are Multimode applicators; the Single-Mode applicators; and the Traveling Wave Devices. The basic requirements of an efficient microwave system are reliability of the system operation, efficient power transfer, and low cost (Orsat, Yang, Changrue, & Raghavan, 2007; Osepchuk, 2002; Venkatesh & Raghavan, 2004). Short descriptions of different microwave applicators are provided in the following section.

Table 5.2 Allocation of frequency for industrial, medical, and scientific (ISM) purposes in the range 433.92 MHz to 40 GHz in different countries (Püschner, 2015)

Frequency (MHz)	Frequency area permitted (tolerance + or −)	Area
433.92	0–2%	Austria, The Netherlands, Portugal, Germany, Switzerland
896	*10 MHz*	*United Kingdom*
915	*13 MHz*	*North and South America*
2,375	50 MHz	Albania, Bulgaria, Hungary, Romania, Russia
2,450	*50 MHz*	*Worldwide except where 2,375 MHz is used*
3,390	0–6%	The Netherlands
5,800	75 MHz	Worldwide
6,780	0–6%	The Netherlands
24,150	125 MHz	Worldwide
40,680		United Kingdom

5.2.1 Multimode applicator

The multimode applicator is the most common type of applicators and it is used in approximately 50% of industrial microwave systems. The principle is similar to a common household microwave system but is designed for handling a larger quantity of material. It is a rectangular conductive metal box where the product is placed. The microwave energy is transmitted into the applicator through the waveguide and undergoes multiple reflections from the cavity walls. The reflecting waves interact with the product and establish a distribution of the electric field within the cavity and the product. This creates stable modes of propagation of the electric field at different points; hence, it is called a multimode applicator.

The dielectric properties of the materials to be processed and the size and location of the sample material within the applicator determine the field distribution in a multimode applicator. Multimode applicators are best suited for products that have high dielectric loss factor and occupy larger volumes of the applicator ($>50\%$). For sample materials having a lower loss factor and occupying a lower volume of the applicator ($<20\%$), the electric field is unevenly distributed resulting in nonuniform temperature distribution which can potentially lead to "hotspots or coldspots" within the sample. In order to improve the uniformity in temperature distribution, a continuously rotating turntable or a mode stirrer (rotating reflector) should be incorporated into the applicator. This helps in achieving more effective and uniform electric field distribution within the sample.

The spatial distribution of the electric field within the sample material cannot be generalized and it is influenced by the dielectric properties of the material and its position and quantity within the cavity. Therefore, the experimental data from a particular scientific investigation may not apply when scaling up to industrial level. In order to map the field distribution within the cavity, water is often used as an experimental model-sample, however, the translation to other liquids or materials yields different temperature distributions, as the field distribution will change as a function of the type of material to be processed within the microwave cavity due to the material's inherent dielectric properties. In a study by Dai, Orsat, and Raghavan (2010), gypsum plates using cobalt chloride as an indicator, were able to map the microwave energy distribution in a multimode microwave cavity.

Numerical modeling (e.g., finite difference time domain) can be used to predict the field distribution within a product loaded within a microwave cavity. Once product parameters are established such as the size, shape, position, dielectric and thermal properties, and uniformity distribution of the materials, the temperature profile within the microwave system, while in operation, can be simulated and validated. This will help the industrial scale up of multimode systems for specific products (see Chapter 1: Introducing microwave-assisted processing of food: fundamentals of the technology).

5.2.2 Single-mode applicator

Within the single-mode cavity, only one mode of propagation is permitted in a defined space with a specific volume. This makes the single-mode resonant cavity applicator the most efficient form of microwave energy distribution. The samples are placed in a rectangular or cylindrical single-mode cavity at predetermined positions where the electric field is concentrated. Single-mode applicators consist of one feeding waveguide, a tuning plunger, and a small microwave resonator with dimensions in the range of the wavelength of the microwaves. The simplest and the most common single-mode cavity operates at TM_{010} mode and is physically the smallest size (see Chapter 1: Introducing microwave-assisted processing of food: fundamentals of the technology). The mode of operation in a single cavity applicator is that the target sample is positioned at regions of high field strength. The reason for the small size relates to the major limitation that the width of the sample must be less than half the wavelength of the incoming microwave. This can be adjusted by moving the tuning plunger in a reciprocating fashion so that the time-average field is distributed across the sample. Otherwise the periodicity of the device will be reflected in a periodic heating pattern across the sample (see Chapter 1: Introducing microwave-assisted processing of food: fundamentals of the technology).

5.2.3 *Traveling wave applicator*

A traveling wave applicator consists of a waveguide arranged in such a way that the product to be processed passes through it from one end while the microwave is placed at the other end. The product absorbs the microwave as it passes along the microwave waveguide. The distance that the material travels, the material's dielectric properties, and size and volume of the product all play a vital role in absorbing the microwaves.

A dummy water load is placed at one end of the waveguide to absorb any microwaves that are not absorbed by the product as a precautionary measure. The speed of the conveyor belt is adjusted so that the product moves at a constant speed through the waveguide and that each part of the product experiences the same microwave field strength. The system is operated in such a way that one pass of product through the waveguide would ensure a uniform and steady-state microwave energy transfer to the product (see Chapter 1: Introducing microwave-assisted processing of food: fundamentals of the technology).

5.3 Current food processing applications

As mentioned earlier, there is a widespread use of domestic microwave ovens and industrial uses of microwaves, e.g., in medical applications, in foods, rubber, textiles, wood products, ceramics, organic/inorganic synthesis, polymerization, and cellular breakdown, are increasing. Microwave technology has proven to be an efficient alternative for sludge processing, medical waste treatment, contaminated soil remediation, wastewater remediation, activated carbon regeneration, and as the source of heat/energy for integrated remedial methodologies (Wu, 2008). This chapter summarizes the current and potential future applications of microwaves in food processing, which are increasingly being optimized for tempering, vacuum drying, freeze-drying, dehydration, cooking, blanching, baking, roasting, rendering, pasteurization, sterilization, and extraction (Krishnaswamy et al., 2013; Lew, Krutzik, Hart, & Chamberlin, 2002; Lidstrom, Tierney, Wathey, & Westman, 2001; Venkatesh & Raghavan, 2004).

The major reasons for the relatively slow development and implementation of industrial applications of microwave processing in the food industry are due to the considerable cost of the microwave equipment, operating costs, and safety issues (to limit microwave leakage) during scaling up (Wray & Ramaswamy, 2015). In order to successfully establish industrial applications of microwave technology in the food industry, either the process or the manufactured product should possess unique features compared to conventional processing. Some industrial applications of microwaves are presented in the following subsections.

The last 10 years have seen the development of several initiatives by companies like Ferrite's Industrial Microwave Systems (Nashua, NH, USA) who acquired Amana Industrial Microwave Division in 2003, EnWave Corporation (EnWave Corp., British Columbia, Canada), Püschner Microwave Power Systems (Schwanewede, Germany), Microdry Incorporated (Crestwood, KY, USA), Sairem (Neyron, France), Industrial Microwave Systems L.L.C. (New Orleans, LA, USA), Defreeze Corporation (Southboro, MA, USA), Stalam (Vicenza, Italy), and Advanced Microwave Technologies (AMT, Edinburgh, UK). These companies are well-recognized commercial suppliers of microwave processing equipment and are making an impact in the food and pharmaceutical sectors.

5.3.1 Cooking and blanching (see also Chapter 9: Microwave-assisted blanching)

The application of microwave cooking at industrial scale was successfully implemented in the processing of bacon, meat patties, and meat balls. The volumetric temperature increase achieved in the microwave process leads to more uniform coagulation of meat proteins and renders high quality fat (Edgar, 1986). Microwave processing of sliced bacon diminishes the risk of nitrosamine formation and provides good product appearance. Microwave rendered fat presents greater stability and higher quality (Ohlsson & Bengtsson, 2000). Combinations of conventional and microwave ovens are available at the industrial scale, where the application of microwaves allows for an acceleration of the cooking process, and the surface water is evaporated through convectional airflow. Microwave cooking of bacon can be achieved at lower energy consumption, less product shrinkage, overall improvement in color, appearance, and taste, which are all key factors for food manufacturers (James, Barlow, James, & Swain, 2006). An industrial-scale continuous microwave oven was used to treat cooked rice in order to control the growth of pests and mildew, and the treated rice was shown to have an improved quality (Zhao, Xiong, Qiu, & Xu, 2007). Many thermolabile vitamins leach out during thermal processing, and this can be overcome (at least to some extent) by microwave processing, which has been demonstrated to have advantages over traditional processing methods, such as reduced nutrient losses

(Begum & Brewer, 2001; Brewer & Begum, 2003; Villanueva, Marquina, Vargas, & Abellan, 2000). In the 1970s, microwave technology was used in food processing specifically for the precooked bacon industry which dominated the early industrial microwave market. Industrial microwave bacon cookers are available from Microdry Incorporated (Crestwood, KY, USA), Defreeze Corporation (Southboro, MA, USA), Ferrite's Industrial Microwave Systems (Nashua, NH, USA).

Blanching is an important process generally carried out before drying, freezing, frying, and canning as enzymes are inactivated, tissues are softened, and color and flavor losses in certain fruits and vegetables are minimized. Microwave blanching is gaining considerable interest, mainly because short processing times and reduced loss of nutrients can be achieved (Adedeji et al., 2008; Kowalska et al., 2008; Walde, Velu, Jyothirmayi, & Math, 2006).

Wang, Zhang, Mujumdar, Mothibe, and Azam (2012) used microwave blanching and conventional blanching methods and investigated their effects on microwave freeze-drying (MFD) in a circular conduit drying chamber and conventional freeze-drying (FD) of stem lettuce cubes. The results showed that the microwave blanched stem lettuce cubes had twice the electrical conductivity than the conventionally blanched samples, and five times more than that of unblanched samples. Dielectric loss factors and electrical conductivity increased in the microwave blanched samples leading to reduced drying times. This study demonstrated that microwave blanching can improve the microwave freeze-drying process (Wang et al., 2012). It has been demonstrated, by Sarang, Sastry, Gaines, Yang, and Dunne (2007), that blanching increases the ionic content of foods, thereby increasing their electrical conductivity for improved heating uniformity in microwave and ohmic heating applications.

Begum and Brewer (2001) compared the microwave blanching of tomatoes with conventional water and steam blanching. Microwave blanched tomatoes retained higher nutritive value with acceptable color and flavor scores (Begum & Brewer, 2001). The effects of microwave blanching on carrot slices were studied to inactivate pectin methylesterase (PME) and peroxidase (POD) enzymes. This study revealed that microwave blanching was more effective in inactivating PME as compared to conventional blanching; furthermore, operating time could be reduced (Başkaya Sezer & Demirdöven, 2015).

It is still a great challenge to design an industrial scale microwave blanching system as for any other equipment for microwave applications. Effective design of applicator, better control of microwave heating cycles, composition of the food product, and nature of heat distribution within the product have to be taken into consideration. The industrial scaling up of microwave blanching remains a mostly unexplored area (Shaheen et al., 2012).

5.3.2 Tempering and thawing (see also Chapter 12: Microwave-assisted thawing and tempering)

Frozen foods are either thawed or tempered before further processing. Thawing is the process when the frozen material is allowed to stand at ambient temperature, i.e., above 0°C to become free from ice so that further mechanical processing such as slicing, cutting, molding can take place. At the industrial level, thawing is done by immersing the frozen products in water for extended periods of time. Taher and Farid (2001) conducted studies on microwave thawing of frozen minced beef, both theoretically and experimentally, and found that the time required for microwave thawing was 20% shorter than that required for conventional thawing.

For the tempering process, the temperature of the frozen material is brought to just below freezing to ease the mechanical chopping and handling. Tempering is an initial process of increasing the product's temperature which is continued by complete thawing (Shaheen et al., 2012). Microwave tempering is used to improve the uniformity of temperature distribution with no temperature abuse of the product and reduced drip losses. Microwave tempering tunnels are commercially available from various suppliers. Ferrite's Industrial Tempering Systems provide continuous throughput of 17,000 lbs (~7.7 t) per hour, but can be modified for small batch throughputs (Ferrite's Industrial Microwave Systems, Nashua, NH, USA). Microdry Incorporated manufactures industrial tempering tunnels, with large tunnel opening to accommodate standard meat packages, which are constructed of stainless steel and polypropylene for easy cleaning (Model TC-75 75 KW—915 MHz—Microdry Incorporated, Crestwood, KY, USA). Tempering and defrosting equipments are also available from Stalam (Vicenza, Italy). Fig. 5.1 shows an image of the AMT4400 continuous microwave tempering system manufactured by AMTek Microwave (Cedar Rapids, IA, USA). The AMT4400 microwave tempering system is powered by four AMT7510 microwave transmitter assemblies. The system provides sophisticated software control with computer display, which provides complete monitoring of the tempering process and facilitates the operation of the equipment (Amtek, 2015).

Figure 5.1 Industrial continuous microwave tempering system AMT4400.
Photo courtesy: AMTek Microwave, Cedar Rapids, IA, USA.

5.3.3 Drying (see also Chapter 8: Microwave-assisted drying)

Drying is one of the oldest methods of food preservation to prolong the storage life of foods. Convective air drying is the most commonly used method of drying in the food industry. Despite the simple operating principles of convective drying systems and low construction cost, there are certain criteria, such as energy inefficiency and undesired product quality which make air-dried food products less desirable. In convective drying, dry air is used to remove the moisture from the product creating a pressure gradient between the inner core and the surface leading to movement of moisture from the inside to the surface. This may lead to case hardening, wherein the outer layer of the product is overdried, preventing the removal of moisture from the less accessible inner core. The equilibrium is reached after an extended time period and with potential overheating of the product surface. Prolonged exposure of the food product to high temperature results in loss of flavor, volatile aromatics, and color, texture deterioration, and it affects the overall quality of the products (Alibas, 2007; Guiné & Barroca, 2012; Mayor, Moreira, & Sereno, 2011; Wray & Ramaswamy, 2015).

The issues related to convectional drying can be addressed by microwave-based drying methods where volumetric heating occurs. There is a possibility of combining different drying methods with microwave technology. Commercial scale microwave-based drying equipment is available for microwave-assisted convection drying, microwave-vacuum drying, and microwave-assisted freeze-drying. Microwave-hot air tunnel dryers are available from suppliers like Ferrite's Industrial Microwave Systems (Nashua, NH, USA), Püschner Microwave Power Systems (Schwanewede, Germany), Microdry Incorporated (Crestwood, KY, USA), Sairem (Neyron, France), Industrial Microwave Systems L.L.C. (New Orleans, LA, USA), and Defreeze Corporation (Southboro, MA, USA).

Microwave/hot air drying is used as a final drying process to finish the complete drying of the product and to level-off moisture content in pasta, cracker, or chip drying (Osepchuk, 2002). The MIP9 dryer from Ferrite's Industrial Microwave Systems (Formerly Amana) is currently the world's most successful microwave drying system used in drying and dewatering applications. Ferrite's patented microwave feed, applicator, and control technologies ensure the uniform distribution of the microwave energy to the product and offer a rapid and predictable drying process (Ferrite, 2015).

Microwave vacuum drying (MVD) is a unique process of combining microwaves and vacuum technology. MVD is a highly efficient process that enables short processing times providing superior product quality. The removal of moisture takes place at temperatures lower than the boiling point of the product at ambient conditions. The lower oxygen concentration prevents oxidation of the products during drying (Regier, Knorzer, & Erle, 2004; Swittra, Therdthai, & Dhamvithee, 2011). A continuous vacuum drying unit using a microwave belt dryer (model μWaveVac1290) was designed by Püeschner Microwave Power Systems (Schwanewede, Germany). It is an automated continuous drying system for temperature sensitive products with low thermal conductivity. The belt in the microwave vacuum dryer is made up of selected Polytetrafluoroethylene (PTFE)- coated glass fibers. While drying wet products, they transition from a viscous and sticky phase to a dry cake at the end of the microwave vacuum drying process. This cake is brittle and can be powdered or granulated at the end of the conveyor belt depending on the final product requirements. The normal working pressure for microwave vacuum drying is from 10 to 50 mbars (Püschner, 2015).

EnWave's Radiant Energy Vacuum (REV) technology uses microwave energy under vacuum to dehydrate food, pharmaceutical, and industrial materials at low temperatures. The two REV platforms designed for food industries are *nutra*REV and *quanta*REV. Commercial scale *nutra*REV equipments are used in several major companies and can be designed to deliver up to 150 kW power to dry fruits, vegetables, herbs, dairy products, meat, and seafood at or below 37°C. The microwave—vacuum-dried cheese snack production unit manufactured by EnWave *nutra*REV at a nominal power of 100 kW is presented in Fig. 5.2.

A pilot scale design of the *quanta*REV system has been tested by EnWave Corporation for high-volume, low-temperature dehydration of solids, liquids, granular, or encapsulated products. The commercial *quanta*REV process equipment was expected to be launched in 2015.

Microwave freeze-drying (MFD) is a microwave-assisted-drying method operating below the freezing point of water. Freeze-drying of foods and pharmaceutical compounds like vaccines is a costly process which may be replaced by MFD technology, which promises shorter processing time and lower energy consumption. Currently, an MFD pilot scale platform called *freeze*REV has been designed by Enwave Corporation for testing in collaboration with industrial partners like Merck (Enwave, 2015).

Figure 5.2 Dehydrated cheese snack production unit, EnWave *nutra*REV, 100 kW.
Photo courtesy: EnWave Corporation, BC, Canada.

5.3.4 Pasteurization (see also Chapter 10: Microwave-assisted pasteurization and sterilization—commercial perspective)

There have been several developments in the past 10 years in industrial scale microwave pasteurization and sterilization systems for solid, semisolid, and liquid foods. The rapid volumetric heating induced by the dissipation of microwaves is proven to deliver a more efficient and gentle process as overall processing times can be reduced compared to those in conventional retorts where slow heat conduction delivers the lethal temperatures to the core of the product. When combined with rapid cooling, this technology can minimize thermal degradation and damage to nutrients like flavor, color, texture, and other functional components (Cañumir et al., 2002; Lau & Tang, 2002). Microwave pasteurization equipment is, for example, manufactured by Advanced Microwave Technologies (AMT, Edinburgh, UK) for liquids at pilot and industrial scales. Stalam (Vicenza, Italy) also manufactures a series of microwave pasteurizing equipment for fresh semolina pasta processing and pasteurization of packaged bread loaves (Stalam, 2015). Additional companies distributing microwave equipment, very successfully on a global scale, for the pasteurization of ready-to-eat meals are Micvac AB (Moelndal, Sweden) and Tops Foods nv (Olen, Belgium).

An industrial scale continuous flow microwave volumetric heating system, by AMT (Edinburgh, UK), comprising of process tank, pump, pressure and temperature sensor, flow meter, rotation device, and microwave volumetric heating unit was used for processing of tomato juice. A schematic line diagram and the pictorial image of the AMT unit are provided in Fig. 5.3. The tomato juice flows through transparent processing tubes at $85 \pm 0.4°C$, flow capacity of 100 L/h with 81.8 ± 1.1 s residence time inside the processing tube (Stratakos, Delgado-Pando, Linton, Patterson, & Koidis, 2016). Six magnetrons are used to produce the microwaves and the microwave pasteurized samples are collected in sterile containers and cooled using ice water (Stratakos et al., 2016).

5.3.5 Baking (see also Chapter 6: Microwave-assisted baking)

Microwaves are used in baking, mostly in combination with conventional and infrared processing to accelerate the baking process. Since 1950, microwave baking has been under research and there have been many investigations on appropriate formulations with various emulsifiers, gums, starches, fats, and enzymes (Keskin, Sumnu, & Sahin, 2004; Seyhun, Sumnu, & Sahin, 2003; Sumnu, 2001) and microwave baking resulted in greater drying rates and uniform internal moisture profiles (Ahmad et al., 2001). Microwave baking drastically reduces the baking time compared to traditional oven baking. Stalam (Vicenza, Italy) manufactures microwave and RF baking equipment as postbaking dryers or for baking white crust bread. These microwave baking machines are marketed as economical systems to produce normal crusted bread, parbaked, or fully baked white crust bread with high energy efficiency and production capacity (Stalam, 2015).

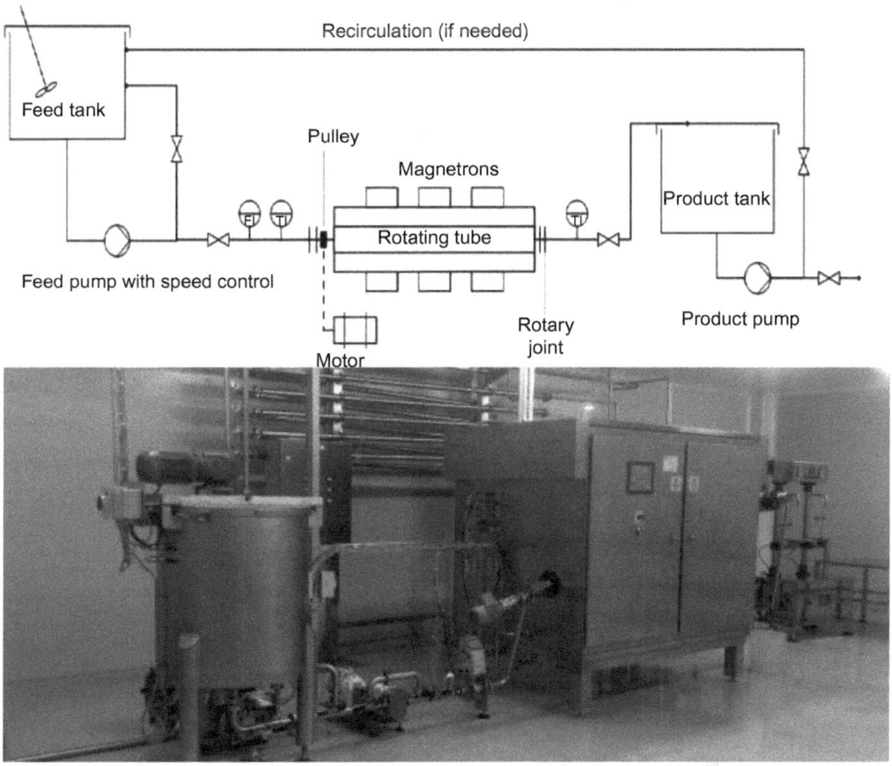

Figure 5.3 Industrial scale continuous flow microwave volumetric heating system by Advanced Microwave Technologies.
Photo courtesy: Stratakos, A., Delgado-Pando, G., Linton, M., Patterson, M. F., & Koidis, A. (2016). Industrial scale microwave processing of tomato juice using a novel continuous microwave system. *Food Chemistry*, *190*, 622–628 and AMT, Edinburgh, UK.

5.3.6 Microwave packaging (see also Chapter 13: Packaging for microwave foods)

Microwave susceptor material consists of a layer of metallized plastic film that laminates paperboard or a stable substrate. The metal film absorbs the microwave energy and converts it into heat. Depending on the thickness of the metal film, absorption or reflection of the microwaves may be dominating (Regier, 2014). Susceptors are used to brown or to make the food crisp. Several commercially available frozen pizza products have susceptors that help in browning and crisping of the pizza crust (Cole & Lafferty, 2002). There are various patents on different microwave susceptors owned by a number of food companies. Innovations in food packaging have led to intelligent packaging systems with the development of new RFID sensors. These sensors allow for the remote and nondestructive monitoring of food

(Vanderroost, Ragaert, Devlieghere, & De Meulenaer, 2014; Yam, Takhistov, & Miltz, 2005).

5.3.7 Microwave-assisted extraction (MAE)

Microwave-assisted extraction (MAE) is a method to enhance the extraction of bioactive compounds. When a dielectric material (sample along with the solvent used for extraction) is placed in a microwave field, both the biomolecules and the solvent try to align themselves with the alternating electromagnetic field either by distortion or molecular movements, or by physical rotation of the molecular dipoles. This leads to rapid heating of both the solvent and sample matrix (Dandekar & Gaikar, 2002; Sun, Liao, Wang, Hu, & Chen, 2007). Microwave extraction rates and the quality of the extract depend on the dielectric properties of the sample and solvent, solubility of the interested compounds in the solvent, and temperature. Temperature is important in the extraction process as the higher the temperature the higher is the diffusion rate and, therefore, the extraction rate. However, this may be undesirable when extracting heat-sensitive compounds over long extraction periods. Nonetheless, MAE is advantageous over conventional extraction techniques with improved extraction yield, reduced extraction time, rapid and volumetric heating of the absorbing medium, low solvent consumption, higher selectivity of target molecules, and high potential for automation (Pallaroni, Von Holst, Eskilsson, & Björklund, 2002; Wang et al., 2010). MAE has been used to extract many different bioactive compounds such as secoisolariciresinol diglucoside from flaxseeds (Nemes & Orsat, 2010), pectin from apple pomace (Wang et al., 2007), phenolic antioxidants from peanut skin (Ballard, Mallikarjunan, Zhou, & O'Keefe, 2009), zearalenone from wheat and corn (Pallaroni et al., 2002), anthocyanin from red raspberries (Sun et al., 2007), tea polyphenols and tea caffeine from green tea leaves (Pan, Niu, & Liu, 2003), grape polyphenols (Krishnaswamy et al., 2013), and phenolic compounds from blueberry leaves (Routray & Orsat, 2014).

Solvent-free microwave extraction is a sustainable technology based on the concept of green chemistry. The sample to be extracted is placed in a microwave cavity without the addition of solvents or water. The internal water present in the sample matrix gets heated, disrupts the cell wall, and aids the extraction process at shorter extraction times (Chemat, Fabiano-Tixier, Abert Vian, Allaf, & Vorobiev, 2015). Currently, solvent free microwave extraction technology is at the laboratory scale and it requires further research to scale up to pilot or industrial processing levels.

5.4 Research and development in novel and future applications

A recent report in October 2015 summarized the emerging novel food processing technologies around the world where high pressure processing (HPP) and microwave processing have been identified as the main technologies for research and development in the next 10 years (Jermann, Koutchma, Margas, Leadley, & Ros-Polski,

2015). There is a geographical difference in the acceptance of novel technologies, where cold plasma and pulsed electric field were projected to be important in Europe, while HPP (80%), microwave (88%), and UV (84%) processing would have more significant trends in North America (Jermann et al., 2015).

Despite the technological advancement and developments that have taken place since the 20th century, a substantial amount of research is still required to apply microwave technologies in the food processing industry more widely. The high cost of initial investments, process optimization, and operation costs related to the nature of the food products and safety measures are some of the current limitations that industrial applications of microwaves are facing. Precise design of processing equipment for specific applications, with safety measures associated with radiation leakage, and trained technical personnel for proper maintenance and operation of the microwave equipment are necessary.

Microwave technologies have a wide scope of applications in functional and novel food processing sectors, sterilization, pasteurization, pretreatment processing, dehydration, and shelf-life extension of food products. The possibility of using microwave energy in innovative ways and identifying novel applications will remain an exciting research area and ensure an ever-growing field of opportunities.

References

Adedeji, A. A., Gachovska, T. K., Ngadi, M. O., & Raghavan, G. S. V. (2008). Effect of pretreatments on drying characteristics of okra. *Drying Technology*, *26*(10), 1251–1256.

Ahmad, S. S., Morgan, M. T., & Okos, M. R. (2001). Effects of microwave on the drying, checking and mechanical strength of baked biscuits. *Journal of Food Engineering*, *50*(2), 63–75.

Alibas, I. (2007). Microwave, air and combined microwave–air-drying parameters of pumpkin slices. *LWT-Food Science and Technology*, *40*(8), 1445–1451.

Amtek. (2015). *Amtek microwaves*. http://www.4amtek.com/index.php/products/microwave-tempering-systems/ Accessed in October 2015.

Ballard, T. S., Mallikarjunan, P., Zhou, K., & O'Keefe, S. F. (2009). Optimizing the extraction of phenolic antioxidants from peanut skins using response surface methodology. *Journal of Agricultural and Food Chemistry*, *57*(8), 3064–3072.

Başkaya Sezer, D., & Demirdöven, A. (2015). The effects of microwave blanching conditions on carrot slices: Optimization and comparison. *Journal of Food Processing and Preservation*. Published Online 12 MAR 2015. Available from: http://dx.doi.org/10.1111/jfpp.12463.

Begum, S., & Brewer, M. (2001). Chemical, nutritive and sensory characteristics of tomatoes before and after conventional and microwave blanching and during frozen storage. *Journal of Food Quality*, *24*(1), 1–15.

Brewer, M. S., & Begum, S. (2003). Effect of microwave power level and time on ascorbic acid content, peroxidase activity and color of selected vegetables. *Journal of Food Processing & Preservation*, *27*(6).

Brittain, J. E. (1985). The magnetron and the beginnings of the microwave age. *Physics Today*, *July 1985*, 60–67.

Cañumir, J. A., Celis, J. E., de Bruijn, J., & Vidal, L. V. (2002). Pasteurisation of apple juice by using microwaves. *LWT-Food Science and Technology*, *35*(5), 389–392.

Chemat, F., Fabiano-Tixier, A. S., Abert Vian, M., Allaf, T., & Vorobiev, E. (2015). Solvent-free extraction of food and natural products. *TrAC Trends in Analytical Chemistry*, *71*, 157–168.

Cole, L. R., & Lafferty, T. P. *Patterned microwave susceptor*. U.S. Patent No. 6,414,290. 2 Jul. 2002.

Dai, J., Orsat, V., & Raghavan, G. S. V. (2010). Visualization of microwave energy distribution in a multimode microwave cavity using $CoCl_2$ on gypsum plates. *Journal of Microwave Power and Electromagnetic Energy*, *44*(3), 144–152.

Dandekar, D. V., & Gaikar, V. (2002). Microwave assisted extraction of curcuminoids from *Curcuma longa*. *Separation science and technology*, *37*(11), 2669–2690.

Edgar, R. (1986). The economics of microwave processing in the food industry. *Food Technology*, *1986*, 106–112.

Enwave. (2015). *Enwave Corporation*. http://www.enwave.net/nutrarev.php: http://www.enwave.net/freezerev.php: http://www.enwave.net/nutrarev.php Accessed in October 2015.

Ferrite. (2015). *Ferrite Incorporated*. http://www.ferriteinc.com/products_drying.html Accessed in October 2015.

Guiné, R. P. F., & Barroca, M. J. (2012). Effect of drying treatments on texture and color of vegetables (pumpkin and green pepper). *Food and Bioproducts Processing*, *90*(1), 58–63.

Hong, X., & Wang, J. (2015). Use of electronic nose and tongue to track freshness of cherry tomatoes squeezed for juice consumption: Comparison of different sensor fusion approaches. *Food and Bioprocess Technology*, *8*(1), 158–170.

James, C., Barlow, K. E., James, S. J., & Swain, M. J. (2006). The influence of processing and product factors on the quality of microwave pre-cooked bacon. *Journal of Food Engineering*, *77*(4), 835–843.

Jermann, C., Koutchma, T., Margas, E., Leadley, C., & Ros-Polski, V. (2015). Mapping trends in novel and emerging food processing technologies around the world. *Innovative Food Science & Emerging Technologies*, *31*, 14–27.

Keskin, S. O., Sumnu, G., & Sahin, S. (2004). Usage of enzymes in a novel baking process. *Nahrung/Food*, *48*(2), 156–160.

Kowalska, H., Lenart, A., & Leszczyk, D. (2008). The effect of blanching and freezing on osmotic dehydration of pumpkin. *Journal of Food Engineering*, *86*(1), 30–38.

Krishnaswamy, K., Orsat, V., Gariépy, Y., & Thangavel, K. (2013). Optimization of microwave-assisted extraction of phenolic antioxidants from grape seeds (*Vitis vinifera*). *Food and Bioprocess Technology*, *6*(2), 441–455.

Lau, M. H., & Tang, J. (2002). Pasteurization of pickled asparagus using 915 MHz microwaves. *Journal of Food Engineering*, *51*(4), 283–290.

Lew, A., Krutzik, P. O., Hart, M. E., & Chamberlin, A. R. (2002). Increasing rates of reaction: Microwave-assisted organic synthesis for combinatorial chemistry. *Journal of Combinatorial Chemistry*, *4*(2), 95–105.

Lidstrom, P., Tierney, J., Wathey, B., & Westman, J. (2001). Microwave assisted organic synthesis – Review. *Tetrahedron*, *57*, 9225–9283.

Lopez-Iturri, P., de Miguel-Bilbao, S., Aguirre, E., Azpilicueta, L., Falcone, F., & Ramos, V. (2015). Estimation of radiofrequency power leakage from microwave ovens for dosimetric assessment at nonionizing radiation exposure levels. *BioMed Research International*, *2015*.

Mayor, L., Moreira, R., & Sereno, A. M. (2011). Shrinkage, density, porosity and shape changes during dehydration of pumpkin (*Cucurbita pepo* L.) fruits. *Journal of Food Engineering*, *103*(1), 29–37.

Narayan, S., Arun, K., Jha, R., & Bommer, J. P. (2012). *RF leakage radiation from microwave oven for aircraft interior applications. Antennas and propagation society international symposium (APSURSI), 2012 IEEE* (pp. 1–2). IEEE.

Nemes, S. M., & Orsat, V. (2010). Screening the experimental domain for the microwave-assisted extraction of secoisolariciresinol diglucoside from flaxseed prior to optimization procedures. *Food and Bioprocess Technology*, *3*(2), 300–307.

Ohlsson, Thomas, & Bengtsson, Nils (2000). Microwave technology and foods. *Advances in Food and Nutrition Research*, *43*, 65–140.

Orsat, V., Yang, W., Changrue, V., & Raghavan, G. S. V. (2007). Microwave-assisted drying of biomaterials. *Food and Bioproducts Processing*, *85*(3), 255–263.

Osepchuk, J. M. (1984). A history of microwave heating applications. *IEEE Transactions on Microwave Theory and Techniques*, *32*(9), 1200–1224.

Osepchuk, J. M. (2002). Microwave power applications. *IEEE Transactions on Microwave Theory and Techniques*, *50*(3), 975–985.

Pallaroni, L., Von Holst, C., Eskilsson, C., & Björklund, E. (2002). Microwave-assisted extraction of zearalenone from wheat and corn. *Analytical and Bioanalytical Chemistry*, *374*(1), 161–166.

Pan, X., Niu, G., & Liu, H. (2003). Microwave-assisted extraction of tea polyphenols and tea caffeine from green tea leaves. *Chemical Engineering and Processing*, *42*(2), 129–133.

Püschner. (2015). *Pueschner microwave power systems*. http://www.pueschner.com/en/microwave-technology/frequency-ranges: http://www.pueschner.com/downloads/vacuumdrying.pdf Accessed October 2015.

Regier, M. (2014). Microwavable food packaging, Chapter 20. In J. H. Han (Ed.), *Innovations in food packaging* (pp. 495–514). Elsevier.

Regier, M., Knorzer, K., & Erle, U. (2004). Microwave- and microwave vacuum-drying of food [Review]. *Chemie Ingenieur Technik*, *76*(4), 424–432.

Rocha, A. M., Facão, M., Sousa, J. P., & Viegas, A. (2008). Optimization of a door seal structure of a microwave oven using a FDTD method. *International Journal of Numerical Modelling: Electronic Networks, Devices and Fields*, *21*(6), 507–513.

Routray, W., & Orsat, V. (2014). MAE of phenolic compounds from blueberry leaves and comparison with other extraction methods. *Industrial Crops and Products*, *58*, 36–45.

Sarang, S., Sastry, S. K., Gaines, J., Yang, T. C. S., & Dunne, P. (2007). Product formulation for ohmic heating: blanching as a pretreatment method to improve uniformity in heating of solid–liquid food mixtures. *Journal of Food Science*, *72*(5), E227–E234.

Seyhun, N., Sumnu, G., & Sahin, S. (2003). Effects of different emulsifier types, fat contents, and gum types on retardation of staling of microwave-baked cakes. *Nahrung/Food*, *47*(4), 248–251.

Shaheen, M. S., El-Ghorab, A. H., Anjum, F. M., & El-Massry, K. F. (2012). *Microwave applications in thermal food processing*. INTECH Open Access Publisher.

Stalam. (2015). http://www.stalam.com/en/rf-applications-for-industrial-bakeries/pasteurisation-equipment.html Accessed in October, 2015.

Stratakos, A., Delgado-Pando, G., Linton, M., Patterson, M. F., & Koidis, A. (2016). Industrial scale microwave processing of tomato juice using a novel continuous microwave system. *Food Chemistry*, *190*, 622–628.

Sumnu, G. (2001). A review on microwave baking of foods. *International Journal of Food Science and Technology*, *36*, 117–127.

Sun, Y., Liao, X., Wang, Z., Hu, X., & Chen, F. (2007). Optimization of microwave-assisted extraction of anthocyanins in red raspberries and identification of anthocyanin of extracts using high-performance liquid chromatography–mass spectrometry. *European Food Research and Technology*, *225*(3), 511–523.

Swittra, B.-N., Therdthai, N., & Dhamvithee, P. (2011). Characterization of microwave vacuum-dried durian chips. *Journal of Food Engineering*, *104*(1), 114–122.

Taher, B., & Farid, M. (2001). Cyclic microwave thawing of frozen meat: Experimental and theoretical investigation. *Chemical Engineering and Processing: Process Intensification*, *40*(4), 379–389.

Vanderroost, M., Ragaert, P., Devlieghere, F., & De Meulenaer, B. (2014). Intelligent food packaging: The next generation. *Trends in Food Science & Technology*, *39*(1), 47–62.

Venkatesh, M. S., & Raghavan, G. S. V. (2004). An overview of microwave processing and dielectric properties of agri-food materials. *Biosystems Engineering*, *88*(1), 1–18.

Villanueva, M. T. O., Marquina, A. D., Vargas, E. F., & Abellan, G. B. (2000). Modification of vitamins B_1 and B_2 by culinary process: Traditional systems and microwaves. *Food Chemistry*, *71*(4), 417–421.

Walde, S. G., Velu, V., Jyothirmayi, T., & Math, R. G. (2006). Effects of pretreatments and drying methods on dehydration of mushroom. *Journal of Food Engineering*, *74*(1), 108–115.

Wang, S., Chen, F., Wu, J., Wang, Z., Liao, X., & Hu, X. (2007). Optimization of pectin extraction assisted by microwave from apple pomace using response surface methodology. *Journal of Food Engineering*, *78*(2), 693–700.

Wang, Y., Xi, G. S., Zheng, Y. C., & Miao, F. S. (2010). Microwave-assisted extraction of flavonoids from Chinese herb Radix puerariae (Ge Gen). *Journal of Medicinal Plant Research*, *4*(4), 304–308.

Wang, Y., Zhang, M., Mujumdar, A. S., Mothibe, K. J., & Azam, S. R. (2012). Effect of blanching on microwave freeze drying of stem lettuce cubes in a circular conduit drying chamber. *Journal of Food Engineering*, *113*(2), 177–185.

WHO. (2015). World Health Day. Food safety: What you should know. http://www.searo.who.int/entity/world_health_day/2015/whd-what-you-should-know/en/ Accessed in October 2015.

Wray, D., & Ramaswamy, H. S. (2015). Novel concepts in microwave drying of foods. *Drying Technology*, *33*(7), 769–783.

Wu, T.-N. (2008). Environmental perspectives of microwave applications as remedial alternatives: Review. *Practice Periodical of Hazardous, Toxic, and Radioactive Waste Management*, *12:2*(102), 102–115.

Yam, K. L., Takhistov, P. T., & Miltz, J. (2005). Intelligent packaging: Concepts and applications. *Journal of Food Science*, *70*(1), R1–R10.

Zhao, S., Xiong, S., Qiu, C., & Xu, Y. (2007). Effect of microwaves on rice quality. *Journal of Stored Products Research*, *43*(4), 496–502.

Microwave-assisted baking

6

E. Turabi Yolacaner[1], *G. Sumnu*[2] *and S. Sahin*[2]
[1]Hacettepe University, Ankara, Turkey, [2]Middle East Technical University, Ankara, Turkey

6.1 Introduction

Saving time, energy, space, and nutrients are the advantages of microwave baking. On the other hand, some quality problems are encountered when only microwave technology is used in baking. Therefore, microwave-baked products have not been fully accepted by consumers. Common quality problems observed in microwave-baked products are firm and tough texture, rapid staling, dryness, and lack of color, flavor, and crust formation. Mechanism and speed of microwave heating are the major reasons for the inferior quality of microwave-baked products. Product development is necessary to obtain microwave-baked products that have the same quality as conventionally baked ones. Studies in recent years have focused on combination or hybrid baking technologies rather than using microwave technology alone to improve the product quality. This chapter first concentrates on the basic principles of microwave heating. Secondly, it discusses the advantages and disadvantages of microwave heating. Then, it gives information about technologies and equipment for microwave-assisted baking. Finally, it summarizes information about recent trends to improve quality of specific baked products such as breads, cakes, biscuits, and gluten-free products.

6.2 Principles of microwave baking

Baking is a complex process, which involves simultaneous heat and mass transfer. Many physical, chemical, and biochemical changes occur in food during baking. These changes are gelatinization of starch, denaturation of protein, liberation of carbon dioxide from leavening agents, volume expansion, evaporation of water, crust formation, and browning reactions.

In conventional baking, heat is transferred to the product surface mainly by convection from the heating media and by radiation from oven walls followed by conduction to the center. Rate and amount of heat transfer, humidity level in a baking oven and baking time are important baking conditions which affect final product quality (Therdthai & Zhou, 2003). On the other hand, in microwave baking, food is surrounded by air at ambient temperature, which is not heated by microwaves. Due to the interaction of microwaves with charged particles and polar molecules, heat generation occurs inside the food and it is conducted through the food material.

The Microwave Processing of Foods. DOI: http://dx.doi.org/10.1016/B978-0-08-100528-6.00006-1

The time of heating is very fast in a microwave oven due to the rapid heat generation. Therefore, there may not be enough time for completion of baking reactions such as starch gelatinization, starch conversion by enzymes, enough expansion of dough/batter, crust formation, browning, and final setting of the dough/batter into a rigid crumb structure during microwave baking. Kinetic reactions such as starch gelatinization and browning reactions depend not only on temperature but also on baking time. For this reason, it is necessary to ensure that gelatinization and browning reactions are completed when baking time is reduced, otherwise the quality of the product can be degraded (Therdthai, Zhou, & Adamczak, 2002).

The inability of microwave ovens to induce browning and crust formation is the biggest difference between microwave and conventional baking. This is due to the fact that the air inside the microwave oven is at ambient temperature and not heated as in conventional baking ovens. Therefore, the surface of foods baked in a microwave oven does not reach temperatures required for browning reactions.

For microwave heating the energy equation can be expressed as follows:

$$\frac{\partial T}{\partial t} = \alpha \nabla^2 T + \frac{\dot{Q}}{\rho C_\mathrm{p}} \tag{6.1}$$

where T is temperature, t is time, α is thermal diffusivity, ρ is density, C_p is the specific heat capacity of the material, and $\dot{Q}$ is the heat generation rate per unit volume of material, which represents the conversion of electromagnetic energy.

The relation between rate of heat generation and electric field intensity (E) at that location can be derived from Maxwell's equations of electromagnetic waves as shown by Metaxas and Meredith (1983):

$$\dot{Q} = 2\pi \varepsilon_0 \varepsilon'' f E^2 \tag{6.2}$$

where ε_0 is the dielectric constant of free space, ε'' is the dielectric loss factor of the food, f is frequency of the oven, and E is the root-mean-squared (rms) value of the electric field intensity.

Mass transfer mechanisms during conventional baking involve diffusion together with evaporation and condensation (Tong & Lund, 1993). Moisture movement in crumb and crust can be described by Fick's Law with apparent diffusion coefficients (Sablani, Marcotte, Baik, & Castaigne, 1998; Thorvaldsson & Janestad, 1999).

Pressure and concentration gradients are the driving forces for moisture transfer during microwave baking. The high pressures developed during microwave heating of high moisture foods have been studied by various researchers (Feng, Tang, Cavalieri, & Plumb, 2001; Ni, Datta, & Orrance, 1999).

A one-dimensional mathematical model of heat and mass transfer with variable thermal and moisture transfer properties and internal heat generation was proposed for microwave heating of dough (Tong & Lund, 1993).

The physical properties that affect microwave heating are dielectric and thermal properties. Dielectric properties are the dielectric constant and loss factor, while thermal properties are thermal conductivity and specific heat capacity. The dielectric constant represents the ability of a material to store microwave energy, while the dielectric loss factor represents the ability of a material to dissipate microwave energy into heat. Dielectric properties of foods depend on food composition, temperature, and frequency (Calay, Newborough, Probert, & Calay, 1995; see also Chapter 2: Microwave heating and the dielectric properties of foods and Chapter 3: Determination of the dielectric properties of foods).

6.3 Strengths and weaknesses of microwave baking

The overall baking time is shortened due to the fact that the temperature of a microwave-baked product rises more rapidly as compared to conventional baking process. This in turn saves energy. Higher nutritive value of the final product is another advantage of microwave baking when compared to conventional baking. Moreover, microwaves penetrate into the dough and inactivate α-amylase thermally, which is an enzyme causing the crumb to be sticky (Chamberlain, 1973). Finally, microwave equipment requires less space, due to shorter processing times.

On the other hand, many quality problems occur in microwave-baked products. The reason for most of these quality problems is short baking time, while generally completion of baking reactions requires a long time. Physicochemical changes, which would normally occur over a lengthy baking period in conventional baking, could not always be completed during the short period of microwave baking (Hegenbert, 1992). Differences between microwave and conventional heating mechanisms and specific interactions of each component in the product with microwave energy are other reasons for inferior quality of microwave-baked products (Goebel, Grider, Davis, & Gordon, 1984). Insufficient gelatinization, microwave-induced changes of the gluten structure, and rapidly generated gas and steam were stated to cause some quality changes in microwave-baked breads (Yin & Walker, 1995).

The quality defects observed in microwave-baked products are firm and tough texture, rapid staling, lack of color and crust formation, and a dry product (Sumnu, 2001). Microwave-induced changes of the gluten structure, extensive leaching of amylose during baking, and insufficient starch gelatinization cause firm and tough texture. In microwave baking, due to nonuniform heating, dramatically different patterns of starch transformation as compared to conventional heating are observed, especially in intermediate to low moisture foods such as baked goods. As a result, poor textures such as tough and leathery crumbs are obtained (Zallie, 1988).

The cause of rapid staling in microwave-baked products is not clear yet. For rapid staling of microwave-reheated breads (Higo & Noguchi, 1987) and microwave-baked cakes (Seyhun, 2002), extensive leaching of amylose during microwave baking as compared to conventional baking may be the reason. The higher moisture loss during microwave baking results in drier baked products than conventionally baked ones. It was shown that cakes and breads baked in a microwave oven lost more moisture than those baked by conventional heating (Keskin, Sumnu, & Sahin, 2004a; Seyhun, Sumnu, & Sahin, 2003; Sumnu, Ndife, & Bayindirli, 1999). Interior heating creates significant internal pressure gradients, which increase the flow of moisture through the food to the boundary (Datta, 1990).

Heating of reducing sugars with proteins or nitrogen-containing substances to form compounds like melanoidins causes the nonenzymatic browning reactions (Maillard reaction). These reactions start at around 160°C. On the other hand, colored substances called caramels are formed when sugars such as fructose, maltose, and dextrose are heated to similar temperatures. Relatively low food surface and surrounding temperatures in microwave baking restrict browning reactions to occur. Additionally, in microwave ovens, evaporated water molecules from the food system directly interact with cold air around the product and condense, which prevents browning reactions and crisping (Schiffmann, 1994). After baking, dough products, which are expected to be crisp and brown, become soggy. When heated for a longer period, they become dry and brittle but never brown when heated for a longer period. High temperature accompanied by dehydration leads to brown surfaces by Maillard reactions and caramelization of sugars. Moreover, time is necessary for completion of these browning reactions. The kinetic rate constant of browning reaction increases with increased temperature (Ibarz, Pagan, & Garza, 2000) and decreased moisture content (Moyano, Rioseco, & Gonzalez, 2002).

Flavors generated as a result of browning reactions do not exist in microwave-baked products. The aroma profile of a microwave-baked cake was shown to be similar to that of batter. Many of the nutty, brown, and caramel-type aromas observed in the conventional cake were lacking in microwave-baked cakes (Whorton & Reineccius, 1990). During microwave baking, individual flavor components are subjected to losses through distillation, flavor binding by starches and proteins, and chemical degradation. Crust formation also provides a barrier against the loss of flavors (Eliasson & Larsson, 1993). Due to absence of crust in microwave-baked products, flavors can easily be released. Changing the food formulation to reduce compound volatility minimizes loss of flavor compounds during microwave baking. This can be achieved by adding an oil phase or increasing the oil content (Yaylayan & Roberts, 2001). Flavoring agents may be encapsulated to reduce the volatility of aroma compounds (Whorton & Reineccius, 1990). On the other hand, during microwave baking of cakes, undesirable flavors such as flour or egg-like flavors develop. Flavoring agents may be added to mask these undesirable flavors and obtain a similar flavor profile to conventionally baked cakes.

Since products baked in microwave ovens have inferior quality, improving this quality represents a challenge to food technologists. Therefore, combining microwave heating with other heating methods or using hybrid technologies may be a solution for obtaining desired quality in bakery products. By this way, process acceleration advantage of microwave heating can be combined with other advantages of several heating mechanisms such as hot air, infrared (IR) heating, or jet impingement technique.

6.4 Technologies and equipment for microwave-assisted baking

A microwave–conventional baking oven for postbaking was developed by APV Baker (UK) as an alternative to radiofrequency (RF) heating in the early 1990s (Bengtsson, 2001). RF heating has the advantage of moisture leveling activity but microwave equipment is more compact and flexible; and the technology is more mature. Therefore, there is a tendency toward microwave and conventional combination heating instead of RF in baking and postbaking processes.

Microwave combination heating appliances currently in use can be grouped into:

1. microwave only,
2. microwave with IR (halogen lamp or grill),
3. microwave with hot air (jet or otherwise),
4. microwave with steam,
5. microwave with induction heating,
6. various combinations of the above (Datta & Rakesh, 2013).

Many disadvantages have been mentioned in the previous section when microwave heating is used in baking alone. Other heating techniques combined with microwaves may be helpful to obtain high quality bakery products.

Microwave heating was combined with IR heating in the Advantium oven which was manufactured by General Electric (Louisville, KT, USA) in 1999. Halogen lamps are placed at the top and bottom of the oven (Fig. 6.1). The halogen lamp–microwave combination oven combines the browning and crisping advantages of halogen lamp heating with the time-saving advantage of microwave heating. A halogen lamp which

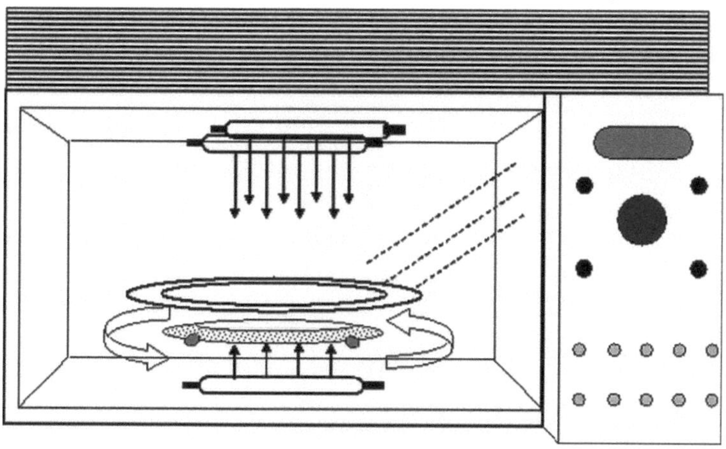

Figure 6.1 Illustration of halogen lamp–microwave combination oven.
Source: Reprinted from Sumnu, G., Sahin, S., & Sevimli, M. Microwave, infrared and infrared-microwave combination baking of cakes. *Journal of Food Engineering*, *71*(2), 150–155. Copyright, © (2005) with permission from Elsevier.

provides near-IR radiation is used to achieve surface browning and to prevent a soggy surface.

Typically forced hot air is provided that emulates simultaneous heating in the microwave oven combined with hot air. Airflow in a typical microwave oven without forced air is low, leading to very low rates of heat and moisture transfer (Verboven, Datta, Anh, Scheerlinck, & Nicolai, 2003). In jet impingement technology, air jets with very high velocities impinge on the food surface, which increase the rates of heat and mass transport. Microwaves with grill or a source of IR are also combined with hot air.

Microwaves combined with steam is a new feature, where steam is generated and fed to the oven cavity. Steam appears to be used not simultaneously with microwaves but as a mode available in the same appliance. No scientific study exists about this combination. In a patent (US 8772685 B2), however, a microwave steam cooking container system is described (Backaert & Demuynck, 2014). This system consists of a water reservoir, a food container with several apertures in its bottom section, and a microwave shielding material in the wall sections. The food is placed in the container, and is therefore shielded from cooking by direct exposure to microwave energy.

In the patent literature, a microwave oven in combination with an induction-heating cooker has been reported (LG Electronics, 1997). A shielding plate is mounted on the bottom surface and an induction coil is provided below the shielding plate in order to selectively choose between microwave and/or induction cooking (transmit a high-frequency magnetic field). There is no known commercially available unit or scientific study about this combination heating.

In a further patent (9006616 B2), a cooking appliance, including combination system, was described (McNamee et al., 2008). In this system, the appliance controls a radiant heating element in an oven cavity, both a blower and a convection heating element in an air plenum, and microwave energy sources to provide for numerous potential operating modes.

6.5 Applications of microwave-assisted baking to particular foods

As discussed before, there are various quality problems when only microwave heating is used in baking. To reduce these problems, there are various studies using microwave-assisted techniques. Most studies are focused on improving the quality of breads, cakes, and some gluten-free baked products. There are limited studies on other baked products.

6.5.1 *Microwave-assisted baked bread*

Hybrid ovens combining impingement with microwaves have been introduced to produce a brown and crisp crust in microwave-baked products

(Walker & Li, 1993). The process of directing jets of fluid at the surface to accelerate the surface heating is referred to as impingement (Ovadia & Walker, 1998). Rapid microwave baking of the product interior can be matched by the rapid impingement baking of the crust. Additionally, it is possible to reduce the higher evaporative moisture loss related to microwave baking by the crust formed by impingement baking.

Recently, combining microwave with IR heating is becoming a more commonly used technology in baking. IR–microwave combination oven combines the browning and crisping advantages of IR heating with the time-saving advantage of microwave heating. This technology has been used in bread baking and reduced the conventional baking time by about 75% (Keskin et al., 2004a). In this study, the effects of IR–microwave combination baking were compared with other baking methods such as conventional, microwave, and IR-heating. Breads baked in a IR–microwave combination oven yielded specific volume and color values comparable with the conventionally baked breads but their weight loss and firmness values were still higher. In IR–microwave combination heating, microwave heating was found to be dominant mechanism in affecting firmness and weight loss. IR–microwave combination ovens can be recommended to be used in bread baking to reduce the problem of lack of browning and crust formation.

Some types of enzymes (protease, α-amylase, xylanase, and lipase) were added to the bread formulation baked in microwave and IR–microwave combination ovens to reduce the firmness (Keskin, Sumnu, & Sahin, 2004b). Enzymes are commonly used in conventional baking to reduce the firmness and staling and to increase the volume of breads. The percentages of reduction in firmness of breads containing different enzymes and baked in different ovens were compared to breads containing no enzymes. As can be understood from the results, in microwave and IR–microwave combination baking, all of the enzymes were effective in reducing both initial crumb firmness and firmness during storage. All of the enzymes were found to be effective in reducing the initial firmness and increasing the specific volume of breads baked in microwave and IR–microwave combination ovens. However, in conventional baking, the effects of enzymes on crumb firmness were seen mostly during storage. The color of protease enzyme-added breads were found to be significantly different from that of the breads produced without enzyme addition and also the color of breads produced using other enzymes.

The addition of α-amylase and lipase enzymes may reduce the initial firmness problem of the interior of bread due to the high amount of amylose leaching while the firmness problem caused by the microwave-induced gluten changes may be reduced by the use of proteases which break down gluten protein, resulting in softer breads. Xylanases can have a specific action on the rate of gluten development and the quality of gluten (Hamer, 1995). The beneficial effects of xylanase on the crumb structure of breads baked in microwave ovens can be the result of this. The effects of enzymes on reducing the firmness of breads during IR–microwave combination baking or storage were found to be similar to those resulting from microwave baking since microwave heating was observed to be more dominant in affecting firmness than IR-heating.

By providing the required humidity within the oven during baking, it was possible to reduce the firmness of breads baked in IR–microwave combination ovens (Demirekler, Sumnu, & Sahin, 2004). Breads baked for 5 min at 70% upper, 50% lower halogen lamp powers, and 20% microwave power in an IR–microwave combination oven (in the IR–combination oven 100% MW-power was measured to be 706 W [IMPI 2-liter test], the initial mass was 50 g, the temperature/power of IR lamps is unfortunately not known) had comparable quality with conventionally baked ones in terms of color, textural characteristics, specific volume, and porosity.

After baking, all bakery products undergo a series of chemical and physical changes which are broadly referred to as staling (Cauvain, 1998). It is known that starch retrogradation implies hardening of the starch gel; therefore it is responsible for the increased firmness of the stale product. The mechanism of rapid staling in breads baked in microwave ovens is not yet fully understood. IR–microwave combination baking may be a promising method to retard staling of baked products. Staling of breads baked in different ovens (microwave, IR–microwave, and conventional) was investigated with the help of mechanical (compression measurements), physicochemical (Differential Scanning Calorimeter, DSC; X-ray; Fourier Transform Infrared Spectroscopy, FTIR), and rheological (Rapid Visco Analyzer, RVA) methods (Ozkoc, Sumnu, Sahin, & Turabi, 2009). All values related to starch retrogradation of microwave-baked samples were found to be the highest among other heating modes. Using IR–microwave combination heating made it possible to produce breads with similar staling degrees as conventionally baked ones in terms of retrogradation enthalpy and FTIR outputs related to starch retrogradation.

Noninvasive method of FITR spectroscopy, which can be used in staling studies, measures the degree of short-range ordering in a system. Conformational changes brought by starch retrogradation can be monitored by analyzing the band-narrowing, which is caused by a reduction in the range of conformations and smaller distributions of bond energies since the system becomes more ordered upon staling (Karim, Norziah, & Seow, 2000; Wilson, Goodfellow, & Belton, 1991). The spectral region 1200–800 cm^{-1} has been shown to be sensitive to the degree of molecular order in starch (Ottenhof, Hill, & Farhat, 2005). The modifications with aging of the absorption values in this spectral region, consisting of the variation of the relative intensities of overlapped bands at $\sim$1000 cm^{-1}, have been observed by other researchers (Cocchi et al., 2005), relating it to the progressive ordering of the amylopectin polymer present in bread. Peaks at 1047 cm^{-1} are related to crystalline regions of starch (Karim et al., 2000).

Water-related variations, such as drying and water redistribution, have also an influence on the measured spectra. It can be seen from Tables 6.1 and 6.2 that as storage time increased the ratio of peaks appearing at 2980–3600 cm^{-1} (A1) and 2810–2970 cm^{-1} (A2) significantly decreased, which was because of the decrease in moisture content of crumb of bread samples during storage and reorganization in the protein–starch network, resulting in the change in band intensities. The ratio of the peak intensities of samples baked in microwave and combination ovens were found to be significantly lower than that of conventionally baked ones. Additionally, the effect of gum addition on decreasing moisture loss during storage was especially seen for breads baked in microwave oven (Table 6.2).

Table 6.1 The integral area ratios of peaks appearing at 2980–3600 cm^{-1} (A1) and 2810–2970 cm^{-1} (A2); appearing around 1060–1070 cm^{-1} (A3), and 1151 cm^{-1} (A4) related to control breads

	Oven type					
	Conventional		Microwave		Infrared–microwave combination	
Peak ratios/ storage time	A1/A2	A3/A4	A1/A2	A3/A4	A1/A2	A3/A4
1 h	5.1 ± 0.3	0.89 ± 0.05	1.8 ± 0.20	1.25 ± 0.32	3.5 ± 0.40	0.96 ± 0.07
120 h	3.5 ± 0.7	1.07 ± 0.09	1.4 ± 0.30	1.27 ± 0.18	2.1 ± 0.50	1.03 ± 0.11

Source: Reprinted from Ozkoc, S. O., Sumnu, G., Sahin, S., & Turabi, E. Investigation of physicochemical properties of breads baked in microwave and infrared-microwave combination ovens during storage. *European Food Research and Technology*, *228*, 883–893, Copyright, © (2009) with permission from Springer.

Table 6.2 The integral area ratios of peaks appearing at 2980–3600 cm^{-1} (A1) and 2810–2970 cm^{-1} (A2); appearing around 1060–1070 cm^{-1} (A3) and 1151 cm^{-1} (A4) related to gum-added breads

	Oven type					
	Conventional		Microwave		Infrared–microwave combination	
Peak ratios/ storage time	A1/A2	A3/A4	A1/A2	A3/A4	A1/A2	A3/A4
1 h	5.0 ± 0.8	0.91 ± 0.09	3.3 ± 0.3	1.22 ± 0.07	3.6 ± 0.2	0.97 ± 0.03
120 h	3.8 ± 0.4	1.09 ± 0.04	2.1 ± 0.4	1.30 ± 0.05	2.4 ± 0.3	1.09 ± 0.06

Source: Reprinted from Ozkoc, S. O., Sumnu, G., Sahin, S., & Turabi, E. Investigation of physicochemical properties of breads baked in microwave and infrared-microwave combination ovens during storage. *European Food Research and Technology*, *228*, 883–893, Copyright, © (2009) with permission from Springer.

Bread crumb structure is one of the major quality attributes of bread. There is an obvious relationship between crumb structure and crumb appeareance. Crumb structure plays also a role on loaf volume and texture. Therefore, it may be concluded that having knowledge on the structure of breads may be helpful to predict many of the quality and transport properties of bread. Heating mechanism and presence of different types of gums also play a role in the formation of pore structure.

Ozkoc, Sumnu, and Sahin (2009) investigated the effects of gums (xanthan, guar, κ-carrageenan, and xanthan–guar gum blend) on the macro- and microstructure of breads baked in different ovens (IR–microwave combination and conventional) of the developed structures were assessed by image and SEM analysis, respectively.

Xanthan–guar gum blend addition improved bread quality with increasing fractions of pore areas according to the image analysis results. SEM analysis results demonstrated that pores in control breads (containing no gum) baked in conventional oven were found to be smaller, and had more spherical, or oval-like shape as compared to the ones baked in IR–microwave combination oven. Moreover, a more homogeneous closed-cell structure was observed for conventionally baked control breads. The pores of breads baked in IR–microwave combination oven were in close proximity to each other which resulted in coalescence of the gas cells to form channels, leading to their being no longer spherical. The starch granules in conventionally baked breads were more distorted and seen as a continuous sheet of gelatinized starch. On the other hand, granular residues and continuous starch structures were observed together in IR–microwave combination heating.

Microwaves have been combined with jet impingement or IR heating to obtain the desired quality characteristics in microwave-baked products. In the study of Sumnu, Datta, Sahin, Keskin, and Rakesh (2007), physical properties of breads during baking were measured using three different heating modes of microwave combined with IR, microwave combined with jet impingement, and jet impingement alone, in two different commercially available microwave ovens. Breads baked in the jet impingement oven were significantly different from the breads baked in the other ovens with respect to their specific volume and moisture content. Transient values of dielectric constant, dielectric loss factor, specific bulk volume, porosity, thermal conductivity, and moisture content were determined. Regression equations were developed to relate dielectric properties and thermal conductivity as a function of moisture content and porosity. The physical property data obtained here can be used as quantitative information on baking processes using novel heating modes and are critical for mathematical modeling studies to optimize these processes.

The porous structure affects the transport and other properties of food and these determine the final quality of a processed food. Therefore, quantifiying the nature of the pores is the first step for obtaining the relationships between the transport properties and the porous structure (Datta, Sahin, Sumnu, & Keskin, 2007). The nature of pore spaces in breads baked using various heating modes (microwave–IR, microwave–jet impingement, and jet impingement) were characterized in terms of total porosity, fraction of closed, blind and flow-through pores, and pore size distributions using several novel and old techniques (liquid extrusion porosimetry, scanned image analysis, pycnometry, volume displacement method, and scanning electron microscopy). According to the results, it appears that the pore size distribution is bimodal and only a combination of techniques can provide comprehensive information. A significant fraction of the pores was found to be closed. Breads baked in jet impingement had the highest total porosity followed by microwave–jet impingement and microwave–IR.

6.5.2 Microwave-assisted baked cakes

Mode of heating was the primary factor in affecting the temperature profile and physical properties of the cakes. In the study of Li and Walker (1996), combining air impingement with microwaves reduced the baking time significantly and produced white layer cakes with acceptable color but the lowest volume and firmest texture.

In another study, combined effects of conventional and microwave baking on white layer cake characteristics were investigated (Bilgen, Coskuner, & Karababa, 2004). Conventional baking was applied to form the cake crust before baking in a microwave oven used to form the crumb of the cakes. Different conventional baking times (8 or 11 min), microwave power (400 or 600 W), and microwave baking times (30, 40, or 50 s) were chosen as baking parameters. Initial mass of the cake batter samples were 100 g each. The bake losses of cakes using the combination of short oven times and low microwave power were smaller than those of the control cakes. Long oven times and low microwave power increased the specific volume of cakes. According to this study, using a combination of conventional and microwave baking produces cakes with qualities equivalent to those of cakes produced with conventional baking.

IR heating as well as conventional baking and air impingement can also be combined with microwave heating in baking cake products. In the study of Sumnu, Sahin, and Sevimli (2005), quality of cakes during baking in microwave, IR, and IR–microwave combination ovens were compared. Cakes baked in the microwave oven had the lowest quality in terms of weight loss, specific volume, firmness, and color. It was possible to achieve surface color formation in cakes by IR and IR–microwave combination baking. Moreover, IR–microwave combination baking reduced weight loss and firmness while it increased the volume of microwave-baked cakes. IR–microwave combination oven produced cakes with similar firmness and color values to that of conventionally baked ones. Conventional baking time was also reduced by about 75%.

Optimization of the conditions of IR–microwave combination baking of cakes by using response surface methodology was performed in the study of Sevimli, Sumnu, and Sahin (2005). Upper IR power, lower IR power, microwave power, and baking time were the independent variables. Weight loss, specific volume, color, and texture profile of the cakes were determined. The initial mass of cake sample was 100 g and the absolute microwave power of the oven 706 W, which was determined by IMPI 2-liter test. The upper IR power, the microwave power, and baking time were found to have significant effects on the weight loss, the specific volume, and the firmness of the cakes. Baking for 5 min at 60% upper IR power, 70% lower IR power, and 30% microwave power yielded comparable quality to that of conventionally baked cakes, except color. By using the IR–microwave combination oven, it was possible to obtain high-quality cakes, while reducing the conventional baking time by about 79%.

Different ingredients also affect the physical properties of cakes during microwave or microwave-assisted baking. The effects of the variation of fat content and addition of emulsifier or fat replacer on color and textural properties of cakes baked in microwave and IR–microwave combinations ovens were investigated in the study of Sakiyan, Sumnu, Sahin, and Meda (2007). They found that the use of a protein-based fat replacer (Simplesse) in microwave and IR–microwave combination ovens is not recommended, as it yielded textures that were considered too hard. The color change of the samples during IR–microwave combination baking increased linearly with baking time. The cakes containing maltodextrin, Lecigran, and Purawave had comparable results.

The starch gelatinization in cakes is important for the final quality of the product. One of the most important drawbacks of microwave baking is insufficient starch gelatinization due to the short baking period. The effects of different baking ovens and also different different cake formulations on the degree of starch gelatinization were investigated in the study of Sakiyan et al. (2011). Baking was performed in microwave, IR–microwave combination, and conventional ovens. DSC and RVA were used to examine starch gelatinization levels of fat-free, 25% fat, and 25% Simplesse-containing cake samples. DSC and RVA results indicated that increasing the baking time led to a significant increase in gelatinization level for all baking ovens. The effect of fat content on starch gelatinization was different depending on the type of baking oven. In the case of microwave and IR–microwave combination baking, addition of fat to the formulation caused higher gelatinization degree, which can be explained by the higher dielectric properties of the samples with higher fat content than the dielectric properties of other formulations (Sakiyan et al., 2007). Although fat is known to have lower dielectric properties, fat-containing cakes were shown to have higher dielectric properties than fat-free cakes since fat-containing cakes had lower porosity (Sakiyan et al., 2007). Addition of fat reduced the degree of starch gelatinization in conventional baking. This may be because of the fact that fats are known to delay starch gelatinization by delaying heat transport and the transport of water into the starch granule due to the formation of complexes between the lipid and amylose during baking (Elliasson, 1985; Ghiasi, Hoseney, & Varriano-Marston, 1982; Larsson, 1980). There was insufficient starch gelatinization in microwave-baked cakes in which the degree of gelatinization ranged from 55 to 78% depending on formulation. On the other hand, combining IR with microwaves increased the degree of starch gelatinization (70–90%).

A two-cycle microwave–toaster oven was used for baking a pound cake in the study of Sánchez-Pardo, Ortiz-Moreno, García-Zaragoza, Necoechea-Mondragón, and Chanona-Pérez (2012). It is a technology that combines the time saving advantage of microwave heating and browning and crisping advantages of an electric toaster. This oven is equipped with an internal electrical heater used to toast the surface of the pound cakes. Optimum microwave intensity and baking time conditions for the cake were determined and the image analysis methodology was applied to analyze the structure. Two-cycle microwave baking was also compared with conventional baking. Weight loss, density, water activity, moisture, luminosity, and seven image analysis parameters were measured as baking quality parameters. It was shown that two baking cycles in microwave ovens were able to eliminate the quality problems observed in microwave-baked products (Sumnu, 2001). The first baking cycle can be applied at low intensity and longer time to improve internal cooking, while the second cycle can be performed at high intensity and shorter time to contribute to the overall cooking. This type of oven is equipped with an internal electrical heater used to toast the surface of the pound cakes. Response surface analysis was used to determine the optimal baking conditions. The best baking conditions for the pound cakes having an initial weight of 45 g resulted in a low power for the first cycle (204 W for 120 s), a high power for the second cycle (937 W for 70 s), and a toasting time of 5.5 min at 200°C. Image analysis showed a 7% reduction in the crumb cell number of the cake baked with microwaves as compared to the conventional process. The two-cycle microwave-toaster oven generated a product with a higher volume (11%) and higher luminosity (2%) than the conventional process.

A further optimization study by using response surface methodology to determine the formulation and baking conditions of a functional cake (soy-cake) baked in an IR–microwave combination oven was performed by Sakiyan (2015). The initial weight of cake sample was 100 g. Baking time (8, 9, 10 min), soy flour concentration (30, 40, 50%), and emulsifier (DATEM, diacetyl tartaric acid ester of mono- and diglycerides) concentration (0.4, 0.6, 0.8%) were the independent variables. The quality parameters that were examined were specific volume, weight loss, total color change, and firmness of the cake samples. According to multiple regression analysis, the model equations and coefficient of determination (r^2) for each dependent variable are given in Table 6.3. The optimum baking time, soy-flour

Table 6.3 Regression equations for cakes with different formulations baked during different baking times in combination oven (X_1 is soy flour concentration (%); X_2 is DATEM content (%); and X_3 is baking time (min))

	Equation	r^2	Lack of fit
Weight loss (%)	$Y_1 = 27.2833^{***} - 0.6936^{**}X_1 - 0.4263^{*}X_2 + 2.6026^{***}X_3 - 0.9962^{**}X_1^2 - 1.0384^{**}X_2^2 - 1.2487^{**}X_3^2 - 0.1037^{ns}X_1X_2 + 0.3010^{ns}X_1X_3 - 0.0262^{ns}X_2X_3$	96.91	0.267^{ns}
Total color change (ΔE)	$Y_1 = 58.3728^{***} + 1.6641^{**}X_1 + 0.0716^{ns}X_2 + 3.9558^{***}X_3 - 0.5407^{*}X_1^2 + 0.1550\text{ns}X_2^2 - 1.8106^{***}X_3^2 - 0.1075^{ns}X_1X_2 + 0.0533^{ns}X_1X_3 - 0.3022^{ns}X_2X_3$	99.11	0.594^{ns}
Specific volume (m^3/kg)	$Y_1 = 1.30497^{***} - 0.04634^{**}X_1 + 0.02243^{*}X_2 + 0.09863^{***}X_3 - 0.05883^{*}X_1^2 - 0.06608^{**}X_2^2 - 0.02389^{ns}X_3^2 - 0.00014^{ns}X_1X_2 + 0.01725^{ns}X_1X_3 + 0.00500^{ns}X_2X_3$	93.32	6.49^{ns}
Hardness (N)	$Y_1 = 0.1528^{***} + 0.0090^{ns}X_1 - 0.0137^{*}X_2 + 0.0395^{***}X_3 + 0.0035^{ns}X_1^2 - 0.0099^{ns}X_2^2 + 0.0036^{ns}X_3^2 - 0.0109^{ns}X_1X_2 + 0.0097^{ns}X_1X_3 - 0.0029^{ns}X_2X_3$	85.51	15.65^{ns}

* The term is significant at $p \leq 0.05$.
** The term is significant at $p \leq 0.01$.
*** The term is significant at $p \leq 0.001$.

Source: Reprinted from Sakiyan, O. Optimization of formulation of soy cake baked in infrared-microwave combination oven by response surface methodology. *Journal of Food Science and Technology*, *52*(5), 2910–2917, Copyright, © (2015) with permission from Springer.

concentration and DATEM concentration were found as 9.5 min, 30%, and 0.72%, respectively. The corresponding responses of the optimum points were almost comparable with those of conventionally baked soy-cakes. Therefore, it was concluded that it was possible to produce high quality soy-cakes in a very short time by using IR–microwave combination oven.

6.5.3 Microwave-assisted baked gluten-free products

Celiac disease (gluten sensitive entropathy) is an intolerance to certain amino acid sequences found in the prolamin fraction of wheat (gliadin), rye (secalin), and barley (hordein). In genetically susceptible individuals, these kind of prolamins damage the mucosa of the small intestine leading to malabsorption of some nutrients. Until now, the only available treatment for celiac disease has been strict adherence to gluten-free diet and avoidance of gluten-containing foods. Since wheat is one of the main components of a daily diet worldwide, as well as the basic ingredient of most baked goods, a gluten-free diet is complex and can easily overwhelm patients (Kupper, 2005). Gluten replacement is nowadays one of the biggest challanges in the food science and technology disciplines, since finding high quality gluten-free food is one of the major issues for celiac disease sufferers.

Microwave-assisted baking was also used in baking of gluten-free products in some studies. Determination of the suitable baking conditions as well as the right formulation for gluten-free bakery products is crucial since obtaining the desired quality is difficult due to the lack of gluten protein. Optimization of the baking conditions and formulation of rice flour-based gluten-free cake by response surface methodology was performed in the study of Turabi, Sumnu, and Sahin (2008). Two types of cake formulations containing different types of gums, xanthan gum and xanthan–guar gum blend, were used in the experiments. The initial weight of cake samples was 100 g. The independent variables were emulsifier content (0, 3, and 6% of flour weight), upper halogen lamp power (50, 60, and 70%), and baking time (7.0, 7.5, and 8.0 min). Specific volume, surface color change, firmness, and weight loss were determined as quality parameters. The independent and dependent variables were fitted to the second-order model equation. Dependent variables at different experimental conditions were also predicted using the model. Cakes formulated with xanthan gum and 5.28% emulsifier and baked using 60% halogen lamp power for 7 min yielded the most acceptable quality. Values for the quality characteristics were calculated at these optimum conditions and are shown in Table 6.4. Cakes formulated by xanthan or xanthan–guar blend using optimum concentration of the emulsifier (Purawave), were also baked in a conventional oven for comparison. Cakes baked by using IR–microwave combination yielded similar color and firmness values to those baked in a conventional oven (Table 6.4). IR–microwave combination baking resulted in cakes with higher volume but lower moisture content. As IR–microwave combination baking decreased conventional baking time by 77%, it may be recommended to be used in production of gluten-free rice cakes. Using only xanthan gum in cakes gave better results with respect to specific volume and firmness (Table 6.4).

Table 6.4 Quality characteristics of the cakes baked in conventional oven and IR–microwave combination oven

Parameters	IR–microwave combination baking		Conventional baking	
	Xanthan–guar blend	Xanthan gum	Xanthan–guar blend	Xanthan gum
Specific volume (mL/g)	1.57	1.75	1.30	1.38
Total color change (ΔE)	32.18	30.56	26.43	25.34
Firmness (N)	2.58	2.32	2.35	2.13
Weight loss (%)	13.69	11.94	10.65	7.70

Source: Reprinted from Turabi, E., Sumnu, G., Sahin, S. Optimization of baking of rice cakes in infrared-microwave combinationo oven by response surface methodology. *Food Bioprocess Technology*, *1*, 64–73, Copyright, © (2008) with permission from Springer.

Chemical and physical changes that occur after baking are called staling and it is a complex process that cannot be explained by a single effect. Baking method and modifications in formulation mainly affect the staling process postbaking. For retardation of staling of bakery products, one of the common strategies is addition of gums into the formulation. In the study of Sumnu, Koksel, Sahin, Basman, and Meda (2010), the effects of different concentrations of xanthan and guar gums and their blends on staling of gluten-free rice cakes baked in an IR–microwave combination oven were determined. Additionally, the results were compared with the cakes baked in a conventional oven. Gums were added at concentrations of 0.3 and 1.0%. Cakes were stored at 22 ± 2°C for 120 h in order to assess the staling behavior of cakes. Xanthan–guar gum blend decreased hardness, weight loss, retrogradation enthalpy, and the change in setback viscosity values of cakes during storage for both types of ovens as compared to the control formulation. It was concluded that oven type was a significant factor in affecting staling parameters of gluten-free cakes. Baking in IR–microwave combination oven resulted in gluten-free cakes with lower retrogradation enthalpies but higher hardness values as compared to the ones baked in conventional oven.

Another study for understanding the effects of chesnut flour and xanthan–guar gum blend–DATEM mixture on staling of gluten-free rice breads baked in conventional and IR–microwave combination ovens was performed by Demirkesen, Campanella, Sumnu, Sahin, and Hamaker (2014). Texture analysis was carried out through mechanical compression (TA), differential scanning calorimetry (DSC), and X-ray diffraction (X-ray diffraction here was used to determine the crystalline structure of starch which helps examining bread staling). During staling starch (amylose and amylopection) shows different crystal structures (V-type for amylose and B-type for amylopectin retrogradation) and FTIR were used for the assessment of staling properties of the bread. Hardness, moisture loss, and retrogradation enthalpy values for all bread samples increased significantly during storage. Since carbohydrate polymers such as starch are extensively hydrogen bonded, conformational changes due to starch retrogradation can be monitored in the FTIR spectra by analyzing band narrowing and changes in band intensities (Wilson et al., 1991). FTIR spectra showed that the integrated area of peaks around 1041 and 1150 cm^{-1} wavelengths, which were related to the structure of starch retrogradation, increased with storage time. According to the X-ray diffractograms, during staling, the peaks observed in fresh breads (at around 2θ of 20°) remained unchanged. The X-ray diffractograms of aged breads indicated a B-type structure with the appearance of peaks at around 17°, 19.5°, and 22°, which showed the amylopectin retrogradation. An additional peak at 24° was observed in breads stored for longer periods. Retrogradation enthalpies and total crystallinity values of breads did not show significant differences with baking process.

Some other types of gluten-free flours can be used in combination with rice flour to design gluten-free products for baking in microwave-assisted ovens. In the study of Demirkesen, Sumnu, and Sahin (2013a, 2013b), the effects of different tigernut flour/rice flour ratios on quality of gluten-free bread formulations baked in IR–microwave combination and conventional ovens were determined. The degrees of starch gelatinization of breads baked in different ovens were also compared. Conventionally baked breads prepared with tigernut flour/rice flour ratio of 10:90 and the ones baked in IR–microwave combination at a tigernut/rice flour ratio of 20:80 had the most acceptable firmness and specific volume values. These breads had also similar color. Breads baked in both conventional and IR–microwave combination ovens had sufficient starch gelatinization degrees ranging from 91 to 94% and from 84 to 85%, respectively.

In a further optimization study, response surface methodology was used to design gluten-free breads containing chestnut and rice flour and xanthan–guar gum blend to be baked in an IR–microwave combination oven (Demirkesen, Sumnu, Sahin, & Uysal, 2011). Gluten-free bread formulations and processing conditions were optimized. Weight loss, firmness, specific volume, and color change of the breads were determined. Rice flour mixed with different proportions of chestnut flour and different emulsifier contents were used to prepare breads. The gluten-free formulations were baked using different upper halogen lamp powers, microwave powers, and baking time which were varied from 40 to 80%, 30 to 70%, and 9 to 17 min, respectively. Four dough samples each having initial weight of 100 g were baked at the same time. The absolute microwave power were measured as 682 W by IMPI 2-L test. For comparison, gluten-free breads and wheat breads baked in conventional oven were used. Breads containing 46.5% chestnut flour and 0.62% emulsifier and baked at 40% IR and 30% microwave power for 9 min had statistically comparable quality with conventionally baked ones.

Understanding the internal macro- and microstructure of bakery products is essential since the physicochemical, sensory, and transport properties of foods are largely dependent on crumb structure and crumb structure affects the appearance of crumb and texture of bakery products. Examining food microstructure has always been a challenge because of the complexity of the material involved. Image analysis and quantification of relative features is the basis of modern food microscopy. This helps understanding the internal macro- and microstructure of bakery products.

In the study of Turabi, Sumnu, and Sahin (2010), the images obtained by a scanner and SEM were used to investigate the effects of different gums on macrostructure of gluten-free rice cakes baked in conventional and IR–microwave combination ovens. Xanthan, guar, locust bean, κ-carragenan, and xanthan–guar blend were the types of gums used in the cake formulations. As control, cake containing no gum was used. It was observed that both addition of different types of gums affected the pore area fraction and number of pores of the rice cakes. The highest pore area fraction was obtained in cakes containing xanthan and xanthan–guar blend. Cakes baked in IR–microwave combination oven had higher porosity than those baked in conventional oven. Microstructure of gluten-free rice cakes was also analyzed (Figs. 6.2 and 6.3). According to these results, conventionally baked cakes showed more starch granule deformations. Both granular starch residues and deformed starch structure were observed together in cakes baked in an IR–microwave combination oven. It was found that not all starch granules lost their identity and did not disintegrate completely.

Image analysis was also performed for a gluten-free bread prepared from chestnut and rice flours and baked in different ovens in the study of Demirkesen et al. (2013a, 2013b). The effects of chestnut flour and xanthan–guar gum blend–emulsifier DATEM mixture addition on macro- and microstructure of rice breads baked in conventional and IR–microwave combination ovens were investigated by using images obtained by a scanner and SEM. Pore area fraction, pore size distribution, and roundness values of pores were determined. Scanned images of different gluten-free bread formulations baked in different ovens can be seen in Fig. 6.4. The highest pore area fraction values were obtained in breads prepared by replacement of 46% of rice flour with chestnut flour containing xanthan–guar gum blend–DATEM mixture and baked in an IR–microwave combination oven. On the other hand, breads containing no additives or chestnut flour had the lowest pore area fraction values. IR–microwave combination baking increased both pore area fraction values and total number of pores. Approximately 23–28% increase in the number of small pores (0–5 mm^2) in rice breads and 71% increase in the number of the large pores (>10 mm^2) in chestnut-rice breads were obtained by IR–microwave combination. The fiber content and larger starch granules of chestnut flour contributed toward the stabilization of gas bubbles resulting in better crumb structure. More homogeneous pore distributions were observed when additives and an IR–microwave combination oven were used. It was seen that starch granules in chestnut–rice breads baked in an IR–microwave combination oven did not disintegrate completely when microstructure of gluten-free breads was investigated.

6.5.4 Crustless bread

Crustless bread is a new concept in bakery industry. Baking crustless bread by using microwave technology is possible and the only difference from regular bread is that there is no crust formation. As bread crust contains acrylamide, which is a

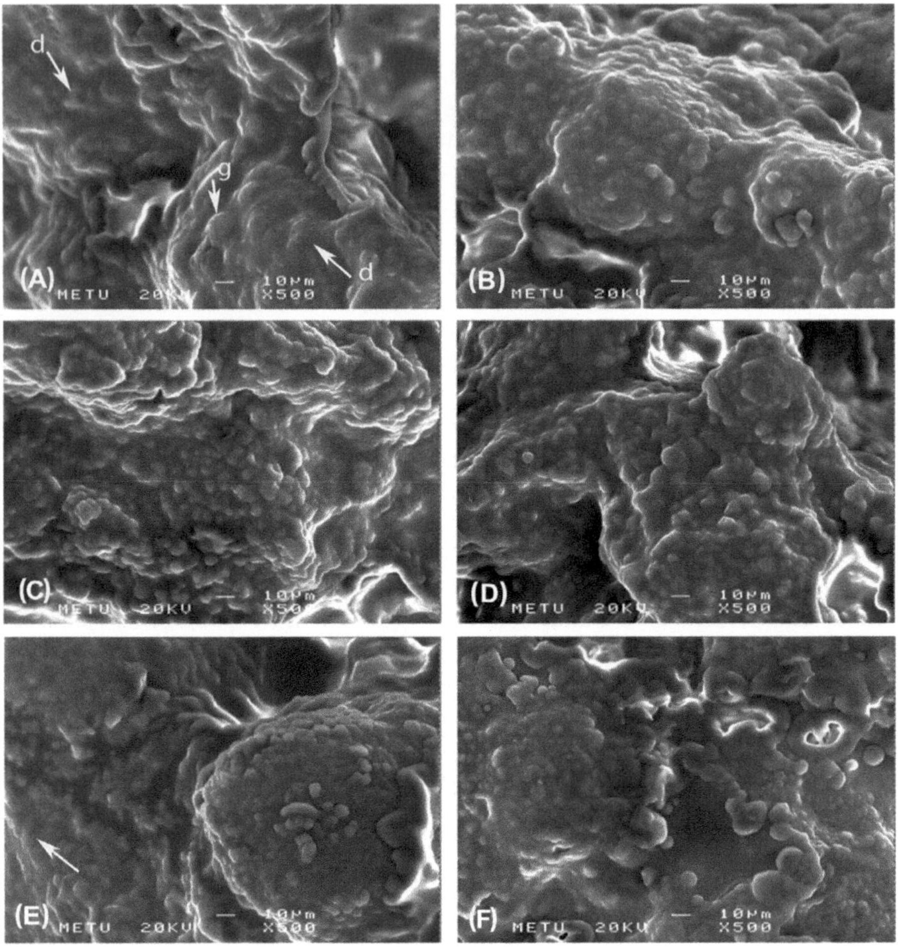

Figure 6.2 SEM micrographs (500 ×) for cakes baked in conventional oven (A. control, B. xanthan, C. guar, D. xanthan–guar, E. locustbean, F. kappa-carrageenan). Arrows show *d*, deformed starch structure; *g*, granular starch residues.
Source: Reprinted from Turabi, E., Sumnu, G., & Sahin, S. Quantitative analysis of macro and micro-structure of gluten-free rice cakes containing different types of gums baked in different ovens. *Food Hydrocolloids*, *24*, 755–762, Copyright, © (2010) with permission from Elsevier.

probable human carcinogen, the application of crustless bread in the industry is likely to be far-reaching (Mondal & Datta, 2011). For children and elderly people, this type of bread may also be an alternative. Generally, crustless bread is produced by removing the crust from the regular bread after the baking process; however, this procedure is highly ineffective in terms of energy consumption and waste of raw materials. Therefore, it is problematic economically as well as ethically (Technical Trend, 2009).

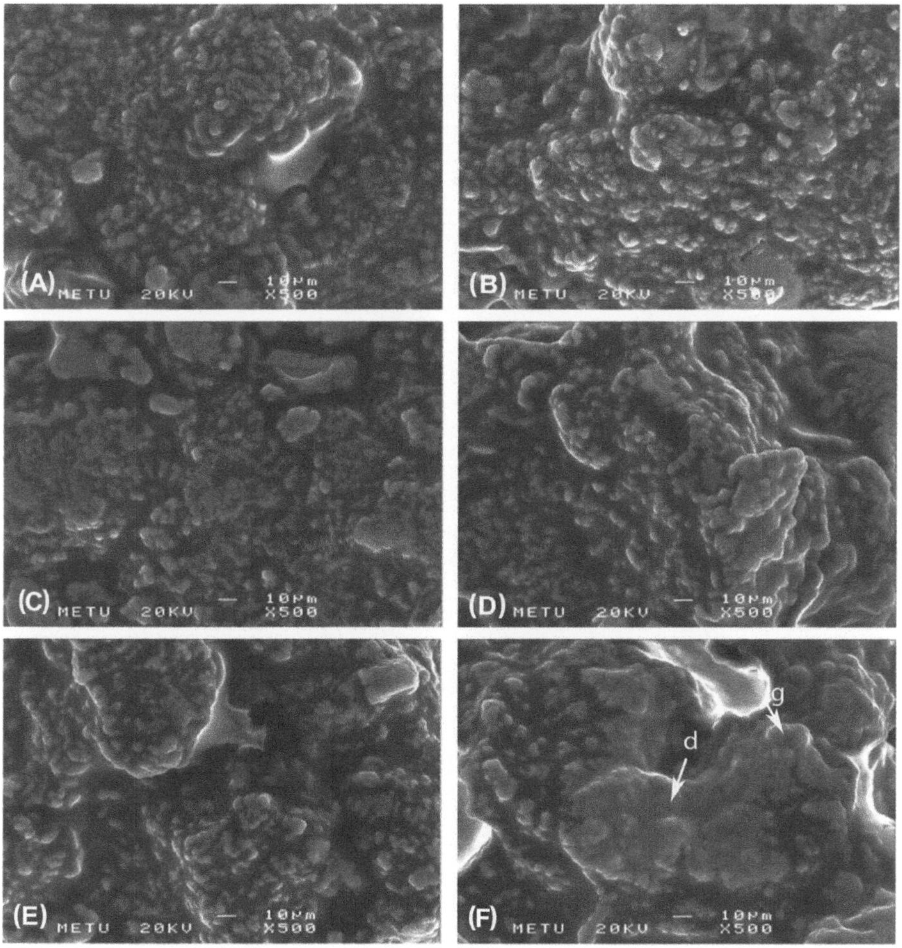

Figure 6.3 SEM micrographs (500 ×) for cakes baked in infrared–microwave combination oven (A. control, B. xanthan, C. guar, D. xanthan–guar, E. locustbean, F. kappa-carrageenan). Arrows show *d*, deformed starch structure; *g*, granular starch residues. *Source*: Reprinted from Turabi, E., Sumnu, G., & Sahin, S. Quantitative analysis of macro and micro-structure of gluten-free rice cakes containing different types of gums baked in different ovens. *Food Hydrocolloids*, *24*, 755–762, Copyright, © (2010) with permission from Elsevier.

The British food group RHM launched crustless bread baked by a proper baking process (www.foodnavigator.com). Crustless bread is also available in other countries including the United States, Italy, and Spain; however, there it is manufactured by cutting the crusts off after baking. In the study of Mondal and Datta (2010), two-dimensional CFD modeling of the crustless bread baking process was performed by using a special kind of baking oven. During gentle baking of the dough, water is sprayed on the surface of the dough. Microwave-assisted baking

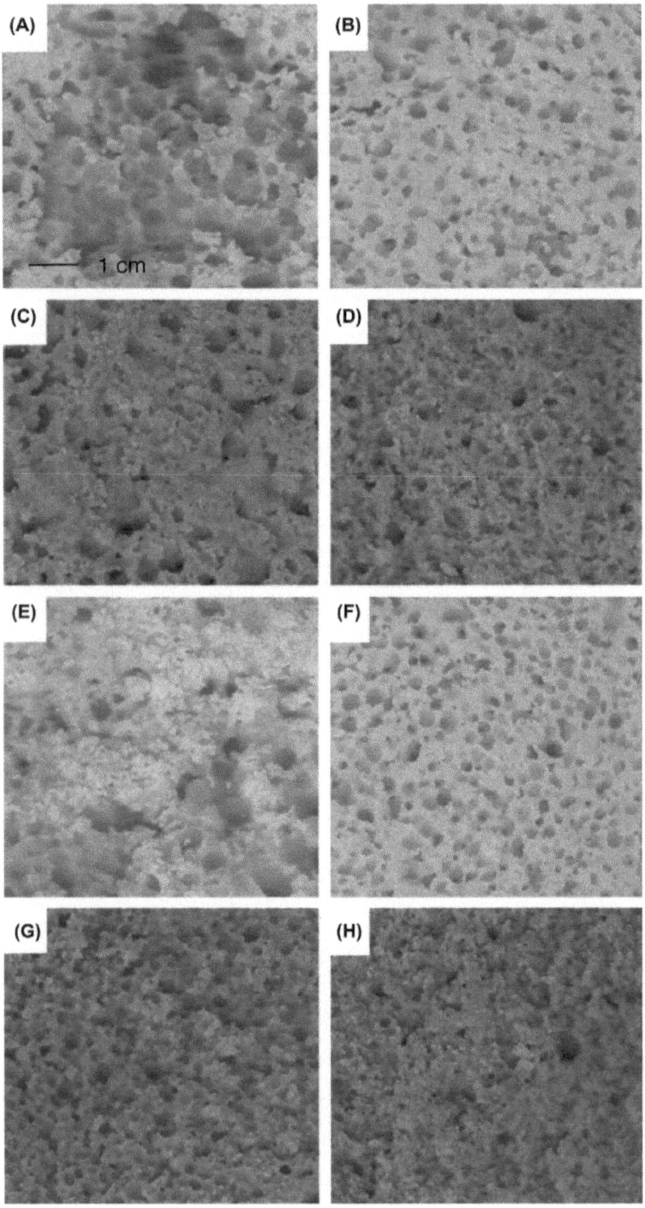

Figure 6.4 Scanned images of different gluten-free bread formulations baked in different ovens. (A) Rice bread baked in a conventional oven. (B) Rice bread containing xanthan–guar gum blend–DATEM mixture and baked in a conventional oven. (C) Chestnut–rice breads baked in a conventional oven. (D) Chestnut–rice bread containing xanthan–guar gum blend–DATEM mixture and baked in a conventional oven. (E) Rice bread baked in an infrared–microwave combination oven. (F) Rice bread containing

may be an alternative method in producing crustless bread. Further research has to be done in designing such an oven and determination of process conditions.

The microwave-baked bread is "normal" bread, but without a crust (Kintsel). The surface of the bread has a thin white film, which is as smooth and elastic as the center of the bread. The bread is baked in microwave transparent (PP, PTFE) boxes. The boxes are placed on a special holding system, which is fed into the microwave chamber on a conveyor belt. Next, the boxes loaded with bread dough travel continuously on the conveyor belt through the microwave furnace. The patented microwave chamber with a round shape and the magnetrons placed helically around the chamber leads to a homogeneous heating, not only from the sides, but also from below.

The French company MO2 has developed a system which turns the possibility of baking of crustless bread by using microwave technology into an economically interesting process that is currently being put into practice in France (Technical Trend in Baking + Biscuit, 2009). Francois Gelé and his staff came up with the idea of exploring microwave technology when the demand for crustless bread increased steadily. They began in the company lab near Nimes, Southern France, to take a closer look into the process of microwave cooking and its use for professional baking processes. The result is a complete line for the production of crustless bread for which the oven, the molds and their use had to be completely reinvented. The bread baked on this line is "common" toast bread, without any crust. The surfaces of the products are covered by a thin, doughy-white skin which is as tender and as elastic as the crumb itself. Microwave baking removes less water from the product than conventional baking. The water collects at the outer surfaces in covered molds and this is when perforation holes of the patented molds come into play.

The addition of conventional oven segment with turbulence air flow would also be a good alternative. Such a combination, however, would provide for a second benefit, namely not only to bake crustless bread with white surfaces but also to brown the product in a short time which would give the bread a tender crust.

6.6 Future trends

Microwave baking when used alone results in products with inferior quality. As mentioned in the previous sections, a recent trend is toward combining microwave heating with other processing methods such as hot air, impingement and IR heating. Although there are studies where microwave-assisted heating is used, these are limited in terms of product variety. Breads, cakes, and some gluten-free products were baked in microwave-assisted ovens. Further research is required to study the effects

◀ xanthan–guar gum blend–DATEM mixture and baked in infrared–microwave combination oven. (G) Chestnut–rice breads baked in an infrared–microwave combination oven. (H) Chestnut–rice bread containing xanthan–guar gum blend–DATEM mixture and baked in an infrared–microwave combination oven.

Source: Reprinted from Demirkesen, I., Sumnu, G., & Sahin, S. Image analysis of gluten-free breads prepared with chestnut and rice flour and baked in different oven. *Food Bioprocess Technology*, 6, 1749–1758, Copyright, © (2013) with permission from Springer.

of these combination ovens on quality of different bakery products such as cookies, bagels, muffins, donuts, crackers, etc.

Phase control, which is a method of microwave heating, can be used to control heating rates and temperature distribution in the product (Bows, Patrick, Andrew, Metaxas, & Dibben, 1999). Since this method offers opportunities for direct control of microwave field and the consequent temperature distribution, it may be used during baking of foods to achieve browning.

Investigation of the interaction of different ingredients with microwave energy is still an area in which further research is needed. Such studies will provide insights into improving the quality of microwave-assisted baked products. Modeling the changes of dielectric, thermal, and functional properties of batter or dough during microwave baking can reduce the quality defects in microwave-assisted baked products. Once the objective of improving the quality of microwave-assisted baked products is achieved, significant savings in time and energy will be possible; furthermore, the manufacture of products with high nutritive quality will be possible.

References

Backaert, D. M. C. J., & Demuynck, M. (2014). *Microwave steam cooking container system*. US Patent US8772685 B2.

Bengtsson, N. (2001). Development of industrial microwave heating of foods in Europe. *Journal of Microwave Power and Electromagnetic Energy*, *36*(4), 227–240.

Bilgen, S., Coskuner, Y., & Karababa, E. (2004). Effects of baking parameters on white layer cake quality by combined use of conventional and microwave ovens. *Journal of Food Processing and Preservation*, *28*, 89–102.

Bows, J. R., Patrick, M. L., Andrew, R. J., Metaxas, C. R., & Dibben, D. C. (1999). Microwave phase control heating. *International Journal of Food Science & Technology*, *34*(4), 295–304.

Calay, R. K., Newborough, M., Probert, D., & Calay, P. S. (1995). Predictive equations for dielectric properties of foods. *International Journal of Food Science & Technology*, *29*(6), 699–713.

Cauvain, S. P. (1998). Improving the control of staling in frozen bakery products. *Trends in Food Science & Technology*, *9*(2), 56–61.

Cocchi, M., Foca, G., Marchetti, A., Sighinalfi, S., Tassi, L., & Ulrich, A. (2005). *Annali di Chimica 95 by Societa Chimica Italiana*.

Datta, A. K. (1990). Heat and mass transfer in the microwave processing of food. *Chemical Engineering Progress*, *86*(6), 47–53.

Datta, A. K., & Rakesh, V. (2013). Principles of microwave combination heating. *Comprehensive Reviews in Food Science and Food Safety*, *12*, 24–39.

Datta, A. K., Sahin, S., Sumnu, G., & Keskin, S. O. (2007). Porous media characterization of breads baked using novel heating modes. *Journal of Food Engineering*, *79*, 106–116.

Demirekler, P., Sumnu, G., & Sahin, S. (2004). Optimization of bread baking in halogen lamp-microwave combination oven by response surface methodology. *European Food Research and Technology*, *219*(4), 341–347.

Demirkesen, I., Campanella, O. H., Sumnu, G., Sahin, S., & Hamaker, B. R. (2014). A study on staling characteristics of gluten-free breads prepared with chestnut and rice flours. *Food and Bioprocess Technology*, *7*, 806–820.

Demirkesen, I., Sumnu, G., & Sahin, S. (2013a). Image analysis of gluten-free breads prepared with chestnut and rice flour and baked in different ovens. *Food and Bioprocess Technology*, *6*, 1749–1758.

Demirkesen, I., Sumnu, G., & Sahin, S. (2013b). Quality of gluten-free bread formulations baked in different ovens. *Food and Bioprocess Technology*, *6*, 746–753.

Demirkesen, I., Sumnu, G., Sahin, S., & Uysal, N. (2011). Optimization of formulations and infrared-microwave combination baking conditions of chestnut-rice breads. *International Journal of Food Science & Technology*, *46*, 1809–1815.

Elliasson, A. C. (1985). Starch gelatinization in the presence of emulsifiers: A morphological study of wheat starch. *Starch*, *37*(12), 411–415.

Elliasson, A. C., & Larsson, K. (1993). *Cereals in breadmaking: A molecular colloidal approach*. New York: Marcel Dekker.

Feng, H., Tang, J., Cavalieri, R. R., & Plumb, O. A. (2001). Heat and mass transfer in microwave drying of porous materials in a spouted bed. *AICHE Journal*, *47*(7), 1499–1511.

Ghiasi, K., Hoseney, R. C., & Varriano-Marston, E. (1982). Effects of flour components and dough ingredients on starch gelatinization. *Cereal Chemistry*, *60*, 58–61.

Goebel, N. K., Grider, J., Davis, E. A., & Gordon, J. (1984). The effects of microwave energy and conventional heating on wheat starch granule transformations. *Food Microstructure*, *3*(1), 73–82.

Hamer, R. J. (1995). Enzymes in the baking industry. In G. A. Tucker, & L. F. J. Woods (Eds.), *Enzymes in food processing* (pp. 199–222). Glasgow: Blackie Academic and Professional.

Hegenbert, S. (1992). Microwave quality: Coming of age. *Food Product Design*, *17*, 29–52.

Higo, A., & Noguchi, S. (1987). Comparative studies on food treated with microwave and conductive heating. I. Process of bread hardening by microwave heating. *Journal of the Japanese Society for Food Science and Technology*, *34*(12), 781–787.

Ibarz, A., Pagan, J., & Garza, S. (2000). Kinetic models of non-enzymatic browning in apple puree. *Journal of the Science of Food and Agriculture*, *80*(8), 1162–1168.

Karim, A. A., Norziah, M. H., & Seow, C. C. (2000). Methods for the study of starch retrogradation. *Food Chemistry*, *71*, 9–36.

Keskin, S. O., Sumnu, G., & Sahin, S. (2004a). Bread baking in halogen lamp-microwave combination baking. *Food Research International*, *37*(5), 489–495.

Keskin, S. O., Sumnu, G., & Sahin, S. (2004b). Usage of enzymes in novel baking process. *Nahrung-Food*, *48*(2), 156–160.

Kintsel, N. *Microwaves in the food industry*. Available from: http://www.linn-high-therm.de/about-us/download/publications.html.

Kupper, C. (2005). Dietary guidelines and implementation for celiac disease. *Gastroenterology*, *128*(4), 121–127.

Larsson, K. (1980). Inhibition of starch gelatinization by amylase-lipid complex formation. *Starch*, *32*, 125–126.

LG Electronics. (1997). *Microwave oven in combination with induction heating cooker*. US Patent No. 5698125.

Li, A., & Walker, C. E. (1996). Cake baking in conventional, impingement and hybrid ovens. *Journal of Food Science*, *61*(1), 188–191.

Metaxas, A. C., & Meredith, R. J. (1983). *Industrial microwave heating*. London: Peter Peregrinus.

Mcnamee, R., Schulte, R. A., Miller, T., Carlson, R. W., Goranson, J. A., & Cook, E. R. (2008). *Cooking appliance including combination heating system*. US Patent 9006619.

Mondal, A., & Datta, A. K. (2010). Two-dimensional CFD modeling and simulation of crustless bread baking process. *Journal of Food Engineering*, *99*, 166–174.

Mondal, A., & Datta, A. K. (2011). Investigation of the process parameters using response surface methodology on the quality of crustless bread baked in a water spraying oven. *Journal of Food Process Engineering*, *34*, 1819–1837.

Moyano, P. C., Rioseco, V. K., & Gonzalez, P. A. (2002). Kinetics of crust color changes during deep-fat frying of impregnated french fries. *Journal of Food Engineering*, *54*(3), 249–255.

Ni, H., Datta, A. K., & Torrance, K. E. (1999). Moisture transport in intensive microwave heating of biomaterials: A multiphase porous model. *International Journal of Heat and Mass Transfer*, *42*(8), 1501–1512.

Ottenhof, M. A., Hill, S. E., & Farhat, I. A. (2005). Comparative study of the retrogradation of intermediate water content waxy maize, wheat and potato starches. *Journal of Agricultural and Food Chemistry*, *53*, 631–638.

Ovadia, D. Z., & Walker, C. E. (1998). Impingement in food processing. *Food Technology*, *52*(4), 46–52.

Ozkoc, S. O., Sumnu, G., & Sahin, S. (2009). The effects of gums on macro and micro-structure of breads baked in different ovens. *Food Hydrocolloids*, *23*, 2182–2189.

Ozkoc, S. O., Sumnu, G., Sahin, S., & Turabi, E. (2009). Investigation of physicohemical properties of breads baked in microwave and infrared-microwave combination ovens during storage. *European Food Research and Technology*, *228*, 883–893.

Sablani, S. S., Marcotte, M., Baik, O. D., & Castaigne, F. (1998). Modeling of simultaneous heat and water transport in the baking process. *Lebensm-Wiss U-Technolology*, *31*(3), 201–209.

Sakiyan, O. (2015). Optimization of formulation of soy-cakes baked in infrared-microwave combination oven by response surface methodology. *Journal of Food Science and Technology*, *52*(5), 2910–2917.

Sakiyan, O., Sumnu, G., Sahin, S., & Meda, V. (2007). The effect of different formulations on physical properties of cakes baked with microwave and near infrared-microwave combinations. *Journal of Microwave Power and Electromagnetic Energy*, *41*(1), 20–26.

Sakiyan, O., Sumnu, G., Sahin, S., Meda, V., Koksel, H., & Chang, P. (2011). A study on degree of starch gelatinization in cakes baked in three different ovens. *Food and Bioprocess Technology*, *4*, 1237–1244.

Sánchez-Pardo, M. E., Ortiz-Moreno, A., García-Zaragoza, F. J., Necoechea-Mondragón, H., & Chanona-Pérez, J. J. (2012). Comparison of pound cake baked in a two cycle microwave-toaster oven and in conventional oven. *LWT-Food Science and Technology*, *46*, 356–362.

Schiffmann, R. F. (1994). Critical factors in microwave generated aromas. In J. McGorrin, T. H. Parlimant, & M. J. Morello (Eds.), *Thermally generated flavors: Maillard, microwave and extrusion processes, ACS symposium series no. 543* (pp. 386–394). Washington, DC: American Chemical Society.

Sevimli, K. M., Sumnu, G., & Sahin, S. (2005). Optimization of halogen lamp–microwave combination baking of cakes: A response surface methodology study. *European Food Research and Technology*, *221*, 61–68.

Seyhun, N. (2002). *Retardation of staling of microwave baked cakes. MS thesis*. Middle East Technical University, Ankara, Turkey.

Seyhun, N., Sumnu, G., & Sahin, S. (2003). Effects of different emulsifiers, gums and fat contents on retardation of staling of microwave baked cakes. *Nahrung-Food*, *47*(4), 248–251.

Sumnu, G. (2001). A review on microwave baking of foods. *International Journal of Food Science & Technology, 36*(2), 117–127.

Sumnu, G., Datta, A. K., Sahin, S., Keskin, S. O., & Rakesh, V. (2007). Transport and related properties of breads baked using various heating modes. *Journal of Food Engineering, 78*, 1382–1387.

Sumnu, G., Koksel, F., Sahin, S., Basman, A., & Meda, V. (2010). The effects of xanthan and guar gums on staling of gluten-free rice cakes baked in different ovens. *International Journal of Food Science & Technology, 45*, 87–93.

Sumnu, G., Ndife, M. K., & Bayindirli, L. (1999). Temperature and weight loss profiles of model cakes baked in the microwave oven. *Journal of Microwave Power and Electromagnetic Energy, 34*(4), 221–226.

Sumnu, G., Sahin, S., & Sevimli, M. (2005). Microwave, infrared and infrared-microwave combination baking of cakes. *Journal of Food Engineering, 71*, 150–155.

Technical Trend (2009). Crustless without waste. *Baking + Biscuit, 3*, 28–30.

Therdthai, N., & Zhou, W. B. (2003). Recent advances in the studies of bread baking process and their impacts on the bread baking technology. *Food Science and Technology Research, 9*(3), 219–226.

Therdthai, N., Zhou, W. B., & Adamczak, T. (2002). Optimisation of the temperature profile in bread baking. *Journal of Food Engineering, 55*(1), 41–48.

Thorvaldsson, K., & Janestad, H. (1999). A model for simultaneous heat, water and vapour diffusion. *Journal of Food Engineering, 40*(3), 167–172.

Tong, C. H., & Lund, D. B. (1993). Microwave-heating of baked dough products with simultaneous heat and moisture transfer. *Journal of Food Engineering, 19*(4), 319–339.

Turabi, E., Sumnu, G., & Sahin, S. (2008). Optimization of baking of rice cakes in infrared–microwave combination oven by response surface methodology. *Food and Bioprocess Technology, 1*, 64–73.

Turabi, E., Sumnu, G., & Sahin, S. (2010). Quantitative analysis of macro and micro-structure of gluten-free rice cakes containing different types of gums baked in different ovens. *Food Hydrocolloids, 24*, 755–762.

Verboven, P., Datta, A. K., Anh, N. T., Scheerlinck, N., & Nicolai, B. M. (2003). Computation of airflow effects on heat and mass transfer in a microwave oven. *Journal of Food Engineering, 59*(2–3), 181–190.

Walker, C. E., & Li, A. (1993). Impingement oven technology-Part III: Combining impingement with microwave. *American Institute of Baking Technology Bulletin, 15*(9), 1–6.

Whorton, C., & Reineccius, G. (1990). Current developments in microwave flavors. *Cereal Foods World, 35*(6), 553–555.

Wilson, R. H., Goodfellow, B. J., & Belton, P. S. (1991). Comparison of Fourier transform mid infrared spectroscopy and near infrared reflectance spectroscopy with differential scanning calorimetry for the study of the staling of bread. *Journal of the Science of Food and Agriculture, 54*, 471–483.

Yaylayan, V. A., & Roberts, D. D. (2001). Generation and release of food aromas under microwave heating. In A. K. Datta, & R. C. Anantheswaran (Eds.), *Handbook of microwave technology for food applications* (pp. 173–189). New York: Marcel Dekker.

Yin, Y., & Walker, C. E. (1995). A quality comparison of breads baked by conventional versus non-conventional ovens: A review. *Journal of the Science of Food and Agriculture, 67*(3), 283–291.

Zallie, J. P. (1988). The role and function of starches in microwaveable food formulation. In: *Proceedings of MW foods, Chicago, Illinois.*

7 Microwave-assisted frying

R. Schiffmann
R.F. Schiffmann Associates, Inc., New York, NY, United States

7.1 Introduction

Microwave heating is rarely used by itself in the industrial processing of materials. This is largely due to the higher capital and operating costs of microwave systems by comparison to conventional heating-mode systems using hot air, steam, among others. There are of course exceptions, two of them are major processes in the food industry: microwave precooking of bacon and microwave tempering of meat, fish, vegetables, etc. However, the addition of microwave energy to a conventional heating system can often have significant benefits, such as:

- Increased speed of manufacture
- Improved product quality
- Improved production of difficult to heat materials
- Reduction in manpower
- Reduction in floor space
- Increased throughput
- Cost savings

In general, microwave energy can be combined with conventional heat transfer means in two ways: either as a sequential system in which the microwave energy is separate from the conventional heating process, i.e., either preceding or following it, as is common in microwave-assisted drying of materials (see Chapter 8: Microwave-assisted drying); or as hybrid systems, in which the microwave energy and the conventional heating modes are combined in a single unit, an example of which is microwave-assisted frying, the subject of this chapter.

7.2 Deep fat frying (deep frying)

Deep fat frying of food is a process in which food is completely submerged (or only partially submerged, i.e., doughnuts) in hot oil whose temperature is typically between 350°F (177°C) and 375°F (191°C). (While lower frying temperatures (140–180°C) are recommended for the prevention of acrylamide formation in rather dry products, this is not useful for doughnuts because they require significantly longer frying time, and the low oil viscosity at that temperature contributes to much greater fat absorption, making the surface oily, causing detrimental effects on eating quality, and poor coating stability.) Foods, such as chicken and fish, are coated with a breading or batter (cornmeal, flour, tempura) prior to being

The Microwave Processing of Foods. DOI: http://dx.doi.org/10.1016/B978-0-08-100528-6.00007-3

submerged in the hot oil. Once the coated food is submerged in the oil, the surface begins to dehydrate and undergoes Maillard reactions that break down sugars and proteins, creating the golden brown or brown exterior of the food. This dehydration forms a crust which prevents or limits oil absorption. The crust also acts as an "ablative shield"[1] that controls and protects overcooking of the product in a manner similar to the protection to overheating and destruction provided by the ablative-shield tiles used on rocket nose cones. In the preparation of doughnuts and similar dough-based items, the products are not coated prior to frying, but form their own coating, i.e., crust, during frying, as will be described below.

The appeal of deep fat frying as an industrial process is due to the efficiency of heat transfer, initially by conduction and then continuously assisted by convection as hotter oil adjacent to the heating elements replaces the oil at the surface of the object being fried that is being cooled due to heat losses and water evaporation. The thermal conductivity for cooking oils is approximately an order of magnitude higher than that of air (Zimmerman, 2006), and therefore, the close contact of the hot oil with the surface of the coating, or dough, provides rapid transfer of heat to that surface. A number of physical and chemical effects prevent the overheating or overcooking of the food. A good synopsis of the deep-frying procedure is provided by Eunice Liu (2015), in an amusing article about "Fair Food," such as deep-fried Twinkies:

- Phase 1: "Initial heating. … the food is completely submerged into the hot oil, until the surface of the food reaches the boiling point of water … While convection uniformly heats the food surface, it doesn't cook the center of the food. Rather, the food center is heated through conduction … (whereby the) heat diffuses from the hot surface into the core…" of the food.
- Phase 2: "Surface boiling. … the hot oil surrounding the food causes water inside the food to evaporate which in turn circulates the currents of frying oil and increases the rate of heat transfer… (at the same time) keeping the hot oil from soaking into the crust or surface of the food."
- Phase 3: "Decreasing heat transfer rate. As the crust continues to dehydrate, it conducts less heat to the rest of the food, so the rate of heat transfer through the escaping steam decreases. The remaining moisture inside of the food is slowly heated to the boiling point of water, which cooks the food…"
- Phase 4: "Bubble end point. This is the last stage of deep-frying … Water from inside the food is no longer evaporating … The fried product should be removed from the oil…"

There are a number of problems with industrial deep fat fryers:

- As the oil heats food, it undergoes a number of chemical and physical changes, manifested as deterioration of the cooking oil. Such changes include:
  - Hydrolysis, oxidation, isomerization, and polymerization. These result in the generation of free fatty acids, aldehydes, ketones, and other compounds (Bordin, 2013).
  - This results in an increase in oil viscosity which has a negative effect upon the heat transfer coefficient; and the smoke point of the oil.
- The frying of the coating or surface creates a barrier to efficient heat transfer, thereby extending the time necessary to complete the cooking of the food.

[1] Ablative shields protect surfaces from being heated excessively, utilizing chemical changes of the surface, and the evolution of gases that reduce the otherwise destructive transfer of heat.

- Water vapor created by the interaction of the hot oil and water in the coating or surface prevents the oil from reaching the interior of the coating or food, but at the same time reduces the heat transfer to the surface of the food.
- Blocking gases in the crust or coating, generally carbon dioxide, formed by the breakdown of the carbohydrates in the surface, also prevent efficient heat transfer from the oil.
- A manifestation of this reduction in heat transfer efficiency is seen when frying doughnuts. In this case, it results from a combination of the water vapor and blocking gas described above, with the ever-expanding dough creating a better and better thermal insulator as the doughnut fries. This will be explained in greater detail in the next section.

7.3 Microwave frying of doughnuts

This is a process whose end product is not achievable by any other means, and with a quality unmatched by the conventional process. In 1961, while employed at DCA Food Industries, Inc., a manufacturer of industrial doughnut processing machinery as well as the dry mixes necessary to produce doughnut dough, the author discovered the concept of frying doughnuts in oil while simultaneously exposing them to microwave energy (Schiffmann et al., 1972; Schiffmann & Roth, 1972). This combination was able to overcome a serious problem that prevents the full expansion of the dough during frying, thereby preventing the doughnut from reaching its full potential volume. In order to explain the importance of this problem and how it was solved by the addition of microwave energy, it is necessary to first describe the conventional frying procedure.

Industrial microwave fryers are essentially long stainless steel kettles containing the heated frying oil (see Fig. 7.1).

The oil is heated either electrically, or through indirect gas-firing, or using heat transfer-fluid tubes. Industrial doughnut fryers might produce up to 24,000 doughnuts

Figure 7.1 Typical doughnut fryer. Raw dough doughnuts are extruded into the left side of the fryer and fried doughnut exit at the right.
Source: Courtesy from W.H. Kemper GmbH, Rietberg, Germany.

per hour. Fryers contain metal conveyors that extend the length of the fryer and employ metal pusher-bars to transport the frying doughnuts through the hot oil. Conventional doughnut frying is a two-stage process in which the doughnut is fried for ~50 s on the first submerged side, and then flipped to the second side for an additional ~50 s of frying. By this point, only the lower half of the doughnuts is fully expanded (see Fig. 7.2). (Completely submerging the doughnut dough to allow for uniform browning is not possible, as it would restrict the overall processing volume.)

Once the second side of the mostly-raw dough is submerged in the hot oil, a crust forms rapidly, preventing this dough from expanding. It results in what is known as the "core"—an unexpanded portion of the doughnut, having roughly twice the density as the already expanded and baked dough. This core has also a firmer texture and tends to stale more rapidly. Above all, it prevents the doughnut from reaching its full potential volume (Schiffmann, 1971).

Through the addition of microwave energy to the doughnut fryer it was possible to bake the upper portion of the doughnut at the same time that the hot oil was frying and baking the lower portion of the doughnut. Thus, when the doughnut was flipped it was merely to brown that side because it had already been fully expanded, resulting in doughnuts that were as much as 25% larger (in volume) than the conventional. This is shown in Fig. 7.3.

7.3.1 The microwave-assisted doughnut fryer

A microwave-assisted doughnut fryer consists of a conventional doughnut fryer upon which a stainless steel multimode microwave applicator is bolted to the frame of the front half of the fryer. One or more microwave generators, each operating at 2.45 GHz, are coupled through waveguide(s) to the applicator. The amount of microwave energy required in doughnut frying systems is quite small. For example, a microwave-assisted fryer with a throughput of approximately 14,000 doughnuts per hour requires approximately 10 kW of microwave power. In order to maintain the best efficiency of microwave coupling to the doughnuts, i.e., keep the microwave energy requirements as low as possible, it was necessary to cover the surface of the cooking

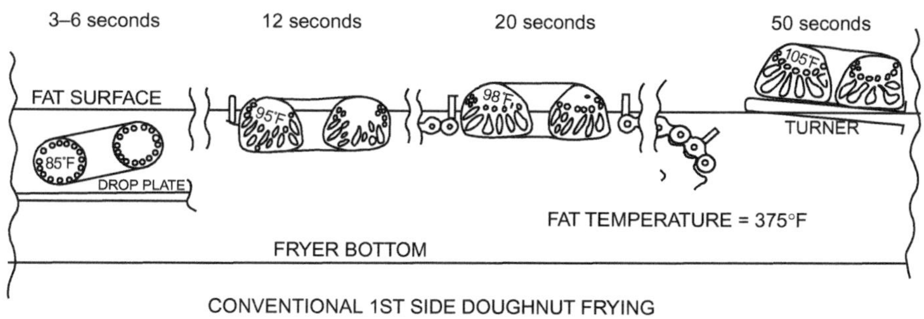

Figure 7.2 Conventionally fried doughnut. Note the cool and still raw dough (105°F) at the turner hand. Time 50 seconds per side.
Source: Courtesy from Institute of Food Technologists.

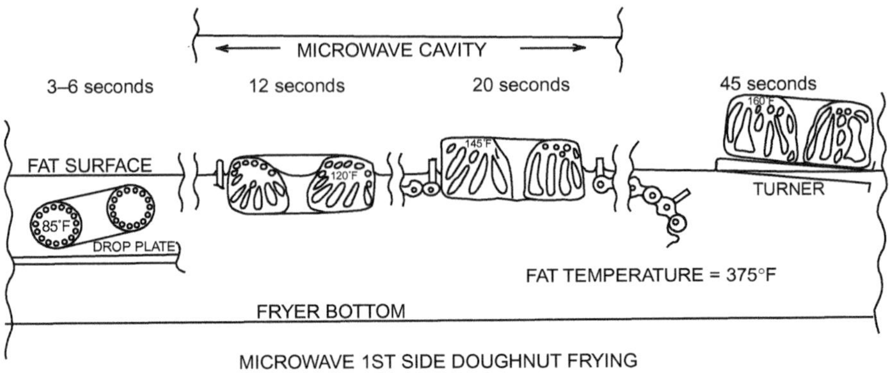

Figure 7.3 Microwave-assisted fried doughnut. Note the near-complete expansion of doughnut at turner and dough temperature 160°F. Time 45 seconds per side.
Source: Courtesy from Institute of Food Technologists.

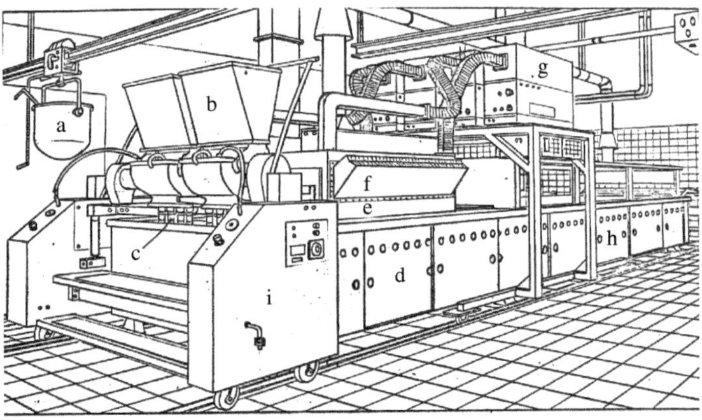

Figure 7.4 Typical microwave-assisted doughnut fryer. The components are: (a) mixed dough being transported to the extruder; (b) dough hoppers feed the dough to the extruder nozzles; (c) extruder nozzles also called cutters; (d) front half of the fryer; (e) microwave applicator; (f) microwave window (see Fig. 7.5); (g) microwave generators plus waveguides coupling them to the applicator (e); (h) second half of the fryer; (i) operator's control panel.
Source: Courtesy Institute of Food Technologists.

oil with doughnuts to its maximum capacity, thereby restricting the exposure of the oil to the microwave energy. In other words, an eight doughnut wide conveying system should contain eight doughnuts, since, if it only had six doughnuts, a great deal of the microwave energy would be spent in heating the oil rather than the doughnuts, a costly way of producing doughnuts and reducing the overall efficiency of the system.

Fig. 7.4 is an illustration of a typical microwave-assisted doughnut fryer. It shows the microwave-assisted doughnut fryer from the point, on the far left, at

Figure 7.5 "Microwave window"—viewing grid consisting of multiple aluminum tubes, 1″ × 1′ cross section and 6′ or more long. There is zero measurable microwave leakage at the viewing end (Belton, 2016).

which the raw toroid-shaped dough is extruded into the fryer, followed by the first half of the doughnut fryer with the microwave applicator bolted to the frame of the fryer, followed by stanchion upon which are located four 2.5 kW microwave generators with waveguides connecting them to the applicator, followed the rest of the non-microwave half of the doughnut fryer, all the way to the exit end at the far right of the illustration. The addition of the microwave hardware to the fryer created a number of unique engineering problems described below along with their solutions.

1. There is a good deal of art involved in the manufacture of doughnuts, so it is important for a baker to see the doughnut throughout its time in the applicator, from the moment it drops into the hot oil up to the point at which it is flipped. In order to do this with a microwave applicator, it was necessary to provide an unobstructed view, a microwave window to the entire surface of doughnuts in the microwave portion of the fryer. This was done by bolting a large number of waveguides, with dimensions small enough to restrict microwaves at 2.45 GHz to be transmitted, i.e., long rectangular tubes of approximately 1″ × 1″ opening and illustrated as a grid, better shown in Fig. 7.5.
2. Due to the deterioration of frying fat as described earlier, it is necessary to periodically empty deep fat fryers, replacing the used frying fat with fresh fat and to clean the fryer. For the microwave-assisted fryer it was necessary to first unbolt the microwave applicator from the fryer frame and then raise the applicator above the fryer with a hoist system. This allowed access to the fryer, and at the same time, permitted removal of congealed polymer (resulting from the vapors given off by the hot oil) from the interior walls of the applicator.
3. It was also necessary to unbolt the waveguides from the applicator. In order to assure that they were properly bolted again onto the applicator when it was restored to the fryer, a system was devised in which a positive air pressure, sensed by the generators, was maintained in each waveguide, so that the generators could not be operated unless the waveguides had been properly bolted to the applicator as evidenced by the proper air pressure having been achieved.
4. As noted above, the microwave generators were located remotely from the fryer, so a master control panel was provided for the baker/operator at the entry end of the fryer, through which he could control both the output powers of the microwave generators, and the fryer conveyor operation.

7.3.2 Advantages to the baker of microwave-assisted doughnut manufacture

- The significantly greater doughnut volume achieved by the use of the microwave-assisted frying process, from 15 to 25%, meant that a bakery was able to produce 15 to 25% more doughnuts from the same lot of microwave doughnut mix compared to the conventional mix. Alternatively, the baker could choose to sell doughnuts of much greater volume than the conventionally fried ones.
- A number of benefits were achieved by eliminating the core:
  - These doughnuts had significantly less fat absorption than the conventionally fried ones, providing a bakery with significant savings in the cost of frying fat.
  - The shelf life of the microwaved doughnuts was several days greater than the conventional ones, i.e., they did not stale as quickly.
    - This is due to the elimination of the core that stales faster than the remainder of the doughnut crumb. Also, the shorter frying time has an effect.
  - The sugar-based coatings employed on these microwave doughnuts had longer shelf life because there was lower fat absorption.
  - Frying time was reduced by as much as 20%, allowing for greater production volume per unit of time.

Fig. 7.6A and B illustrates the differences between the conventionally fried doughnut with its core (A) and the microwave-assisted fried doughnut in which the core is eliminated (B).

7.3.3 Commercialization of the microwave-assisted doughnut process

After over 2 years of in-plant testing, 25 microwave-assisted doughnut fryers were sold in the United States, Canada, and Europe beginning in the late 1960s and into the first half of the 1970s. In developing the dough to be used in the microwave doughnut fryer, and because microwaves provided a different

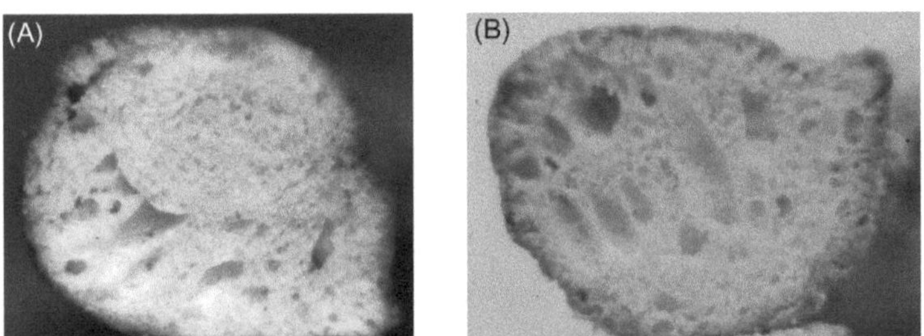

Figure 7.6 Comparison of conventionally fried doughnut (A) on left, with microwave-assisted fried doughnut (B) on right. Note the dense core in (A) that has been completely eliminated in (B).
Source: Courtesy from Institute of Food Technologists.

mechanism of heat transfer, we discovered that it was necessary to modify the amount and type of leavening acid; we also needed to strengthen the dough with additional gluten in order to control the degree of expansion in the microwave applicator section of the fryer. As a result, and since no conventional doughnut mix was capable of working successfully in the microwave environment, proprietary doughnut mixes were developed and sold by the company, to accompany the microwave-assisted fryers.

7.3.4 Current status of microwave-assisted doughnuts fryers

Currently there are no microwave-assisted doughnut fryers in operation anywhere in the world. They ceased operating sometime in the mid-1970s when, despite the satisfaction of many bakeries, DCA Food Industry, Inc. ceased to manufacture the microwave-doughnut mixes. This is an example of an innovation that was far ahead of its time. There were almost no domestic microwave ovens at this time, so the operators were generally unfamiliar and sometimes frightened by the concept of microwave heating. Furthermore, the company manufacturing the microwave-assisted doughnut machinery and mixes (DCA Food Industries, Inc.) also manufactured and sold conventional doughnut processing equipment which the salespeople were more familiar with and more willing to sell instead of the microwave equipment.

7.4 Other microwave-assisted frying processes

While microwave-assisted doughnut production has been the only successful application of this technology to date, there have been investigations of this microwave application for other food products. The author has done significant work on microwave-assisted frying of breaded chicken and breaded fish.

- Microwave-assisted frying of breaded chicken: in this case the desire was to increase the throughput of conventional continuous breaded chicken frying systems. It was found that it was possible to increase the throughput, i.e., the yield, of breaded chicken parts by as much as 50%. In other words, 50% more breaded chicken parts could be fried in the same amount of time by adding a microwave system to the line whose purpose was to cook the interior of the chicken while it was frying. Although the manufacturer was impressed by the demonstration of the technique, they were reluctant to adopt the new process.
- Microwave-assisted frying breaded fish products: once again, there was a desire by a manufacturer to increase the production volume by the addition of microwave energy to a conventional breaded fish frying process. The difficulty in this case was that it was necessary to employ a hold-down conveyor commonly used in this kind of fish frying process, and this presented some design challenges for the microwave process. As a result, the manufacturer was reluctant to continue investigating this new procedure.

7.5 The future of microwave-assisted frying in the food industry

It is surprising that the technique of microwave-assisted frying has not been adopted by the food industry despite the success of its original manufacturer. This lack of adoption is instructive:

- The timing of innovation is critical, i.e., innovation that appears too soon is unlikely to meet with great success. In this case, the use of microwave energy was so limited within industry that the salesmen were reluctant to sell the equipment. Users also had concerns as illustrated when one large baked goods supply company had great success with this process in its West Coast operation but failed in its East Coast operation because of roadblocks put up by the plant personnel. Today, given the wide use of domestic microwave ovens throughout the world, this should no longer be a barrier to adoption of this microwave-assisted frying process.
- Perhaps the major barrier to adoption is the requirement for intensive and extensive involvement by the user to achieve a successful microwave-assisted frying process. The most successful microwave processing systems are those in which the user is actively involved in their development. Microwave equipment manufacturers are generally incapable of developing innovative systems or processes on their own. In the case of the microwave-assisted doughnut fryers, most of the product and process development was done over several years by the personnel at DCA, while the supplier of the microwave generators and associated microwave hardware had a much smaller process development role. It was the laboratory personnel at DCA who understood the requirements for successful doughnut production and were able to utilize their skills to define not only the hardware requirements, but also to produce the "software," i.e., the unique doughnut dry mix, followed by two years of extensive in-plant testing and further development required to produce successful systems and product.

Microwaves have the unique ability of heating the interior of material, and when combined with another source of heat, such as hot oil, may produce products of equal or greater quality, at costs that are equal or less than the conventional procedures. This combination may also provide a greater production volume at a reasonable capital cost. Further, when the process involves a foam-like or similar structure that is developed during the heating process, microwave energy is often the best way to heat, bake, cook, or dry the interior in a time that is shorter than achievable in conventional processes, and usually much less destructive, i.e., avoiding case hardening or overheating the surface in order to heat the interior. Should companies wish to adopt microwave-assisted frying as a way of producing their products, there will likely be extensive research and development involved that will likely require the efforts of a dedicated team and may also require modification of the existing product or its components. However, the potential rewards are great in terms of product quality, process efficiency, and return on investment.

(The author wishes to thank Bob Swientek, Editor-in-Chief of Food Technology, for allowing the use of several figures from his paper (Schiffmann, 1971).)

References

Belton, J. (2016). *personal communication*, June 9, 2016.

Bordin, K. (2013). Changes in food caused by deep fat frying. *ALAN*, *63*(1).

Liu, E. (2015). *Fair food: The science of deep-frying*. https://scienceandfooducla.wordpress.com/2015/08/25/fair-food-deep-frying/.

Schiffmann, R. F. (1971). Applications of microwave energy to doughnut production. *Food Technology*, *25*, 58–62.

Schiffmann, R. F., et al. (1972). *Apparatus for producing cooked products*. U.S. Patent 3.633, 490.

Schiffmann, R. F., & Roth, H. (1972). *Method of producing cooked products*. U.S. Patent 3,679,432.

Zimmerman, B. (2006). *Heat transfer and cooking*. http://www.cookingforengineers.com/article/224/Heat-Transfer-and-Cooking.

Further reading

Farkas, B. E., et al. (1996). Modeling heat and mass transfer in immersion frying. I. Model development. *Journal of Food Engineering*, *29*(2), 211–226.

8 Microwave-assisted drying

V. Gaukel[1], T. Siebert[1] and U. Erle[2]
[1]Karlsruhe Institute of Technology (KIT), Karlsruhe, Germany,
[2]Nestlé R&D Center, Solon, OH, United States

8.1 Introduction

Drying is among the oldest processing techniques for food and its original purpose is to conserve perishable food in an edible state for a longer time. The deterioration of foods is based on chemical and biological processes that are mainly driven by free or unbound water inside the product and/or the presence of oxygen (Barbosa-Cánovas & Vega-Mercado, 1996).

However, prolongation of shelf-life is not the only reason for food drying. Sokhansanj and Jayas (2015) list several other aspects, such as the development of certain product quality parameters, e.g., color and taste. Certain changes in those quality parameters can be beneficial for the overall customer acceptance of the product. These changes can be summarized as "Quality Enhancement." The "Ease of Handling" is another factor that is beneficially influenced by drying processes. Packaging, storing, handling, and transportation of dried goods are often easier and/or less expensive compared to fresh products. Reasons for this are reduced weight and volume and, especially for powders, better flow characteristics. "Further Processing" procedures, such as milling, mixing, and segregation can also be improved by drying processes. Dried products are less sticky during mixing processes and milling of dried goods consumes less energy. Furthermore, some microorganisms, especially meso- and cryophilic organisms, as well as insects can be partly inactivated, depending on the process conditions.

Due to the fact that drying has been conducted for thousands of years all over the world, many different drying processes have been developed. The most important ones are hot air-drying (AD), freeze-drying (FD), vacuum-drying (VD), solar drying (SD), and osmotic drying (OD). Most of those processes can be assisted by microwaves, and especially microwave vacuum-drying (MVD) has become an important industrial drying process. To understand the particular benefits of microwave drying processes, this chapter will first discuss the principles of drying processes using the example of AD.

8.2 Principles of drying processes

AD is a common and well-established industrial drying process. As idealized in Fig. 8.1, a typical drying curve of a foodstuff can be subdivided into three periods.

The Microwave Processing of Foods. DOI: http://dx.doi.org/10.1016/B978-0-08-100528-6.00008-5

After an initial phase (IP), a steady state is reached in which the drying rate (evaporated water over time) is constant. Therefore, this period is called "constant rate period" (CRP). In this period the product surface is still wet and the passing airflow transfers the heat for water evaporation to the surface. Simultaneously, the air is collecting the water vapor and is transporting it away from the product surface out of the dryer. This is a simultaneous heat and mass transfer. To keep the surface wet, enough water has to flow from the inside of the food to its surface. In the constant rate period this water flow is mainly capillary-driven and the surface temperature of the product is theoretically constant. As the water evaporation consumes energy, the surface temperature is typically significantly lower than the air temperature. The temperature profile in Fig. 8.1A is adopted from a hot air-drying process of carrot slices and it shows that this constant temperature period is not necessarily exactly reflected in real data. The factors that determine and limit the drying rate in the CRP are the state of the air (temperature and relative humidity) as well as air velocity. By changing any of these parameters the drying process may be accelerated (or delayed) considerably.

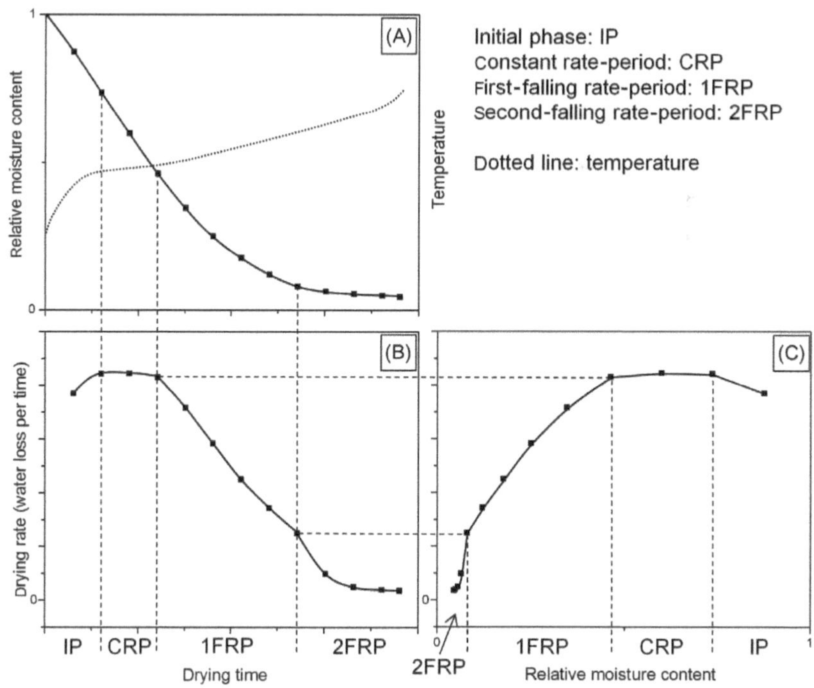

Figure 8.1 Drying curves adopted from experimental hot air-drying data of carrots to illustrate typical drying curves with initial-heating-, constant-, first-falling- and second-falling rate-period in three typical diagrams. (A) (relative) moisture content over drying time together with typical temperature curve. (B) drying rate over drying time, and (C) drying rate over (relative) moisture content.

This changes in the next drying period, the so-called "first falling rate period" (1FRP). This period starts when the surface becomes dry, as water from the inside of the product cannot be transported fast enough to the surface. Therefore heat must be transported inside the product to enable water evaporation and the water vapor must be transported through the dry product area to its surface. This means that the resistance for heat and mass transfer increases, and consequently the drying rate decreases and the surface temperature rises slowly. As the area where evaporation takes place retracts more and more to the center of the product, the drying rate slows down continuously. It is not easy to increase the drying rate in this period. To enhance heat transfer, higher temperatures or higher air velocities have to be applied, both leading to a higher product surface temperature. The product may then suffer from overheating as there is no cooling effect from water evaporation, as seen in the constant rate period. The product may also dry out too fast in the outer layers, causing a phenomenon called "case hardening." This leads to the formation of a hard skin around the product that may hinder the further drying process or lead to product cracks.

The "second falling rate period" (2FRP) starts when all free water is evaporated and only bound water is left. For the drying of bound water, an additional desorption energy is necessary, which is a further resistance for the drying process. Whether or not this period is reached in industrial drying processes depends on the hygroscopicity of the product and the target moisture of the process. A 2FRP is only visible when the drying rate is analyzed (see Fig. 8.1B and C) and not from Fig. 8.1A. The slope of the curve in Fig. 8.1C can be different depending on the dried product (Mujumdar, 2015). If the process continues, the temperature increases to the temperature of the hot air.

For industrial food drying processes it is difficult to determine the absolute length of the individual drying periods, because there are no sharp limits between them. As every piece of food is individual and the process conditions are not the same at all positions in the drying chamber, not all food particles are in the same drying period at the same time. Therefore, it is a rather fluent transition between the drying periods, and the characteristic value to compare drying processes is commonly the time for the whole process until the product has reached its final moisture content. The drying time varies with process conditions and product. At which water content an acceptable dry state is reached depends on the intended purpose of the product (Jangam, Law, & Mujumdar, 2010).

8.3 Specific aspects of MW-assisted drying

As described in Chapter 1, Introducing microwave-assisted processing of food: fundamentals of the technology, microwaves can transmit energy in different ways. Their unique feature is that they provide heating inside the product. The extent of MW absorption depends on dielectric properties, size and shape of the product, and the existence of other materials which absorb the microwaves in competition and therefore influence the local field strength.

8.3.1 Heating mechanism and selective drying

As described in Chapter 2, Microwave heating and the dielectric properties of foods, the extent to which microwaves are absorbed by a material depends on its dielectric properties. It is the imaginary part of the dielectric constant, ϵ'', that determines the heat generated in the dielectric by an electromagnetic field of a certain strength and frequency. As water molecules—together with dissolved ions—have the main influence on ϵ'', microwave absorption usually decreases with decreasing water content of the material. The real part of the dielectric constant, ϵ', is also governed by water. Fig. 8.2 shows this trend for apple pieces, but the curves for foodstuffs all show more or less the same behavior. Reading from right (high moisture) to left (low moisture), we can see a slight increase of ϵ''. It seems strange that the samples lose a lot of water without a loss in ϵ'' but this effect is leveled out due to the rising concentration of dissolved ions and the shrinkage. At lower moisture levels (below 0.6), the absence of water causes ϵ'' to drop considerably. The material is now absorbing the microwave energy to a significantly lesser extent. When dry and wet food parts are coexisting, e.g., in the falling rate periods, it leads to a selective heating of water rich parts and is referred to as "leveling effect" (Zhang, Tang, Mujumdar, & Wang, 2006). Besides other reasons, such as focusing effects, this may lead to hot and cold spots or even a reverse temperature gradient in the product (center warmer than the surroundings). The heating with microwaves overcomes the heat transfer limitations of other heat sources, accelerates the evaporation of water and consequently the mass transfer rate. In microwave drying, the drying rate is proportional to the absorbed microwave power in all phases of the process. This is what makes microwave drying interesting: the larger the products, the more important becomes the differences in heat and mass transfer and the more obvious are the key advantages of microwaves.

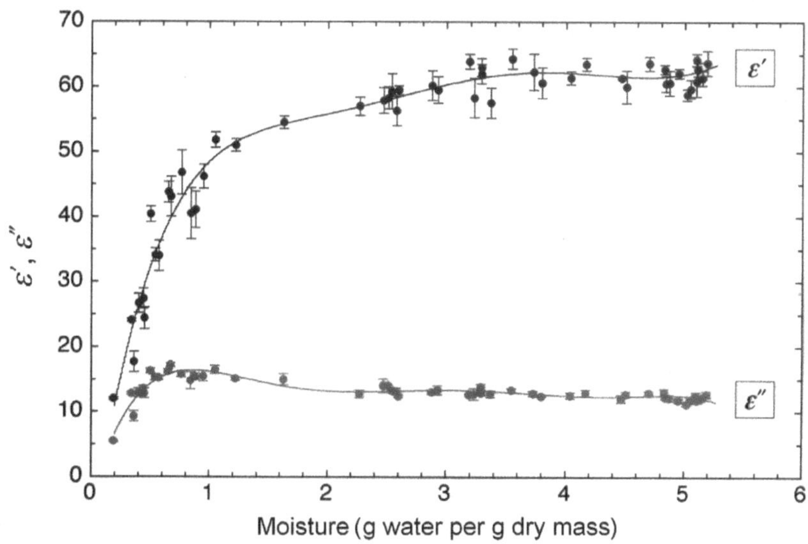

Figure 8.2 Dielectric properties of apples as a function of moisture (Erle, 2001).

As the amount of available water decreases over the course of the drying process, at a certain point, the efficiency of the microwave drying process will decrease. If the microwave power is not adjusted at this point of the drying process, the electromagnetic field strength increases, because the less microwaves are absorbed in the drier, the higher the field strength becomes. This can create other problems, such as thermal runaway, arcing, or plasma formation.

8.3.2 Thermal runaway

Like other polymers, many biopolymers in food show a steep increase in ϵ'' when exceeding a critical temperature. As the substance gets soft with increasing temperature, the molecules can move more freely and thereby absorb more microwaves, which leads to a self-accelerating process. This can also be seen for melting ice during heating of frozen products (see Chapter 12: Microwave-assisted thawing and tempering). This phenomenon is known as "thermal runaway." An unstable system, which bears the risk of sudden overheating and burning of the product, is created. Without vacuum, it may even cause a fire. As a consequence, one is limited with the microwave intensity toward the end of the drying process, because the evaporation of water may not cool the product enough to prevent a thermal runaway. An alternative way to deal with the low dielectric properties at the end of the drying process is to increase the amount of product exposed to the microwaves, for instance by creating a thicker product layer on a belt.

8.3.3 Plasma formation and arcing

Another reason to limit microwave power during drying has to be considered, especially if a vacuum is applied: a breakdown or discharge of the electrical field. This is basically a flow of electrons or air ions, which may even result in visible phenomena such as plasma or arcing. The field strength necessary for plasma formation at 2.45 GHz is reaching its minimum at a pressure of approximately 2.5 mbar (see Chapter 17: Improving the heating uniformity in microwave processing). While effects like plasma are desired in some technological fields, they are usually unwanted in microwave drying. Plasma formation consumes energy and may damage the equipment as well as the product. For instance, Teflon parts can turn black and erode due to plasma formation. Foods can get thin black crusts, making them unsuitable for consumption. Metaxas and Meredith (1983) have provided a quite detailed description of breakdown effects along with their causes. Important factors are electrical field strength, pressure, and the type of gas in the atmosphere. Local field strength is of course a function of microwave power setting, drier design, mass and dielectric properties of the product, etc. As poorly absorbing products lead to high field intensities, it is often necessary to reduce microwave power toward the end of the process. Once a plasma has been created, it requires much less electrical field strength to be maintained. That is why it is so important to prevent peaks in the microwave intensity. The use of "low ripple" magnetrons with a steady power supply or a solid state microwave source may reduce the problem.

8.3.4 Microwave field distribution

Another reason not to exaggerate microwave power during drying is the distribution of the microwave field. Although considerable progress has been achieved in drier design, the microwave field is never completely even. The occurrence of "hot" and "cold" spots is described in Chapter 18, Simulation of microwave processes. Some manufacturers use mode stirrers to distribute the waves, but it is always a good idea to move the product, so that all parts of it get approximately the same amount of energy. This is why household microwave ovens usually have a turntable. Using microwaves in a "modest" way means that there is more time for heat conduction to level out any unbalanced microwave absorption. A simple way to improve the uniformity of the microwave field is to increase the size of the drying apparatus as the number of modes increase with the apparatus size.

8.3.5 Product size

Fig. 8.3 illustrates one more potential pitfall in microwave drying of particulate foods. This curve for Mie's absorption coefficient stems from a calculation based on spheres of various diameters, which have the dielectric properties of water in an electromagnetic field with the commonly used frequency of 2.45 GHz; the corresponding wavelength is around 12.2 cm. A value of 1 for this coefficient means that 100% of the incident electromagnetic energy is being absorbed by the object. The calculation reveals that the absorption of electromagnetic waves is increasing slightly with decreasing radius down to approximately 4 cm. In smaller spheres resonant interaction and focusing effects lead to fluctuating values and even values higher than 1. But for small spheres with a radius less than 0.5 cm, the absorption is very poor.

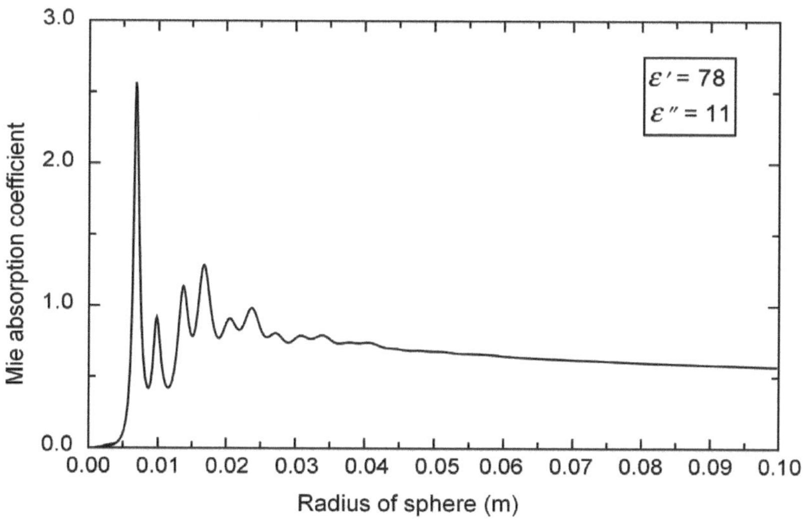

Figure 8.3 Mie's absorption coefficient calculated for spheres of water (Erle, 2001).

This means that a small wet object may not receive enough microwave power and will not be heated. This effect is of practical relevance if the design and operation of the drier allows for single particles. As long as the particles are in contact with each other and form a bulk, the effect is not significant. Nevertheless, due to the focusing effects, objects within a radius size of 0.5 and 3 cm may experience uneven heating or overheating. These effects influence the microwave absorption especially during the falling rate period of a drying process when the amount of water decreases and the wet part of the product shrinks. In this phase, the water-rich parts decrease, and the microwave absorption will undergo all the limitations discussed above.

8.4 MW in drying processes

8.4.1 General aspects of microwave application

Microwaves are used as single or additional heat source in drying processes. The most important use of microwaves as heat source in drying processes are in AD, VD, and FD. In the following subsections these three processes will be discussed.

Most publications in this field investigate the drying time and report one or two further product parameters like rehydration, shrinkage, ingredient retention, color retention, and hardness, before and after the drying process. The fact, that there is no standard procedure leads to numerous possible ways to investigate quality and process parameters and often makes the direct comparison of results from different authors not possible. In addition, microwaves can be applied in a drying process in various ways, i.e., *Continuous application*, *intermittent/pulsed,* or *temperature controlled.*

Continuous application means that a certain power is used, regardless of any parameters other than the end-conditions of the process. These can be time, weight, or temperature.

Intermittent or pulsed application describes a process which introduces the microwaves in periodic repetitions. The frequency of the pulses can be constant or changing.

Temperature controlled systems regulate the power of the introduced microwaves depending on the product surface temperature. The temperature is hereby commonly measured with infrared temperature sensors. This makes it possible to keep the product temperature more or less constant.

In most of the available literature, the way of microwave application is not described in detail. This is the reason why the influence of the above described ways of microwave application on the drying process cannot be discussed further.

8.4.2 Microwave-assisted hot air-drying

Conventional hot air-drying (AD) is the most common drying process, because of its low investment and operating cost (Soysal, Ayhan, Estürk, & Arkan, 2009) and ease of operation (Megías-Pérez, Gamboa-Santos, Soria, Villamiel, & Montilla, 2014). It accounts for approximately 85% of all industrial drying operations

(Mujumdar & Beke, 2003). Relatively long drying times (roughly 1–7 h, depending on temperature, moisture content, and product structure) and high air temperatures (up to 180°C) lead to a degradation of valuable ingredients (Gamboa-Santos et al., 2014; Kowalski, Szadzinska, & Lechtanska, 2013; Mclaughlin & Magee, 1998). Product structure is another characteristic that is drastically changed during hot air-drying. Volume retention of hot air-dried foodstuffs reaches values between 15 and 30% (Aversa, Curcio, Calabrò, & Iorio, 2012; Bruijn & Bórquez, 2014; Witrowa-Rajchert & Rzaca, 2009). The formation of a hard crust (case hardening) at the outside of the product can be mentioned as a disadvantage. Taste and color are changing as well and may be beneficial or detrimental depending on the product and the consumer's taste.

When *microwaves* are used as heat source for an air-drying process, the airflow enables the removal of outside moisture of the product, while the volumetric heating of the microwaves leads to an accelerated moisture-transport from the inside of the product to the outside. This is especially advantageous in the falling rate period where the heat transfer is limited in a classical air-drying process. This significantly decreases drying time. Several research groups found a reduction of drying time for many different products, like carrot slices (25–90%) (Prabhanjan, Ramaswamy, & Raghavan, 1995), kiwifruit (40–89%) (Maskan, 2001), bananas (64.3%) (Maskan, 2000), cranberries (60–82.5%) (Beaudry, Raghavan, & Rennie, 2003), garlic (up to 80%) (Sharma & Prasad, 2006), oyster mushrooms (75%) (Bhattacharya, Srivastav, & Mishra, 2015), and collard greens (up to 70%) (Alibas, 2009). The most investigated quality parameters were rehydration, which could be improved for carrots (Prabhanjan et al., 1995), kiwifruit (Maskan, 2001), bananas (Maskan, 2000), and garlic (Sharma & Prasad, 2006) and color retention, which was better for banana (Maskan, 2000), garlic (Sharma & Prasad, 2006), and collard greens (Alibas, 2009). The air used in these processes can be either hot (e.g., 80°C) to prevent surface cooling or cold (e.g., 20°C) to prevent overheating of the surface.

8.4.3 Microwave vacuum-drying

Vacuum-Drying is generally recognized as a method for gentle drying of foodstuffs (Durance & Wang, 2002). Reduced pressures lead to evaporation of water at lower temperatures, which causes only few chemical, sensorial, and nutritional changes (Drouzas & Schubert, 1996). Orikasa et al. (2014) found that VD of kiwi fruit is superior compared to AD in terms of ascorbic acid and color retention. The drying times of vacuum-drying (up to 48 h) are among the longest of all drying processes. The main limitation is the energy transmission. As hot air is not available, contact plates, radiation heat, or microwaves are commonly used.

Microwaves are an effective way to transmit energy in this system, because of the volumetric character of the heating process. This helps again in particular in the falling rate drying periods and results in one of the shortest process times of all drying processes.

Fig. 8.4 shows experimental curves from an MVD process of carrots at three different power levels at a pressure of 50 mbar using 250 g of carrot slices with a thickness of

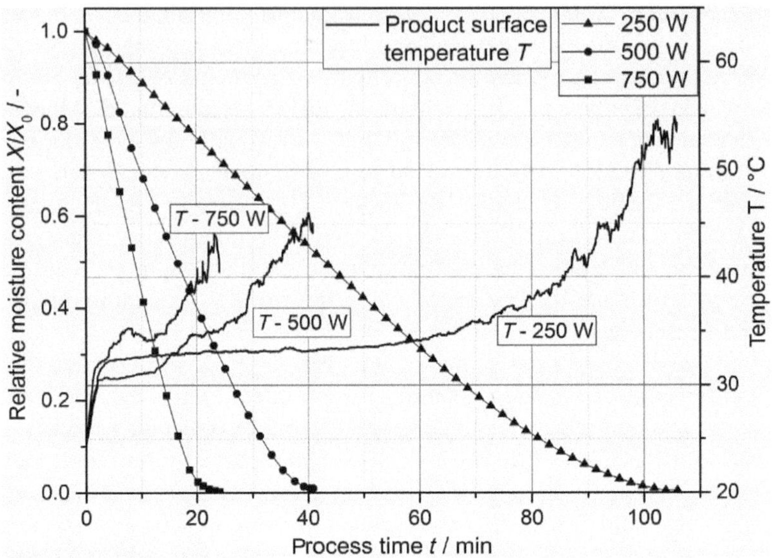

Figure 8.4 Drying curves with surface temperatures in microwave vacuum-drying (MVD) of carrots at a pressure of 50 mbar and three power levels (Siebert, Gall, Schuchmann, & Gaukel, 2016).

5 mm. Drying experiments were conducted in a modular drying processor with a rotating drum as described in Rother, Steimle, Gaukel, and Schuchmann (2011). After a short heating time to reach the boiling point, the surface temperature stays between 30 and 35°C until more than half of the water has been removed (the boiling point of pure water would be 32.9°C at this pressure). When the product surface becomes dry, the temperature starts to rise until the process is finished.

The key difference from an air-drying curve is that the drying rate in MVD is proportional to the absorbed microwave power in all phases of the process. In some literature (Rother & Schuchmann, 2009), it is mentioned that the drying rate is proportional to the applied microwave power but this is only valid if all the applied microwave power is absorbed by the product and the product temperature is comparable, which is not always the case especially during the end of the drying process.

Fig. 8.5 shows the drying rates of the carrot-drying experiments shown in Fig. 8.4. The basic shape appears similar to typical hot air-drying curves (see Fig. 8.1C), but the underlying physical principles are not. Reading them from right to left (as the process takes place) the curves have a slight initial increase, because some energy is needed to reach the boiling point. This is followed by a long slight decline of the drying rate. Thus showing that we can speed up the process even in the "falling rate period" simply by the application of more microwave power. As long as there is enough water left, the product temperature is controlled only by the pressure level and the corresponding boiling temperature. As described in Section 8.3.1, at lower moisture levels, the absence of water causes ϵ'' to drop considerably. The material is now much less inclined to absorb the offered energy, and

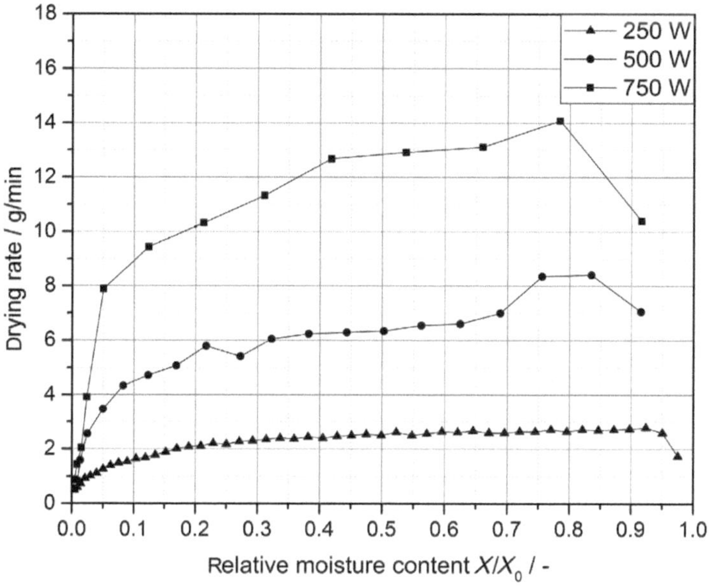

Figure 8.5 Drying rate over relative moisture content during MVD of carrots at three different power levels.

the drying rate decreases. The excess energy may be absorbed by the apparatus, or go back to a safety water load or be absorbed by the dry product parts leading to a fast temperature increase (see Fig. 8.4) or cause other side effects described in Section 8.3.1. Nevertheless, the drying rate is still higher than in an air-drying process as the water-rich parts are still selectively heated.

As described above, not all of the applied microwave energy is necessarily absorbed by the product. Fig. 8.6 shows the absorbed microwave energy versus relative moisture content during the above shown drying process. First, it is interesting that even in the beginning of the drying process only approximately 70% of the applied microwave power is absorbed by the product. This absorption rate depends strongly on the relation of product (water) volume to the volume of the cavity and can also be adopted to values close to 100% by tuning the cavity (see Chapter 1: Introducing microwave-assisted processing of food: fundamentals of the technology). But during the drying process the water-load will change and the absorption rate decreases. This is especially pronounced at higher microwave power levels where at the end of the drying process only 30% of the applied energy is absorbed. For reduction of energy losses it would be advantageous to reduce the microwave power at the end of the process or to adjust the tuning of the cavity automatically.

Higher drying rates at comparatively low temperatures contribute to a higher retention of valuable ingredients, which was shown by Cui, Li, Song, and Song (2008) for the ascorbic acid and carotenoid content of carrots (Mayer-Miebach, Behsnilian, Regier, & Schuchmann, 2005). Both ingredients had a retention ≥89%, which was in both cases over 20% higher compared to AD.

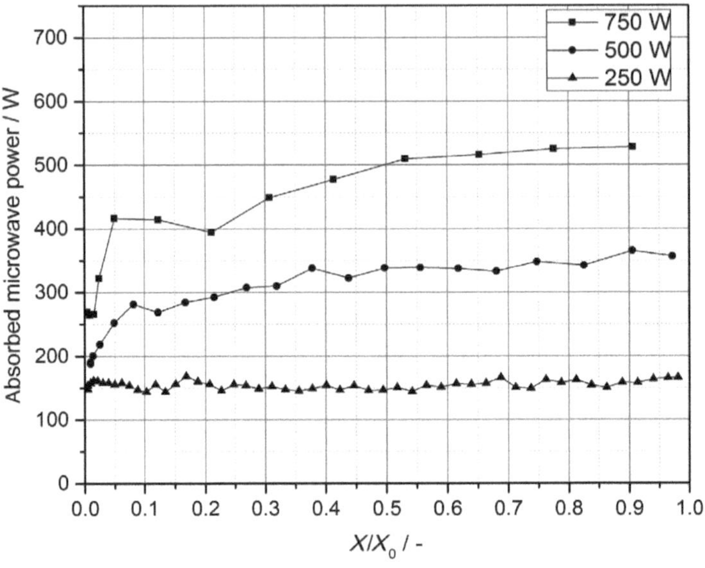

Figure 8.6 Absorbed microwave energy versus relative moisture content during MVD of carrots at three different power levels (applied microwave energy).

Volume retention is better compared to hot air-drying which was already shown for strawberries (Bruijn & Bórquez, 2014), carrots (Cui et al., 2008; Rother et al., 2011), and apples (Sham, Scaman, & Durance, 2001). One reason for that is an effect called puffing, which occurs when the evaporation rate inside the product is higher than the transportation rate of steam to the outside of the product. The resulting overpressure leads to a product expansion which results in a higher volume retention of the final product. This is illustrated in Fig. 8.7 where microwave vacuum-dried and puffed carrot cubes are compared with air-dried and freeze-dried ones. As expected, FD produces very porous, though a little pale products, whereas microwave puffing yields a stronger color, but the porous pieces look more like pillows, sometimes with only one air bubble inside. The AD control samples exhibit the usual problem of extensive shrinkage. The rate of rehydration is best for freeze-dried products. This is because they have an open pore structure and, due to capillary effects, draw up water within seconds. Puffed particles, although quite porous, usually have a closed surface and cannot make use of the capillary effect. Still they are superior to air-dried qualities, not necessarily for a higher rehydration rate, but because they get soft quicker in hot water. This is described in detail in Rother et al. (2011).

Furthermore, oxidation reactions are diminished, due to the low pressures and thus the low partial pressure of oxygen. This circumstance is beneficial for the color, which was proven by Yongsawatdigul and Gunasekaran (1996) for cranberries and Cui et al. (2008) for carrots.

Figure 8.7 Carrot cubes dried in a freeze-drying (FD), microwave vacuum-drying (MVD) and an air-drying (AD) process.

In general, the previously mentioned attributes may lead to a product quality which is sometimes on par with freeze-drying, but in most cases somewhere in between freeze- and hot air-drying. Special effects on aroma retention during MVD and microwave processes in general are addressed in Section 8.5.

8.4.4 Microwave freeze-drying

Freeze-drying (FD) can be described as a special form of vacuum-drying. The product is frozen in a first process step and subsequently dried in a vacuum chamber using a pressure below the triple point of water. This enables the water to sublimate directly from the solid phase into the gaseous phase. Commonly, heat is transferred like in vacuum-drying processes by contact heating plates or radiant heat or both. Usually, plate temperatures between 30°C and 40°C are being used for contact heating or up to 150°C for radiant heating. This results in low product temperatures throughout the whole process, which resolves in a gentle process. This is why freeze-drying in general has the highest retention rates for valuable ingredients of all drying processes. Special quality aspects are discussed in Section 8.5. In addition the supportive character of the still frozen parts of the product is responsible for the highest volume retention possible. This goes along with a very porous structure, which lets the products appear a little pale, compared to fresh ones. When it comes to preserving the fresh product in its current state, freeze-drying is generally acknowledged as the best drying process. Retention rates for ascorbic acid and carotenoids in carrots were reported to be above 95% when compared to the fresh value (Cui et al., 2008). The volume retention has been found to be between 70% (Rother et al., 2011) or even 80% (Cui et al., 2008), again with carrot as the investigated product. For other products, like quince (Koc, Eren, & Ertekin, 2008) or pumpkin (Nawirska, Figiel, Kucharska, Sokol-Letowska, & Biesiada, 2009), values of over 90% are documented. These values are representative also for other products. Because of this fact, FD is mainly used for high-value fruit and vegetable foods

(Huang, Zhang, Yan, Mujumdar, & Sun, 2009). The reason for that is the high energy demand due to the need for a continuous vacuum and many phase transitions (freezing, sublimating, and condensing/freezing) throughout the whole process and the long drying times of commonly between 12 and 48 hours. FD is also widely used for the drying of pharmaceutical products (Liapis & Bruttini, 2015).

The main drawback of FD is the limited heat transfer rate as there is no constant rate drying period. As water is frozen and not liquid, it is not transported to the surface by capillary forces but only by diffusion. In addition, a dry and porous structure is built up in the outer product parts directly after surface drying. This leads to a low and decreasing heat transfer coefficient. It is obvious that microwaves instead or in addition to heating plates can be useful. This method was already invented in the late 1950s, and it has shown that an acceleration of the process is possible compared to standard FD (Copson, 1958). Regrettably, some major drawbacks have prevented microwave FD from replacing ordinary FD. Apart from uneven heating due to the nonuniform distribution of microwaves, especially plasma formation (see Section 8.3.3) is a massive problem at the low pressures, because it can lead to localized thawing of the product (Wray & Ramaswamy, 2015). When unfrozen parts are present besides frozen parts, the unfrozen parts are preferentially heated due to the high differences in the dielectric properties of water and ice. This leads to a self-enforcing melting and structure collapse of the product.

Nevertheless, at low energy levels microwave-assisted FD can be much faster than conventional FD. Experiments of FD of starter cultures (*Lactobcillus paracaseii*) showed that a conventional FD of *L. paracaseii* took 840 minutes to reach a final water content of 4.5% in a 100 g batch at 1 mbar and 30°C contact heating plate temperature compared to 340 minutes in a microwave-assisted FD at the same conditions where the product was initially heated with 200 W (2 W/g) microwave power to a surface temperature of 30°C. The bacterial survival ratio in the MW-assisted process was 86% of the one in the FD process at less than half of the process time. These results were achieved in the frame of a research project (AIF 17477) financed by the German Ministry of Economics in cooperation with Arbeitskreis industrieller Forschung AIF and Forschungskreis der Ernährungsindustrie e.V. (FEI) using the drying processor described in Rother et al. (2011).

Recently, the problem of uneven heating in FD has been tackled by moving the product in rotating drums. This, however, is not recommended for products that are sensitive to abrasion (see also Chapter 17: Improving the heating uniformity in microwave processing).

8.4.5 Microwave drying in serial combinations

MVD may also be used as one step in a chain of dehydration processes. For example, when combined with an osmotic pretreatment, much of the water can already be removed by keeping fruits or vegetables in a concentrated solution for a few hours (Zhang et al., 2006). Meanwhile, some of the solutes, like sugar or calcium ions, may penetrate the tissue and have beneficial effects on the quality of the final product. For example, it has been found that sucrose and calcium chloride from an osmotic treatment reduce shrinkage of apples and strawberries during subsequent MVD (Erle & Schubert, 2001).

Furthermore, microwave-drying processes, especially MVD, is being used in serial combination with AD and FD processes. The combination of MVD and AD is perhaps the most obvious option: as AD is relatively cheap, the first part of the process is preferably an air-drier. During the constant rate period, the drying rate is acceptable and the product remains cool due to the cooling effect of the evaporation. When the falling rate period starts and AD becomes inefficient, microwaves and vacuum can be used to overcome the heat and mass transfer limitations as described above (Cui, Xu, & Sun, 2003; Figiel, 2010). Such a process was developed by Battelle Ingenieurtechnik GmbH (Eschborn, Germany) and implemented by Zifru (Zittau, Germany). In this case, the air-drying step was designed to create a barrier on the outside of the particles so that the following microwave process produced a puffing effect.

Apart from this combination, especially a combination of MVD with FD is part of more recent research (Huang et al., 2009; Kumar, Radhakrishna, Nagaraju, & Rao, 2001; Pei et al., 2014; Rother et al., 2011; Xu et al., 2005). FD is, as mentioned above, a rather expensive and time-consuming process. By combining it with another drying process, like AD or MVD, it would be possible to reduce the amount of water before FD, and this may lead to a cost and drying time reduction. A crucial factor for this is the development of the product quality during the individual drying processes, because such a serial combination is only relevant if quality parameters, such as volume retention, rehydration properties, color and ingredient retention are better, or energy consumption is lower compared to single-drying processes.

Rother et al. (2011) found that the order of combination is of great importance. They combined FD of carrots either with AD or with MVD. FD could be the predrying or the finish-drying step. A predrying with hot air or microwave vacuum-drying led to a significant shrinkage of the product, which could not be made up for in the finish-drying step. Furthermore, achieved time-savings during the first phase of the process were more than compensated by a longer finish-drying phase. The reason for that is the formation of an outer crust, which acted as a diffusion barrier for the escaping moisture. It was found that it is much more beneficial for the product, in terms of volume retention, rehydration properties, and drying time, when freeze-drying is used as a predrying step. In this combination, the structure of the product was preserved in the predrying step, which allowed an even faster finish-drying step due to a very porous, already dried outer layer, which allowed the moisture to escape more quickly. This was investigated at fixed cross-over points (relative moisture ratios at which the change between the processes is conducted) between the processes. However, the cross-over point is very important in terms of product quality and drying time. Table 8.1 shows the influence of cross-over points between FD and AD as well as between FD and MVD on volume retention, rehydration ratio, and drying time of carrot cubes. The experimental procedures were the same as described in detail in Rother et al. (2011). 250 g carrot cubes of 9 mm length were dried in the drying processor at 50 mbar and 515 W. AD was

Table 8.1 Influence of cross-over points (relative moisture ratios in %) between FD/MVD and FD/AD on selected quality attributes and drying time for drying of carrot cubes

Process	Drying time (*t*/min)	Volume retention (V/V_0/%)	Rehydration ratio after 10 min
Pure FD	551	68.4	0.68
Pure AD	266	12.2	0.3
Pure MVD	68	24.4	0.37
FD10%-AD	493	57.2	0.5
FD27%-AD	442	39.2	0.44
FD44%-AD	376	24	0.38
FD61%-AD	350	16.2	0.3
FD10%-MVD	441	67.5	0.46
FD27%-MVD	336	48.7	0.41
FD44%-MVD	269	31.7	0.38
FD61%-MVD	204	22.2	0.41

conducted at 95°C air temperature. FD was performed under plate heating at a temperature of 30°C and a pressure of 2 mbar. The volume retention was measured as a bulk volume in a measuring cylinder before and after the drying, the rehydration ratio is the product weight after 10 minutes in 80°C water relative to the fresh weight before drying. The finishing point for all experiments was when the drying rate was less than 0.03 g/min. FD was always used as predrying step and the % value after FD gives the relative moisture content at which the cross-over between the processes was conducted. This means the higher the value, the shorter is the predrying FD process.

Considering the drying time, it can be seen that as expected for the pure drying techniques, FD takes significantly longer than AD, and MVD is by far the fastest process. In addition, it can be derived in general that as more FD is involved in a particular process combination, the longer takes the drying process. As pure FD develops the best product structures in terms of volume retention and rehydration ratio, also the combination processes with the longest FD part produce the best qualities. For volume retention this was the FD10%-MVD process and the best value for rehydration ratio is observed for FD10%-AD. The results show also, that a puffing effect of MVD is not observed when FD is used prior to MVD. This is due to the fact that FD creates a porous microstructure from which moisture can easily migrate from the core to the surface, hence a pressure build up is not possible.

Also, Kumar et al. (2001) investigated a combination of a FD predrying step with an AD process, instead of a MVD finish drying step with various, time-dependent cross-over points. Their results showed that an increase in pre-FD entailed a higher volume retention, a higher rehydration ratio, and a shorter rehydration time for carrots as well as for pumpkins. Xu et al. (2005) found that the opposite combination (AD, followed by FD) had an even worse overall quality than single AD. Nevertheless, a combination of FD and MVD has proven by many authors to be superior to a combination of FD and AD in terms of drying time (Huang, Zhang, Mujumdar, et al., 2009; Pei et al., 2014; Rother et al., 2011), structure (Rother et al., 2011), and ingredient retention (Pei et al., 2014).

In summary, serial combinations of drying processes are mostly beneficial for product quality parameters when compared to AD or MVD as single processes. In terms of energy consumption, they are more effective than single FD (Huang, Zhang, Mujumdar, et al., 2009; Kumar et al., 2001; Xu, Zhang, Mujumdar, Duan, & Jin-Cai, 2006; Xu et al., 2005), but the drying time is not necessarily shorter.

8.5 Special quality attributes of microwave-assisted (vacuum) drying

If we look specifically at the retention of aroma, it becomes necessary to distinguish between two basic cases. In most foods, the aroma molecules are present in very small amounts so that they are likely dissolved in the aqueous phase. In this situation, the volatility of the aroma molecules in water is essential. Considering the fact that we perceive aroma—as opposed to taste—with our noses, it is quite clear that aroma molecules are generally volatile. Otherwise they would stay in the food during eating and not contribute to the aroma. In other words: If there is an interface between a water phase (i.e., a food) and a gas phase, the aroma molecules tend to choose the gas phase. In AD, the surface, where the aroma molecules can escape, is mainly the outer surface of the particles. This is also where the water molecules evaporate. So the aroma molecules already on the surface of the food particles will disappear into the gas phase, but more aroma molecules only reach the surface with the capillary water flow from within. As a result, the loss of aroma cannot happen at a higher rate than the loss of water, based on the respective starting contents.

This theory is supported by the measured curve in Fig. 8.8, in which the aroma loss (hexanol) of apple slices in AD and MVD are presented as a function of the water loss. For AD the measured values are initially on the diagonal, that means the aroma loss is proportional to the water loss. Later, as the product becomes drier, the capillary flow stops and other effects, such as binding with carbohydrates, come into play. Here, the aroma retention is even better than proportional to the remaining water.

In MVD, the situation is different. Due to the combination of volumetric heating and lowered pressure, the boiling point is reached throughout the material. Consequently, steam bubbles are created in the apple slices. As mentioned before, the aroma tends to accumulate in the gas phase, and when there is more surface area, the losses are higher. Fig. 8.8 shows the corresponding measurements. The losses of hexanol are in fact much more pronounced than for air-drying. Only toward the end of the process (on the left hand side of Fig. 8.8), some of the aroma is bound to the almost dry matrix. The calculated curve, using Eq. (8.1), is an approximation based on the assumption of an infinite inner surface and thus

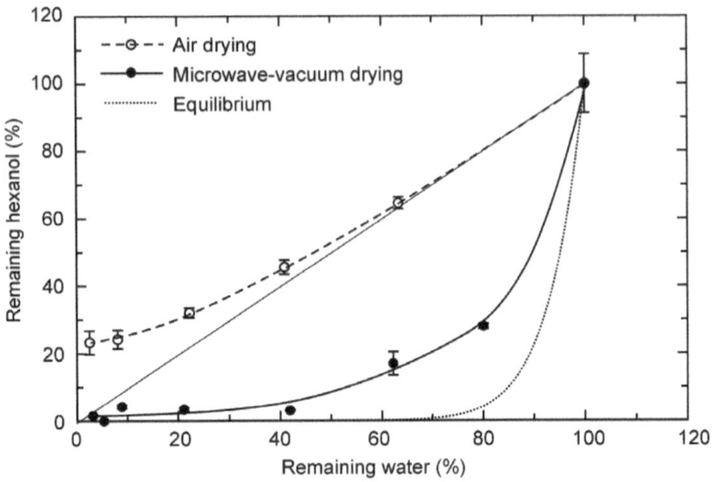

Figure 8.8 Hexanol losses during MVD and AD of apple slices. Equilibrium according to Eq. (8.1).

represents the worst case of thermodynamic equilibrium. It shows how high values for Henry's constant, H_{aroma}, (high volatility) lead to high aroma losses (Erle, 2001):

$$\frac{X_{\text{aroma}}}{X_{\text{aroma},0}} = \left(\frac{X}{X_0}\right)^{\frac{H_{\text{aroma}}}{p*} - 1} \tag{8.1}$$

with relative aroma loading ($X_{\text{aroma}}/X_{\text{aroma},0}$) called "remaining aroma" and relative water loading (X/X_0) called "remaining water" and p^* as vapor pressure of pure water.

For a more precise calculation, the amount of the inner surface and the diffusion coefficients of the aroma molecules would be needed. These are not available, but the simple, equilibrium-based curve has the right tendency.

In herbs, the aroma is much more intense, because the portion of aroma compounds, called "essential oils," is higher. Considering the poor water solubility of these compounds, we can assume that they form a phase of their own. Only very small amounts are dissolved in the aqueous phase. Under these circumstances it is not so much the volatility, i.e., the tendency to escape from the aqueous phase, but more the boiling point that determines the aroma retention. As essential oils mostly have a higher boiling point than water, it is possible to evaporate water without losing much of the aroma. This is supported by the results shown in Fig. 8.9, which compares microwave drying at various power levels. As demonstrated, it is better to apply high microwave power levels in order to evaporate the water quickly at its boiling point. In this example, there was practically no loss of essential oils at the two highest power levels. At lower power levels we give the oils more time to escape, either directly into the air or via the aqueous phase.

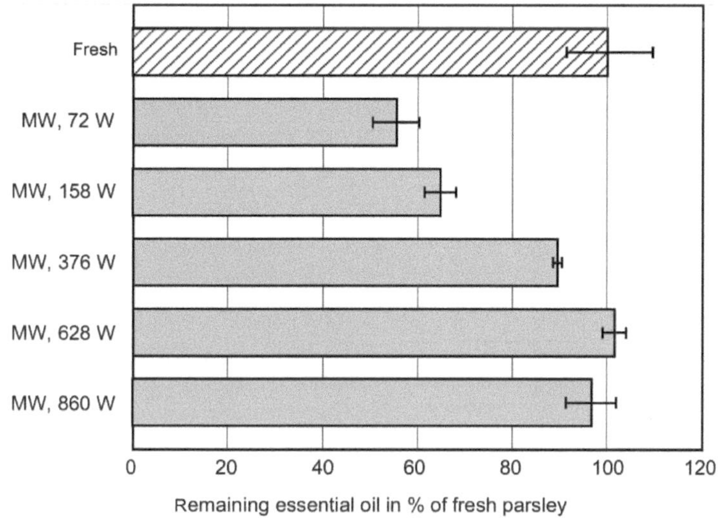

Figure 8.9 Essential oil losses of herbs after MVD at different power levels (Erle, 2001).

8.6 Microwave drying applied in food industry

Microwave drying is still not common in the food industry. There are many reasons for its limited use: the technical problems described above were not well understood in the past. This has led to some failures, which have surely discouraged other potential users. Schiffmann (2001) has listed a number of formerly successful applications that have been discontinued. Among these are the finish drying of potato chips, pasta drying, snack drying, and the finish drying of biscuits and crackers. It is apparently not always the microwave process itself but rather changes in the circumstances of production that make competing technologies more successful.

In spite of these difficulties, there are some applications. Schiffmann (2001) mentions cereal cooking and drying with a production rate of nearly 1 ton/h. Pasta drying with microwaves is carried out in Italy. MVD is being used for meat extract and, at least for a number of years, for the production of a powder made from orange juice concentrate (Attiyate, 1979). The combination of air-drying and microwave vacuum-puffing is being used in Germany and Poland for fruits and vegetables. As the food industry does not disclose all its production processes, we cannot expect this list to be complete but additional actual aspects of industrial microwave applications are given in Chapter 17, Improving the heating uniformity in microwave processing.

Hauri (1989) has provided values for the necessary investment and the specific energy requirements of five different drying methods (see Table 8.2). Based on the same throughput, the investment needed for MVD is rather high, while the energy figures are more favorable than for AD.

Table 8.2 Specific energy demand and investment costs for selected drying processes according to Hauri (1989)

Type of drying process	Specific energy demand kWh/kg water	Specific investment costs for equal throughput
Air band-drying	1.9	100%
Spray-drying	1.6	120%
Vacuum contact-drying	1.3	150%
Microwave vacuum-drying	1.5	190%
Freeze-drying	2.0	230%

8.7 Modeling microwave drying

What is modeling of microwave drying? The first thing most scientists would think about is the prediction of drying time, temperature, and moisture distribution in dependency of process parameters like MW-power, air temperature, and velocity or pressure in case of vacuum processes. Consequently, this "process modeling" is also the focus in literature. But it has to be pointed out, that especially concerning the drying of foods, there are other important parameters, mostly related to product quality, like rehydration, color, ingredient retention, volume retention, sensory aspects, and more, which are also determined by the process parameters. Consequently, it would be obvious to develop models for the prediction of product quality in dependency of process parameters ("product modeling"). The latter is very rare in literature and most studies are limited to the monitoring of product quality parameters and just show qualitatively how they are influenced by process parameters; like "increasing air temperature leads to decreasing color values." Quantification of these effects is rare and only possible with empirical equations. This was for instance recently shown with response surface methods for the rehydration ratio of jerusalem artichokes in dependency of microwave power (Karacabey, Baltacioglu, Cevik, & Kalkan, 2015) or for color change, rehydration ratio, and hardness of okra in dependency of microwave power and air temperature during a microwave air-drying process (Kumar, Prasad, & Murthy, 2014). Nevertheless, the correlation coefficients range from good to fair. To the best knowledge of the authors, there is no publication showing the attempt to develop a physical-based model for the simulation of product quality parameters in dependency of process parameters. After having a closer look on the physical background and the efforts in "process modeling" it will get clear why "product modeling" is so to speak not existing.

Like for every modeling attempt, *process modeling* of microwave drying can be achieved using two fundamentally different methods for the model development. There are empirical models that are simply based on correlation equations which describe the observed effects, and models based on physical principles. Sometimes, these methods are also mixed, and physical-based relationships are used when known and other unknown relationships are described with empirical equations. A common way to describe an unknown relationship is also to neglect it. This is not necessarily negative, as validation experiments can be performed in a model-supporting suitable way like "thin-layer" drying. In addition, it is the way how models developed over many years of research from very simplified models to more complex ones, after the simplified models helped to understand basic drying principles. Not to forget the increase in computational power, which makes it possible to solve complex implicit equation systems in a short period of time.

Feng, Yin, and Tang (2012) recently reviewed models for microwave drying of foods. They provide an overview about which parts of the drying process are addressed by the model and they also include the special aspects of microwave heating. The first part is the basic theory of the drying model:

Simple *empirical equations* are often used to correlate the drying rate with moisture content, microwave power density, microwave frequency, temperature, air flow rate, and sample geometry (Feng et al., 2012). Frequently used equations are the page model (Page, 1949) along with other exponential equations, e.g., shown in McMinn (2006) for drying of lactose, which has also been applied by Khraisheh, Cooper, and Magee (1995), Tulasidas, Raghavan, and Mujumdar (1995), Diaz, Martinez-Monzo, Fito, and Chiralt (2003), Bhattacharya et al. (2015), and many others. MVD drying has been modeled for apples, kiwi fruits, and pears by a statistical regression method in order to show the influence of microwave power and pressure on a drying constant (Kiranoudis, Tsami, Maroulis, & Marinos-Kouris, 1997). From a scientific point of view, these empirical models are very limited in use, because the physical principles are not included. But for an industrial process, these are useful and easy to handle tools for the control of known processes.

Another approach is the *diffusion model* which limits the drying to a mass transfer problem by calculating an effective diffusion coefficient (D_{eff}) from the measured drying rates. This is possible from Fick's second law for several geometries (Crank, 1976) and the temperature dependency of D_{eff} may be attributed by the law of Arrhenius (Sharma & Prasad, 2004). Several studies have been done underlying these principles (see Feng et al., 2012), but especially for the modeling of microwave-assisted drying, this method is limited, as a part of the idea of microwave heating is to overcome diffusion barriers, e.g., by steam creation inside the product. This changes the physical principles underlying Fick's second law and turns the model to a quasiempirical one. For example, Erle (2001) showed that models on MVD can be simplified by assuming that the absorbed energy is just used for the evaporation of water, which leaves the object as a convective flow of steam. This makes the calculations of mass transfer obsolete and has yielded a good correlation with measured drying curves of apple cuboids.

The most physically based approach for the modeling of microwave drying processes is the application of *heat and mass transfer* balances. There are different models developed from different assumptions like the "irreversible thermodynamics approach" (Luikov, 1964), the "unsaturated porous media theory" (Philip & de Vries, 1957), the "volume average technique" (multiphase model) (Whitaker,

1977), the "two region model" (Peishi & Pei, 1989), and the "continuous-thermo-hydro-mechanical model" (Kowalski, Musielak, & Banaszak, 2009). Several simplifications of these models exist for special geometries, like sliced or spherical food (Lu, Tang, & Liang, 1998; Lu, Tang, & Ran, 1999), one-dimensional models for infinite slab or cylinder (McMinn, Khraisheh, & Magee, 2003), or two-dimensional models (Lian, Harris, Evans, & Warboys, 1997). Additional to the principles of the calculation approach, different conditions can be taken into account. There are the assumptions for free water transport, which may be capillary- or diffusion-controlled, the consideration of bound water and the consideration of vapor and/or air pressure differences (e.g., for vacuum-drying) (Feng et al., 2012).

A general challenge for microwave-assisted processes is the calculation of the energy source in the simulation. It ranges from empirical calculations, over values based on the applied MW-power evenly distributed in the product, or changing in the product according to Lambert's law, to a full electrical field analysis distribution in the cavity and the sample according to the equations of Maxwell (Feng et al., 2012). From the local electric field strength, the local heat dissipation can be calculated with the knowledge of the dielectric properties (see Chapter 2: Microwave heating and the dielectric properties of foods). However, as well as the local transfer coefficients for heat and mass transfer and the local heat and mass properties of the product, the dielectric constants are functions of local moisture, temperature, and structure of the product. A deep understanding on the dependencies between these parameters and their local distribution would be necessary for an accurate simulation. In addition, all these parameters are not only dependent on each other, but also changing with time. Nevertheless, to improve the existing models, several empirical correlations are available between moisture content or temperature and material properties or transfer coefficients (Jia, Islam, & Mujumdar, 2003), e.g., for carrot and potato samples. The influence of temperature, moisture content, and porosity on the dielectric properties are described in literature for several foods (exemplarily shown in Jha et al., 2011; Knoerzer, Regier, Erle, Pardey, & Schubert, 2004; Ozturk, Kong, Trabelsi, & Singh, 2016; Piyasena, Dussault, Koutchma, Ramaswamy, & Awuah, 2003; Solyom et al., 2013; Sosa-Morales, Valerio-Junco, López-Malo, & García, 2010; Venkatesh & Raghavan, 2004), see also Chapter 2, Microwave heating and the dielectric properties of foods. This shows the complexity of an accurate physical principles-based simulation of a microwave-assisted drying process even if only a "process modeling" is considered.

Addressing "product modeling" would mean to transfer the simulated data of local product temperature, pressure, and moisture into product quality changes. These are mainly structural aspects which determine volume retention, hardness, and rehydration rate, and ingredient-dominated aspects leading, e.g., to color and concentration changes, not to forget the sensorial quality. The latter are often complex chemical reactions which are not fully understood. Even if the physical/chemical principles underlying these product changes would be clear (which they are not), the heterogeneous multiphase nature of many foods, consisting of cellular structure, fibers, suspensions, or emulsions, makes "product modeling" to date impossible. As described above, only empirical correlations are reported (Bal, Kar,

Satya, & Naik, 2011; Karacabey et al., 2015; Kumar et al., 2014; Qi, Zhang, Mujumdar, Meng, & Chen, 2014; Wang, Zhang, Mujumdar, & Sun, 2009).

A major drawback for all drying models and simulations is their validation. The first challenge is a proper measurement of the part of microwave power which was absorbed by the product and not lost in the cavity or reflected back to the magnetron. Cavity tuning and diodes measuring the returning microwave power are necessary. As the recent models deliver a lot of information on a microscale dimension (e.g., finite-elements/differences methods) the validation of temperature, moisture, and pressure data is already demanding for integral values and even more for 3D resolved measurements. Possible process variables to measure are temperature and pressure with, e.g., optical fibers, which are limited due to the fiber size, cost, and resolution (Feng et al., 2012), see also Chapter 15, Measuring temperature distributions during microwave processing. Magnetic resonance imaging (MRI) has provided the opportunity to investigate the distribution of water and temperature inside the product online and nondestructive during the process. This is a valuable tool to verify 3D calculations, comprising heat capacity, thermal conductivity, and diffusion coefficients or moisture distribution measurements. With 3D-MRI data on product composition together with a microwave field simulation, it was possible to simulate and validate the heating of a model food and a chicken wing (Knoerzer, Regier, & Schubert, 2008).

In a recent study (Gulati, Zhu, Datta, & Huang, 2015; Zhu, Gulati, Datta, & Huang, 2015), a drying model was established using the volume averaging method for heat and mass transfer calculations as well as a full microwave field calculation, capillary driven water flux, and consideration of vapor and air pressure. The authors validated the model for potato spheres of different sizes. Some changes of material properties were introduced based on empirical calculations. Measured were the surface temperature during drying, the spatial temperature distribution in the middle plane offline via infrared camera and one fiber optical temperature probe as well as a fiber optical pressure probe inside the potato and an integral moisture measurement with an online balance. Focusing effects were highest for the 3 cm diameter potato and very low for small spheres of 1.2 cm diameter. The small spheres experienced a very even heating. In summary, good agreements were found between the model predictions and the measurements concerning absorbed power, temperature, and pressure.

A different strategy on "process and product modeling" is the use of *artificial neural networks (ANN).* The basic idea is to have a self-learning system based on empirical correlations between the input and output parameters. A comprehensive review on these techniques used in drying modeling is given by Aghbashlo, Hosseinpour, and Mujumdar (2015). In the concluding remarks, the authors state that these methods principally have a high potential to model, optimize, monitor, and control drying processes and even outperform other available techniques, but they also advise that ANNs are preferentially used as a complement and not as alternative techniques. A major drawback from a scientific point of view is that even if an ANN can predict a wanted parameter in dependency of several process parameters very well, we cannot learn anything on the physical principles of this correlation as it is completely empirical and not extractable from the ANN in detail.

This aspect may jeopardise the use of ANNs in the real-world drying technology (Aghbashlo et al., 2015).

From a practical point of view, none of the physical or ANN-based modeling attempts has achieved much significance in food industry. So far it has always been easier to use trial and error, based on former practical experience, and to use simple empirical calculations which correlate well with the observed parameter changes than to collect all necessary information on thermophysical properties, dielectric properties, field distribution in the drier, etc.

References

Aghbashlo, M., Hosseinpour, S., & Mujumdar, A. S. (2015). Application of artificial neural networks (ANNs) in drying technology. A comprehensive review. *Drying Technology*, *33*, 1397–1462.

Alibas, I. (2009). Microwave, vacuum, and air drying characteristics of collard leaves. *Drying Technology*, *27*, 1266–1273.

Attiyate, Y. (1979). Microwave vacuum drying: first industrial application. *Food Engineering International*, *51*, 30–31.

Aversa, M., Curcio, S., Calabrò, V., & Iorio, G. (2012). Experimental evaluation of quality parameters during drying of carrot samples. *Food and Bioprocess Technology*, *5*, 118–129.

Bal, L. M., Kar, A., Satya, S., & Naik, S. N. (2011). Kinetics of colour change of bamboo shoot slices during microwave drying. *International Journal of Food Science & Technology*, *46*, 827–833.

Barbosa-Cánovas, G. V., & Vega-Mercado, H. (1996). *Dehydration of foods*. New York: Chapman & Hall.

Beaudry, C., Raghavan, G. S. V., & Rennie, T. J. (2003). Microwave finish drying of osmotically dehydrated cranberries. *Drying Technology*, *21*, 1797–1810.

Bhattacharya, M., Srivastav, P. P., & Mishra, H. N. (2015). Thin-layer modeling of convective and microwave-convective drying of oyster mushroom (*Pleurotus ostreatus*). *Journal of Food Science and Technology*, *52*, 2013–2022.

Bruijn, J. de, & Bórquez, R. (2014). Quality retention in strawberries dried by emerging dehydration methods. *Food Research International*, *63*, 42–48.

Copson, D. (1958). Microwave sublimation of foods. *Food Technology*, *12*, 270–272.

Crank, J. (1976). *The mathematics of diffusion* (2nd ed.). Oxford: Clarendon Press.

Cui, Z. W., Li, C. Y., Song, C. F., & Song, Y. (2008). Combined microwave-vacuum and freeze drying of carrot and apple chips. *Drying Technology*, *26*, 1517–1523.

Cui, Z. W., Xu, S. Y., & Sun, D. W. (2003). Dehydration of garlic slices by combined microwave-vacuum and air drying. *Drying Technology*, *21*, 1173–1184.

Diaz, G. R., Martinez-Monzo, J., Fito, P., & Chiralt, A. (2003). Modelling of dehydration-rehydration of orange slices in combined microwave/air drying. *Innovative Food Science and Emerging Technologies*, *4*, 203–209.

Drouzas, A. E., & Schubert, H. (1996). Microwave application in vacuum drying of fruits. *Journal of Food Engineering*, *28*, 203–209.

Durance, T. D., & Wang, J. H. (2002). Energy consumption, density, and rehydration rate of vacuum microwave- and hot-air convection- dehydrated tomatoes. *Journal of Food Science*, *67*, 2212–2216.

Erle, U. (2001). *Untersuchungen zur Mikrowellen-Vakuumtrocknung von Lebensmitteln*. Aachen: Shaker Verlag.

Erle, U., & Schubert, H. (2001). Combined osmotic and microwave-vacuum dehydration of apples and strawberries. *Journal of Food Engineering*, *49*, 193–199.

Feng, H., Yin, Y., & Tang, J. (2012). Microwave drying of food and agricultural materials. Basics and heat and mass transfer modeling. *Food Engineering Reviews*, *4*, 89–106.

Figiel, A. (2010). Drying kinetics and quality of beetroots dehydrated by combination of convective and vacuum-microwave methods. *Journal of Food Engineering*, *98*, 461–470.

Gamboa-Santos, J., Megías-Pérez, R., Soria, A. C., Olano, A., Montilla, A., & Villamiel, M. (2014). Impact of processing conditions on the kinetic of vitamin C degradation and 2-furoylmethyl amino acid formation in dried strawberries. *Food Chemistry*, *153*, 164–170.

Gulati, T., Zhu, H., Datta, A. K., & Huang, K. (2015). Microwave drying of spheres. Coupled electromagnetics-multiphase transport modeling with experimentation. Part II: Model validation and simulation results. *Food and Bioproducts Processing*, *96*, 326–337.

Hauri, F. W. (1989). Vakuumbandtrocknung in der Lebensmittelindustrie. *Die Ernährungsindustrie*, *10*, 32–34.

Huang, L. L., Zhang, M., Mujumdar, A. S., Sun, D. F., Tan, G. W., & Tang, S. (2009). Studies on decreasing energy consumption for a freeze-drying process of apple slices. *Drying Technology*, *27*, 938–946.

Huang, L. L., Zhang, M., Yan, W. Q., Mujumdar, A. S., & Sun, D. F. (2009). Effect of coating on post-drying of freeze-dried strawberry pieces. *Journal of Food Engineering*, *92*, 107–111.

Jangam, S. V., Law, C. L., & Mujumdar, A. S. (Eds.) (2010). *Drying of foods, vegetables and fruits* (Vol. 1).

Jha, S. N., Narsaiah, K., Basediya, A. L., Sharma, R., Jaiswal, P., Kumar, R., & Bhardwaj, R. (2011). Measurement techniques and application of electrical properties for nondestructive quality evaluation of foods – A review. *Journal of Food Science and Technology*, *48*, 387–411.

Jia, L. W., Islam, M. R., & Mujumdar, A. S. (2003). A simulation study on convection and microwave drying of different food products. *Drying Technology*, *21*, 1549–1574.

Karacabey, E., Baltacioglu, C., Cevik, M., & Kalkan, H. (2015). Optimization of microwave-assisted drying of Jerusalem artichokes (*Helianthus tuberosus* L.) by response surface methodology and genetic algorithm. *Italian Journal of Food Science*, *28*, 121–131.

Khraisheh, M. A. M., Cooper, T. J. R., & Magee, T. R. A. (1995). Investigation and modeling of combined microwave and air-drying. *Food and Bioproducts Processing*, *73*, 121–126.

Kiranoudis, C. T., Tsami, E., Maroulis, Z. B., & Marinos-Kouris, D. (1997). Drying kinetics of some fruits. *Drying Technology*, *15*, 1399–1418.

Knoerzer, K., Regier, M., Erle, U., Pardey, K. K., & Schubert, H. (2004). Development of a model food for microwave processing and the prediction of its physical properties. *Journal of Microwave Power and Electromagnetic Energy*, *39*(3 + 4), 67–78.

Knoerzer, K., Regier, M., & Schubert, H. (2008). A computational model for calculating temperature distributions in microwave food applications. *Innovative Food Science & Emerging Technologies*, *9*, 374–384.

Koc, B., Eren, I., & Ertekin, F. K. (2008). Modelling bulk density, porosity and shrinkage of quince during drying: The effect of drying method. *Journal of Food Engineering*, *85*, 340–349.

Kowalski, S. J., Musielak, G., & Banaszak, J. (2009). Heat and mass transfer during microwave-convective drying. *AIChE Journal*, *56*, 24–35.

Kowalski, S. J., Szadzinska, J., & Lechtanska, J. (2013). Non-stationary drying of carrot: Effect on product quality. *Journal of Food Engineering*, *118*, 393–399.

Kumar, D., Prasad, S., & Murthy, G. S. (2014). Optimization of microwave-assisted hot air drying conditions of okra using response surface methodology. *Journal of Food Science and Technology*, *51*, 221–232.

Kumar, H. S. P., Radhakrishna, K., Nagaraju, P. K., & Rao, D. V. (2001). Effect of combination drying on the physico-chemical characteristics of carrot and pumpkin. *Journal of Food Processing and Preservation*, *25*, 447–460.

Lian, G., Harris, C. S., Evans, R., & Warboys, M. (1997). Coupled heat and moisture transfer during microwave vacuum drying. *Journal of Microwave Power & Electromagnetic Energy*, *32*, 34–44.

Liapis, A. I., & Bruttini, R. (2015). Freeze drying. In A. S. Mujumdar (Ed.), *Handbook of industrial drying* (4th ed., pp. 259–282). Boca Raton: CRC Press.

Lu, L., Tang, J., & Liang, L. (1998). Moisture distribution in spherical foods in microwave drying. *Drying Technology*, *16*, 503–524.

Lu, L., Tang, J., & Ran, X. (1999). Temperature and moisture changes during microwave drying of sliced food. *Drying Technology*, *17*, 414–431.

Luikov, A. V. (1964). *Advances in heat transfer*. New York: Pergamon Press.

Maskan, M. (2000). Microwave/air and microwave finish drying of banana. *Journal of Food Engineering*, *44*, 71–78.

Maskan, M. (2001). Drying, shrinkage and rehydration characteristics of kiwifruits during hot air and microwave drying. *Journal of Food Engineering*, *48*, 177–182.

Mayer-Miebach, E., Behsnilian, D., Regier, M., & Schuchmann, H. P. (2005). Thermal processing of carrots: Lycopene stability and isomerisation with regard to antioxidant potential. *Food Research International*, *38*, 1103–1108.

Mclaughlin, C. P., & Magee, T. R. A. (1998). The effect of shrinkage during drying of potato spheres and the effect of drying temperature on vitamin C retention. *Food and Bioproducts Processing*, *76*, 138–142.

McMinn, W. A. M. (2006). Thin-layer modelling of the convective, microwave, microwave-convective and microwave-vacuum drying of lactose powder. *Journal of Food Engineering*, *72*, 113–123.

McMinn, W. A. M., Khraisheh, M. A. M., & Magee, T. R. A. (2003). Modelling the mass transfer during convective, microwave and combined microwave-convective drying of solid slabs and cylinders. *Food Research International*, *36*, 977–983.

Megías-Pérez, R., Gamboa-Santos, J., Soria, A. C., Villamiel, M., & Montilla, A. (2014). Survey of quality indicators in commercial dehydrated fruits. *Food Chemistry*, *150*, 41–48.

Metaxas, A. C., & Meredith, R. J. (1983). *Industrial microwave heating*. London: Peregrinus.

Mujumdar, A. S. (2015). *Principles, classification and selection of dryers. Handbook of industrial drying* (4th ed.). Boca Raton: CRC Press.

Mujumdar, A. S., & Beke, J. (2003). Grain drying: Basic principles. In A. Chakraverty, A. Mujumdar, G. Raghavan, & H. Ramaswamy (Eds.), *Handbook of postharvest technology: Cereals, fruits, vegetables, tea, and spices*. New York, USA: Marcel Dekker, Inc.

Nawirska, A., Figiel, A., Kucharska, A. Z., Sokol-Letowska, A., & Biesiada, A. (2009). Drying kinetics and quality parameters of pumpkin slices dehydrated using different methods. *Journal of Food Engineering*, *94*, 14–20.

Orikasa, T., Koide, S., Okamoto, S., Imaizumi, T., Muramatsu, Y., Takeda, J. I., ... Tagawa, A. (2014). Impacts of hot air and vacuum drying on the quality attributes of kiwifruit slices. *Journal of Food Engineering*, *125*, 51–58.

Ozturk, S., Kong, F., Trabelsi, S., & Singh, R. K. (2016). Dielectric properties of dried vegetable powders and their temperature profile during radio frequency heating. *Journal of Food Engineering*, *169*, 91–100.

Page, G. E. (1949). *Factors influencing the maximum rates of air drying shelled corn in thin layers*. West Lafayette/US.

Pei, F., Yang, W. J., Shi, Y., Sun, Y., Mariga, A., Zhao, L. Y., & Hu, Q. H. (2014). Comparison of freeze-drying with three different combinations of drying methods and their influence on colour, texture, microstructure and nutrient retention of button mushroom (*Agaricus bisporus*) slices. *Food and Bioprocess Technology*, *7*, 702–710.

Peishi, C., & Pei, D. C. (1989). A mathematical model of drying processes. *International Journal of Heat and Mass Transfer*, *32*, 297–310.

Philip, J. R., & de Vries, D. A. (1957). Moisture movement in porous materials under temperature gradients. *Transactions, American Geophysical Union*, *38*, 222.

Piyasena, P., Dussault, C., Koutchma, T., Ramaswamy, H. S., & Awuah, G. B. (2003). Radio frequency heating of foods: Principles, applications and related properties – A review. *Critical Reviews in Food Science and Nutrition*, *43*, 587–606.

Prabhanjan, D. G., Ramaswamy, H. S., & Raghavan, G. (1995). Microwave-assisted convective air drying of thin layer carrots. *Journal of Food Engineering*, *25*, 283–293.

Qi, L.-L., Zhang, M., Mujumdar, A. S., Meng, X.-Y., & Chen, H.-Z. (2014). Comparison of drying characteristics and quality of shiitake mushrooms (*Lentinus edodes*) using different drying methods. *Drying Technology*, *32*, 1751–1761.

Rother, M., & Schuchmann, H. P. (2009). Mikrowellen-gestützte Erwärmungsprozesse – Erwärmung unter Konkurrenz. *CIT*, *81*, 1131.

Rother, M., Steimle, P., Gaukel, V., & Schuchmann, H. P. (2011). How to meet the freeze drying standard in combined drying processes: Pre and finish drying of carrot dice. *Drying Technology*, *29*, 266–277.

Schiffmann, R. F. (2001). Microwave processes in the food industry. In A. Datta & R. Anantheswaran (Eds.), *Handbook of microwave technology for food applications* (pp. 299–339). New York: Marcel Dekker.

Sham, P. W. Y., Scaman, C. H., & Durance, T. D. (2001). Texture of vacuum microwave dehydrated apple chips as affected by calcium pretreatment, vacuum level, and apple variety. *Journal of Food Science*, *66*, 1341–1347.

Sharma, G., & Prasad, S. (2004). Effective moisture diffusivity of garlic cloves undergoing microwave-convective drying. *Journal of Food Engineering*, *65*, 609–617.

Sharma, G. P., & Prasad, S. (2006). Optimization of process parameters for microwave drying of garlic cloves. *Journal of Food Engineering*, *75*, 441–446.

Siebert, T., Gall, V., Schuchmann, H. P., & Gaukel, V. (2016). Determination of change-over points of serial combination drying processes: A measure to improve quality of dried carrot discs and process efficiency. In: *Proceedings of the 20th international drying symposium (IDS 2016), Gifu Japan, accepted.*

Sokhansanj, S., & Jayas, D. S. (2015). *Drying of foodstuffs*. In *Handbook of industrial drying*, (4th ed.). pp. 521–544). Boca Raton: CRC Press.

Solyom, K., Kraus, S., Mato, R. B., Gaukel, V., Schuchmann, H. P., & Cocero, M. J. (2013). Dielectric properties of grape marc: Effect of temperature, moisture content and sample preparation method. *Journal of Food Engineering*, *119*, 33–39.

Sosa-Morales, M. E., Valerio-Junco, L., López-Malo, A., & García, H. S. (2010). Dielectric properties of foods. Reported data in the 21st century and their potential applications. *LWT-Food Science and Technology*, *43*, 1169–1179.

Soysal, Y., Ayhan, Z., Estürk, O., & Arkan, M. F. (2009). Intermittent microwave-convective drying of red pepper: Drying kinetics, physical (colour and texture) and sensory quality. *Biosystems Engineering*, *103*, 455–463.

Tulasidas, T. N., Raghavan, G. S. V., & Mujumdar, A. S. (1995). Microwave drying of grapes in a single mode cavity at 2450 MHz – I: Drying kinetics. *Drying Technology*, *13*, 1949–1971.

Venkatesh, M. S., & Raghavan, G. (2004). An overview of microwave processing and dielectric properties of agri-food materials. *Biosystems Engineering*, *88*, 1–18.

Wang, R., Zhang, M., Mujumdar, A. S., & Sun, J.-C. (2009). Microwave freeze–drying characteristics and sensory quality of instant vegetable soup. *Drying Technology*, *27*, 962–968.

Whitaker, S. (1977). *Simultaneous heat, mass, and momentum transfer in porous media: A theory of drying, . Advances in heat transfer* (Vol. 13, pp. 119–203). New York, San Francisco, London: Academic Press.

Witrowa-Rajchert, D., & Rzaca, M. (2009). Effect of drying method on the microstructure and physical properties of dried apples. *Drying Technology*, *27*, 903–909.

Wray, D., & Ramaswamy, H. S. (2015). Novel concepts in microwave drying of foods. *Drying Technology*, *33*, 769–783.

Xu, Y. Y., Zhang, M., Mujumdar, A. S., Duan, X., & Jin-Cai, S. (2006). A two-stage vacuum freeze and convective air drying method for strawberries. *Drying Technology*, *24*, 1019–1023.

Xu, Y. Y., Zhang, M., Tu, D. Y., Sun, J. C., Zhou, L. Q., & Mujumdar, A. S. (2005). A two-stage convective air and vacuum freeze-drying technique for bamboo shoots. *International Journal of Food Science and Technology*, *40*, 589–595.

Yongsawatdigul, J., & Gunasekaran, S. (1996). Microwave-vacuum drying of cranberries. 1. Energy use and efficiency. *Journal of Food Processing and Preservation*, *20*, 121–143.

Zhang, M., Tang, J., Mujumdar, A. S., & Wang, S. (2006). Trends in microwave-related drying of fruits and vegetables. *Trends in Food Science & Technology*, *17*, 524–534.

Zhu, H., Gulati, T., Datta, A. K., & Huang, K. (2015). Microwave drying of spheres. Coupled electromagnetics-multiphase transport modeling with experimentation. Part I: Model development and experimental methodology. *Food and Bioproducts Processing*, *96*, 314–325.

Microwave-assisted blanching

9

L. Dorantes-Alvarez[1], A. Ortiz-Moreno[1], R. Guzmán-Gerónimo[2], and L. Parada-Dorantes[3]
[1]Instituto Politécnico Nacional, Mexico City, Mexico, [2]Universidad Veracruzana, Veracruz, Mexico, [3]Universidad del Caribe, Cancún, Mexico

9.1 Introduction

Blanching of vegetal materials is a pretreatment to preserve food quality mainly through inactivation of enzymes, and also to reduce the volume of the material by expelling intracellular trapped air, reduce the microbial load, and eliminate undesirable odors and flavors (Binsi et al., 2014). The target enzymes are mainly peroxidase, polyphenol oxidase, and pectinases. Traditional techniques that employ steam, hot solutions, or hot water inactivate enzymes by convective and conductive heat transfer, and may cause leaching of vitamins and other soluble components to the water/solutions (Chavez-Reyes, Dorantes-Alvarez, Arrieta-Baez, Osorio-Esquivel, & Ortiz-Moreno, 2013).

Microwave-assisted blanching (MAB) is recognized as a technology that reduces the time and energy required to achieve the inactivation of enzymes since it is a method that allows high temperature/short time treatment of solid foods through volumetric heating. This is important to preserve thermolabile nutrients, vitamins, and other bioactive compounds. Dorantes-Alvarez, Jaramillo-Flores, González, Martinez, and Parada (2011) calculated from various reports including different commodities that the energy density for microwave blanching varies from 0.15 in strawberries to 2.55 KJ/g in potatoes. Some publications showing the advantages of the use of microwave-assisted blanching in the last years are presented in Table 9.1. Microwave heating takes place in dielectric materials such as in foods due to the polarization effect of electromagnetic radiation at frequencies between 300 MHz and 300 GHz. Dielectric properties of foods play a critical role in determining the interaction between the microwave field and the foods, and they are dependent on composition, temperature, and microwave frequency (see Chapter 2: Microwave heating and the dielectric properties of foods). Hence, there is an important influence of the dielectric properties of the material on the efficiency of electromagnetic energy coupled into the materials, electromagnetic field distribution, and conversion of electromagnetic energy into thermal energy within the material (Tang, 2005). Matrix must contain dipolar or ionic species to enable heating to occur. Microwave and conventional blanching may lead to changes of physicochemical properties. The following are some examples: the dielectric and other properties of *Agaricus bisporus* (mushroom) slices were compared when blanched with microwaves and

The Microwave Processing of Foods. DOI: http://dx.doi.org/10.1016/B978-0-08-100528-6.00009-7

Table 9.1 **Comparison of traditional versus microwave blanching in fruits and vegetables**

Material/reference	Blanching techniques that are compared	Parameters studied	Conclusion	Conditions of microwave blanching
Agaricus bisporus slices Jiang et al. (2015)	Before and after microwave vacuum-drying	Dielectric properties and microstructure	The microstructure of microwave slices was more uniform and the time was reduced	2450 MHz microwave oven 800 W Galaxy Inc. Shunde, China 100°C/1 min
Sutchi catfish (*Pangasianodon hypophthalmus*) 100 g samples Binsi et al (2014)	Quality of microwave blanched fish vs non blanched and chilled	Color and texture of fillets catfish	Combination of microwave blanching and quick chilling improve color and texture	Home-model microwave oven, Panasonic, 2450 MHz/18 s
Beta vulgaris L. Latorre et al. (2013)	Hot water and microwave-assisted blanching	Cell wall polymers hydrophobicity, microstructure, and functional properties	A better mechanical performance observed at 350 watts	Microwave system ETHOS Milestone SRI, Sonsole, Italy 2450 MHz 350 W, 900 W
Centella asiática Trirattanapikul and Phoungchandang (2014)	Microwave blanching and heat pump-assisted drying	Total phenolics, percentage inhibition and moistures diffusivities	Microwave blanching retained the highest total phenolics and the highest inhibition	Microwave oven LG Korea 2450 MHz 800 W
Stem lettuce cubes Wang et. al. (2012)	Effects of blanching methods on freeze-drying of stem lettuce cubes	Dielectric properties, electrical conductivity and microstructure	Electrical conductivity of sample blanched by microwave was higher than those blanched by boiling water. Time of freeze-drying was shorter	Microwave oven Panasonic NN-GF339M, China 2450 MHz 800 W Fiber optic thermometer 100°C/1 min

Dried potatoes Krokida, Kiranoudis, Maroulis, and Marinos-Kouris (2000)	Osmotic-, microwave-, sulfite-, water-, and steam-blanching on the color of dried potatoes	Browning as shown by the L*, a*, and b* parameters.	Untreated and microwave pretreated materials showed extensive browning, while osmotic, sulfite, water-, and steam-blanching suppressed browning during drying	
Strawberries preserved by osmotic treatment Moreno, Chiralt, Escriche, and Serra (2000).	Steam, microwave blanching	Polyphenol oxidase activity and texture	Both treatments lead to about 80% of residual enzyme activity, microwaves preserved cells better at the microstructural level	Microwave oven 400 W
Frozen carrots Kidmose and Martens (1999)	Microwave, steam and water blanching	Inactivation of peroxidase	Higher contents of dry matter, minerals, ascorbic acid, total sugars and carotene in the microwaved product	Continuous conveyer microwave oven 10 kW, M25, with four magnetrons of 1.25 kW each
Frozen endive and spinach leaves Ponne, Baysal, and Yuksel (1994)	Microwaves, a combination of microwaves and steam, steam, water, infrared and radio frequency energies	Texture, color, dry matter, vitamin C content, nitrate content, and sensory characteristics	The leaves had better texture and sensory evaluation when blanched with microwaves or the microwave-steam combination	Microwave tunnel (6 kW, 2450 MHz) for 90 s
Banana slices Cano, Marin, and Fu' Ster (1990)	Microwaves and boiling water	Inactivation of polyphenol oxidase and peroxidase	Even though it accomplished a great enzyme inactivation, it resulted in a poor quality product from the sensory point of view	650 W for 2 min

conventional methods. The authors concluded that the microwave treated samples showed a more uniform microstructure, highest dielectric loss factor ε'' and the lowest dielectric constant ε', probably due to a higher ionic conductivity and a lower amount of dissolved air (Jiang, Liu, Li, & Zhou, 2015). Effects of blanching methods on freeze-drying stem lettuce cubes were studied, and it was reported that the electrical conductivity of samples blanched by microwave was two times higher than that of the samples blanched by boiling water, probably by a different salt concentration due to loss of water in the microwaved samples; also, the time required to achieve the same level of residual moisture in the freeze-drying experiments was shorter and a better microstructure was observed (Wang, Zhang, Mujumdar, Mothibe, & Azam, 2012). The texture of vegetables is related to cell wall polymers and the hydrophilicity, rehydration, and mechanical properties of food commodities may change after blanching. Latorre, De Escalada-Plá, Rojas, and Gerschenson (2013) found that cell wall composition of red beets *Beta vulgaris* was similar in blanched samples compared to the control, however, when comparing the microstructure, the microwave treated beets (36 g) at a nominal power of 350 W showed no alterations in the middle lamella, which were evident in hot water and microwave treated beets at powers exceeding 900 W. Evaluation and optimization of the impact of microwave power on different fruits and vegetables is therefore imperative. Some authors have also evaluated the retention of bioactive compounds and water-soluble vitamins as well as antioxidant properties, generally in plant materials. Tirattanapikul and Phoungchandang (2014) studied the combined effect of microwave blanching and heat-pump drying of *Centella asiatica*, concluding that the samples retained the highest total phenolics content and the highest antioxidant activity when compared to traditional methods. Other successful applications of MAB in animal tissue are the improvement of color and texture of catfish fillets, when a combination of microwave blanching and quick chilling was assessed (Binsi et al., 2014). The conditions to achieve a successful microwave-assisted blanching depend on the particular vegetable, the power of the microwave emitters, and the amount of vegetable material in the microwave oven. Most of the studies utilize 2450 MHz ovens, with a power ranging from 350 to 900 watts, sample weights from 10 to 200 grams, time of microwave treatment from 0.25 to 6 min, and temperature usually ranges from 80 to 100°C.

9.1.1 Blanching and enzyme inactivation

Enzymatic browning is a prominent deteriorative reaction in fruits. It causes discoloration associated with increased concentration of polymeric derivatives of *o*-quinones, which derive from phenolic substrates through oxidative reactions catalyzed by polyphenol oxidase (PPO) in the presence of atmospheric oxygen (Del Valle, Ara'nguiz, & Leo'n, 1998). Polyphenol oxidase is a generic term for a group of enzymes. The brown pigments they produce also lead to the development of off flavors and losses in nutritional quality. These enzymes are relatively heat labile: temperatures of more than 50°C and proper time of treatment allow a decrease in activity, whereas they are inactivated at temperatures of 80°C.

Peroxidase is widely distributed in higher plants, with high concentrations in fig sap and horseradish. It is also found in some animal tissues and in microorganisms. This enzyme controls the level of peroxides in the tissue. However, its catalytic action produces a deteriorative effect in raw vegetable products, since it contributes to browning action and other oxidative reactions. In fact, peroxidase catalysis is associated with four types of activity: peroxidatic, oxidatic, catalatic, and hydroxylation. Under the usual assay conditions in vitro, where a phenolic substrate is used, only the peroxidatic reaction is of importance (Whitaker, 1994). Peroxidatic activity on phenolic compounds, such as ferulic acid, can generate phenolic cross-links that connect polymer chains; this action affects the mechanical properties of cell walls and therefore the texture of vegetables. This enzyme system has gained much attention due to its role in modulating the mechanical properties of cell walls during extension and cell adhesion, as related to the thermal stability of texture (Tijskens et al., 1997). The assay most frequently used to detect peroxidatic activity is the peroxidatic reaction with guaiacol as the substrate. This assay is very simple and readily followed in a continuous fashion in a recording spectrophotometer. Complete inactivation of peroxidase assures that all other enzymes have been destroyed. Hence loss of peroxidase activity is generally used as an index of proper blanching of vegetables.

Lipoxygenase is found in a wide variety of plants, particularly the legumes. There are at least three detrimental effects of the action of lipoxygenase in foods: (1) destruction of the essential fatty acids (linoleic, linolenic, and arachidonic); (2) production of free radicals, which damage other compounds including vitamins and proteins; and (3) development of off flavor and odor. In beans and peas this is characterized as a hay-like flavor. This is a serious problem in unblanched frozen peas, beans, and corn since lipoxygenase can continue its action even at low temperatures (Whitaker, 1994).

The pectic enzymes are a group of catalysts that degrade the pectin substances and have been found in higher plants and in microorganisms. They are useful for treatment of fruit juices and beverages to facilitate filtration and clarification, to increase juice yields, and in the production of low-methoxyl pectin and galacturonic acids. They are considered deteriorative enzymes since they cause excessive softening of many fruits and vegetables, as well as "cloud" separation in such products as tomato and citrus juices. There are three types of pectic enzymes that catalyze three different types of reactions. The enzyme pectin methyl esterase (PME) demethylates the carboxymethyl groups of the pectic polysaccharide chains, which triggers different processes related to texture and firmness. These processes may comprise cross-linking by Ca^{2+}, increasing the hydration at the demethylated sites, enhancing shielding and repulsion forces by the electric charges within the biopolymer matrix of the cell wall, as well as decreasing the susceptibility to polygalacturonase activity (Tijskens, Rodis, Hertog, Proxenia, & Van Dijk, 1999). This enzyme has been claimed to be activated by low-temperature long-time (LTLT) blanching treatments (Del Valle et al., 1998), hence to prevent changes in texture due to PME activity, a high temperature blanching process is recommended. Other pectic enzymes are the polygalacturonases and the pectate lyases that hydrolyze glycosidic linkages in pectic substances, causing viscosity reduction in fruit products.

9.1.1.1 D, z, and thermal inactivation time values

The thermal inactivation curve for enzymes is generated by subjecting an enzyme extract or a food sample to a series of heat treatments at a specific temperature, and then testing for residual enzyme activity. The range of temperature tested is beyond the optimal temperature, where a rather steep decline in activity is observed, due to denaturation. In general, the shape of the decrease with temperature is exponential, indicating a first-order reaction. From the resulting plot of log of residual enzyme activity versus time, the *D* value of an enzyme may be calculated. It represents the heating time in minutes required to inactivate 90% of the total enzyme activity at a particular temperature (Ramaswamy & Abbatemarco, 1996).

The *z*-value is a temperature sensitivity indicator, and it is obtained by plotting the logarithm of *D*-values against temperature. The *z*-value indicates the temperature range between which the *D*-value curve passes through one logarithmic cycle. Svensson (1977) reported the *z*-values for the thermal inactivation of potato enzymes, peroxidase being the enzyme most resistant to temperature with a value of 35°C, followed by polyphenoloxidase (7.8°C), lipoxygenase (3.6°C), and lipolytic acid hydrolase (3.1°C).

9.1.1.2 Thermal inactivation time

An enzyme is considered inactivated when there is no measurable residual activity, and then the corresponding heating time is taken. This is known as the thermal inactivation time. The calculation of blanching time for a specific product is based on the heating time of its most heat-resistant enzyme. Alternatively, the enzyme that causes commercial deterioration in a particular commodity may also be considered.

The thermal inactivation time of a particular enzyme is strongly dependent on the processing temperature: at higher temperatures, there are lower thermal inactivation time values. The same applies for the thermal death time of microorganisms. However, at relatively low thermal processing temperatures, the destruction rate for enzymes is greater than that for microorganisms, but as process temperature increases, the destruction rate for microorganisms increases faster than that for enzymes. This is why the inactivation of some enzymes is used to verify the efficiency of a thermal process, e.g., pectin methyl esterase activity has been used to determine the adequacy of pasteurization of fruit juices, and alkaline phosphatase inactivation is used to verify the pasteurization of milk. Thermal inactivation times of enzymes are also affected by different blanching media, type of vegetable, weight and size of samples, and in general by all the factors that influence the energy transfer in the food or the stability of enzymes.

High-temperature short-time (HTST) processes were developed upon the knowledge that enzymes and vegetative cells of pathogenic bacteria may be inactivated at high temperatures and short times. One of the advantages of HTST is that it results in greater nutrient retention (Lund, 1977). As HTST treatments may be accomplished in solid foods by the application of microwaves, a greater nutrient retention is expected.

9.1.2 Microwave-assisted blanching patents

Often the studies are done comparing microwave-assisted blanching to traditional techniques, and include the whole food processing steps, such as cooling, freezing, or drying. As a result of this research, some patents were generated and are presented in Table 9.2. A combination of blanching techniques where the vegetable material is subjected to microwave heating prior to hot air treatment was registered as a patent US 2005/0037118 A1 (Panaioli, Rotunno, & Simeone, 2005), and is reported to be applicable to a wide range of vegetables suitable for freezing, such as potatoes, carrots, cauliflower, broccoli, paprika, and leaf vegetables such as spinach. The preliminary exposure to microwave heating rapidly raises the temperature of the vegetable material and helps to assure the core is heated. After microwave treatment, the vegetables are subjected to hot air from 100 to 150°C to complete the blanching process. Following blanching, the vegetable material is preferably frozen. The invention is registered as a novel blanching technique, claimed to retain the flavor and nutritional components of the vegetable. Cammaroto et al. (2014) presented an invention (Patent EP 2692245A2) about a continuous process applying microwaves and hot water to food products in order to save time and to assure heating of the core in the blanching process. The food product may be fish, fruits, vegetables, herbs, spices, and meat, preferably packed into a steam bag. The system energy may be transferred to the food product in different ways: forced water convection, conduction, and irradiation with microwaves, which increases the efficiency of the system and reduces the time of blanching. Huijun, Chunli, and Chunyu (2012) describe a microwave technology for blanching leafy vegetables using sodium alginate and calcium chloride solutions, followed by a vacuum packaging of the product (Patent CN102499278A). The time of inactivation of peroxidase is shorter than in a conventional blanching, improving the texture, better preservation of vitamin C, and more acceptable appearance of leafy vegetables. Deng (2012) is the inventor of the method published in Patent CN202184107U, about a microwave machine for blanching fruits and vegetables that comprises a microwave-light wave vaporizing and blanching chamber, a rotary row-knife slicing machine, a continuous pressing unit, and a chamber to store the pulp of high quality because of shorter times of the blanching process. Patent CN102150927B describes a microwave blanching multistage tandem rolling fruit and vegetable pulping machine (Deng, 2013). The fruit is blanched quickly in the microwave and optical-wave blanching chamber, and then are sliced into flakes in the rotating knife slicing machine; the flakes are squeezed into pulp through a two-stage trapezoidal tear roller. The claims include saving time and energy in the blanching and pulping process of fruits and vegetables, as well as improved nutritional value of the products.

9.2 Advantages and disadvantages of microwave-assisted blanching

General advantages of MAB include speed of operation, energy savings, precise process controls, and faster startup and shutdown times (Kidmose & Martens, 1999).

Table 9.2 Summary of microwave blanching applications to particular foods based on patents

Patent, authors, and date of publication	Conditions of microwave application	Claims
WO 2005016012 A1 Process for blanching vegetables Panaioli et al. (2005)	Blanching of the vegetable material consisting in microwave heating prior to be heated with water vapor at 120°C from 3 to 6 min before freezing	Vegetable material is heated with microwaves and in a jet stream oven. The method reduces loss of water of spinach and legumes and preserves flavor
EP 2692245 A2 Food product preparation method Cammaroto et al. (2014)	The food product is treated with microwaves and hot water simultaneously to inactivate peroxidase before freezing. The temperature used was 98–99°C at atmospheric pressure for 3 min	The enzymes are rapidly deactivated and preserves taste and appearance relative to a product conventionally blanched
CN202184107 Microwave blanching multistage continuous pressing pulping machine for fruits and vegetables. Deng (2012)	A continuous system which comprises a microwave blanching chamber a rotary row-knife slicing devices pressing unit and storage chamber	Applicable to the production of pulp and sauce from fruits and vegetables efficient blanching and pulping. It saves time and energy
CN102150927B Microwave blanching multistage tandem rolling fruit and vegetable pulping machine Deng (2013)	The fruit rotating in the microwave for fast light and steam in a multistage blanching machine and then is sliced and pressed to obtain juice and sauce. The control system ensures a temperature within 97–103°C range, preventing over-heating	The energy conservation, emission reduction and high quality fruit and vegetable juice and pulp are realized, through a high efficiency blanching and pulping process
CN102499278A Fresh-keeping technology of microwave enzyme inactivation for leafy vegetables Huijun et al. (2012)	Microwave enzyme inactivation in alkaline solution and alginate is achieved. A vacuum packing of leafy vegetables is included	An inactivation of chlorophyllase is realized, and the green color and texture is better preserved in leafy vegetables

Since MAB requires no additional water, lower leaching of vitamins and other soluble nutrients is achieved, and the generation of waste water is eliminated or greatly reduced. One of the main advantages of microwave-assisted blanching is achieving temperatures to denature enzymes of fruits, vegetables, and other commodities in a relatively short time. Besides energy savings, this fact prevents or at least minimizes the action of enzymes in the temperature range where they have highest activity, which may occur in a slow increase of temperature.

As seen in Table 9.2, two patents (Deng, 2012, 2013) cover equipment in which the size reduction of the fruit or vegetables is carried out in the same microwave cabinet with knives made of materials compatible with microwaves. This is particularly useful in the preparation of pulps, purées, and juices. The retention of properties of phytochemicals is another advantage of this type of blanching.

The main disadvantage of microwave-assisted blanching is the temperature nonuniformity often associated with microwave heating and the occurrence of cold spots. This becomes more evident when the process is scaled up and larger pieces of the commodities are treated; the treatment time is short and, therefore, equilibration of temperatures is generally not possible due to slow thermal conduction. This may to some extent be solved by the use of suitable devices that rotate or move the product in the microwave cabinet. Other inconvenient effects of traditional blanching are losses in product quality (texture and turgor), environmental impact, and energy costs. Leaching and degradation of nutritive components, such as sugars, minerals, and vitamins, may occur when blanching with water or steam. The blanching process should assure enzyme inactivation, preventing discoloration and degradation of other sensory characteristics, while minimizing the negative effects, taking into account the interdependence of every aspect (Arroqui, Rumsey, Lopez, & Virseda, 2002; Gaiser, Rathjen, & Spiess, 1996). The optimization of this unit operation relies on heat capacity properties of the foodstuff and the heat transfer. Another limiting factor for the commercial uptake of the technology is the associated capital costs for industrial scale microwave equipment; these may be acceptable, however, in cases where premium quality and retention of valuable components is required, e.g., in the pharmaceutical or nutraceutical industry.

9.3 Case studies/examples of application of microwave-assisted blanching

9.3.1 Inactivation of oxidoreductases and preservation of antioxidants

Enzymes catalyzing oxidoreduction reactions belong to the oxidoreductases class. The substrate oxidized is the electron donor which may be ascorbic acid, or natural antioxidants, or bioactive compounds, present in fruits and vegetables, such as flavonoids, anthocyanins, and other phenolic substances. Therefore, one of the main purposes of blanching is to inactivate the enzyme systems that may cause color, flavor, and textural changes. The oxidoreductases involved in deterioration of color, flavor, and loss of bioactive compounds are: peroxidase, polyphenol-oxidase, lipoxygenase, and ascorbic oxidase. The efficiency of the blanching process is usually based on the inactivation of one of the more heat resistant enzymes, i.e., peroxidase or polyphenoloxidase. Their activity causes discoloration by the oxidation of phenolic compounds and other antioxidants. The following are examples of applications of microwave-assisted blanching of fruit and vegetables, in which the target enzymes to be inactivated are oxidoreductases, and the effect of MAB on other sensory characteristics, as well as in preservation of bioactive compounds and antioxidant activity (e.g., ascorbic acid, flavonoids, carotenoids, phenolic compounds) is assessed. Comparisons with conventional blanching will be shown, and a summary is given in Table 9.3.

Table 9.3 Inactivation of oxidoreductases and preservation of antioxidants

Product	Conditions	Compounds	Results
Capsicum annuum Jalapeño type Dorantes-Alvarez et al. (2011)	Microwave pretreatment 10 g/10, 15, 20, 25, and 30 s	Phenolic antioxidant activity	Microwave inactivation of PPO of peppers was achieved applying an energy density of 0.38 KJ/g. Antioxidant activity was enhanced from 29 to 42 μM Trolox/g peppers (dry weight basis)
Mamey fruit (*Pouteria sapota*) Palma-Orozco et al. (2012)	Microwave (MT) treatment: 937 W/165 s Conventional blanching: 92°C/300 s	Polyphenol oxidase (PPO) inactivation and microstructure	PPO was completely inactivated when conventional blanching treatment was performed but required a higher temperature (92°C/300 s). Microwave-assisted blanching resulted in a negligible damage in microstructure of mamey pulp, while other blanching resulted in large damaging effects on tissue organization and shape
Green beans (*Phaseolus vulgaris* L.) Ruiz-Ojeda and Peñas (2013)	Conventional hot-water and microwave blanching	Ascorbic acid Enzyme activities: Peroxidase (POD)	No significant differences in product quality were found between hot water blanched and microwaved pods at optimal processing conditions. Shorter processing times and higher ascorbic acid retention were found

(*Continued*)

Table 9.3 (Continued)

Product	Conditions	Compounds	Results
Carrot slices Başkaya Sezer and Demirdöven (2015)	Conventional Blanching (CB) 300 s at 94°C Microwave Blanching (MB) 360 W/ 300 s/75 mL water	Peroxidase activity, pectin and carotenoid contents	POD was inactivated by 100% in 300-s blanching time of CB at 94°C. Under the same conditions, it is verified that PME is inactivated too. Also the pectin content is higher (24%) in MB in comparison with CB. The carotenoid content remained the same in both treatments

Dorantes-Alvarez et al. (2011) reported on the blanching of peppers using microwaves. The purpose of this work was to evaluate changes in antioxidant activity of *Capsicum annuum* Jalapeño type as induced by microwave-assisted blanching to inactivate polyphenoloxidase (PPO). The whole fresh peppers (85% moisture) were blended until a paste was obtained and 10 g samples were heated with microwaves. The energy density to achieve the complete inactivation of PPO, was 0.38 KJ/g. Changes in the content of phenolic compounds were confirmed using high performance liquid chromatography (HPLC), and the emergence of other phenol derivatives with enhanced antioxidant activity was detected in blanched samples. It may therefore be concluded that blanching Jalapeño peppers with microwaves can induce the formation of derivatives of phenolics which may exhibit enhanced antioxidant activity.

Sapote mamey (*Pouteria sapota)* is a tropical fruit native to Mexico and Central America; when used for commercial purposes, the quality of mamey fruit rapidly deteriorates during handling, storage, and processing. This has prevented the use of mamey fruit as a potential alternative in the development of new products in the food industry. Therefore, Palma-Orozco, Sampedro, Ortiz-Moreno, and Nájera (2012) studied the effect of microwave blanching on the polyphenol inactivation, color changes, and microstructural damage of mamey fruit. It was observed that polyphenol oxidase was completely inactivated when the temperature reached 79°C for 165 s. The optimum energy intensity for polyphenol oxidase inactivation by microwave treatment of 300 g of mamey was 0.51 kJ/g; and the microwave treatment was found to result in negligible damage of the mamey pulp's microstructure.

Ruiz-Ojeda and Peñas (2013), reported on a comparative study of conventional hot-water and microwave blanching and the effect of the processes on the quality of green beans (*Phaseolus vulgaris* L.). Batches of raw pods were treated similarly to an industrial process employing a hot-water treatment, but using a microwave oven for blanching instead. The effects of microwave processing time and nominal output power on physical properties (shrinkage, weight loss, texture, and color), enzyme activities (guaiacol peroxidase, L-ascorbate peroxidase, and catalase), and ascorbic acid content of pods were measured and modeled by first-order kinetics. Inactivation of peroxidase (POD) was the best indicator to assess the efficiency of microwave blanching of green beans. No significant differences in product quality were found between hot-water blanched and microwave treated pods at optimum processing conditions. Furthermore, since shorter processing times and higher ascorbic acid retention were found, microwave processing of green beans may be a suitable alternative to conventional blanching methods. Microwave blanching of green bean pods has been proven as a reliable alternative method to the conventional heating process used in the vegetable canning industry. The overall quality of the product processed by microwave heating under optimum conditions was comparable to that of the current industry process; in addition to an effective enzyme inactivation in a shorter processing time, the microwave process retained ascorbic acid better than what was found in the conventional blanching process.

Başkaya and Demirdöven (2015) studied the effects of microwave blanching conditions on carrot slices. The objectives of this study were: (1) to determine conventional blanching (CB) conditions providing 100% peroxidase (POD) inactivation and adapt these inactivation conditions to microwave-assisted blanching (MAB) for inactivation of pectin methylesterase (PME); (2) to optimize the MAB conditions by response surface methodology to ensure almost 100% PME inactivation; and (3) to compare the properties of the samples at the optimum points with those of conventionally blanched and unblanched carrot samples. Conventional blanching conditions for 100% peroxidase inactivation were found to be 50 g of sliced carrots for 300 s with 150 mL of blanching water at 94°C. MAB conditions for blanching 50 g of sample were optimized by using independent variables: microwave power (360–900 W), blanching time (10–300 s), and blanching water volume (0–150 mL). It was determinated that for the MAB exist three optimal process points for the PME inactivation: (1) 900 W, 170 s, 75 mL water; (2) 630 W, 190 s, 75 mL water; and (3) 360 W, 300 s, 75 mL water. In these points, PME inactivation extent was determined as almost 100%. Therefore, pectin was highly retained, carotenoid contents and color values were better protected as well.

9.3.2 Microwave-assisted blanching and extraction of pigments

Generally, the elaboration of juices from fruit and vegetables, includes extraction of pigments and other soluble compounds from the matrix. One of the consumer demands is a minimum addition or no addition of additives in food, including pigments. Therefore, the extraction of pigments is important since color is one of the most important quality attributes of juices, and due to the fact that some of the plant pigments have been reported to have biological activity such as phenolics and carotenoids (Hasler & Blumberg, 1999). In order to obtain a longer shelf life and a better stability of color and flavor in juices it is necessary to have a blanching process, since inactivation of enzymes and a reduction of microbial load is required. Several reports suggest that microwaves may be used to obtain fruit juices with a higher quantity of polyphenolic compounds (Gerard & Roberts, 2004), since they produce a quick and uniform heating, minimize changes in flavor and color, as well as extract a higher amount of these compounds. Zelong and Zhang (2010) registered a patent about a rapid extraction of lycopene, which is a carotenoid that has shown anticancer activity. The method included a pretreatment equivalent to a blanching step using ultrasound disruption of the fruit matrix and by microwave heating. Microwave-assisted extraction employs microwave energy to facilitate partition analytes from the sample matrix into the solvent. Jianli, Zhongqing, and Liang (2015) registered a patent in which they describe the use of microwave treatment to extract algae polyphenols using water as a solvent followed by a macroporous resin separation. The mass of seaweed powder and water are in a ratio of 1:2. The temperature range was between 60 to 80°C using 300 W, and the resultant extract was rich in polyphenols. This suggests potential for use as a pretreatment in diverse processes looking to maintain or increase the bioavailability of pigments. An application of microwave-assisted blanching and extraction of pigments is the processing of beet juice, whose color is due to betalains (which are thermolabile substances). The application of 450 W power for 12 min to 640 g of peeled red beet produced a sample temperature of 50–55°C, which minimized the degradation of the pigments and showed a higher quantity of betaxanthins and betacyanins compared to the control juice, although the betacyanins content was reduced (Slavov, Karagyozov, Denev, Kratchanova, & Kratchanov, 2013).

Other studies (Guzmán-Gerónimo et al., unpublished results) studied the effect of blanching of 250 g blackberry purée as a pretreatment in the preparation of the juice. The heating was performed using a microwave oven (Panasonic NN-6468, Secaucus NJ, USA) with an operating frequency of 2450 MHz and a power of 453 W for 10–60 s in which the temperatures reached 50–77°C. It was observed that microwave blanching produced an increase in total anthocyanins content in the juice with increasing treatment time. In the same way, an enhancement in color density was observed in juice when blackberry purées were heated with microwaves (Fig. 9.1).

The increase in color and in anthocyanins may be attributed to changes in the microstructure that cause liberation of the pigment that is accumulated in the vacuoles of the fruit cells. To prove this hypothesis, changes in microstructure of blackberry purée induced by microwave heating were studied using scanning electron microscopy. The sample of blackberry without treatment is shown in Fig. 9.2A; where the polyhedral shape of the cells is well defined. As can be seen in Fig. 9.2B, C, and D, upon heating with microwaves the size of the pores increases and a major disruption of the cellular structure is obtained.

These findings suggest that microwave heat treatment of the purée can be used prior to juice extraction to obtain a better extractability of anthocyanins and to

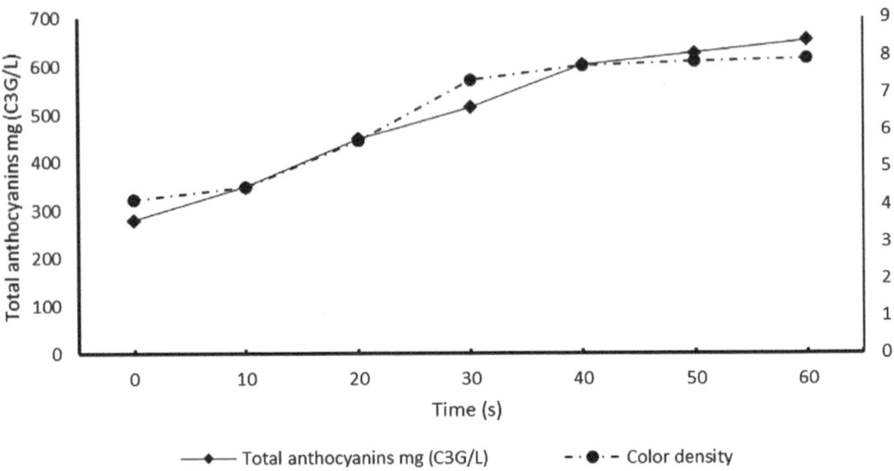

Figure 9.1 Effect of microwave pretreatment of blackberry purée on the total anthocyanins content and color density of blackberry juice.

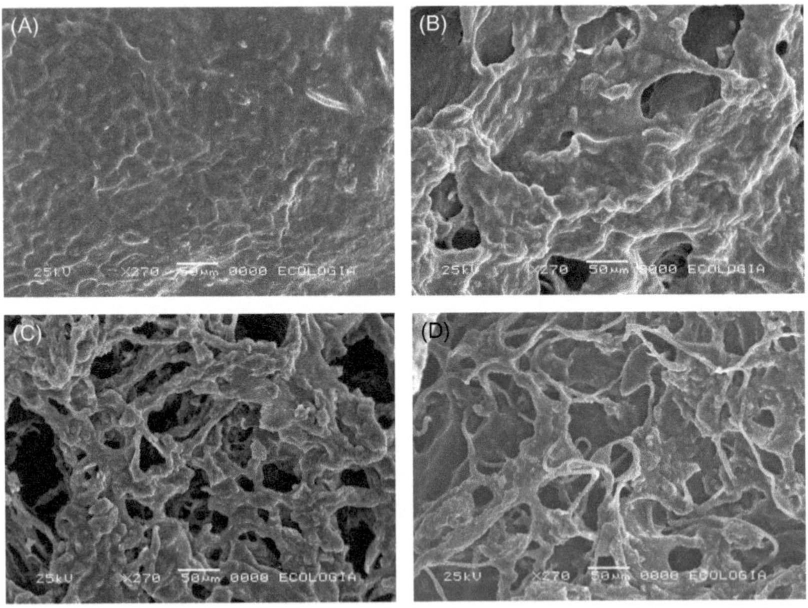

Figure 9.2 Scanning electron micrographs of blackberry purée after microwave treatment at 0 (A), 20 (B), 40 (C), and 60 s (D).

improve the color in blackberry juice. Parameters that should be considered in the use of microwave blanching pretreatment to obtain products of a better pigment content are the applied power and treatment time, the variety, matrix and size of the material, water content, and dimensions of the container.

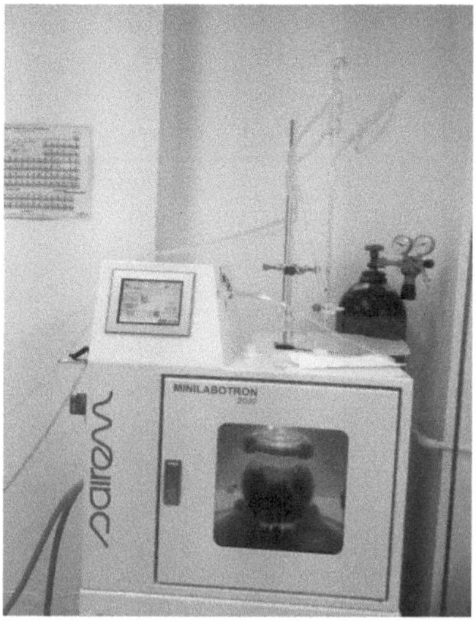

Figure 9.3 Focused microwave-assisted extraction—FMAEon the Minilabotron 2000.
Source: Courtesy of Sairem SAS France and Chemspeed SRL, Romania.

One example of microwave-assisted extraction equipment to obtain nutraceuticals and phytochemicals from food materials and residues, which may be equivalent to a microwave blanching step, is shown in Fig. 9.3.

9.3.2.1 Microwave-assisted blanching and preservation of bioactive compounds

Currently, the food industry's interest in the application of technologies that help maintain the level of bioactive compounds is increasing. However, various reports indicate that in fruits and vegetables blanched by conventional methods there is a greater loss of nutrients and bioactive compounds caused by leaching. Thus, recent research has been focused on the application of microwave-assisted blanching as a pretreatment in various processes to maintain or increase the bioactive compounds in products such as purées, juices, snacks, and among others. Table 9.4 shows the literature reports which studied the effect of microwave-assisted blanching and other techniques on the extractability and retention of bioactive compounds such as polyphenols and ascorbic acid.

It was found that the application of microwaves improved tea quality due to enzyme inactivation. Therefore, Huang, Sheng, Yang, and Hu (2007) studied the effect of microwaves and oven heating on the preservation of bioactive compounds, such as polyphenols and ascorbic acid, which are some of the components responsible for green tea quality, on tea harvested and stored in two different processing seasons (spring and autumn). The results show that the tea from spring and autumn tea processed by microwaved heating had higher content of vitamin C (4.74 g/kg and 5.51 g/kg dry mass, respectively) in comparison with those treated by oven

Table 9.4 Microwave-assisted blanching and preservation of bioactive compounds

Product	Conditions	Compounds	Results
Green tea autumn and spring processing season Huang et al. (2007)	Microwave: 6 KW for 2 min. Oven heating: 3 KW at 350°C	Polyphenols Vitamin C Chorophyll	Microwave heating increased the content of vitamin C (4.74 g/ kg and 5.51 g kg, respectively) compared to oven heating (4.46 g/ kg and 3.96 g/ kg, respectively) and the sensory quality was also higher There was no significant difference in both tea polyphenols and total chlorophylls
Loquat (*Eriobotrya japonica*) Chavez-Reyes et al. (2013)	478 W for 210 s	Phenolic compounds Antioxidant activities Polyphenol oxidase	Total phenolic content of water/ methanol extract significantly increased (295%) after microwave treatment. Methanolic extract of microwave-treated mesocarp showed higher antioxidant activity (250%) than that of fresh mesocarp. It was also noted that phenolics were more abundant in the microwaved samples than in the fresh samples
Apple purées Oszmianskia et al. (2008)	Microwave 2 min at 80°C Blanched in water at 90°C for 4 min	Procyanidins, epicathequin, chlorogenic acid Antioxidant activity	Microwave blanching maintained a higher content of phenolic substances such as procyanidins, (-) epicathequin, and chlorogenic acid when compared with sample blanched in water at 90°C for 4 min

(Continued)

Table 9.4 **(Continued)**

Product	Conditions	Compounds	Results
Apple purée varieties Fuji and McIntosh at Gerard and Roberts (2004)	Microwave blanching 1500 W for 4 to 16.2 min at 40–70°C	Phenolic compounds	The phenolic compound content increase of 5 to 34% in the apple juice, depending on the variety
Grapes varieties Barbera and Nebbiolo Segade et al. (2014)	1 and 2 W/g for 60 s	Anthocyanin	Applying microwaves of 1 W/g for 60 s maintained the disubstituted anthocyanins content and a fraction of malvidin derivatives increased as compared to untreated samples. When 2 W/g was used for 60 s, increase trisubstituted anthocyanins

heating (3.96 g/kg and 4.46 g/kg dry mass, respectively). There was no significant difference in both tea polyphenol and total chlorophyll between microwave heating and oven heating. They concluded that microwave treated tea provided higher ascorbic acid content, which decreased significantly slower during storage compared to that treated by conventional heating. The sensory quality was also improved in tea produced by microwave blanching.

The loquat (*Eriobotrya japonica* Lindl) plant is grown in subtropical areas of China, Japan, India, Israel, and the Mediterranean. Enzymatic browning greatly depreciates the potential of loquat as a food product. Chavez-Reyes et al. (2013) studied the application of microwave technology to inactivate polyphenol oxidase of loquat fruit and evaluated its effect on the phenolic profile. The antioxidant in 8 g dry mass significantly increases after microwave treatment to 1230 ± 0.36 mg GAE/100 g dry mass. Antioxidant activity also increases from 20 μM Trolox/g dry mass (fresh sample) to 82 μM Trolox/g dry mass, in the microwaved sample (225 g treated al 478 watts for 210 s). Five glycoside phenolics, 3-caffeoylquinic acid, 3-p-coumaroylquinic acid, 5-caffeoylquinic acid, and quercetin-3-O-sambubioside, were identified using a high-performance liquid chromatography system equipped with diode array detector coupled to a mass detector (HPLC-DAD-MS). The authors concluded that microwave treatment inactivated polyphenol oxidase and preserved the integrity of phenolic compounds. It was also noted that phenolics were more extractable after microwave treatment.

Studies on the application of microwave blanching for 1 kg of Idared and Shampion apple purées for a duration of 2 min at temperatures of 80°C showed that the Shampion apple purée maintained a higher content of phenolic substances such as procyanidins, (-) epicathequin, and chlorogenic acid when compared with sample blanched in water at 90°C for 4 min, while the antioxidant activity was slightly higher in the microwave processed samples. In addition, a decrease of procyanidin polymerization (DP) was observed during the storage of Shampion purée (Oszmianskia, Wolniak, Wojdyloa, & Wawerb, 2008).

Other studies on microwave blanching of 3 kg apple purées made from apple varieties Fuji and McIntosh using a nominal power of 1500 W, for 4 to 16.2 min in which the temperatures reached were in the range from 40 to 70°C, resulted in an increase from 5 to 34% phenolic compound content in the apple juice, depending on the variety (Gerard & Roberts, 2004).

Recently, microwave-assisted blanching has been explored in processes such as winemaking from grapes using varieties Barbera and Nebbiolo with the aim of increasing the anthocyanin content in the must prior to maceration. It was reported that, when applying microwaves of 1 W/g for 60 s to the Barbera grape variety, the disubstituted anthocyanin content was maintained and the fraction of malvidin derivatives increased as compared to the untreated samples. When 2 W/g was used for 60 s, a marked increase in trisubstituted anthocyanins was found. Meanwhile, in the Nebbiolo variety, the application of 1 W/g for 30–60 s did not produce changes in the anthocyanins profile as compared to the unprocessed sample (Segade et al., 2014). It is worth mentioning that in order to maintain the bioactive substances in processed fruits and vegetables, the right conditions for each individual product should be determined. This will assist in making health claims around the intake that corresponds to the established recommended daily requirement, such as 2.5 mg/kg/day and 1 g/day for anthocyanins and polyphenols, respectively (Giusti & Wrolstad, 2001). While research has been focused on establishing processing conditions that favor the content of these compounds, new research on the effect of microwave-assisted blanching for the purpose of increased bioavailability of bioactive compounds should be undertaken.

9.4 Concluding remarks and future trends

The references that are presented in this chapter as well as our experience of microwave-assisted blanching of fruits, vegetables, and catfish fillets show its advantages when compared to conventional blanching methods. In addition to the savings of time and energy, the retention and extractability of phytochemicals increased. Pigments and bioactive compounds, such as antioxidants, are preserved. The combination with other techniques, e.g., ultrasound, has been explored. While currently there is no equipment available for continuous microwave-assisted blanching at the commercial scale, such developments are expected to occur in the near future.

With the ever increasing understanding about phytochemicals and their properties, such as anticancer, antidiabetes, anticholesterol, and antioxidant activities, the interest to preserve them during processing will continue and, therefore, it is expected that applications of microwave-assisted blanching will continue to be an interesting and active research topic. With the further development of microwave technology, and investigations on combination treatments, new approaches may be identified and developed assuring the retention of biological activities of phytochemicals by microwave-assisted blanching and subsequent operations, such as freezing and drying. Equipment scale-up is challenging, but will be an important step in making such processes a commercial reality.

References

Arroqui, C., Rumsey, T. R., Lopez, A., & Virseda, P. (2002). Losses by diffusion of ascorbic acid during recycled water blanching of potato tissue. *Journal of Food Engineering*, *52*, 25–30.

Başkaya, S. D., & Demirdöven, A. (2015). The effects of microwave blanching conditions on carrot slices: Optimization and comparison. *Journal of Food Processing and Preservation*, *39*(6), 2188–2196. Available from: http://dx.doi.org/10.1111/jfpp.12463.

Binsi, P. K., Ninan, G., Zynudheen, A. A., Neethu, R., Ronda, V., & Ravishankar, C. N. (2014). Compositional and chill storage characteristics of microwave blanched sutchi catfish (*Pangasianodon hypophthalmus*) fillets. *International Journal of Food Science and Technology*, *49*, 364–372.

Cammaroto, A., Di Principe, A., Rotunno, C., Esti, M., Liburdi, K., & Benucci, I. (2014). *Food product preparation method*. Patent number EP2692245A2, Classification International A23B7/06; A23L1/01; A23L1/025; A23N12/00, applicant: CSI CIE Surgelati Italiana SPA.

Cano, P., Marin, M. A., & Fu' Ster, C. (1990). Freezing of banana slices. Influence of maturity level and thermal treatment prior to freezing. *Journal of Food Science*, *55*(4), 1070–1072.

Chavez-Reyes, Y., Dorantes-Alvarez, L., Arrieta-Baez, D., Osorio-Esquivel, O., & Ortiz-Moreno, A. (2013). Polyphenol oxidase inactivation by microwave oven and its effect on phenolic profile of loquat (*Eriobotrya japonica*) fruit. *Food and Nutrition Sciences*, *4*(9), 8.

Del Valle, J. M., Ara'nguiz, V., & Leo'n, H. (1998). Effects of blanching and calcium infiltration on PPO activity, texture, microstructure and kinetics of osmotic dehydration of apple tissue. *Food Research International*, *31*(8), 557–569.

Deng, H. (2012). *Microwave blanching multistage continuous pressing pulping machine for fruits and vegetables*. Patent number CN202184107. Classification International A23N1/02l, applicant: Haibo Deng.

Deng, H. (2013). *Microwave blanching multi-stage tandem rolling fruit and vegetable pulping machine*. Patent number CN102150927B. Classification International A23N1/02, applicant: Haibo Deng.

Dorantes-Alvarez, L., Jaramillo-Flores, E., González, K., Martinez, R., & Parada, L. (2011). Blanching peppers using microwaves. *Procedia Food Science*, *1*, 178–183.

Gaiser, M., Rathjen, A., & Spiess, W. E. L. (1996). Nitrate extraction during blanching of spinach. *Lebensm-Wiss u -Technol*, *30*(4), 432–435.

Gerard, K. A., & Roberts, J. S. (2004). Microwave heating of apple mash to improve juice yield and quality. *Society of Food Science and Technology*, *37*, 551–557.

Giusti, M. M., & Wrolstad, R. E. (2001). Unit F1.2. Anthocyanins. Characterization and measurement with UV-visible spectroscopy. In R. E. Wrolstad, T. E. Acree, E. A. Decker, M. H. Penner, D. S. Reid, S. J. Schwartz, C. H. F. Shoemaker, D. M. Smith, & P. Sporns (Eds.), *Current protocols in food analytical chemistry* (pp. 1–13). New York: Wiley & Sons.

Hasler, C. M., & Blumberg, J. B. (1999). Symposium on phytochemicals: Biochemistry and physiology. *Journal of Nutrition*, *129*, 756S–757S.

Huang, Y., Sheng, J., Yang, F., & Hu, Q. (2007). Effect of enzyme inactivation by microwave and oven heating on preservation quality of green tea. *Journal of Food Engineering*, *78*, 687–692.

Huijun, Z., Chunli, S., & Chunyu, G. (2012). *Fresh-keeping technology of microwave enzyme inactivation for leafy vegetables*. Patent number CN102499278A. Classification International: A23B7/00; A23B7/015; A23B7/154; A23B7/157, applicant: Univ Qiqihar.

Jiang, N., Liu, C., Li, D., & Zhou, Y. (2015). Effect of blanching on the dielectric properties and microwave vacuum drying behavior of *Agaricus bisporus* slices. *Innovative Food Science and Emerging Technologies*.

Jianli, S., Zhongqing, Y., & Liang, W. (2015), *Extraction method of seaweed polyphenols*. Patent number CN102935093, Classification international A61K36/02, applicant: Haiersoft Corporation, Qingdao Wenchuang Technology CO., LTD.

Kidmose, U., & Martens, H. J. (1999). Changes in texture, microstructure and nutritional quality of carrot slices during blanching and freezing. *Journal of the Science of Food and Agriculture*, *79*, 1747–1753.

Krokida, M. K., Kiranoudis, C. T., Maroulis, Z. B., & Marinos-Kouris, D. (2000). Effect of pretreatment on color of dehydrated products. *Drying Technology*, *18*, 1239–1250.

Latorre, M. E., De Escalada-Plá, M. F., Rojas, A. M., & Gerschenson, L. N. (2013). Blanching of red beet (*Beta vulgaris* L. var. conditiva) root. Effect of hot water or microwave radiation on cell wall characteristics. *LWT-Food Science and Technology*, *50*, 193–203.

Lund, D. B. (1977). Effects of commercial processing and storage on nutrients. In R. S. Harris, & E. Karnas (Eds.), *Nutritional evaluation of food processing* (pp. 203–240). Westport, CT: AVI.

Moreno, J., Chiralt, A., Escriche, I., & Serra, J. A. (2000). Effect of blanching/osmotic dehydration combined methods on quality and stability of minimally processed strawberries. *Food Research International*, *33*, 609–616.

Oszmianskia, J., Wolniak, B. M., Wojdyloa, A., & Wawerb, I. (2008). Influence of apple puree preparation and storage on polyphenol contents and antioxidant activity. *Food Chemistry*, *107*, 1473–1484.

Palma-Orozco, G., Sampedro, J. G., Ortiz-Moreno, A., & Nájera, H. (2012). In situ inactivation of polyphenol oxidase in mamey fruit (*Pouteria sapota*) by microwave treatment. *Journal of Food Science*, *77*, C359–C365.

Panaioli, S., Rotunno, C., & Simeone, V. (2005). *Blanching vegetables*. Patent number US20050037118A1, applicant: Unilever Bestfoods North America, Division Of Conopco, Inc.

Ponne, C. T., Baysal, T., & Yuksel, D. (1994). Bleaching leafy vegetables with electromagnetic energy. *Journal of Food Science*, *59*(5), 1037–1041.

Ramaswamy, H. S., & Abbatemarco, C. (1996). Thermal processing of fruitsIn H. S. Somogyi Ramaswamy, & Y. H. Hui (Eds.), *Biology, principles and applications, processing fruits science and technology* (Vol. 1, pp. 31–33). Lancaster, PA: Technomic Publishing.

Ruiz-Ojeda, L. M., & Peñas, F. J. (2013). Comparison study of conventional hot water and microwave blanching on quality of green beans. *Innovative Food Science and Emerging Technologies*, *20*, 191–197.

Segade, R. S., Torchio, F., Giacosa, S., Ricauda, A. D., Gay, P., Lambri, M., & Rolle, L. (2014). Impact of several pre-treatments on the extraction of phenolic compounds in winegrape varieties with different anthocyanin profiles and skin mechanical properties. *Journal of Agricultural and Food Chemistry*, *62*, 8427–8451.

Slavov, A., Karagyozov, V., Denev, P., Kratchanova, M., & Kratchanov, C. (2013). Antioxidant activity of red beet juices obtained after microwave and thermal pretreatments. *Czech Journal of Food Sciences*, *31*(2), 139–147.

Svensson, S. (1977). Inactivation of enzymes during thermal processing. In T. Hoyern, & O. Kvä le (Eds.), *Physical, chemical and biological changes in food caused by thermal processing* (pp. 202–217). Barking, UK: Applied Science Publishers.

Tang, J. (2005). Dielectric properties of foods. In H. Schubert, & M. Regier (Eds.), *The microwave processing of foods* (pp. 22–40). Cambridge: Woodhead Publishing limited.

Tijskens, L. M. M., Rodis, P. S., Hertog, M. L. A. T. M., Proxenia, N., & Van Dijk, C. (1999). Activity of pectin methyl esterase during blanching of peaches. *Journal of Food Engineering*, *39*, 167–177.

Tijskens, L. M. M., Rodis, P. S., Hertog, M. L. A. T. M., Waldron, K. W., Ingham, L., Proxenia, N., & Van Dijk, C. (1997). Activity of peroxidase during blanching of peaches, carrots and potatoes. *Journal of Food Engineering*, *34*, 355–370.

Trirattanapikul, W., & Phoungchandang, S. (2014). Microwave blanching and drying characteristics of *Centella asiatica* (L.) urban leaves using tray and heat pump-assisted de humidified drying. *Journal Food Science Technology*, *51*(12), 3623–3634. Available from: http://dx.doi.org/10.1007/s13197-012-0876-8.

Wang, Y., Zhang, M., Mujumdar, A. S., Mothibe, K. J., & Azam, S. M. R. (2012). 'Effect of blanching on microwave freeze-drying of stem lettuce cubes in a circular conduit drying chamber'. *Journal of Food Engineering*, *113*, 177–185.

Whitaker, J. R. (1994). *Principles of enzymology for the food sciences* (Vol. 425–433, pp. 565–573). New York: Marcel Dekker.

Zelong, L., & Zhang, L. (2010). *Method for fast extracting lycopene*. Patent number CN101121631B, Classification international C07C11/21, C07C7/10, C09B61/00, applicant: Jiangnan University.

10 Microwave-assisted pasteurization and sterilization—commercial perspective

R.A. Stanley[1] *and K. Petersen*[2]
[1]University of Tasmania, Hobart, TAS, Australia, [2]915 Labs, Denver, CO, United States

10.1 Introduction

This chapter reviews industrial developments in microwave-assisted continuous thermal pasteurization and sterilization of foods over the last 20 years and considers factors that impact on their commercial implementation.

The industrial sterilized processed foods industry started with foods canned in metal and glass replacing traditional smoked, salted, pickled, or dried products. Applications for meats, fruit, and vegetables were initially industrialized for military use but became prevalent from the late 19th century (Zanchi, 2012). While these products are still commonly made and used today, their role as stabilized convenience foods was largely supplanted in Westernized countries by the advent of frozen foods. This was enabled in the last half of the 20th century by the widespread establishment of infrastructure for distribution, retailing, and consumer appliances to support it. Although the nutritional quality of canned, and especially frozen foods, is similar to that of fresh (Rickman, Bruhn, & Barrett, 2007), the eating quality is compromised. Extended thermal processing negatively influences the fresh taste qualities of foods (Ortega-Rivas, 2009). Consumer demand has therefore fostered the rise of chilled and fast food technologies that can deliver higher levels of convenience and eating experience.

In order to meet consumer needs new food stabilization technologies must deliver to the quality, convenience, and price expectations of the changing markets. In addition, efficiencies in growing, processing, distribution, and food formulation have resulted in the relative price of food falling over the last 30 years. There has also been a trend toward globalization of markets with the development of longer national and international food supply chains putting downward pressures on prices of domestic products. This has been accompanied by an increasing proportional spend on food not prepared at home (Gambrell, 2013). Consumers are therefore more able to pay for convenience foods such as ready-meals.

From the 1960s to about 1980, Swedish Alfa-Laval scientist Lennart Stenstrom invented and built a commercial system in which food was sterilized by a combination of microwave energy, pressured hot water, and the steam within a sealed microwave permeable pack (Stenstrom, 1974a, 1974b). He termed this Multitherm

The Microwave Processing of Foods. DOI: http://dx.doi.org/10.1016/B978-0-08-100528-6.00010-3

processing (Brody, 2012). Foods were sterilized in thin microwave permeable packages using surrounding hot water to prevent edge burning at the corners due to microwave overheating. The product was conveyed through a lock to a separate pressured cold water chamber to cool the food and stop further heat-induced changes. These microwaved foods were produced and marketed commercially in Sweden but ultimately the production of machines was commercially abandoned in the 1980s (Brody, 2012; Yang, 2012). However, the potential of microwave-assisted continuous thermal pasteurization and sterilization of packaged foods was established. Improvements in packaging and processing technology, and the development of clearer regulatory pathways, are now starting to meet the growing market needs for ready-to-eat meals.

10.2 Market applications for extended shelf life foods

Continuous microwave processing is targeted to the production of packaged foods in ready-to-use forms. The global market for consumer, food service, and institutional ready-meals has grown by 3% p.a. to US$53.6 billion and 8.1 billion kg per year in 2013. Europe accounts for the largest share at 42% of the global ready-meals market with its market projected to grow to US$51.4 billion by 2018 (MarketLine Industry Profile Report, 2014).

The requirements of markets for new food processing technologies varies amongst the different market segments depending on supply chain characteristics and customer needs. However, for all market sectors, advances in meal quality, convenience, and stability are being enabled by better microwave and packaging technology. These advances will continue to generate further product improvements in the future. Resulting improvements are shifting food stabilization technologies from cook-chill, frozen, and retort canned products toward higher quality chilled and shelf stable ready-to-eat meal options (Hunter, 2011).

10.2.1 Chilled fast moving consumer goods meals

Supermarkets specialize in Fast Moving Consumer Goods (FMCG), which are small consumer products that have high volume turnover and generally low profit margin. Supermarkets are the predominate FMCG food products distribution channel with 77% of the global ready-meal markets (Marketline Industry Profile Report, 2014). They compete in the processed foods market on range, quality, and price of chilled, frozen, and shelf stable meal products. Chilled FMCG products are usually stocked in open refrigerators. They have higher quality attributes and command a higher price point compared to frozen or ambient stable products (MarketLine Industry Profile Report, 2015a, 2015b).

Many FMCG meal products, such as chilled ready-to-heat-and-serve meat and/or vegetable dishes, pastas, and casseroles, are highly perishable. If not chilled and vacuum-packed they can support rapid microbial growth and increasing oxidative rancidity. Appropriate technology is needed to deliver products that will have an adequate shelf life under commercial storage conditions while maintaining eating quality. The shelf life must cover the period required for manufacture, distribution, warehousing, display, and consumer use while maintaining food quality and safety (Chilled Foods Association, 2010).

Cook-chill meals will have a shelf life usually between 5 and 20 days depending on process factors and the presence of preservatives. Microwave processing within a sealed package can achieve longer shelf lives of up to 30 days as the product is not exposed to potential atmospheric contamination and oxygen after heating (MicVac, 2015; Pendrous, 2011). However, in general, the longer shelf life requires more intense processing which can be at the expense of the "fresh" qualities of the food. Pasteurization processing that minimizes damage to quality has advantages for chilled FMCG products (Gibbons, 2012; Hunter, 2011). Microwave-assisted pasteurization technology that can apply more rapid and even heating will maximize retention of "fresh" attributes in chilled products and have an advantage over traditional heating processes (Ortega-Rivas, 2007; Pasha, Saeed, Sultan, Khan, & Rohi, 2014).

10.2.2 Shelf stable FMCG meals

Temperature stable FMCG meal products must not deteriorate within the specified shelf life. Typically, for FMCG meals this is 12 months. The longer shelf life compared to fresh chilled items enables distributers, retailers, and the consumer more flexibility to stock items for much later sale or use. The ability to ship ambient stable or frozen meals internationally by sea-freight also opens up market competition to producers from all parts of the world.

Food products that have traditionally been canned in metal containers, such as meat meals and soups, are usually perceived to be of lower quality due to the thermal damage to eating attributes that occurs in processing. These foods are also less convenient in that they must also be extracted from the cans and reheated out of their containers. Retort pouch technologies, combined with mixing to ensure heat distribution during sterilization processing, can considerably minimize thermal processing times and therefore thermal damage (Angalet, 2003). Technologies that process meals in packaged formats by microwaves also enables the customers to reheat the meal in the same container using domestic microwave systems. This raises the convenience values of such microwave sterilized products, but quality factors such as taste and texture nevertheless remain important issues. Minimization of any degradation during processing and sterilization to within limits acceptable to consumers is still needed.

The shelf stable ready-meals market has seen an increasing growth of microwavable convenience meals over the last 20 years as technology has become available. The trend has been led by Europe where microwave processing technology was adopted faster, possibly due to more amenable pathways for regulatory approval (Tang, 2015). The quality of microwavable convenience meal offerings also appears to be improving with many new microwavable snack meals and multi-component meals being developed. Despite microwave-assisted thermal pasteurization and sterilization systems offering potential advantages for the production of FMCG microwavable convenience meals such systems are not yet widely applied.

10.2.3 Institutional and hospitality industry meals

Institutional and hospitality preprepared foods are a rapidly growing market segment. The centralized preparation of food service meals or meal components and corporate catering distribution to restaurants, cafes, schools, hospital, and age-care is progressively displacing on-site kitchens as services are outsourced. The bulk buying, centralized quality control, and nutritional management systems enable cheaper and more efficient preparation of foods and food components. These meals are then distributed to facilities such as hospitals, aged-care, and large venue catering facilities. The foods are subsequently reheated, plated with any fresh food accompaniments, and served at the point of use. This centralized processing and distribution model negates the need for multiple smaller kitchens and preparation areas. It therefore improves efficiencies and generates cost savings. Currently, most institutional preprepared meals are made using cook-chill or cook-freeze systems (IBISWorld, 2015). Rapid microwave meal pasteurization or microwave meal sterilization systems offer shelf life and quality advantages over cook-chill or cook-freeze systems for the institutional and hospitality market sector.

10.2.4 Military, remote, and emergency foods

Extended shelf life military foods, emergency foods, and foods for remote feeding situations, such as on board vessels, oil rigs, and in remote mines, are further markets that need extended shelf life foods (IBISWorld, 2015). Institutional or hospitality catering operations also need contingency provision for catastrophic failures of their meal production systems. In these instances, power requirements or reliability of power for cold storage can also be an issue. Shelf stable meals can therefore have advantages over chilled or frozen products. While current canned/retorted and dried products offer solutions with adequate shelf life, the quality of such products can be below the acceptable standards for people to routinely eat for an extended period. For military and remote feeding, high food quality can also represent a psychological compensation for the confinement or isolation of people in these situations. The requirement for improved quality shelf stable foods for the US military has driven the development of microwave-assisted thermal sterilization (MATS) technology (Tang, 2015). The critical military and specialized applications of microwave stabilized foods can therefore enable the development, demonstration, and improvement of the technology to a point that it can be adopted for lower value mass-scale commercial industry needs.

In summary, there are drivers of cost, quality, and convenience that present opportunities for new microwavable meal technologies. Developments undertaken over the last 20 years are now reaching commercial implementation and wider application. These developments could have very significant impacts on future food processing across a range of markets.

10.3 Industrial continuous microwave processing systems

Industrial food processing technology must deliver products that are safe, of high consumer quality, and at a throughput and cost that is appropriate to the market being served. While technically microwave systems are able to do this, the development and regulatory validation of successful commercial applications has been slow. Continuous thermal heating systems, such as microwave tunnels, must deliver a uniform input and distribution of energy in response to the changes in the mass and dielectric properties of foods moving through the processing apparatus. This is needed in order to uniformly heat the products to the desired temperature and times (Tang & Liu, 2015). The presence of any cold spots within the food has to be compensated for by extended thermal processing allowing time for heat transfer and temperature equilibration to ensure all parts of the food get the minimum treatment (Lee, Shin, & Yam, 2002). Nonuniform heating is thus potentially damaging to consumer quality by requiring, on average, more thermal treatment. Additionally, an effective method for determining the location and temperature of cold spots in foods during continuous microwave sterilization processing is required in order to be able to validate the safety of the food. The difficulties in determining the cold spot and demonstrating the thermal history at the cold spot have contributed to delays in obtaining regulatory approvals for continuous microwave compliance in the United States. Methods for visual location by heat activated changes in model foods and for correlation with the predicted cold spot by energy calculations have been developed (Tang, 2015).

Industrial systems must also be able to operate reliably and typically with high levels of automation in order to reduce labor costs. The integration of the mechanical, electrical, and control systems necessary for successful commercial microwave-assisted tunnel oven operation requires high levels of investment and time. Nevertheless, continuous pasteurization and continuous sterilization systems have been developed and the technology improved to the point where advantages over current conventional thermal pasteurization and retort systems are becoming apparent (Brody, 2012; Gander, 2011).

10.3.1 Microwave-assisted continuous pasteurization

10.3.1.1 Process description

Microwave pasteurization is a process that rapidly microwave heat-treats and extends the shelf life of foods while minimizing quality damage compared to conventional externally applied heat sources. Microwave processing, which heats the food directly through absorption of electromagnetic energy by the food components, can be potentially more spatially uniform than conventional heating processes that rely on thermal diffusivity from the food surface. The ability to apply microwave energy to the interior of the food can thus lower the overall time required to fully treat a food product (Datta & Davidson, 2000). Pasteurized foods, however, are usually limited in their ability to withstand heat treatment. This can be due to labile components that will degrade, such as textural characteristics or chlorophyll in green vegetables. Pasteurized foods also need to be of higher quality closer to that of freshly cooked foods in order to meet market expectations and deliver a premium over shelf stable foods. The heat is minimized to levels sufficient to control potentially pathogenic bacteria for the duration of the shelf life. The actual heat treatment required depends on the level of initial contamination. For most foods, where only low levels of bacterial cells are present before cooking, process temperatures between 65 and 90°C for times that result in reductions in bacterial numbers by a factor of 6 $\log_{10}$, are often used (Tang, 2015). For vacuum-packed low acid chilled foods, a minimum of 90°C for 10 minutes is specified to achieve a 6 $\log_{10}$ inactivation of psychrotophic nonproteolytic *Clostridium botulinum* (FDA Regulation Guidance, 2016a, 2016b; Food Standards Agency Guidance, 2008). For packaged foods, where the organism of concern is *Listeria monocytogenes*, a treatment of 72°C for 2 minutes is used to achieve a 6 $\log_{10}$ bacterial inactivation for chilled products (Chilled Foods Association, 2010).

Pasteurization treatment conditions alone, however, are insufficient to ensure elimination of all microorganisms. Spoilage and food safety hazards can result, particularly from heat-resistant spores in low acid ($pH > 4.5$), high water activity ($a_w > 0.85$) foods. The products therefore have a limited shelf life and must be kept under refrigeration for a maximum of 10 days (Food Standards Agency Guidance, 2008). Extended shelf life products, sometimes known as Refrigerated Processed Foods of Extended Durability (REPFEDs), can be made using a variety of further control steps (also referred to as hurdles) to increase storage time and assure microbial safety. In addition to mild heat treatment for pasteurization, pH, water activity, modified atmosphere packaging, and preservatives can be manipulated to enhance shelf life (Daelman, Jacxsens, Devlieghere, & Uyttendaele, 2013; Food Standards Agency Guidance, 2008). The actual shelf life achievable by additional preservation steps is governed by the treatment conditions, product parameters, and storage temperatures. It is commonly between 10 and 30 days at storage temperatures at $\leq 4°C$.

Continuous microwave pasteurization processing is possible for liquids such as juices, beverages, and milks or products such as soups that can flow through a microwave-permeable tube. The microwave becomes a supplementary or alternative source of energy to control the temperature of the fluid as it flows through the pasteurizer. This arrangement has advantages of lower heat exchange surface temperatures compared to pasteurization systems using external hot water or steam as the energy source. In these cases, the high temperature difference between the surface of the heat exchanger and contact liquid risks burn-on of temperature sensitive products such as dairy proteins fouling the surface and inhibiting heat transfer

(Coronel, Simunovic, Sandeep, Cartwright, & Kumar, 2008; Saltiel & Datta, 1999). For microwaves, the energy input in response to thermocouple readings is also more rapidly variable than the corresponding steam or hot water energy sources. This allows a precise level of tailoring of in-process heat treatment by microwave systems. The disadvantage of continuous flow pasteurizers is the need to subsequently aseptically package the foods. Aseptic packaging technology was developed beginning in the 1950s. The Tetra Pak technology method (Tetra Laval, Pully, Switzerland) passes flat, unformed packaging material through a heated 30% hydrogen peroxide bath at 70°C for 6 seconds. The hydrogen peroxide is then eliminated from the packaging material, which is then aseptically filled with product and sealed (TetraPak, Website).

Continuous pasteurization processing of meals in trays or pouches is also possible. This is needed for meal products that contain solid pieces, such as meat, pasta, or vegetables. Meals may also have multiple food components in a tray or be comprised of a viscous solid such as mashed potato. In these circumstances it is not possible to flow or mix the food during pasteurization and it must be processed within the sealed tray or pouch. The energy needs to be applied to the food with adequate input and treatment times for all the meal components to be pasteurized. The critical element becomes the ability to evenly heat all the components of the meal. In early pilot-scale trials at a microwave frequency of 896 MHz, single-mode rather than multimode microwave systems were shown to be superior in this aspect (Burfoot, Griffin, & James, 1988).

Microwave pasteurization is carried out by passing hermetically sealed food in a barrier plastic tray or pouch down a conveyer belt through the microwave tunnel. The food is heated by the microwave to the desired temperature, up to the boiling point near 100°C at atmospheric pressure, and held for the desired time. Acidified foods such as vegetables must reach a temperature of 74°C for in-pack pasteurization (Koskiniemi, Truong, Simunovic, & McFeeters, 2011). Slotted microwave waveguides have been used to distribute the energy more evenly along the heating part of the tunnel (Meredith, 1989). External heating may also be used to supplement the microwave energy and to maintain the temperature to ensure that the entire interior food mass is heated uniformly for a sufficient time. Finally, the containers pass into a cooling section to arrest temperature-induced changes caused by thermal exposure (Fig. 10.1).

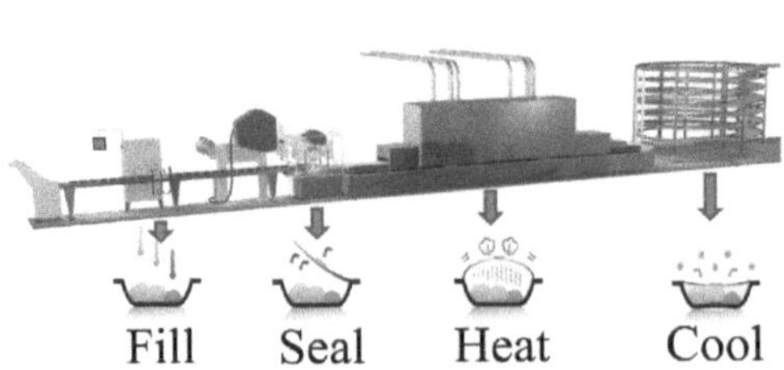

Figure 10.1 MicVac continuous microwave pasteurization process.
Source: Courtesy of MicVac, Molndal, Sweden.

As the microwave tunnels are not sealed and the microwave chamber remains open during operation to allow product to be continuously passed through the chambers, measures are needed to suppress radiation from exiting the microwave area and causing thermal injury or microwave communications interference. Metallic baffles have been used to suppress radiation from exiting the microwave (Kaufman, Schiffmann, & Stein, 1972). A dielectric liquid such as water flowing through a microwave permeable coil or slot can be used in the attenuation zone to absorb excess energy (Jeppson, 1968).

10.3.1.2 Industrial examples

There are various continuous flow microwave systems under development, such as the UHT/HTST Microthermics model (Microthermics, Raleigh, NC, USA). The Microthermics' pilot-scale unit can process liquid foods at flow rates between 48 and 180 L/h. The system uses a focused microwave heater operating at 2.45 GHz and 6 kW with a product temperature range from 60 to 150°C making sterilization as well as pasteurization possible. Integration of the time–temperature and flow rate along the heating, holding and cooling stages of the tube provides the time–temperature history of the product.

Other commercially available and industrially used microwave flow pasteurizers and sterilizers include the EnbioJet (Enbio Technology, Kosakowo-k-Gdyni, Poland), a system described as being able to control liquid temperatures to within 0.5°C (Enbiojet Brochure, 2016). Trials using the EnbioJet Microwave Flow Pasteurizer at 90–130°C showed that microbiological purity was achieved without the rapid decline of anthocyanin content observed with the same batches of juices submitted to heating at 100°C (Piasek, Kusznierewicz et al., 2011).

Pilot-scale microwave processing of nonflowable foods is possible with microwave tunnels such as the Sairem Labotron 8000 (Sairem, Neyron Cedex, France). This system operates at a frequency of 2.45 GHz and nominal power of 8 kW and can be used to develop continuous microwave-assisted thermal processes (Sairem Company Website, 2015).

There are only a limited number of industrial continuous systems specifically for pasteurization of packaged ready-meals or food ingredients currently commercialized. However, there have been many past industrial developments on microwave processing to produce packaged meals. An early commercial technology for microwave processing of pasta meals came from Officine Meccaniche Attrezzature per Ceramiche (OMAC). Engineer Giuseppe Ruozi devised a multistage in-pack microwave pasteurization and sterilization system. This system was based on a 2.45 GHz continuous microwave sterilization process and was patented with variations from 1989 to 1998 (Ruozi, 1989, 1991, 1998, 2000). The machines were capable of processing up to 2 tons of food per hour. The system was used for filled fresh pasta. Products became common in Italy in the 1990s and the technology was also applied to a variety of other foods throughout Europe, Japan, and parts of South America. In this period OMAC sold and installed over 200 microwave heat processing systems worldwide. While OMAC ceased manufacturing new machines in 1995, the

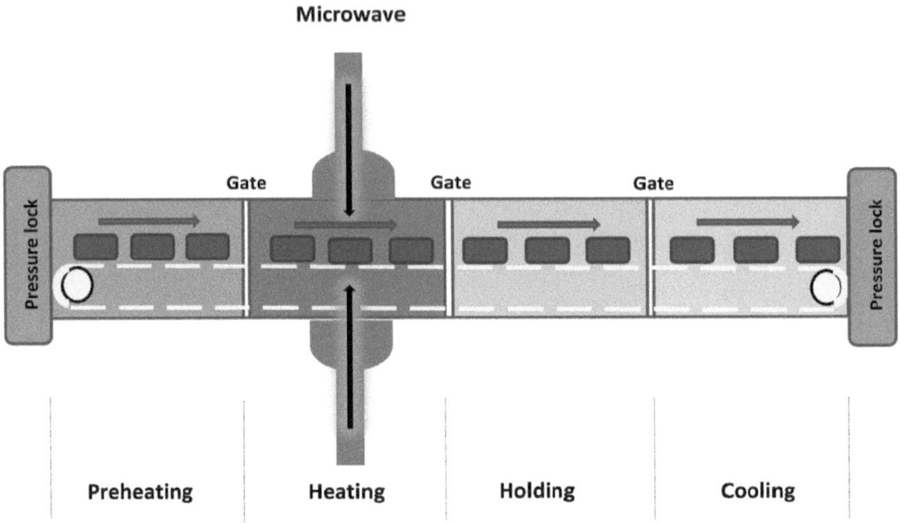

Figure 10.2 Continuous microwave sterilization schematic. Food packages are loaded through the entry lock and pass through the system on a conveyer belt to exit via another pressure lock. In the OMAC (Tops Foods) and the Gustosi systems the microwave is at 2.45 GHz and the heat transfer medium is air. In a MATS system the microwave is at 915 MHz and the pressure chamber uses water as the heat transfer medium.

Classica Group (Classica Microwave Technologies (CMT), Sayreville, NJ, USA) acquired OMAC's assets in 2000 with continued system development by Ruozi (Higgins, 2003). CMT ceased operations in 2004.

An industrial continuous in-pack meal pasteurization system at ambient atmospheric pressure (Fig. 10.2) is marketed by MicVac (Molndal, Sweden). This was originally developed by Joel Haamer for processing mussels into vacuum processed meals (Gander, 2011). Meal pack sizes range from 150 g up to 1 kg. The process utilizes a one-way steam valve that creates a slight overpressure-temperature within the pack when the microwave energy input is sufficient to create steam. The vent releases excess energy as steam in addition to flushing out oxygen. When the microwave energy is removed, the product falls below the boiling point and the steam valve seals aseptically, creating a vacuum-packaged and preserved tray of food that is cooled in a spiral chiller (Hunter, 2011). The valve is also used for reheating the food in the tray. On reaching the boiling point, the consumer hears the steam "whistle" venting as an indication that the reheat process is finished (Gilbert, 2011; Luke, 2011). Since 2006 the technology has been applied in many countries. A plant operating in Australia at Saizeriya (Melton, Victoria) processes vacuum-sealed pasta dishes for distribution to their restaurant chain (Turner, 2013). The cooking and pasteurization device is claimed to offer a shelf life of at least 30 days at 8°C (Gilbert, 2011; Hunter, 2011) although microbiological testing has shown microbial stability at ≤ 4°C of at least 6 weeks (MicVac, 2015).

Another industrial microwave pasteurization development was reported from Washington State University (WSU). The microwave-assisted pasteurization system developed at WSU by Tang and coworkers is currently at pilot-scale. The pasteurization system can semicontinuously process 8 to 20 oz (227–570 g) prepackaged chilled meals with 2 to 4 minutes of heating the product to 90°C to achieve a 6 $\log_{10}$ reduction in numbers of pathogenic bacteria. A shelf life exceeding one month at refrigeration temperatures has been demonstrated for several formulated food items, including stroganoffs, curries, burritos, and hors d'oeuvres (Zearott, 2014). An industrial version at 50 meals a minute (MAPS-50) concept is being developed by 915 Laboratories (Manning, 2015).

10.3.2 Microwave-assisted sterilization

Conditions required for sterilization depend on the pH of the food product. High acid foods such as canned fruit juice, fruits, and pickled vegetables are defined by having a pH less than 4.5. Vegetative cells of *C. botulinum* are killed in a few minutes at 60°C in high acid foods and germination of *C. botulinum* spores is suppressed at low pH. Commercial sterility requires the absence of microorganisms capable of growing in the food at normal nonrefrigerated conditions. Commercial sterility for high acid foods can be achieved by treatment at temperatures of 65–70°C for extended periods of time, unless the removal of spoilage mesophiles such as *Bacillus polymyxa* and *Bacillus macerans* from the food is necessary (Hersom & Hulland, 1980). Commercial sterility for high acid foods is usually achieved industrially by 10 min heating at 90°C (FDA Regulation Guidance, 2016a, 2016b). To preserve eating quality, more lightly processed acid foods must also use additional strategies such as lowered water activity in validated processes to prevent microbial growth (FDA Regulation Guidance, 2016a, 2016b). As commercial sterilization of high acid foods does not require temperatures exceeding 100°C, it can be achieved without a pressured vessel.

The use of temperatures above 100°C is required to achieve sterilization within feasible process times with low acid products, such as plain pasta, rice, meat, or vegetable dishes that include any one component with a pH of greater than 4.5. This can only be achieved within pressured retorts or Ultra High Temperature (UHT) rapid heat exchangers. Saturated steam, commonly at 116–121°C, can inactivate microbial cells and spores to achieve a 12 $\log_{10}$ bacterial spore reduction when held within the required time–temperature treatment conditions. Conventional methods for the determination of commercial sterility are based on first order reaction kinetics (primary model) and use extrapolation of the log-linear survival ratio versus treatment time at a constant temperature in case of an isothermal microbial heat inactivation. The temperature dependence is traditionally described by the Arrhenius equation (secondary model). Peleg (2006) has demonstrated that both models are largely doubtful to describe survival curves during microbial inactivation. Advanced models for the isothermal and nonisothermal inactivation of microorganisms, such as the Weibull distribution, are explained in the book of Peleg (2006) and will be needed to ensure an accurate prediction of food safety in microwave pasteurization and sterilization applications, where minimization of heat treatments to preserve quality is targeted.

Minimization of thermal treatment times to preserve quality has been targeted by other retort systems. Shaking retorts such as the Allpak SHAKA (Covington, LA, USA) can speed up the cycle time through movement to minimize cold spots but the products being treated must be sufficiently fluid to be mixable (Angalet, 2003). This limits the applications to soups and sauce-based meals. Conventional static industrial retorts are usually batch processes and therefore have high energy requirements due to the heating and cooling cycles. Microwave energy penetration and therefore internal heating of containerized foods offers an approach to more rapidly and efficiently sterilize meals in pack without mixing and the associated potential damage to ingredient shapes and textures.

10.3.2.1 Process description

Microwave-assisted continuous flow sterilization technology has been developed by North Carolina State University, USDA–ARS South Atlantic Area Food Science Research Unit, and Industrial Microwave Systems, L.L.C. The technology uses a 60 kW, 915 MHz continuous flow system to heat the product to sterilization temperatures (>121°C), followed by aseptic packaging into flexible containers to obtain a shelf stable product. Pumpable purees from purple-flesh sweet potatoes showed that total phenolic content increased (5.9%) and total monomeric anthocyanin content decreased (14.5%) with microwave application. However, there was high retention of other food quality attributes (Steed et al., 2008). Strawberry puree processed through continuous microwave (CM) heating at 120°C for 10 second compared to conventional thermal processing (CTP) at 90°C for 15 minutes showed that CTP caused higher degradation of polyphenols (c. 7%), anthocyanins (c. 20%), and vitamin C (c.48%) compared to CM at 120°C (Marszalek, Mitek, & Skapska, 2015).

The pressurized continuous microwave sterilization of food in pouches was patented by the US Army in 1974 (Kenyon, Berkowitz, & Ayoub, 1976). Microwave-assisted sterilization technology of packaged foods, such as defense Meals-Ready-To-Eat (MRE), is built on the packaging and processing innovations of overpressure retorting using microwave permeable flexible plastic containers. Gas pressure is applied to the autoclave to keep the pressure above the boiling point at all times in the process cycle. This enables flexible sealed plastic containers to be aseptically processed without internal steam pressures rupturing the seal integrity (Darecchio, 1983).

10.3.2.2 Industrial examples

As with continuous pasteurization systems, there are only a few known commercial continuous in-pack food sterilization systems. By design, most continuous sterilization systems can also be operated at lower temperatures to achieve only pasteurization.

Tops Foods (Olen, Belgium) (Tops Foods Company, 2014) began processing high-end, ambient stable and prolonged shelf life chilled ready-meals to the European private label retail markets from 1993 using an OMAC meal sterilization/pasteurization process. This system, which operates at a frequency of 2.45 GHz and provides a continuous microwave sterilization process for in-pack foods, was patented with variations from 1989 to 1998 (Ruozi, 1989, 1991, 1998). It uses the microwave to heat the food trays at up to 2.5 bar (250 kPa) pressurized internal atmosphere. The system consists of four linearly connected metal tunnel sections, each with a gate between to separate the thermal processes (Fig. 10.2). Multiple 2.45 GHz magnetrons are used in each of the first two sections to provide microwave heating. The third section uses heated air for holding and the fourth section is for cooling with chilled air under pressure (Ruozi, 1991; Tang, 2015).

Gustosi (Bondone, Italy) is an Italian private label manufacturer which also uses a 2.45 GHz continuous microwave sterilization system operating under pressure to produce a range of ready-meals with a 12-month ambient shelf life. A range of Italian foods, such as pastas and meats, are manufactured as ready-meals for microwave reheating (Gustosi Company Website). The food packages are arranged in multilanes down the microwave tunnel (Tang, 2015).

A development program on microwave food processing at WSU started in 1997, partially financed by the US Army Natick Soldier Center and Kraft Foods. The work was continued from 2001 with a grant from the US Department of Defense Dual Use Science and Technology (DUST) program and a microwave consortium consisting of seven food processing, packaging, and equipment companies and the US Army Natick Soldier Center (Tang, 2015). The research on 915 MHz microwave in-pack sterilization has generated the Microwave-assisted Thermal Sterilization (MATS) technology, now licensed and manufactured by 915 Laboratories (Press Release 915 Labs, 2015a, 2015b; Tang, 2015).

The MATS technology, patented in 2006, uses a 915 MHz single-mode cavity and water immersion (Tang, Liu, Pathak, & Eugene, 2006). The 915 MHz microwaves have a wavelength long enough for single-mode cavities to accommodate single-meal portion food packages which, when coupled with the water immersion, provide a more uniform heating across the meal. The 915 MHz microwaves with their longer wavelength can also penetrate further into food products, giving additional means for enhancing heating uniformity (Tang, 2015).

The WSU pilot-scale system has four zones for processing: preheating, microwave heating, holding, and cooling. Each zone has its own water circulation and external temperature with overpressure from compressed air connected to the buffer tanks of the water circulation systems. The microwave heating is generated from four 915 MHz microwave single-mode cavities. The first two cavities have relatively high microwave power at 5 kW each but the last two cavities use 3 and 2 kW, respectively, to provide for individualized process control of the final product temperature. Single-meal packages are moved on a microwave transparent conveyor through the four cavities in a single file, then on to the holding and cooling zones. Transition gates between adjacent zones separate the circulating water while allowing food packages to be conveyed on the moving belt, as illustrated in Fig. 10.2 (Tang et al., 2006).

Industrial pilot versions manufactured by 915 Laboratories (Denver, CO, USA) and labeled MATS-B have been installed at two commercial premises. These companies, Wornick Foods (Cincinnati, OH, USA) and AmeriQual (Evansville, IN, USA) run pilot contract testing researching product applications of MATS technology in collaboration with potential users. The systems have an approximate $\sim 30 \times 15$ ft^2 (9.1×4.6 m^2) footprint and can process up to 150 meals per hour (Anderton, 2015; Lingle, 2015). The MATS facility at Wornick Foods began operation in 2011 with a batch processing plant as part of the WSU industry consortium (Wornick Foods, 2011). In 2015, it was upgraded to a semicontinuous process (Wornick Foods, 2015). The AmeriQual research and development facility also began as a MATS development centre for the WSU consortium of food manufacturers and the Department of Defense Combat Feeding Directorate. MATS is being used to expand the range of high-quality components for both ration MREs and Group Feeding Heat and Serve ration entrees (Foran, 2012). The MATS pilot technology operated by Wornick Foods and AmeriQual uses a carrier tray system to convey the food through the microwave in 10.5 oz (297 g) trays. Containers are limited to a depth of about 2 inches (5.1 cm) although developments are being undertaken on 3 to 3.5 inches (7.6–8.9 cm) packages and on group ration steam trays up to 6.5 lbs (2.95 kg) (Lingle, 2015).

A full scale version of the technology, labeled as a MATS-150, and an initial purchase order for installation in 2017 has been announced by 915 Laboratories. The system has a footprint of $\sim 150 \times 15$ ft^2 (45.7×4.6 m^2) depending on auxiliary options added. The MATS system can also be used for continuous pasteurization processing by programing lower temperature and pressure conditions (Lingle, 2015; Fig. 10.3).

Figure 10.3 MATS-B plant.
Source: Courtesy of 915 Labs, Denver, CO, USA.

10.4 Issues and advantages of microwave-assisted continuous thermal processing

10.4.1 Product quality in comparison to traditional retort and microwave-assisted sterilization

The more rapid pasteurization or sterilization of foods by microwave can improve product quality over conventional retorts (Datta & Davidson, 2000). Continuous industrial systems using separated heating, holding, and collating sections enable more precise control and minimization of heat damage beyond that needed for pasteurization or sterilization. The development of intelligent systems enabling the interaction of complex heating and power parameters to be accounted in continuous processing is also enabling great progress in the resultant food quality of products (Datta & Rakesh, 2013) (see also Chapter 16: Microwave plant requirements and process control for advanced applications). However, the principal benefit derives from overcoming the limitation of heat transfer from the outside of a solid food to the inside. Raising the applied steam temperature in a steam retort above 121°C to accelerate sterilization at the center of a food can increase overall nutritional and textural damage. Microwave energy penetration of the food can therefore be advantageous by enabling High Temperature Short Time (HTST) processing that improves overall quality. For example, HTST microwave processes using 128°C final product temperature for 3 minutes cooking time improved the final texture in carrots compared with controls retorted in cans at 120°C retort temperature for 45 minutes (Lenz & Lund, 1977; Ohlsson, 1987, 1989, and others) in Tang (2015).

There is little other research on direct comparisons of microwave and steam retort products processed under identical levels of sterilization. Potential benefits can be calculated from the cumulative fraction of food that achieves a given temperature under conventional and microwave heating. The spatial heating effects from microwaves allows uniformity of temperatures to be achieved faster and thus decreases processing times (Datta & Davidson, 2000). Benefits can also be inferred from the extended processing times of static retort processing of cans compared to mixing the contents in the can such as in the SHAKA process. Total batch processing retort times for soups and sauces at 121°C of more than 123 minutes required for sterilization of 30 × 41 cm metal cans, or 54 minutes for a pouch could be reduced to less than 10 minutes using the SHAKA process at 130°C (Angalet, 2003). Kinetic data can also be found and compared in Chapter 11: Microwaves for microbial inactivation—efficiency and inactivation kinetics.

10.4.2 Throughput, processing cost, and energy considerations

Many parameters need to be taken into account when assessing the suitability and advantages of new food preservation processes for industrial implementation. Capital and operating costs as well as production capacity, reliability, and product quality for the intended target markets need to be assessed. Continuous microwave pasteurization and sterilization technology is still being refined and commercial information on costs has not been published. However, it is expected that the cost of the equipment will be reduced as increasing uptake of technology and purchase of new systems repays development costs. Technology and packaging refinements will also reduce process costs. Additionally, the capacity of the new technologies to provide ready-meal foods with new functional and sensory properties will allow new markets and values to be gained as the technology becomes more mature.

In a similar manner there is very little information to compare energy efficiencies from microwave-assisted processing compared to steam retort processes. For both processes the proportion of energy from microwaves or steam that is used to heat the food is small being in the range of 5–10% of the total thermal and electrical energy used. Most of the energy is lost to the surroundings from the water and equipment as well as the energy required to cool product. The overall energy efficiency (energy used to heat the product/total energy used in the process) of a continuous MATS industrial-scale operation was projected to be 30% compared to 13–29% for commercial batch retort processing (Peterson, 2016).

10.4.3 Product packaging

The properties of microwave energy prevent the use of metal or foil food packaging. The development of microwaved foods has driven the need for new microwave permeable packaging systems (Ohlsson & Bengtsson, 2001). There are many packaging options, in terms of high-barrier plastic pouches, trays, and lidding for pasteurization processes that are suitable for processing operations that do not exceed 100°C. In contrast sterilization processes run under changing pressures and temperatures up to 130°C pressure have provided challenges to packaging technology. The lack of available heat-resistant microwavable polymers has required development of improved systems to ensure integrity of the seal and the barrier properties of the films. Ethylene Vinyl Alcohol (EVOH) is the benchmark barrier polymer for the tray and resists retort shock better in continuous microwave processing compared to retorting (Lingle, 2015). The microwave technology allows much shorter processing times of 10–15 minutes maximum. The longer processing times of retort sterilization result in increased plasticization of the EVOH layer, causing a decrease in the oxygen barrier properties of the films (Dhawan et al., 2014; Mokwena, Tang, Dunne, Yang, & Chow, 2009).

10.4.4 Dimensional and control restrictions of microwave processing

Higher frequency 2.45 GHz microwaves have more limited ability to penetrate into foods than 915 MHz (Ohlsson & Bengtsson, 2001; Tang, 2015) (see also Chapter 2: Microwave heating and the dielectric properties of foods). Systems such as the MATS that use 915 MHz are therefore able to heat more uniformly but also cope with larger food package dimensions. While normal package systems used for the MATS have a depth up to about 5 cm, food service steam trays with a depth of up to 8.9 cm were reported as being trialed (Lingle, 2015). The MATS system has microwave windows above and below the meal package, so energy is being more uniformly applied, allowing the greater depth of energy penetration.

There are also issues in applying reliable means to monitor product temperatures for continuous microwave processing (see Chapter 15: Measuring temperature distributions during microwave processing and Chapter 17: Improving the heating uniformity in microwave processing). Ensuring every package has been uniformly heat treated is a fundamental critical issue for continuous processing compared to batch retorts. Energy monitoring directly via temperature readings or indirectly via energy adsorption is needed to assure uniform treatment levels.

10.4.5 Regulatory approval of novel pasteurization and sterilization processes

Novel thermal treatment processes must be demonstrated by research to be safe. Industrial-scale microwave-assisted pasteurization and sterilization processes developed faster in Europe than America. A factor that contributed has been the time required for validation of the safety of continuous microwave processes to the satisfaction of US federal regulators (Higgins, 2003). New technologies and protocols for ensuring the reproducibility and effectiveness of the microwave heat treatment were required. This included being able to predict and determine the location of the "cold spot" position in the food, which is the point that is slowest to reach the critical temperature. Safety validation includes putting known levels of challenge microorganisms at the cold spot within the food packages so that efficacy for thermal treatment of worse case contamination can be demonstrated.

Starting in 2009, the MATS microwave consortium at WSU sought and obtained US Food and Drug Administration acceptance of safety for processing mashed potatoes, then salmon and Alfredo sauce, and finally chicken and dumplings. Foods such as mashed potatoes, seafood, and meat are difficult to process to achieve an adequate heat treatment without impacting on food quality (Orr, 2012). In practical industrial terms however, the development of technologies to assist routine industrial validation of meal processing is still a matter for ongoing research.

10.5 Conclusions: impacts and prospects for continuous microwave processing of stabilized meals

Convenience in the form of high quality ready-to-use meals will continue to be a market driving factor for the development of continuous microwave thermal processing of meals. Over the next 10 years, the developments currently underway will have a substantial impact on product formats and marketing channels as more high quality preserved food offerings replace current canned or frozen foods. Additionally, the food hospitality and institution service sectors will seek the efficiencies and quality controls that centralized processing and distribution of meals or meal components can provide. Further improvements to the technologies and to the product quality will be developed through enhancements in process control and formulations of meals.

Sources of further information and advice

1. US Food and Drug Administration Kinetics of Microbial Inactivation for Alternative Food Processing Technologies http://www.fda.gov/Food/FoodScienceResearch/SafePracticesforFoodProcesses/ucm100158.htm.
2. Washington State University http://microwaveheating.wsu.edu/.

References

Anderton, J. (2015). *New microwave system could revolutionize food production*. www.engineering.com. Retrieved 26.02.16, from http://www.engineering.com/AdvancedManufacturing/ArticleID/10277/New-Microwave-System-Could-Revolutionize-Food-Production.aspx.

Angalet, S. (2003). *SHAKA: A new and novel processing technology to produce commercailly sterile canned foods*. Master of Science. Kansas State University.

Brody, A. L. (2012). The coming wave of microwave sterilization and pasteurization. *Food Technology*, *66*(3), 78–80.

Burfoot, D., Griffin, W. J., & James, S. J. (1988). Microwave pasteurisation of prepared meals. *Journal of Food Engineering*, *8*(3), 145–156.

Chilled Foods Association. (2010). *Shelf life of ready to eat food in relation to L. monocytogenes – Guidance for food business operators*. Retrieved 28.02.16, from http://www.chilledfood.org/wp-content/uploads/2015/08/Shelf-life-of-RTE-foods-in-relation-to-Lm-FINAL-v1.1.1-23-3-10-with-worked-examples.pdf.

Coronel, P., Simunovic, J., Sandeep, K. P., Cartwright, G. D., & Kumar, P. (2008). Sterilization solutions for aseptic processing using a continuous flow microwave system. *Journal of Food Engineering*, *85*, 528–536.

Daelman, J., Jacxsens, L., Devlieghere, F., & Uyttendaele, M. (2013). Microbial safety and quality of various types of cooked chilled foods. *Food Control*, *30*, 510–517.

Darecchio, A. (1983). *Apparatus for providing overriding pressure in heat processing sealed containers*. Patent. US4397814.

Datta, A. K., & Davidson, P. M. (2000). Microwave and radio frequency processing. *Journal of Food Science*, *65*, 32.

Datta, A. K., & Rakesh, V. (2013). Principles of microwave combination heating. *Comprehensive Reviews in Food Science & Food Safety*, *12*(1), 24.

Dhawan, S., Varney, C., Barbosa-Cánovas, G. V., Tang, J., Selim, F., & Sablani, S. S. (2014). The impact of microwave-assisted thermal sterilization on the morphology, free volume, and gas barrier properties of multilayer polymeric films. *Journal of Applied Polymer Science*, *131*(12), 40376.

Enbiojet Brochure. (2016). *EnbioJet compact flow pasteurizer*. Retrieved 18.02.16, from http://www.enbiotechnology.com/en/pliki/EnbioJet_en.pdf.

FDA Regulation Guidance. (2016a) *Chapter 16: Pathogenic bacteria survival through cooking or pasteurization*. Retrieved 26.02.16, from http://www.fda.gov/downloads/Food/GuidanceRegulation/UCM252435.pdf.

FDA Regulation Guidance. (2016b). *Chapter 17: Pathogenic bacteria survival through processes designed to retain raw product characteristics*. Retrieved 22.02.16, from http://www.fda.gov/downloads/Food/GuidanceRegulation/UCM252437.pdf.

Food Standards Agency Guidance. (2008). *Food Standards Agency guidance on the safety and shelf-life of vacuum and modified atmosphere packed chilled foods with respect to non-proteolytic Clostridium botulinum.* Retrieved 22.02.16, from http://www.food.gov.uk/sites/default/files/multimedia/pdfs/publication/vacpacguide.pdf.

Foran, A. (2012). *MATS-B microwave installed at AmeriQual assists CFD at Natick.* ARNews. Retrieved 22.02.16, from http://www.army.mil/article/87943/MATS_B_microwave_installed_at_AmeriQual_assists_CFD_at_Natick.

Gambrell, D. (2013). *Food prices: America's Shrinking Grocery Bill.* Bloomberg Business Week Bloomberg Business Week 2013 February 28.

Gander, P. (2011). Micro-wave or macro-trend? *Food Manufacture, 86*(7), 46.

Gibbons, L. (2012). Chilled foods join shelf-life revolution: Sustainable Shelf-Life Extension (SUSSLE) project. *Food Manufacture, 87*(12), 4–5.

Gilbert, H. (2011). Whistling curries: A step change in ready meals. *Grocer, 234*(8004), 11.

Gustosi Company Website. *Italian ready meals.* Retrieved 28.02.16, from http://www.gusto-si.it/.

Hersom, A. C., & Hulland, E. D. (1980). Chapter 13: Microbiology of sound canned foods – 'Commercial sterility'. In: *Canned foods: Thermal processing and microbiology* (7th ed.). Churchill Livingstone, Edinburgh.

Higgins, K. T. (2003). Microwave muscles into processing mainstream. *Food Engineering, 75*(11), 71–72.

Hunter, J. (2011) *Insight – Ready meals: The convenience factor.* Packaging Today June/July 17–18.

IBISWorld. (2015). *Corporate catering services in Australia.* Report OD5504: 1–27.

Jeppson, M. R. (1968). *Apparatus and process for microwave treatment.* Patent US3365562.

Kaufman, H. B., Schiffmann, R. F., & Stein, E. W. (1972). *End trap for microwave oven.* Patent. US3665141.

Kenyon, E. M., Berkowitz, D., & Ayoub, J. A. (1976). *Apparatus for continuous microwave sterilization of food in pouches.* Patent. US3961569.

Koskiniemi, C. B., Truong, V.-D., Simunovic, J., & McFeeters, R. F. (2011). Improvement of heating uniformity in packaged acidified vegetables pasteurized with a 915MHz continuous microwave system. *Journal of Food Engineering, 105*, 149–160.

Lee, D. S., Shin, D.-H., & Yam, K. L. (2002). Improvement of temperature uniformity in microwave-reheated rice by optimizing heat/hold cycle. *Food Service Technology, 2*(2), 87–93.

Lenz, D., & Lund, D. (1977). The lethal-Fourier number methods: Experimental verification of a model for calculating average quality factor retention in conductive heating canned foods. *Journal of Food Science, 42*(4), 997–1001.

Lingle, R. (2015). *Food packaging identified for new MATS processing.* Packaging Digest. Retrieved 28.02.16, from http://www.packagingdigest.com/food-packaging/food-packaging-id-4-new-mats-process-1506.

Luke, S. (2011). *Meal that whistles when it's cooked.* Daily Mail, Solo Syndication Ltd.: 1.

Manning, L. (2015). *4 Questions with 915 Labs' CEO after $3m seed round for sterilization and food processing technology.* AgFunder News. Retrieved 20.02.16, from https://agfundernews.com/5-questions-with-915-labs-ceo-after-3m-seed-round-for-sterilization-and-food-processing-technology5110.html.

MarketLine Industry Profile Report. (2014). *Global ready meals.* Ready Meals Industry Profile, 1–41. Progressive Digital Media Ltd.

MarketLine Industry Profile. (2015a). *Ready meals in Asia-Pacific.* Ready Meals Industry Profile,1–37. Progressive Digital Media Ltd.

MarketLine Industry Profile Report. (2015b). *Ready meals in the United States*. Ready Meals Industry Profile, 1–42. Progressive Digital Media Ltd.

Marszalek, K., Mitek, M., & Skapska, S. (2015). Effect of continuous flow microwave and conventional heating on the bioactive compounds, colour, enzymes activity, microbial and sensory quality of strawberry puree. *Food and Bioprocess Technology*, *8*(9), 1864–1876.

Meredith, R. J. (1989). *Apparatus for heating discrete packages of products using microwaves*. Patent US 4889966 A.

MicVac. (2015). *Quality without compromising on safety*. Retrieved 19.02.16, from http://www.micvac.com/customers-partners/success-stories/quality-without-compromising-on-safety.

Mokwena, K. K., Tang, J., Dunne, C. P., Yang, T. C. S., & Chow, E. (2009). Oxygen transmission of multilayer EVOH films after microwave sterilization. *Journal of Food Engineering*, *92*, 291–296.

Ohlsson, T. (1987). Sterilization of foods by microwaves. Presented at *International Seminar on New Trends in Aseptic Processing and Packaging of Foodstuffs*. Munich, Oct. 22–23.

Ohlsson, T. (1989). Dielectric properties and microwave processing. In A. G. Medina & R. P. Single (Eds.), *Food properties and computer-aided engineering of food processing systems*. Dordrecht, The Netherlands: Kluwer Academic Publishers.

Ohlsson, T., & Bengtsson, N. (2001). Microwave technology and foods. *Advances in Food and Nutrition Research*, *43*, 65–140.

Orr, S. (2012). *AmeriQual involved in "game-changer" food processing research*. Retrieved 22.02.16, from http://www.courierpress.com/business/ameriqual-involved-in-game-changer-food-processing-research-ep-444039898-324557971.html.

Ortega-Rivas, E. (2007). Processing effects for safety and quality in some non-predominant food technologies. *Critical Reviews in Food Science and Nutrition*, *47*(2), 161–173.

Ortega-Rivas, E. (2009). Processing effects on safety and quality of foods. In E. Ortega-Rivas (Ed.), *Food Preservation Technologies* (pp. 3–9). Boca Raton: CRC Press.

Pasha, I., Saeed, F., Sultan, M. T., Khan, M. R., & Rohi, M. (2014). Recent developments in minimal processing: A tool to retain nutritional quality of food. *Critical Reviews in Food Science & Nutrition*, *54*(3), 340.

Peleg, M. (2006). *Advanced quantitative microbiology for foods and biosystems: Models for predicting growth and inactivation*. Boca Raton: CRC Press.

Pendrous, R. (2011). Microwave in-pack pasteurisation gives chilled meals 30-day shelf-life. *Food Manufacture*, *86*(5), 19.

Peterson, K. (2016). Personal communication 3rd March, 2016 of results from BPA Technology Innovation Project TIP 261.

Piasek, A., Kusznierewicz, B., Grzybowska, I., Malinowska-Panczyk, E., Piekarska, A., Azqueta, A., et al. (2011). The influence of sterilization with EnbioJet (R) Microwave Flow Pasteurizer on composition and bioactivity of aronia and blue-berried honeysuckle juices. *Journal of Food Composition and Analysis*, *24*(6), 880–888.

Press Release 915 Labs. (2015a). *915 Labs leads the food processing revolution fueled by $3 million in funding*. Retrieved 26.02.16, from http://www.enhancedonlinenews.com/news/eon/20151104005478/en.

Press Release 915 Labs. (2015b). *915 Labs introduces breakthrough in food processing, paving the way for healthier eating and clean food labels*. 915-LABS-Processing, PR Newswire US. Retrieved 26.02.16, from http://www.prnewswire.com/news-releases/

915-labs-introduces-breakthrough-in-food-processing-paving-the-way-for-healthier-eating-and-clean-food-labels-300096868.html.

Rickman, J. C., Bruhn, C. M., & Barrett, D. M. (2007). Nutritional comparison of fresh, frozen, and canned fruits and vegetables II. Vitamin A and carotenoids, vitamin E, minerals and fiber. *Journal of the Science of Food and Agriculture*, *87*(7), 1185–1196.

Ruozi, G. (1989). *A method of pasteurizing or sterilizing foodstuffs utilizing microwaves, and an oven for the implementation of such a method*. Patent. EP 0347623 A1.

Ruozi, G. (1991). *Method of pasteurizing or sterilizing foodstuffs utilizing microwaves*. Patent. US5066503.

Ruozi, G. (1998). *Plant for pasteurizing or sterilising solid or liquid food products using microwaves*. Patent. US5750966 A.

Ruozi, G. (2000). *Method and apparatus for sanitizing minced meat*. Patent. US6039991 A.

Sairem Company Website. (2015). *SAIREM SAS*. Retrieved 18.02.16, from http://www.sairem.com/the-tunnels-53.html.

Saltiel, C., & Datta, A. K. (1999). Heat and mass transfer in microwave processing. *Advances in Heat Transfer*, *33*, 1–94.

Steed, L. E., Truong, V. D., Simunovic, J., Sandeep, K. P., Kumar, P., Cartwright, G. D., et al. (2008). Continuous flow microwave-assisted processing and aseptic packaging of purple-fleshed sweetpotato purees. *Journal of Food Science*, *73*(9), E455–E462.

Stenstrom, L. (1974a). *Heating of products in electro-magnetic field*. Patent. US3809845.

Stenstrom, L. (1974b). *Method and apparatus for treating heat-sensitive products*. Patent. US3809844.

Tang, J. (2015). Unlocking potentials of microwaves for food safety and quality. *Journal of Food Science*, *80*(8), E1776–E1793.

Tang, J., & Liu, F. (2015). *Microwave sterilization or pasteurization*. Patent WO2015171763A1.

Tang, J., Liu, F., Pathak, S. K., & Eugene, E. E. I. I. (2006). *Apparatus and method for heating objects with microwaves*. Patent. US7119313.

TetraPak Website. *Aseptic solutions keeping food safe and fresh*. Retrieved 22.02.16, from http://www.tetrapak.com/packaging/aseptic-solutions.

Tops Foods Company. (2014). *Revolutionizing the airline market with ovenable high-end quality ambient ready meals*. Belgium: Tops Foods Brochure. Retrieved 22.02.16, from http://www.food.be/public/uploads/company-files/790/brochure.pdf.

Turner, S. (2013). *Saizeriya Australia ready meal line*. Video Retrieved 28.02.16, from https://www.youtube.com/watch?v = HN_PoEeO4S8.

Wornick Foods. (2011). *Wornick Foods launches microwave sterilization initiative*. Retrieved 26.02.16, from http://www.prnewswire.com/news-releases/wornick-foods-launches-microwave-sterilization-initiative-117641743.html.

Wornick Foods. (2015). *Wornick Foods expands MATS Center capacity for product development projects*. Retrieved 26.02.16, from http://wornick.com/blog/2015/07/22/.

Yang, T. C. S. (2012). Thermal prcoessing of rations. In A. H. Barrett & V. Cardello (Eds.), *Military food engineering and ration technology* (pp. 41–68). Lancaster, PA: DEStech Publications Inc.

Zanchi, J. (2012). In A. H. Barrett & A. V. Cardello (Eds.), *An overview of US military field feeding and combat rations. Military food engineering and ration technology* (p. 489). Lancaster, PA: DEStech Publications Inc.

Zearott, S. (2014). *Microwave pasteurization improves food safety, flavor*. Retrieved 28.02.16, from https://news.wsu.edu/2014/02/10/microwave-pasteurization-improves-food-safety-flavor.

11

Microwaves for microbial inactivation—efficiency and inactivation kinetics

H. Bozkurt-Cekmer and P.M. Davidson
University of Tennessee, Knoxville, TN, United States

11.1 Principles of microwave inactivation

Absorption of microwave energy in the food involves two primary mechanisms: ionic interaction and dipolar rotation (Rosen, 1972). Ionic polarization occurs when an electrical field is applied to food solutions containing ions; the ions move at an accelerated pace due to their inherent charge and collide with other ions. The resulting collisions cause the conversion of kinetic energy of ionic movement into thermal energy. Water in food is the primary component responsible for dipolar rotation. Due to their dipolar nature, water molecules try to follow the electric field associated with electromagnetic radiation as it oscillates at microwave frequency (Datta & Davidson, 2000). The absorbed energy from microwaves may raise the temperature of the food high enough to inactivate microorganisms. Even though a large number of studies have proven that the thermal effect is the essential contributor to the destruction of microorganisms, some researchers have attempted to provide an explanation for non-thermal effects of microwaves on microbial inactivation.

11.2 Critical process factors influencing microbial inactivation through microwave application

11.2.1 Food

Food related factors that affect the energy absorption are dielectric properties, size, shape, orientation of food in relation to the oven, physical state of water in the product, the presence of bone in a meat product, and moisture content (Berek & Wickersheim, 1988). Following are brief descriptions of some of the more studied factors during microwave inactivation.

11.2.2 Dielectric properties

The dielectric properties of foods are the principal parameters that determine the coupling and distribution of electromagnetic energy during dielectric heating (Tang,

The Microwave Processing of Foods. DOI: http://dx.doi.org/10.1016/B978-0-08-100528-6.00011-5

2015). The dielectric properties of foods are often temperature dependent (Nelson, 1991), and therefore, must be known over the full range of temperatures experienced by the product to allow prediction of dielectric heating behavior (Wang, Wig, Tang, & Hallberg, 2003). For further information about dielectric properties of food materials, measurement techniques, and applications, the reader is referred to Chandrasekaran, Ramanathan, and Basak (2013) and Chapter 2: Microwave heating and the dielectric properties of foods and Chapter 3: Determination of the dielectric properties of foods.

11.2.3 Moisture content

The moisture content significantly affects the dielectric properties of a food and, consequently, the penetration depth of microwaves is affected (Hill, 1994). Uneven heating is observed in high moisture foods because of low microwave penetration depth. Low moisture foods have a more uniform heating rate because of deeper microwave penetration (Mudgett, 1989). In addition, in high moisture foods, heat loss through surface cooling can occur as a result of evaporation (Berek & Wickersheim, 1988).

Vela and Wu (1979) exposed various microorganisms including bacteria, actinomycetes, fungi, and bacteriophages to microwaves (2450 MHz, 1500 W, 250 g) in the presence and absence of water in soil. Each microorganism was placed into the soil (250 g) with the desired water content. They found that microorganisms were inactivated only in the presence of water and dry or lyophilized organisms were not affected even by extended exposure. They concluded that microorganisms were inactivated by thermal effects and, most likely, there were no other nonthermal effects.

11.2.4 Size and shape

The size of a food product affects the depth of microwave penetration and affects the heating rate and uniformity (Heddleson & Doores, 1994). It is also important to avoid sharp corners and thin edges on the food and containers in order to improve heating uniformity. Irregular shaped products are subjected to nonuniform heating due to the differences in product thickness (Mudgett, 1989). If the diameter of spherical or cylindrical food is 20–60 mm, the center will become much warmer than the surface because of a concentration effect, e.g., in meat balls (Hill, 1994). In rectangular or square products with negligible depth, the slab geometry, while for other depths cuboidal geometry determines the heating rates throughout the product (Decareau & Mudgett, 1985).

11.2.5 Packaging materials

It has been reported that high-density polypropylene (HDPP) is a suitable packaging material for microwave processing, and because of the associated costs, preferred over other materials that can withstand the target temperature. In the case of sterilization, in addition to glass, PET, HDPP, and various polyester-based materials are available as well as high-quality trays, pouches, and bags (Ahmed & Ramaswamy, 2007). The reader is referred to the Chapter 13: Packaging for microwave foods and Tang (2015) for detailed information about packaging materials suitable for microwave processing.

11.2.6 Process and equipment

11.2.6.1 Frequency

There are a limited number of frequency bands that are assigned by International Telecommunications Union (ITU) for industrial, scientific, and medical (ISM) applications. The important microwave frequencies allocated for ISM use are 2450 MHz (worldwide), 896 MHz (United Kingdom), and 915 MHz (United States; only used in large-scale industrial food processing systems). A detailed discussion about ISM bands can be found in several chapters of this book.

The shorter frequency of 896 or 915 MHz has a wavelength in air of approximately 33 cm and this frequency is useful for industrial applications while the longer frequency of 2450 MHz has a wavelength of approximately 12 cm and is commonly used for domestic microwave oven in worldwide (Hill, 1994). The wavelength has special significance as most interactions between energy and materials take place in that region and generate instantaneous heat due to molecular friction. With the exception of moisture, lipids, and ash, food constituents are relatively inert to the prescribed microwave frequencies. In addition, frequency (or wavelength) dictates equipment components such as magnetron, waveguide, and to some extent heating volume.

In the current literature, the most common frequency used in microwave inactivation studies was 2450 MHz. A few studies used 905 MHz (Vrhovac, Hrascan, & Franekic, 2010), 915 MHz (Heddleson, Doores, & Anantheswaran, 1994; Song & Kang, 2016), 41,600–41,800 MHz (Grundler, Keilmann, Putterlik, & Strube, 1982), or 99 GHz (Cohen, Cahan, Shani, Cohen, & Abramovich, 2010). Application of microwaves at 915 MHz has been reported to result in a more uniform temperature distribution due to greater penetration depth (Tong, Lentz, & Rossen, 1994).

11.2.6.2 Power level

Power level (and time) is one of the most critical factors that determines microbial inactivation efficiency in microwave heating. Several studies have been done to investigate the effect of power level on the inactivation of microorganisms. In a recent study, Rougier, Prorot, Chazal, Leveque, and Leprat (2014) investigated the effect of power level (200–2000 W) at a constant frequency (2450 MHz) on the membrane stability of *Escherichia coli* (5 mL) and they proved an increased loss of membrane integrity with increased power level, caused by the faster heating rate and temperatures achieved. Benlloch-Tinoco, Pina-Perez, Martinez-Navarrete, and

Rodrigo (2014) also evaluated the effect of power level (600, 900, and 1000 W) on the inactivation kinetics of *Listeria monocytogenes* during microwave heating and concluded that the higher the power level, the faster the inactivation. Similar findings were reported by several other researchers with *Listeria* spp. (Benlloch-Tinoco et al., 2014; Farber, D'aoust, Diotte, Sewell, & Daley, 1998), *Salmonella* spp. (De La Vega-Miranda, Santiesteban-Lopez, Lopez-Malo, & Sosa-Morales, 2012; Song & Kang, 2016), *Escherichia* spp. (Canumir, Celis, De Bruijn, & Vidal, 2002; Kindle, Busse, Kampa, Meyerkonig, & Daschner, 1996; Rougier et al., 2014; Yaghmaee & Durance, 2005), *Bacillus* spp. (Wang et al., 2003), *Thermoactinomyces vulgaris* and *Streptomyces albus* (Vela & Wu, 1979), and parasites (Ortega & Liao, 2006).

11.2.6.3 Others

Design (size, geometrical configuration, and other factors) of the microwave oven can significantly affect the magnitude and/or spatial variation of the power absorption in the product. In addition, presence or absence of devices added to improve uniformity, such as mode stirrers and turntables, are major factors affecting temperature distribution. The detailed information about critical process factors during microwave application can be found in Section 11.3 of this book (see also Chapter 15: Measuring temperature distributions during microwave processing, Chapter 16: Microwave plant requirements and process control for advanced applications, Chapter 17: Improving the heating uniformity in microwave processing, and Chapter 18: Simulation of microwave processes).

11.3 Inactivation of microorganisms in foods by microwave application

The effectiveness of microwaves on inactivating microorganisms in foods has been studied in dairy products (milk, skim milk), fruit products (kiwifruit puree, apple juice, puree, and cider, pineapple juice), vegetable products (tomato juice, potatoes, jalapeno peppers, asparagus, corn), meat and seafood products (frankfurters, chicken, turkey, beef loaf /minced beef, catfish fillets, and oysters), peanut butter, and coriander (Tables 11.1 and 11.3). Microorganisms tested in these studies have included: pathogenic vegetative bacteria (*Cronobacter sakazaii, E. coli, L. monocytogenes* (and its surrogate *Listeria innocua*), *Salmonella* serovars, *Staphylococcus aureus*), endospore-forming bacteria (*Bacillus* spp., *Clostridium* spp.), vegetative spoilage bacteria (*Alicyclobacillus* spp., *Pediococcus* spp., *Lactobacillus* spp., *Enterococcus* spp., *Corynebacterium* spp., *Pseudomonas* spp., *Acinetobacter calcoaceticus, Proteus vulgaris*), parasites (*Cryptosporidium parvum, Cyclospora cayetanensis*), fungi (*Brettanomyces custersii, Candida albicans*), and bacteriophages (Tables 11.1–11.3). There is no evidence that there are any pathogenic bacteria with a unique resistance to microwave heating.

Table 11.1 Inactivation of microorganisms in various foods or microbiological media by microwave heating at 50–93°C (pasteurization) using 2450 MHz

Target microorganism		Food matrix	Power (W)	*T* (°C)	Time	Log reduction (CFU/mL or g)	Reference
Gram Positive bacteria	*Listeria monocytogenes*	Frankfurters	550	65 75 85	2–19 min	0.41/min 0.65/min 0.94/min	Huang and Sites (2007)
			1100	NR	75 s	3.7	Rodriguez-Marval et al. (2009)
	Mycobacterium terrae	Milk	600	82–93	85–100 s	5	Kindle et al. (1996)
	Staphylococcus aureus	Milk	600	82–93	85–100 s	5	Kindle et al. (1996)
	Alicyclobacillus spp.	Asparagus	900	50	3–7 min	5	Giuliani, Bevilacqua, Corbo, and Severini (2010)
			300–900	< 50	0–30 min	1.11–1.74	Wang, Zhao, Liao, and Hu (2010)
	Lactobacillus plantarum	MRS broth	650	50	30 min	2–4	Shin and Pyun (1997)
Gram negative bacteria	*Cronobacter sakazaii*	Milk	600	82–93	85–100 s	5	Kindle et al. (1996)
			-	-	40–50 s	4–5	Al-Holy, Lin, Abu-Ghoush, Al-Qadiri, and Rasco (2009)
			900	78.8–88.1	0–120 s	5	Pina-Perez et al. (2014)
	Escherichia coli	Buffer	100–500	60–90	0–4 min	4–6	Fujikawa et al. (1992)
		Milk	270–900	82–93	85–100 s	5	Kindle et al. (1996)
		Apple juice	600	83	30 s	2–4	Canumir et al. (2002)

		Peptone water	510–711	53–64	0–300 s	5–6	Yaghmaee and Durance (2005)
		Catfish fillets	1250	80–90	2 min	4–5	Sheen, Huang, and Sommers (2012)
	Salmonella sv.	Corn-soy milk	60,000	56.7–82.2	3.9–10 min	3	Bookwalter, Shukla, and Kwolek (1982)
		Milk[a]	700	60	30–45 s	6	Heddleson et al. (1994)
		Catfish fillets	1250	80–90	2 min	4–5	Sheen et al. (2012)
		Jalapeno peppers	950	63	25 s	4–5	De La Vega-Miranda et al. (2012)
		Coriander			10 s		
		Potato omelet	300–800	70	40–80 s	4	Valero, Cejudo, and Garcia-Gimeno (2014)
		Peanut butter[a]	2000–4000		5 min	3.24–4.26	Song and Kang (2016)
	Klebsiella pneumoniae	Milk	600	82–93	85–100 s	5	Kindle et al. (1996)
	Pseudomonas aeruginosa	Milk	600	82–93	85–100 s	5	Kindle et al. (1996)
Yeast	*Candida albicans*	Milk	600	82–93	85–100 s	5	Kindle et al. (1996)

[a]915 MHz.

Table 11.2 Inactivation of spore-forming bacteria in microbiological media by microwave heating at 90–140°C (sterilization) using 2450 MHz

Target microorganism	Heating menstruum	Power (W)	*T* (°C)	Time (min)	Log reduction (CFU/mL or g)	Reference
Bacillus subtilis	Glass vial		140	2.7	6	Lohman and Manique (1986)
	DI Water	6000	117	45	5	Jeng, Kaczmarek, Woodworth, and Balasky (1987)
			130		5	
			137		5	
Geobacillus Bacillus stearothermophilus	Phosphate buffer	-	130	6.05	1	Perng, Yang, and Fu (1994)
			140	0.85	1	
Clostridium sporogenes	Phosphate buffer	900	90	0–30	<1	Welt et al. (1994)
			100		<1	
			110		2–3	

Table 11.3 Inactivation of microorganisms in various foods media by microwave heating (2450 MHz) using simulated cooking conditions

Target microorganism			Food matrix	Power (W[a])	*T* (°C)	Time (min)	Log reduction (CFU/mL or g)	Reference
Bacteria	Mesophiles		Minced beef	650	57.6–72.6	7–11	4	Hollywood et al. (1991)
	Psychrotrophs						4	
	Coliforms						2	
	Gram positive	*Listeria* spp.	Chicken	700	65.6	NR[b]	0.82	Harrison and Carpenter (1989)
					71.1		1.95	
					73.9		2.5	
					76.7		3.77	
					82.2		3.26	
				650	70	15	2–3	Lund et al. (1989)
				500–800	52.1–93.9	0–10	NR	Farber et al. (1998)
		Staphylococcus aureus	Minced beef	650	57.6–72.6	7–11	<1	Hollywood et al. (1991)
		Lactobacillus spp.			57.6–72.6	7–11	3–4	
		Enterococcus spp.			57.6–72.6	7–11	1	
		Brochothrix thermosphacta			57.6–72.6	7–11	3–4	
	Gram negative	*Escherichia coli*	Chicken portions	650	69.8–73.7	5–35	2–6	Apostolou et al. (2005)
			Whole chicken		60.2–92	22	NR	
		Salmonella sv.	Meat loaves	780	-	4	5	Levre and Valentini (1998)
			Corn	700	130	2	6	Anaya, Aguirrezabal, Ventura, Comellas, and Agut (2008)
			Chicken	480–800	70–90	95–140 s	5–6.4	Pucciarelli and Benassi (2005)
		Pseudomonas spp.	Minced beef	650	57.6–72.6	7–11	3–4	Hollywood et al. (1991)
Bacterial Sporeformers		Anaerobic sporeformers	Minced beef	650	57.6–72.6	7–11	<1	Hollywood et al. (1991)
		Bacillus spp.			57.6–72.6	7–11	<1	
		Bacillus subtilis	Chicken breast/liver/gizzard	900	NR	3	Inhibition zone	Javadi (2011)
Yeasts and molds			Minced beef		57.6–72.6	7–11	2–3	Hollywood et al. (1991)

[a]Watts.
[b]Not reported.

11.4 Mechanism of microwave inactivation of microorganisms

The mechanism of inactivation of microorganisms by microwave heating has been extensively studied and debated for more than half a century (Tables 11.4–11.6). There is universal agreement that the major mechanism for microbial inactivation by microwaves is heat. Contributing mechanism(s) of inactivation, so-called "nonthermal" effects caused by microwave fields have been discussed.

Kozempel, Annous, Cook, Scullen, and Whiting (1998) developed a system to evaluate nonthermal effects of microwaves. The system (5–5.4 kW) used multiple passes through a microwave generator. The flow rate was 0.96 to 1.26 kg/min and the dwell time per pass was 1.1 to 1.5 min. Microwave energy was applied but thermal energy was removed by a cooling tube within the process line in the microwave generator to maintain the temperature below 40°C. They used the microwave system at 40°C to investigate the inactivation of *Pediococcus* spp. in skim milk, water, 10% glucose solution, apple juice, pineapple juice, tomato juice, apple cider, and beer. They found that there was significant reduction in *Pediococcus* spp. in all liquids except skim milk. In order to explain the nonthermal inactivation by microwaves, they proposed four hypotheses: selective heating, electroporation, cell membrane rupture, and magnetic field coupling. They associated the differences in the inactivation of *Pediococcus* spp. in different liquids with protective effects of insoluble solids in apple cider, tomato juice, and pineapple juice but not in water, glucose solution, or apple juice. However, in a later study, Kozempel, Cook, Scullen, and Annous (2000) reported that microwave energy had no nonthermal effect on the inactivation of *Enterobacter aerogenes, E. coli, L. innocua*, or yeast in various liquids including water, liquid egg, apple juice, apple cider, or tomato juice. These studies were done in a system modified from their prior design, which used a double tube that allowed input of microwave energy but removed thermal energy with cooling water. They claimed that their prior design was not practical to achieve turbulent flow within the microwave chamber, which meant the dwell time was not uniform and there was an undefined temperature distribution. Further, they reported that, in the absence of heat, microwaves had no lethal effect on microorganisms. They did however suggest that microwaves might enhance thermal inactivation. In a similar study oinactivation of of *Enterococcus faecalis*, *S. aureus*, and *E. coli* (10 mL of stock culture was placed into 100 mL of distilled deionized water) using microwave (2450 MHz, 130 W) and conventional heating at 55 or 65°C for 1–5 min (Benjamin, Reznik, Benjamin, & Williams, 2007). A good correlation (0.84–0.99) was observed between conventional heating and microwave heating. It was concluded that no other effect, other than heat, played a role in microbial inactivation.

Further research on nonthermal effects has been done by comparing conventional and microwave heating. Hollywood, Varabioff, and Mitchell (1991) investigated the effect of microwave and conventional heating on inactivation of microorganisms including gram-positive and gram-negative bacteria, bacterial spores, yeast,

Table 11.4 **Inactivation mechanisms of bacteria by microwaves at 2450 MHz**

Target microorganism		Power (W)	Mechanism(s)	Reference
Mesophiles Psychrotrophs		650	Thermal only	Hollywood et al. (1991)
Gram positive	*Lactobacillus* spp.	650	Thermal only	Hollywood et al. (1991)
		650	Irreversible damage to cell membrane, increase of permeability of cell components	Shin and Pyun (1997)
	Staphylococcus aureus	1500	Thermal only	Vela and Wu (1979)
		NR	Increased metabolic activity	Dreyfuss and Chipley (1980)
		650	Thermal only	Hollywood et al. (1991)
		750	Thermal only	Shekarforoush and Firozi (2003)
		130	Thermal only	Benjamin et al. (2007)
		650	Thermal only	Mima et al. (2008)
		800	Thermal only	Tahir et al. (2009)
		16	Inactivation <45°	Shamis et al. (2009)
	Listeria monocytogenes	650	Thermal only	Hollywood et al. (1991)
		750	Thermal only	Shekarforoush and Firozi (2003)
	Listeria innocua	7000	Thermal only	Kozempel et al. (1998)
	Brochothrix thermosphacta	650	Thermal only	Hollywood et al. (1991)
	Enterococcus spp.	650	Thermal only	Hollywood et al. (1991)
		7000	Thermal only	Kozempel et al. (1998)
		130	Thermal only	Benjamin et al. (2007)
	Pediococcus spp.	7000	Selective heating, electroporation, cell membrane rupture, and magnetic field coupling	Kozempel et al. (1998)
		7000	Thermal only	Kozempel et al. (2000)

(*Continued*)

Table 11.4 **(Continued)**

Target microorganism		Power (W)	Mechanism(s)	Reference
Gram negative	*Escherichia* spp.	1500	Thermal only	Vela and Wu (1979)
		900	Increase in the amount of DNA and released protein, no significant reduction in cell density	Woo et al. (2000)
		7000	Thermal only	Kozempel et al. (1998)
		750	Thermal only	Shekarforoush and Firozi (2003)
		130	Thermal only	Benjamin et al. (2007)
		800	Thermal only	Tahir et al. (2009)
		16	Inactivation <45°	Shamis et al. (2009)
		NR	Increase in proliferation but not in metabolic activity	Cohen et al. (2010)[a]
		NR	Openings of pore in the membrane	Shamis et al. (2011)
		200–2000	Loss of cell membrane integrity > 400 W	Rougier et al. (2014)
	Salmonella sv.	1500	Thermal only	Vela and Wu (1979)
		700	Thermal only	Heddleson et al. (1994)
		750	Thermal only	Shekarforoush and Firozi (2003)
		700	Enhanced inactivation	Anaya et al. (2008)
	Acinetobacter calcoaceticus	700	Disintegration of cell wall	Zhou et al. (2010)
	Coliforms	650	Thermal only	Hollywood et al. (1991)
	Pseudomonas spp.	1500	Thermal only	Vela and Wu (1979)
		750	Thermal only	Shekarforoush and Firozi (2003)
		650	Thermal only	Mima et al. (2008)
		700	Disintegration of cell wall	Zhou et al. (2010)
	Corynebacterium	800	Thermal only	Tahir et al. (2009)
	Proteus vulgaris	800		
	Cronobacter sakazaii	800–900	Inactivation and cell damage	Pina-Perez et al. (2014)

[a]915 MHz.

Table 11.5 Inactivation mechanisms of bacterial spore by microwaves at 2450 MHz

Target microorganism	Power (W)	Results	Reference
Bacillus subtilis	750	Thermal only	Shekarforoush and Firozi (2003)
	100	Low DPA release and structural damage on the surface compare to conventional heating	Celandroni et al. (2004)
	2000	Aggregation of cytoplasmic proteins	Kim et al. (2008)
	650	Inactivation	Mima et al. (2008)
	700	Disintegration of cell wall	Zhou et al. (2010)
Bacillus licheniformis	2000	Cortex hydrolysis	Kim et al. (2009)
Bacillus cereus	1500	Thermal only	Vela and Wu (1979)
	650	Thermal only	Hollywood et al. (1991)
	800	Thermal only	Tahir et al. (2009)
Clostridium sporogenes	900	Thermal only	Welt et al. (1994)
Clostridium perfringens	750	Thermal only	Shekarforoush and Firozi (2003)
Streptomyces albus	1500	Thermal only	Vela and Wu (1979)
Thermoactinomyces vulgaris	1500	Thermal only	Vela and Wu (1979)

Table 11.6 Inactivation mechanisms of microorganisms other than bacteria by microwaves at 2450 MHz

Target microorganism		Frequency	Power	Effect	Reference
Molds	*Avicularia versicolor* *Penicillium brevicompactum*	2450 MHz	1500 W	Thermal only	Vela and Wu (1979)
Yeasts	*Saccharomyces cerevisiae*	41.6–41.8 GHz	–	Thermal only	Grundler et al. (1982)
		905 MHz	–	Thermal only	Vrhovac et al. (2010)
	Candida albicans	2450 MHz	650 W	Thermal only	Mima et al. (2008)
	Brettanomyces custersii	2450 MHz	0–2.2 W/g	Thermal only	Zeng, Huang, and Zhao (2014)
Parasites	*Cryptosporidium parvum*	915 MHz	2100 W	Thermal only	Collins et al. (2005)
		2450 MHz	650–1100 W	Thermal only	Ortega and Liao (2006)
	Cyclospora cayetanensis	2450 MHz	650–1100 W	Thermal only	Ortega and Liao (2006)
Bacteriophage	*Lactobacillus caseii* bacteriophage PL-1	2450 MHz	–	Thermal only	Kakita et al. (1995)
	Escherichia coli K 12-bacteriophage	2450 MHz	1500 W	Thermal only	Vela and Wu (1979)
	E. coli-bacteriophage T4	2450 MHz	1000 W	Thermal only	Bryant, Rahmanian, Tam, and Zabetian (2007)

and molds in minced beef. They observed and reported that microbial inactivation occurred only due to heating. Other researchers (Heddleson et al., 1994; Mima et al., 2008; Tahir, Mateen, Univerdi, Karagoban, & Zengin, 2009) also reported similar findings.

It is however possible that microbial injury may play a role in the differences seen in inactivation between conventional and microwave heating. Shin and Pyun (1997) exposed *Lactobacillus plantarum* cells in MRS broth to microwave and conventional heat at 50°C for 30 min and reported greater injury in cells treated with microwave radiation. Khalil and Villota (1988) compared microwave and conventional heating at 50°C for 30 min on inactivation and injury of *S. aureus* FRI-100 in phosphate buffer. They found greater injury, time in lag phase, leakage of UV absorbing material, and time for production of enterotoxin in microwave-treated cells compared with water bath-treated cells. One hypothesis was that microwaves caused greater lethality because of the instantaneous distribution of thermal energy to heat sensitive cellular components. They also observed that anaerobic conditions enhanced the recovery of microwave-treated *S. aureus*. They proposed that microwaves might catalyze certain oxidative reactions, possibly in membrane lipids, the by-products of which adversely affect cells during microwave heating. Yaghmaee and Durance (2005) also concluded that the impact of temperature on *E. coli* destruction was different when microwaves were the medium of heat transfer and suggested the existence of factors other than heat contributed to the lethal effect.

Celandroni et al. (2004) developed a microwave waveguide applicator to generate a uniform and measurable distribution of the microwave electric-field amplitude and investigated the inactivation of *B. subtilis* spores by microwaves. Microwave heating was compared with conventional heating at 100°C. There were similar survival kinetics for spores with both microwave and conventional heat. However, the cortex layer of spores subjected to conventional heat was significantly thicker than untreated spores while the cortex of microwaved-treated spores did not change. Conventionally heated spores were also found to release dipicolinic acid (DPA) upon treatment while extracellular DPA was completely undetectable in supernatants of microwave-treated spores. These observations suggested that microwave heating may promote the formation of stable complexes between DPA and other spore components (i.e., calcium ions), thus making release of DPA undetectable (Celandroni et al., 2004). *B. licheniformis* spores (2 mL) treated with microwaves (2450 MHz, 2 kW, 1 min) exhibited spore cortex hydrolysis, swelling, and finally rupture, as well as rupture of the inner membrane (Kim et al., 2009). *B. subtilis* vegetative cells (2 mL) exposed to the same conditions (2450 MHz, 2 kW, for 1 min) showed disruption of cell walls and aggregation of cytoplasmic proteins (Kim, Jo, Kim, Bai, & Park, 2008). They considered the microwave effects as nonthermal because exposure of spores and vegetative cells to conventional heat at the same temperature did not cause similar changes.

Shamis et al. (2009) used microwave heating (18 GHz, 16 W at 45 °C for 1 min) to sterilize transplant biomaterial of pathogenic bacteria (*E. coli* and *S. aureus*; 2 mL) without compromising tissue functionality and durability. They reported ~4 log CFU/mL inactivation of both pathogenic bacteria from biomaterial. This effect was suggested to be nonthermal in nature because the processing temperature was maintained at c.45°C and, based on the previously reported data, temperatures <45°C were nonlethal for *E. coli* (Heddleson & Doores, 1994). In later work, Shamis et al. (2011) investigated the electromagnetic effects of microwaves (18 GHz, 1500 kW/m^3 for 1 min) on *E. coli* (2 mL) at temperatures below 40°C. They found that the cells exhibited different morphology immediately following exposure to microwaves (cells appeared shrunken and dehydrated) compared to that observed 10 min after the exposure had ceased, at which time the morphology resembled that of the control (thermallyheated cells). Confocal laser scanning microscopy also revealed that the microwave-treated cells could take up fluorescein isothiocyanate (FITC) conjugated dextran molecules, suggesting that temporary pores had formed within the cell membrane. The formation of temporary pores within the bacterial membrane led them to conclude that the microwave-cell interaction was electrokinetic in nature.

Zhou, Shin, Hwang, Ahn, and Hwang (2010) investigated the effect of microwaves on cellular disintegration of gram-positive (*Bacillus cereus*) and gram-negative cells (*A. calcoaceticus* and *P. aeruginosa*) in municipal secondary sludge. They reported that, compared to untreated controls, exposure to microwave heating significantly increased the soluble chemical oxygen demand (1.8–4.0 fold), soluble protein concentration (1.1–1.8 fold), and soluble carbohydrate concentration (3.2–14.1 fold) in cell suspensions, with a larger increase for the gram-negative bacteria. Using scanning electron microscopy, they observed disintegration of the cell membrane in *A. calcoaceticus* and *P. aeruginosa*. Consistent with the findings of this study, other researchers also observed similar changes in the cell membrane in *C. sakazaii* (Rougier et al., 2014) and *E. coli* (Pina-Perez, Benlloch-Tinoco, Rodrigo, & Martinez, 2014).

In addition to potential nonthermal effects on cell structure, changes in metabolic activity have also been investigated (Cohen et al., 2010; Dreyfuss & Chipley, 1980). Dreyfuss et al. (Dreyfuss & Chipley, 1980) reported that microwave stressed cells of *S. aureus* exhibited a greater metabolic imbalance than conventionally heated cells. In contrast, Cohen et al. (2010) concluded that exposure of *E. coli* (1.5 mL) to microwave (99 GHz, 2 W/m^2) for 19 h, showed no change in metabolic activity, and slight increase in proliferation of the bacteria compared with control groups. Since the applied frequency affects the penetration depth, the differences between these studies might be associated with the applied frequency. Woo, Rhee, and Park (2000) observed both aggregation of cytoplasmic proteins and swelling of the cell wall in *E. coli* and *B. subtilis* (500 mL) after exposure to microwave heating at 40°C for 25 s, 60°C for 50 s, and 80°C for 80 s (2450 MHz, 600 W). Since the use of a conventional heat source (circulating water bath at 40, 60, and 80°C) did not cause similar changes, they considered the changes as a consequence of other effects.

In summary, while microwave heating has been shown to lead to potential structural changes, metabolic changes, and injury in exposed cells, these changes do not account for the bactericidal effects demonstrated with microwave heating. Therefore, when developing methods for describing inactivation kinetics of microwave heating, it is strongly recommended that only thermal effects should be included in the model (FDA, 2000).

11.5 Application of microwaves in food processing

The applications of microwaves in food processing include pasteurization and sterilization of prepackaged foods, cooking of meat products, tempering of frozen meat and poultry products, drying of various foods, baking of bread, biscuits, and confectionery products, thawing of frozen products, and blanching of vegetables (Ahmed & Ramaswamy, 2007).

11.5.1 Pasteurization

Pasteurization is a heating process that uses relatively mild heat treatment for foods to kill key pathogens and inactivate vegetative spoilage bacteria and enzymes to make food safe for consumption and increase shelf life. Microwave pasteurization systems typically take 1.5 to 2 min to raise product temperatures to 70°C. This is considered sufficient to control *L. monocytogenes* ($\geq$6 log reduction) for a shelf life of 7 days under refrigeration. For a shelf life of $\sim$9 weeks, heating at 90°C, for a minimum of 10 min in the coldest spot, is required to control nonproteolytic *C. botulinum* (Tang, 2015).

Microwave pasteurization has been tested on many food products and has been shown to be effective in inactivating a number of different microorganisms (Table 11.1). Microwave pasteurization is a process that is a viable alternative for liquid food products where temperature homogeneity is easily achieved. However, with solid products, the most important limiting factor is the nonuniform temperature distribution, which is responsible for hot and cold spots in pasteurized products (Hamoud-Agha, Curet, Simonin, & Boillereaux, 2013). The presence of a cold spot could lead to survival of pathogenic microorganisms, which could be a major obstacle to the widespread development of industrial microwave pasteurization systems especially for solid food products (Vadivambal & Jayas, 2010).

11.5.2 Sterilization

Sterilization of food by heating has been designed to achieve commercial sterility (i.e., nonrefrigerated shelf stability) of products. The magnitude of thermal treatment is a function of pH which is interrelated with the thermal resistance of certain bacterial sporeformers. Foods with high pH (low-acid foods, pH $>$4.6) support the growth of *C. botulinum* which produces a potentially lethal neurotoxin when it is allowed to grow. For a low-acid food, a thermal process sufficient to eliminate toxin-producing *C. botulinum* spores makes it commercially sterile in conjunction with packaging in a hermetically sealed container (Ahmed & Ramaswamy, 2007). Table 11.2 illustrates the resistance of several sporeforming bacteria to microwave heating including the surrogate for *C. botulinum, Clostridium sporogenes.*

Commercialization of microwave sterilization has faced several problems including nonuniform and often unpredictable electromagnetic fields in multimode microwave cavities, a lack of high-barrier microwave transparent packaging materials, and reliable means to monitor product temperature (Tang, 2015). In addition, 2450 MHz microwaves could only process foods of shallow thicknesses, due to limited penetration in foods. Recently, Tang (2015) published an extensive literature review in which he described the development of a 915 MHz single-mode microwave-assisted thermal sterilization (MATS) technology which has received regulatory acceptance.

11.5.3 Cooking

Meat and poultry products have been studied extensively using microwave heating for microbial inactivation (Table 11.3). Among these studies, several researchers have reported incomplete inactivation of microorganisms during microwave cooking (Apostolou, Papadopoulou, Levidiotou, & Ioannides, 2005; Farber et al., 1998; Harrison & Carpenter, 1989; Hollywood et al., 1991; Javadi, 2011; Levre & Valentini, 1998; Lindsay, Krissinger, & Fields, 1986; Lund, Knox & Cole, 1989). Incomplete inactivation is most likely associated with the nonuniform temperature distribution within the food sample. For example, Pucciarelli and Benassi (2005) reported that microwave cooking caused temperature gradients between the outside and inside of chicken that varied from 3 to 11°C. Nonuniform temperature distribution throughout the products being heated leads to possible pathogen survival in contaminated food products. Similarly, Levre and Valentini (1998) investigated the inactivation of *Salmonella* during microwave cooking (2450 MHz, 780 W, 3.5–4 min) of sixty meat loaves (c.600 g). They reported that 43.3% of samples exposed to microwave cooking for 3.5 min were positive for *Salmonella* while complete inactivation (5 log CFU/g) took 4 min. Apostolou et al. (2005) also reported that microwave heating of whole chickens did not eliminate *E. coli* O157:H7 because of nonhomogeneous heating and wide temperature variations ranging from 60.2 to 92.0°C. These results emphasize the need to use recommended procedures to determine the endpoint of microwave cooking of foods, such as use of a thermometer and the use of proper standing times to allow temperature equilibration and expose the microflora to elevated temperatures for an adequate period of time.

11.6 Microbial inactivation kinetics

The equations for calculations of thermal inactivation kinetics used to design and optimize conventional heating process are well established and documented. However, when it comes to microwave heating, it becomes more problematic. The main concern lies in the particular form of heating that takes place during microwave exposure (Banik, Bandyopadhyay, & Ganguly, 2003). In conventional heating, a holding period is expected, but in the case of microwaves, the heating that takes place is exclusively nonisothermal (Matsui, Gut, Oliveira, & Tadini, 2008). Furthermore, it is usually not possible with microwave heating to fix parameters that affect the heating process, such as heating rate, range of temperatures at which the samples are exposed, or appropriate sample homogenization. At present, little is known kinetically about the general basic relationship between microbial inactivation and microwave exposure.

Relatively few studies have investigated the kinetics of microbial destruction by microwave heating and even fewer have compared it with conventional heating inactivation kinetics (Table 11.7). According to the results of published studies, microbial inactivation by microwaves can be fitted using first-order kinetics. This has been successfully employed to describe destruction of *Salmonella* spp. *B. subtilis*, *B. stearothermophilus*, *C. sporogenes*, *L. plantarum*, *S. cerevisae*, *E. coli*, *L. monocytogenes*, and *C. sakazaii* (Table 11.7). However, there were no studies found on application of nonlinear models to determine microwave inactivation behavior of microorganism.

The traditional approach to describe the change in number of survivors over time for the first-order kinetic model is as follows:

$$lnN(t) = lnN_0 - kt \tag{11.1}$$

where $N(t)$ is the number of survivors after an exposure time (t) in CFU/mL and the initial population is N_0 (CFU/mL), and k as the first-order rate constant ($1/s$). This equation may then be rearranged to:

$$\log_{10}\frac{N(t)}{N_0} = -\frac{t}{D} \tag{11.2}$$

where D is the decimal reduction time (time required to reduce 90% of the existing microbial population under isothermal conditions) (Bigelow, 1921). This step is followed by the determination of the z-value which is the temperature increase (or decrease) necessary to reduce (or increase) the D-value by 90% and is indicative of the temperature dependence of microbial inactivation.

$$z = \frac{T_1 - T_2}{logD_2 - logD_1} \tag{11.3}$$

This is the conventional method for inactivation kinetics. Peleg (2006) has demonstrated that these models are largely doubtful to describe survival curves during microbial inactivation. Advanced models for the isothermal and nonisothermal inactivation of microorganisms such as the Weibull distribution are explained in the book of Peleg (2006) and will be needed to ensure an accurate prediction of food safety in microwave pasteurization and sterilization applications, where minimization of heat treatments to preserve quality are targeted.

Knowledge of the thermal inactivation parameters (D- and z-values) for a particular microorganism makes it possible to design effective thermal processes.

Based on the available data for microbial inactivation kinetics, a comparison of microwave and conventional methods on the survival of the microorganism indicates that D-values obtained for microwave heating are considerably lower than those for conventional treatment (Benlloch-Tinoco et al., 2014; Koutchma, Bail, & Ramaswamy, 2001; Ramaswamy et al., 2000; Tajchakavit, Ramaswamya, & Fustierb,

Table 11.7 **Experimental *D*- and *z*-values of microorganisms during microwave and conventional heating**

Microorganism	Food matrix		Conventional heating			Microwave heating					Reference
			T (°C)	*D*	*z*-value (°C)	Power level (W)		*T* (°C)	*D*	*z*-value (°C)	
G. stearothermo-philus	Phosphate buffer		–	–	–	–		130	6.05 s	11.73	Perng et al. (1994)
								140	0.85 s		
Bacillus subtilus	Milk		100	5.5	–	–		100	0.65	–	Wu (1996)
	0.1% NaCl solution		100	2.5		750		100	2.5	–	Celandroni et al. (2004)
Clostridium sporogenes	Phosphate buffer		90	370.17	13	900		90	370.17	13	Welt et al. (1994)
			100	62.98				100	62.98		
			110	10.71				110	10.71		
Cronobacter sakazaii	Milk		–	–	–	400		58	1.90	–	Pina-Perez et al. (2014)
						500		58	0.72		
						600		72	0.50		
						700		78	0.38		
						800		79	0.24		
						900		88	0.23		
Escherichia coli	Apple juice		55	173 s	5.15	700		55	12.98 s	8.18	Ramaswamy, Koutchma, and Tajchakavit (2000)
			60	18 s				60	6.31 s		
			65	1.99 s				65	0.78 s		
	DI water	Water bath	55	170 s	5.18	700	MW 1	55	20 s	10	Koutchma et al. (2001)
			60	18 s				60	8.3 s		
			65	2 s				65	2 s		
		Steam	55	73 s	7.21		MW 2	55	13 s	8.18	
			60	18 s				60	6.3 s		
			65	2 s				65	0.78 s		
			56	351.60	5	100		60	76.77	22.79	Fujikawa et al. (1992)
								70	15.35		
			58	135.47				90	3.29		
						200		60	46.06	24.64	
			60	49.00				70	11.52		
								90	2.56		
			62	22.80		300		60	32.90	–	
								70	10.01		
			64	8.56				90	–		
	Apple juice		83	–	–	270		38.3	3.88	–	Canumir et al. (2002)
						450		58.5	0.99		
						720		76.2	0.48		
						900		70.3	0.42		

	Apple cider	–	–	–	2000	80	0.755	–	Gentry and Roberts (2005)
	Peptone water	50.4	5.10	9.1	711	49.2	13	5.9	Yaghmaee and Durance (2005)
						51.7	6.9		
		53.1	4.10			51.8	4.1		
						55.2	0.7		
		54.7	2.60			56.5	0.6		
						57.1	0.8		
		55	1.30			57.3	0.5		
						58.4	0.5		
		57.6	0.60			61.1	0.1		
					510	48.9	28	6.2	
		58.6	0.70			52.8	9.8		
						53.3	3.8		
		60.0	0.50			54.2	2.2		
						55.6	0.5		
		60.1	0.24			56.3	0.4		
						58.1	0.4		
		63.5	0.19			58.2	0.7		
						60.3	0.5		
		65.7	0.18			60.5	0.3		
						60.6	0.1		
Listeria monocytogenes	Kiwifruit juice	50	3.40	13.59	600	60	0.71	–	Benlloch-Tinoco et al. (2014)
		55	1.18		900		0.29		
		60	0.62		1000		0.28		
Lactobacillus plantarum	Apple juice	60	26 s	–	700	60	3.7 s	–	Tajchakavit et al. (1998)
Pediococcus spp.	Water	–	–	–	7000	35	4.2	–	Kozempel et al. (1998)
	%10 glucose					35	1.5		
	Apple juice					39	0.33		
	Pineapple juice					38	10.8		
	Tomato juiceApple cider					39	1.78		
						39.2	6.15		
	Beer					39	6.05		
Salmonella sv.	Milk	–	–	–	700	60	0.69	–	Heddleson et al. (1994)
Saccharomyces cerevisae	Apple juice	60	10 s	–	700	60	0.41 s	–	Tajchakavit et al. (1998)
Yeast	Beer	–	–	–		35.7	1.62	–	Kozempel et al. (1998)

1998; Wu, 1996; Yaghmaee & Durance, 2005). In the case of inactivation of *B. subtilis* in milk, $D_{100°C}$ of microwave heating was 0.65 min but for conventional heating, it was 5.5 min. The destruction of *L. plantarum* and *Saccharomyces cerevisiae* in apple juice with microwave heating was about 6 and 25 times faster than conventional heating, respectively. These observations coincide closely with other published results for *E. coli* in apple juice (Canumir et al., 2002; Ramaswamy et al., 2000), DI water (Koutchma et al., 2001), and peptone water (Yaghmaee & Durance, 2005), and *L. monocytogenes* in kiwifruit puree (Benlloch-Tinoco et al., 2014). In particular, the *D*- values for *E. coli* in DI water in a continuous-flow microwave ranged from 13 s at 55°C to 0.78 s at 65°C and were considerably lower than both continuous flow steam heating (73 and 3 s at 55 and 65°C, respectively) or batch heating (173 and 2 s at 55 and 65°C) (Koutchma et al., 2001). Ramaswamy et al. (2000) reported similar *D*- values for *E. coli* inoculated into apple juice with the same microwave conditions. The exposure of *E. coli* inoculated apple juice was also investigated by Canumir et al. (2002) who reported *D*- values ranging from $D_{70.3°C} = 0.42$ min to $D_{38.3°C} = 3.88$ min for 900 and 270 W, respectively. Yaghmaee and Durance (2005) reported *D*- values for microwave inactivation of *E. coli* in peptone water at 510 W, with $D_{55.6°C} = 0.5$ min and $D_{60.5°C} = 0.3$ min.

Even though the microbial inactivation kinetics of *Listeria* spp. To conventional thermal treatments have been studied extensively in food products, there is limited information about survival with microwave treatments. According to Benlloch-Tinoco et al. (2014), the inactivation of *L. monocytogenes* with microwave heating at 900 W and ($D_{60°C} = 0.29$ min) and 1000 W ($D_{60°C} = 0.28$ min), but not at 600 W ($D_{60°C} = 0.71$ min), was lower than for a conventional thermal process (0.62 min). Thus, it could be concluded that the power level could affect the *D*- values for target pathogens when the different power levels and treatment times were applied to the same mass. Consistent with these findings, Fujikawa, Ushioda, and Kudo (1992) reported that application of microwave heating at 300 W led to considerably faster inactivation of *E. coli* than at 100 W for the same mass.

Only a few researchers have demonstrated that microbial inactivation kinetics for microwave heating were essentially the same as the inactivation kinetics of conventional heating (Fujikawa et al., 1992; Welt, Tong, Rossen, & Lund, 1994). Fujikawa et al. (1992) reported similar inactivation kinetics for microwave and conventional heating. However, when the available inactivation data were used, unlike their conclusion about having same *z*-values for microwave and conventional heating, the *z*-values obtained for microwave heating were considerably higher than those obtained for conventional heating (Table 11.7). Similarly, Welt et al. (1994) reported the similar inactivation kinetics for *C. sporogenes* spores. However, Welt et al. (1994) ignored the nonisothermal nature of the microwave, and used the traditional *z*-value concept to calculate *D*- value of microorganism at 121 °C during microwave. By using the *z*-value of 13°C, and the *D*- values ($D_{90°C} = 370.17$ min, $D_{100°C} = 62.98$ min, $D_{110°C} = 10.71$ min) at 90, 100, 110°C, the calculated *D*- value was 1.5 min at 121°C. Even though these studies have made a significant contribution to the literature, there were problems associated with their methodologies. Thus, predicted *D*- values at higher temperatures require further validation and verification.

Once the thermal resistance of a microorganism is known, and when the sample is subjected to a nonisothermal heating, it is possible to calculate total lethality (F, also referred to as thermal death time) for thermal treatment by integration of the time–temperature history:

$$F = \int 10^{(T-T_{\text{ref}})/z} dt \tag{11.4}$$

Using the time–temperature profile and microbial inactivation kinetics data for *E. coli* in apple cider during microwave pasteurization, a reported total lethality ($F_{80°C}$) was 3.9 min and a 5.13 log CFU/mL reduction was achieved (Gentry & Roberts, 2005). Using the same approach, an $F_{65°C}$ of 12.83 min resulted in a 5.77 log CFU/mL reduction in DI water (Koutchma et al., 2001). Other than *E. coli*, the only reported lethality was for *C. sakazaii* in milk (Pina-Perez et al., 2014). They calculated the amount of *C. sakazaii* reduction in 120 s at different power levels (600, 800, and 900 W), and reported that they achieved 2.7 log CFU/mL at 600 W, and >8 log CFU/mL at 800, and 900 W, respectively. The calculated $F_{72°C}$, $F_{79°C}$, and $F_{88°C}$ values were 1.35, 1.92, and 1.84 min, at 600, 800, and 900 W, respectively.

In conclusion, it is difficult to precisely evaluate effectiveness, and the potential of nonthermal effects, of microwave heating from the literature because of the techniques employed or a lack of detail in the methods or materials, especially in relation to temperature monitoring. Lack of temperature measurement, uneven heating due to microwave field distributions, and inability to control temperatures of microwave-treated samples during process have been cited as reasons for the difficulty resolving the thermal versus nonthermal effects controversy (Koutchma et al., 2001). A further aspect to be taken into considerations is that the ability to properly understand and carry out inactivation kinetics is important not only for the accurate design of the system but also for the establishment of the appropriate comparisons between microwave and conventional methods (Latorre, Bonelli, Rojas, & Gerschenson, 2012). An accurate comparison of microwave and conventional heating might be the basis of many studies dealing with microwave processing. Nevertheless, poor correction of processing time values for come-up times periods prior to kinetic data analysis owing to the nonisothermal nature of microwave heating may lead to erroneous interpretations, hinder comparison of different research, and cause conflicting opinions regarding to potential differences of microwave and conventional heating methods.

11.7 Microbial validation studies

Knowledge of heat penetration in foods and reliable measurement of temperature histories at desired locations are critical in developing a thermal process for the production of safe foods. The heat penetration pattern is used to evaluate heating uniformity and identify locations of slowest heating zones within a food packages. The temperature profile records the changes in food temperature over processing time to provide data for calculations of projected reduction in targeted food pathogen population or possible thermal degradation of food qualities (Holdsworth, 1985). US food safety regulations require that a thermal process is developed based on the temperature history in the slowest heating zone in food packages (FDA, 1977). As with conventional heating, time–temperature history at the coldest point also determines the microbiological safety of a microwave process (Datta & Davidson, 2000). Once the heating pattern and cold spots are determined, temperature sensors could be directly placed at the locations of cold spots to record temperature profiles for developing a thermal process to ensure food safety using Eq. (11.4) (Luan et al., 2013).

11.7.1 Microbial or chemical surrogates/indicators for validation studies

The validation of safe microwave systems is based on two major factors: (1) the selection of the microbial surrogate that has similar or higher *D*-values than the actual pathogen; (2) determination of the nature of heat-penetration characteristics of the food system, which also means temperature mapping (Awuah, Ramaswamy, & Economides, 2007). Once temperature mapping and suitable microbial surrogate are determined, then regulatory acceptance steps can be initiated.

11.7.1.1 Microbial surrogates

Since microwave processing is primarily considered as a heat process, microbiological validation tests should be designed using procedures that parallel those used for thermal processing. It has been reported that there was no microorganism with unique resistance to microwave processing, suggesting that classical surrogates (vegetative cells or spores) would be appropriate for process determination and validation.

For commercially sterilized low-acid (pH > 4.6) foods, putrefactive anaerobe *C. sporogenes* strain PA 3679 is used as a nontoxigenic surrogate for proteolytic *Clostridium botulinum* in the validation of conventional heat processes. It too would be used for derivation of thermal processes for microwave sterilization systems as a surrogate to produce a safe product (Guan, Cheng, Wang, & Tang, 2004; Tang, 2015; Welt et al., 1994). The reported $D_{121^\circ C}$ of 0.6 min for *C. sporogenes* PA 3679 is higher than that of proteolytic *C. botulinum* ($D_{121^\circ C} = 0.24$ min) and both have a *z*-value close to 10°C. Thus, thermal processes based on sufficient inactivation of spores of PA 3679 typically provide a considerable safety margin with respect to the destruction of spores of proteolytic *C. botulinum* (Brown, Tran-Dinh, & Chapman, 2012). Detailed information about the use of *C. sporogenes* PA 3679 in thermal processing as surrogate for *C. botulinum* has been discussed in detail by Brown et al. z(2012).

L. monocytogenes has been commonly used as a target pathogen for designing microwave pasteurization processes (Tang, 2015) because of its greater thermal resistance compared to *E. coli, S. enteritidis,* or *S. typhimurium* (Gabriel & Nakano, 2009; Guerrero-Beltrán & Barbosa-Cánovas, 2005). Nonpathogenic *L. innocua* strains have been used as surrogates for *L. monocytogenes* because of their behavioral and resistance similarities. However, in a recent study, Bozkurt, D'souza, and Davidson (2014) found that the thermal resistance of hepatitis A virus (HAV) ($D_{72°C} = 0.88-0.91$ min, z-value = 12.51−13.92°C) is greater than that of *L. monocytogenes* which calls for a reexamination of target pathogens and surrogates.

11.7.1.2 Chemical indicators

Food safety regulations require that a thermal process be developed based on the temperature history of the slowest heating zone of a food package (Food & Drug Administration, 1977). Thus, a fast and reliable method to monitor and predict microwave-heating pattern in foods during microwave is needed for successful development of commercial microwave processes. In order to design an effective thermal process to ensure adequate sterility for shelf-stable foods, it is essential to determine the location of cold and hot spots in packaged foods (Pandit, Tang, Mikhaylenko, & Liu, 2006).

Microwave heating is different from conventional heating in which the heating patterns are dependent upon the direct interaction between microwave energy and food and are difficult to predict. Thus, assessment of temperature distribution within packaged foods during microwave sterilization is essential, but it cannot be determined with single point each or even various points temperature measurements (Decareau & Mudgett, 1985).

Direct measurement of time−temperature history during microwave heating can be obtained by fiber optic sensors, because metal probes could not be used in microwave due to their interaction with electromagnetic field. Even though fiber optic sensors are expensive, unstable, and fragile, they are well suited to monitor temperature changes during microwave processing (Tang et al., 2008).

On the other hand, chemical marker offers an alternative as a time−temperature integrator to determine heating patterns. For this purpose, chemical markers (M-1; [2,3-dihydro-3,5-dihydroxy-6-methyl-4(H)-pyran-4-one], M-2; [4-hydroxy-5-methy-3(2 H)-furanone], and M-3; [2,3-dihydro-3,5-dihydroxy-6-methyl-4(h)-pyran-4-one]) developed by the US Army Natick Soldier Center are effective in evaluating heating patterns in selected model foods. Those markers were used to assess cumulative time−temperature effects in different processing technologies (Kim & Taub, 1993; Kim, Taub, Choi, & Prakash, 1996; Prakash, Kim, & Taub, 1997). In those studies, marker precursors were added in natural foods to enhance marker formation. It was reported that the M-2 marker is particularly suited for short-time microwave sterilization processes (Lau et al., 2003; Wang, Lau, Tang, & Mao, 2004). Lau et al. (2003) reported that M-2 yield changed in whey protein gels consisting of 20% whey protein and 1% D-ribose. At temperatures from 116 to 131°C, the chemical marker M-2 yield increased sharply with time within 10 min heating. The formation of M-2 followed 1st order kinetics, which was similar to thermal inactivation kinetics of foodborne microbial pathogens (Lau et al. 2003). First order reaction kinetics were also reported for M-2 formation in mashed potatoes made from dehydrated flakes (Pandit et al., 2006). Based on their results, standard whey protein gels and mashed potatoes with chemical marker precursors for M-2 were developed that cover the range of dielectric and thermal properties similar to those of different types of foods. They reported that the color of those model foods were pale white before thermal processing, while localized brown color was observed after microwave. They also concluded that the color intensity was dependent on the intensity of heating (Pandit et al., 2006).

Pandit, Tang, Liu, and Mikhaylenko (2007) reported a positive correlation between local thermal lethality F_0 value (equivalent lethality at 121°C with a $z = 10°C$), and color intensity as a result of M-2 formation during microwave sterilization. In their following study, they developed a computer aided vision method for rapid heating pattern determination after each microwave sterilization run (Pandit, Tang, Liu, & Pitts, 2007). In order to determine M-2 yields from multiple locations in different layers, the use of the computer vision method required only a few minutes while using HPLC took much longer (Pandit, Tang, Liu, & Pitts, 2007). This method has been used to identify cold spots in model foods as a critical step in developing microwave sterilization processes for a wide range of products. The vision method has also been used to verify 3D transient computer simulation models based on equations for microwave propagation, microwave heating, and heat transfer (Luan, Tang, Pedrow, Liu, & Tang, 2015; Resurreccion et al., 2013). The intensity of brown color developed in a whey protein gel during the microwave sterilization process was converted to values in the RGB (red, green, and blue) scale to help visualize heating pattern. According to their results, both computer simulation and chemical maker result from experiments confirm no edge heating in food (Pandit, Tang, Liu, & Pitts, 2007). Tang (2015) reviewed the use of chemical markers in microwave processing and reported that the M-2 marker is particularly suited for short-time microwave sterilization processes.

Direct measurements of temperature after microwave heating include thermocouples, thermal imaging and magnetic resonance imaging. Thermocouples are commonly used to measure the temperature distribution in foods to validate computer simulation results (Wang, Tang, Rasco, Kong, & Wang, 2008). Thermal imaging of temperature fields has also become popular method to map the surface temperature in microwave treatments. To determine the interior temperature distribution, the heated food is cut into different layers where the surface temperature is then measured by the thermal imaging method (Sakai, Mao, Koshima, & Watanabe, 2005). Magnetic resonance imaging temperature mapping has been developed for food processing research based on the sensitivity of water protons to temperature, including relaxation times, diffusion coefficients, and chemical shifts. Magnetic resonance imaging has been used to map temperature distribution after microwave treatment. The detailed information about temperature measurement distribution can also be found in Chapter 15: Measuring temperature distributions during microwave processing.

11.8 Concluding remarks and future trends

A critical examination of the use of microwave heating for microbial inactivation demonstrates the challenges in determining precise microbial inactivation kinetics, heat penetration data, and even mechanism(s) of inactivation. These challenges have proven to be major hurdles for regulatory approval of industrial microwave heating processes for commercially sterile and pasteurized products. One positive is that there have been no pathogenic microorganisms found that have any unique resistance to microwave heating. Thus, there is no reason that conventionally derived thermal inactivation kinetics for foodborne pathogens or spoilage bacteria could not be used when designing processes that utilize microwave technology. In general, there remains a definite need for further studies to bridge the gap between laboratory research and industrial applications.

References

Ahmed, J., & Ramaswamy, H. S. (2007). Microwave pasteurization and sterilization of foods. In M. S. Rahman (Ed.), *Handbook of food preservation*. Boca Raton, FL: CRC Press.

Al-Holy, M. A., Lin, M., Abu-Ghoush, M. M., Al-Qadiri, H. M., & Rasco, B. A. (2009). Thermal resistance, survival and inactivation of enterobacter sakazakii (*Cronobacter* Spp.) in powdered and reconstituted infant formula. *Journal of Food Safety*, *29*, 287–301.

Anaya, I., Aguirrezabal, A., Ventura, M., Comellas, L., & Agut, M. (2008). Survivability of Salmonella cells in popcorn after microwave oven and conventional cooking. *Microbiological Research*, *163*, 73–79.

Apostolou, I., Papadopoulou, C., Levidiotou, S., & Ioannides, K. (2005). The effect of short-time microwave exposures on *Escherichia coli* O157:H7 inoculated onto chicken meat portions and whole chickens. *International Journal of Food Microbiology*, *101*, 105–110.

Awuah, G. B., Ramaswamy, S., & Economides, A. (2007). Thermal processing and quality: Principles and overview. *Chemical Engineering Process*, *46*, 584–602.

Banik, S., Bandyopadhyay, S., & Ganguly, S. (2003). Bioeffects of microwave – A brief review. *Bioresource Technology*, *87*, 155–159.

Benjamin, E., Reznik, A., Benjamin, E., & Williams, A. L. (2007). Mathematical model of manganese ion catalyzed microwave deactivation of *Enterococcus faecalis*, *Staphylococcus aureus* and *Escherichia coli*. *Cellular and Molecular Biology*, *53*, 49–54.

Benlloch-Tinoco, M., Pina-Perez, M. C., Martinez-Navarrete, N., & Rodrigo, D. (2014). *Listeria monocytogenes* inactivation kinetics under microwave and conventional thermal processing in a kiwifruit puree. *Innovative Food Science & Emerging Technologies*, *22*, 131–136.

Berek, H. E., & Wickersheim, K. A. (1988). Measuring temperatures in microwavable packages. *Journal of Packaging Technology*, *2*, 164–168.

Bigelow, W. D. (1921). The logarithmic nature of thermal death time curves. *Journal of Infectious Diseases*, *29*, 528.

Bookwalter, G. N., Shukla, T. P., & Kwolek, W. F. (1982). Microwave processing to destroy Salmonellae in corn-soy-milk blends and effect on product quality. *Journal of Food Science*, *47*, 1683–1686.

Bozkurt, H., D'souza, D. H., & Davidson, P. M. (2014). A comparison of the thermal inactivation kinetics of human norovirus surrogates and hepatitis A virus in buffered cell culture medium. *Food Microbiology*, *42*, 212–217.

Brown, J. L., Tran-Dinh-, L., & Chapman, B. (2012). *Clostridium sporogenes* PA 3679 and its uses in the derivation of thermal processing schedules for low-acid shelf-stable foods and as a research model for proteolytic *Clostridium botulinum*. *Journal of Food Protection*, *75*, 779–792.

Bryant, S., Rahmanian, R., Tam, H., & Zabetian, S. (2007). Effects of microwave irradiation and heat on T4 bacteriophage inactivation. *Journal of Experimental Microbiology and Immunology*, *11*, 66–72.

Canumir, J. A., Celis, J. E., De Bruijn, J., & Vidal, L. V. (2002). Pasteurisation of apple juice by using microwaves. *Lebensmittel-Wissenschaft Und-Technologie-Food Science and Technology*, *35*, 389–392.

Celandroni, F., Longo, I., Tosoratti, N., Giannessi, F., Ghelardi, E., Salvetti, S., et al. (2004). Effect of microwave radiation on *Bacillus subtilis* spores. *Journal of Applied Microbiology*, *97*, 1220–1227.

Chandrasekaran, S., Ramanathan, S., & Basak, T. (2013). Microwave food processing – A review. *Food Research International*, *52*, 243–261.

Cohen, I., Cahan, R., Shani, G., Cohen, E., & Abramovich, A. (2010). Effect of 99 GHz continuous millimeter wave electro-magnetic radiation on *E. coli* viability and metabolic activity. *International Journal of Radiation Biology*, *86*, 390–399.

Collins, M. V., Flick, G. J., Smith, S. A., Fayer, R., Rubendall, E., & Lindsay, D. S. (2005). The effects of E-beam irradiation and microwave energy on Eastern Oysters (*Crassostrea virginica*) experimentally infected with *Cryptosporidium parvum*. *Journal of Eukaryotic Microbiology*, *52*, 484–488.

Datta, A. K., & Davidson, P. M. (2000). Microwave and radio frequency processing. *Journal of Food Science*, *65*, 32–41.

Decareau, R., & Mudgett, R. E. (1985). *Microwaves in the food processing industry*. Orlando, FL: Academic Press.

De La Vega-Miranda, B., Santiesteban-Lopez, N. A., Lopez-Malo, A., & Sosa-Morales, M. E. (2012). Inactivation of *Salmonella typhimurium* in fresh vegetables using water-assisted microwave heating. *Food Control*, *26*, 19–22.

Dreyfuss, M. S., & Chipley, J. R. (1980). Comparison of effects of sublethal microwave-radiation and conventional heating on the metabolic-activity of *Staphylococcus aureus*. *Applied and Environmental Microbiology*, *39*, 13–16.

Farber, J. M., D'aoust, J. Y., Diotte, M., Sewell, A., & Daley, E. (1998). Survival of *Listeria* spp. on raw whole chickens cooked in microwave ovens. *Journal of Food Protection*, *61*, 1465–1469.

Food and Drug Administration (FDA). (1977). *Code of regulations-Title 21*. Washington, DC: Food and Drug Administration, Government Services Administration.

Food and Drug Administration (FDA) (2000). *Kinetics of microbial inactivation for alternative food processing technologies – microwave and radio frequency processing*. Washington, DC: Food and Drug Administration, Government Services Administration.

Fujikawa, H., Ushioda, H., & Kudo, Y. (1992). Kinetics of *Escherichia coli* destruction by microwave irradiation. *Applied and Environmental Microbiology*, *58*, 920–924.

Gabriel, A. Z., & Nakano, H. (2009). Inactivation of Salmonella, *E. coli* and *Listeria monocytogenes* in phosphate-buffered saline and apple juice by ultraviolet and heat treatments. *Food Control*, *20*, 443–446.

Gentry, T. S., & Roberts, J. S. (2005). Design and evaluation of a continuous flow microwave pasteurization system for apple cider. *LWT-Food Science and Technology*, *38*, 227–238.

Giuliani, R., Bevilacqua, A., Corbo, M. R., & Severini, C. (2010). Use of microwave processing to reduce the initial contamination by *Alicyclobacillus acidoterrestris* in a cream of asparagus and effect of the treatment on the lipid fraction. *Innovative Food Science & Emerging Technologies*, *11*, 328–334.

Grundler, W., Keilmann, F., Putterlik, V., & Strube, D. (1982). Resonant-like dependence of yeast growth rate on microwave frequencies. *British Journal of Cancer. Supplement*, *5*, 206–208.

Guan, D., Cheng, M., Wang, Y., & Tang, J. (2004). Dielectric properties of mashed potatoes relevant to microwave and radio-frequency pasteurization and sterilization processes. *Journal of Food Science*, *69*, E30–E37.

Guerrero-Beltrán, J. A., & Barbosa-Cánovas, G. V. (2005). Reduction of *Saccharomyces cerevisiae*, *Escherichia coli* and *Listeria innocua* in apple juice by ultraviolet light. *Journal of Food Process Engineering*, *28*, 437–452.

Hamoud-Agha, M. M., Curet, S., Simonin, H., & Boillereaux, L. (2013). Microwave inactivation of *Escherichia coli* K12 CIP 54.117 in a gel medium: Experimental and numerical study. *Journal of Food Engineering*, *116*, 315–323.

Harrison, M. A., & Carpenter, S. L. (1989). Survival of *Listeria monocytogenes* on microwave cooked poultry. *Food Microbiology*, *6*, 153–157.

Heddleson, R. A., & Doores, S. (1994). Factors affecting microwave heating of foods and microwave induced destruction of food borne pathogens – A review. *Journal of Food Protection*, *11*, 952–1037.

Heddleson, R. A., Doores, S., & Anantheswaran, R. C. (1994). Parameters affecting destruction of *Salmonella* spp. by microwave-heating. *Journal of Food Science*, *59*, 447–451.

Hill, M. A. (1994). Vitamin retention in microwave cooking and cook-chill foods. *Food Chemistry*, *49*, 131–136.

Holdsworth, S. D. (1985). Optimization of thermal processing. *Journal of Food Engineering*, *4*, 89–116.

Hollywood, N. W., Varabioff, Y., & Mitchell, G. E. (1991). The effect of microwave and conventional cooking on the temperature profiles and microbial-flora of minced beef. *International Journal of Food Microbiology*, *14*, 67–75.

Huang, L. H., & Sites, J. (2007). Automatic control of a microwave heating process for in-package pasteurization of beef frankfurters. *Journal of Food Engineering*, *80*, 226–233.

Javadi, A. (2011). Effect of roasting, boiling and microwaving cooking method on doxycline residues in edible tissues of poultry by microbial method. *African Journal of Pharmacy and Pharmacology*, *5*, 1034–1037.

Jeng, D. K. H., Kaczmarek, K. A., Woodworth, A. G., & Balasky, G. (1987). Mechanism of microwave sterilization in the dry state. *Applied and Environmental Microbiology*, *53*, 2133–2137.

Kakita, Y., Kashige, N., Murata, K., Kuroiwa, A., Funatsu, M., & Watanabe, K. (1995). Inactivation of *Lactobacillus bacteriophage* PL-1 by microwave irradiation. *Microbiology and Immunology*, *39*, 571–576.

Khalil, H., & Villota, R. (1988). Comparative study on injury and recovery of *Staphylococcus aureus* using microwave and conventional heating. *Journal of Food Protection*, 51.

Kim, H. J., & Taub, I. A. (1993). Intrinsic chemical markers for aseptic processing of particulate foods. *Food Technology*, *47*, 91–97.

Kim, H. J., Taub, I. A., Choi, Y. M., & Prakash, A. (1996). Principles and applications of chemical markers of sterility in high temperature short-time processing of particulate foods. *Chemical Markers for Processed and Stored Foods*, *631*, 54–69.

Kim, S. Y., Jo, E. K., Kim, H. J., Bai, K., & Park, J. K. (2008). The effects of high-power microwaves on the ultrastructure of *Bacillus subtilis*. *Letters in Applied Microbiology*, *47*, 35–40.

Kim, S. Y., Shin, S. J., Song, C. H., Jo, E. K., Kim, H. J., & Park, J. K. (2009). Destruction of *Bacillus licheniformis* spores by microwave irradiation. *Journal of Applied Microbiology*, *106*, 877–885.

Kindle, G., Busse, A., Kampa, D., Meyerkonig, U., & Daschner, F. D. (1996). Killing activity of microwaves in milk. *Journal of Hospital Infection*, *33*, 273–278.

Koutchma, T., Bail, A. L., & Ramaswamy, H. S. (2001). Comparative experimental evaluation of microbial destruction in continuous-flow microwave and conventional heating systems. *Canadian Biosystem Engineering*, *43*, 1–8.

Kozempel, M., Cook, R. D., Scullen, O. J., & Annous, B. A. (2000). Development of a process for detecting nonthermal effects of microwave energy on microorganisms at low temperature. *Journal of Food Processing and Preservation*, *24*, 287–301.

Kozempel, M. F., Annous, B. A., Cook, R. D., Scullen, O. J., & Whiting, R. C. (1998). Inactivation of microorganisms with microwaves at reduced temperatures. *Journal of Food Protection*, *61*, 582–585.

Latorre, M. E., Bonelli, P. B., Rojas, A. M., & Gerschenson, L. N. (2012). Microwave inactivation of red beet (*Beta vulgaris* L. var. conditiva) peroxidase and polyphenoloxidase and the effect of radiation on vegetable tissue quality. *Journal of Food Engineering*, *109*, 676–684.

Lau, M. H., Tang, J., Taub, I. A., Yang, T. C. S., Edwards, C. G., & Mao, R. (2003). Kinetics of chemical marker formation in whey protein gels for studying high temperature short time microwave sterilization. *Journal of Food Engineering*, *60*, 397–405.

Levre, E., & Valentini, P. (1998). Inactivation of Salmonella during microwave cooking. *Zentralblatt Fur Hygiene Und Umweltmedizin*, *201*, 431–436.

Lindsay, R. E., Krissinger, W. A., & Fields, B. F. (1986). Microwave vs conventional oven cooking of chicken – relationship of internal temperature to surface contamination by *Salmonella typhimurium*. *Journal of the American Dietetic Association*, *86*, 373–374.

Lohman, S., & Manique, F. (1986). Microwave sterilization of vials. *Journal of Parenteral Science and Technology*, *40*, 25–30.

Luan, D., Tang, J., Pedrow, P. D., Liu, F., & Tang, Z. (2013). Using mobile metallic temperature sensors in continuous microwave assisted sterilization (MATS) systems. *Journal of Food Engineering*, *119*, 552–560.

Luan, D. L., Tang, J. M., Pedrow, P. D., Liu, F., & Tang, Z. W. (2015). Performance of mobile metallic temperature sensors in high power microwave heating systems. *Journal of Food Engineering*, *149*, 114–122.

Lund, B. M., Knox, M. R., & Cole, M. B. (1989). Destruction of *Listeria monocytogenes* during microwave cooking. *Lancet*, *1*, 218.

Matsui, K. N., Gut, J. A. W., Oliveira, P. V., & Tadini, C. C. (2008). Inactivation kinetics of polyphenol oxidase and peroxidase in green coconut water by microwave processing. *Journal of Food Engineering*, *882*, 169–176.

Mima, E. G., Pavarina, A. C., Neppelenbroek, K. H., Vergani, C. E., Spolidorio, D. M., & Machado, A. L. (2008). Effect of different exposure times on microwave irradiation on the disinfection of a hard chairside reline resin. *Journal of Prosthodontics*, *17*, 312–317.

Mudgett, R. E. (1989). Microwave food processing. *Food Technology*, *43*, 117.

Nelson, S. O. (1991). Dielectric properties of agricultural product. *IEEE Transaction on Electrical Insulation*, *26*, 845–869.

Ortega, Y. R., & Liao, J. Y. (2006). Microwave inactivation of *Cyclospora cayetanensis* sporulation and viability of *Cryptosporidium parvum* oocysts. *Journal of Food Protection*, *69*, 1957–1960.

Pandit, R. B., Tang, J., Mikhaylenko, G., & Liu, F. (2006). Kinetics of chemical marker M-2 formation in mashed potato – A tool to locate cold spots under microwave sterilization. *Journal of Food Engineering*, *76*, 353–361.

Pandit, R. B., Tang, J., Liu, F., & Mikhaylenko, G. (2007). A computer vision method to locate cold spots in foods in microwave sterilization processes. *Pattern Recognition*, *40*, 3667–3676.

Pandit, R. B., Tang, J., Liu, F., & Pitts, M. (2007). Development of a novel approach to determine heating pattern using computer vision and chemical marker (M-2) yield. *Journal of Food Engineering*, *78*, 522–528.

Peleg, M. (2006). Advanced Quantitative Microbiology for Foods and Biosystems: Models for Predicting Growth and Inactivation, Boca Raton, CRC Press.

Perng, R. S., Yang, K. P., & Fu, M. H. (1994). Heat resistance of *Bacillus stearothermophilus* spores at ultra high temperature. *Food Science*, *21*, 267–274.

Pina-Perez, M. C., Benlloch-Tinoco, M., Rodrigo, D., & Martinez, A. (2014). *Cronobacter sakazakii* inactivation by microwave processing. *Food and Bioprocess Technology*, *7*, 821–828.

Prakash, A., Kim, H. J., & Taub, I. A. (1997). Assessment of microwave sterilization of foods using intrinsic chemical markers. *Journal of Microwave Power and Electromagnetic Energy*, *32*, 50–57.

Pucciarelli, A. B., & Benassi, F. O. (2005). Inactivation of *Salmonella enteritidis* on raw poultry using microwave heating. *Brazilian Archives of Biology and Technology*, *48*, 939–946.

Ramaswamy, H. S., Koutchma, T., & Tajchakavit, S. (2000). Enhanced thermal effects under microwave heating conditions. In: *International conference of engineering and food (ICEF 8), 2000 Puebla, Mexico*.

Resurreccion, F. P., Tang, J., Pedrow, P., Cavalieri, R., Liu, F., & Tang, Z. (2013). Development of a computer simulation model for processing food in a microwave assisted thermal sterilization (MATS) system. *Journal of Food Engineering*, *118*, 406–416.

Rodriguez-Marval, M., Geornaras, I., Kendall, P. A., Scanga, J. A., Belk, K. E., & Sofos, J. N. (2009). Microwave oven heating for inactivation of *Listeria monocytogenes* on frankfurters before consumption. *Journal of Food Science*, *74*, M453–M460.

Rosen, C. G. (1972). Effects of microwaves on food and related materials. *Food Technology*, *26*, 36–37.

Rougier, C., Prorot, A., Chazal, P., Leveque, P., & Leprat, P. (2014). Thermal and nonthermal effects of discontinuous microwave exposure (2.45 gigahertz) on the cell membrane of *Escherichia coli*. *Applied and Environmental Microbiology*, *80*, 4832–4841.

Sakai, N., Mao, W. J., Koshima, Y., & Watanabe, M. (2005). A method for developing model food system in microwave heating studies. *Journal of Food Engineering*, *66*, 525–531.

Shamis, Y., Patel, S., Taube, A., Morsi, Y., Sbarski, I., Shramkov, Y., et al. (2009). A New sterilization technique of bovine pericardial biomaterial using microwave radiation. *Tissue Engineering Part C-Methods*, *15*, 445–454.

Shamis, Y., Taube, A., Mitik-Dineva, N., Croft, R., Crawford, R. J., & Ivanova, E. P. (2011). Specific electromagnetic effects of microwave radiation on *Escherichia coli*. *Applied and Environmental Microbiology*, *77*, 3017–3022.

Sheen, S., Huang, L. H., & Sommers, C. (2012). Survival of *Listeria monocytogenes*, *Escherichia coli* O157:H7, and *Salmonella* spp. on catfish fillets exposed to microwave heating in a continuous mode. *Journal of Food Science*, *77*, E209–E214.

Shekarforoush, S. S., & Firozi, R. (2003). The effect of sodium chloride, pH and acid type on destruction of *Listeria monocytogenes* by microwave heating. *Journal of Faculty of Veterinary Medicine*, *58*, 395–400.

Shin, J. K., & Pyun, Y. R. (1997). Inactivation of *Lactobacillus plantarum* by pulsed-microwave irradiation. *Journal of Food Science*, *62*, 163–166.

Song, W. J., & Kang, D. H. (2016). Inactivation of *Salmonella senftenberg*, *Salmonella typhimurium* and *Salmonella tennessee* in peanut butter by 915 MHz microwave heating. *Food Microbiology*, *53*, 48–52.

Tahir, A., Mateen, B., Univerdi, S., Karagoban, O., & Zengin, M. (2009). Simple method to study the mechanism of thermal and non thermal bactericidal action of microwave radiations on different bacterial species. *Journal of Bacteriology Research*, *1*, 58–63.

Tajchakavit, S., Ramaswamya, H. S., & Fustierb, P. (1998). Enhanced destruction of spoilage microorganisms in apple juice during continuous flow microwave heating. *Food Research International*, *31*, 713–722.

Tang, J. (2015). Unlocking potentials of microwaves for food safety and quality. *Journal of Food Science*, *80*, E1776–E1793.

Tong, C. H., Lentz, R. R., & Rossen, J. L. (1994). Dielectric-properties of pea puree at 915 Mhz and 2450 Mhz as a function of temperature. *Journal of Food Science*, *59*, 121–122.

Tang, Z. W., Mikhaylenko, G., Liu, F., Mah, J. H., Pandit, R., Younce, F., et al. (2008). Microwave sterilization of sliced beef in gravy in 7-oz trays. *Journal of Food Engineering*, *89*, 375–383.

Vadivambal, R., & Jayas, D. S. (2010). Non-uniform temperature distribution during microwave heating of food materials – A review. *Food and Bioprocess Technology*, *3*, 161–171.

Valero, A., Cejudo, M., & Garcia-Gimeno, R. M. (2014). Inactivation kinetics for *Salmonella enteritidis* in potato omelet using microwave heating treatments. *Food Control*, *43*, 175–182.

Vela, G. R., & Wu, J. F. (1979). Mechanism of lethal action of 2,450-Mhz radiation on microorganisms. *Applied and Environmental Microbiology*, *37*, 550–553.

Vrhovac, I., Hrascan, R., & Franekic, J. (2010). Effect of 905 MHz microwave radiation on colony growth of the yeast *Saccharomyces cerevisiae* strains FF18733, FF1481 and D7. *Radiology and Oncology*, *44*, 131–134.

Wang, J., Zhao, G. H., Liao, X. J., & Hu, X. S. (2010). Effects of microwave and ultrasonic wave treatment on inactivation of Alicyclobacillus. *International Journal of Food Science and Technology*, *45*, 459–465.

Wang, Y., Lau, M. H., Tang, J., & Mao, R. (2004). Kinetics of chemical marker M-1 formation in whey protein gels for developing sterilization processes based on dielectric heating. *Journal of Food Engineering*, *64*, 111–118.

Wang, Y., Tang, J., Rasco, B., Kong, F. B., & Wang, S. J. (2008). Dielectric properties of salmon fillets as a function of temperature and composition. *Journal of Food Engineering*, *87*, 236–246.

Wang, Y. F., Wig, T. D., Tang, J., & Hallberg, L. M. (2003). Dielectric properties of foods relevant to RF and microwave pasteurization and sterilization. *Journal of Food Engineering*, *57*, 257–268.

Welt, B. A., Tong, C. H., Rossen, J. L., & Lund, D. B. (1994). Effect of microwave-radiation on inactivation of *Clostridium sporogenes* (Pa-3679) spores. *Applied and Environmental Microbiology*, *60*, 482–488.

Woo, I. S., Rhee, I. K., & Park, H. D. (2000). Differential damage in bacterial cells by microwave radiation on the basis of cell wall structure. *Applied and Environmental Microbiology*, *66*, 2243–2247.

Wu, Q. (1996). Effect of high-power microwave on indicator bacteria for sterilization. *Ieee Transactions on Biomedical Engineering*, *43*, 752–754.

Yaghmaee, P., & Durance, T. D. (2005). Destruction and injury of *Escherichia coli* during microwave heating under vacuum. *Journal of Applied Microbiology*, *98*, 498–506.

Zeng, S. W., Huang, Q. L., & Zhao, S. M. (2014). Effects of microwave irradiation dose and time on Yeast ZSM-001 growth and cell membrane permeability. *Food Control*, *46*, 360–367.

Zhou, B. W., Shin, S. G., Hwang, K., Ahn, J. H., & Hwang, S. (2010). Effect of microwave irradiation on cellular disintegration of Gram positive and negative cells. *Applied Microbiology and Biotechnology*, *87*, 765–770.

12 Microwave-assisted thawing and tempering

S.J. James, C. James and G. Purnell
Food Refrigeration and Process Engineering Research Centre, Grimsby, United Kingdom

12.1 Introduction

Thawing and tempering have received much less attention in the literature than most other food processing operations. In commercial practice there are relatively few controlled thawing systems compared with chilling and freezing systems.

Frozen meat, fish, vegetables, fruit, butter, and juice concentrate are common raw materials for many food-manufacturing operations. Frozen meat, as supplied to the industry, ranges in size and shape from complete hindquarters of beef to small breasts of lamb and poultry portions, although the majority of the material is "boned-out" and packed in boxes approximately 15 cm thick weighing between 20 and 40 kg. Fish is normally in plate frozen slabs; fruit and vegetables in boxes, bags, or tubs; and juice in large barrels. Few processes can handle the frozen material and it is usually either thawed or tempered before further processing (Anon, 2008; James & Crow, 1986).

Thawing is usually regarded as complete when the temperature of all points within the frozen food has been raised to 0°C and no free ice is present. This is the minimum temperature at which the meat can be boned or other products cut or separated by hand. Lower temperatures (e.g., −5 to −2°C) are acceptable for product that is destined for mechanical chopping, but such material is referred to as "tempered" rather than thawed (Anon, 2008). The two processes should not be confused because tempering only constitutes the initial phase of a complete thawing process.

Thawing is often considered as simply the reversal of the freezing process. However, inherent in thawing is a major problem that does not occur in the freezing operation. The majority of the bacteria that cause spoilage or food poisoning are found on the surfaces of food. During the freezing operation, surface temperatures are reduced rapidly and bacterial multiplication is severely limited, with bacteria becoming completely dormant below −10°C. In the thawing operation, these same surface areas are the first to rise in temperature and bacterial multiplication can recommence. On large objects subjected to long uncontrolled thawing cycles, surface spoilage can occur before the center regions have fully thawed (Anon, 2008; Baush, Caren, Stephen, & Tuemy, 1980).

Conventional thawing and tempering systems supply heat to the surface and then rely on conduction to transfer that heat into the center of the product. A few, including microwave, use electromagnetic radiation to generate heat within the food.

The Microwave Processing of Foods. DOI: http://dx.doi.org/10.1016/B978-0-08-100528-6.00012-7

In selecting a thawing or tempering system for industrial use, a balance must be struck between thawing time, appearance, and bacteriological condition of the product, processing problems such as effluent disposal, and the capital and operating costs of the respective systems. Of these factors, thawing time is the principal criterion that often governs selection of the system. Appearance, bacteriological condition, and weight loss are important if the material is to be sold in the thawed condition but are less so if it is for processing (Anon, 2008).

There are a number of reviews of meat (James & James, 2002; Merts, 1999) and fish (Archer, Edmonds, & George, 2008; Haugland, 2002; Jason, 1974a, 1974b) thawing. The main detrimental effect of freezing and thawing meat is the large increase in the amount of proteinaceous fluid (drip) released on final cutting, yet the influence of thawing rate on drip production is not clear. James and James (2002) reported that studies have shown that there was no significant effect of thawing rate on the volume of drip in beef or pork. Several authors concluded that fast thawing rates would result in increased drip, while others showed the opposite. Hergenreder et al. (2011) found that slow thawing rates resulted in higher drip loss from beef subprimals independent of initial freezing rate. However, this difference did not affect the final eating quality of cooked steaks. Similarly, in a comparison of immersion (<2 h) or air thawing (24 h) of pork by Linares, Saavedra, Serrano, Les, and Sosa (2005), immersion thawing resulted in lower drip loss, 5.2% compared with 10.3% for slow frozen and 7.8% compared with 10.8% for fast frozen. Drip being the free liquid emerging from foods, especially meat and fish, during chilling, chilled storage, and especially after thawing and during subsequent storage. It is normally expressed as a percentage of the initial weight of the food item. However, James, Nair, and Bailey (1984) reported that thawing times from −7 to 0°C of less than 1 minutes or greater than 2000 minutes led to increased drip loss). The results are therefore conflicting and provide no useful design data for optimizing a thawing system. With fish, fruit, and vegetables, ice formation during freezing and frozen storage (Ostwald ripening) breaks up cell structure and fluids are released during thawing.

Fik and Macura (2001) suggested from their results that bread that underwent microwave thawing had generally better quality in comparison with air blast thawing at 50°C. Riihonen and Linko (1990) studied the effect of thawing conditions on the quality of mechanically deboned beef (MDB) and mechanically deboned pork (MDP). All samples were analyzed chemically and microbiologically immediately after thawing. Microwave thawing resulted in a better quality product ($P < 0.05$) than thawing at 4 or 21°C. Different methods (air, immersion, microwave, and high-pressure) of thawing $2 \times 2 \times 2\ cm^3$ cubes of hami melon were investigated by Xin, Rui, and Jin-Hong (2015). Less drip loss, textural damage, and discoloration were reported when 700 W of microwave power were used. Holzwarth, Korhummel, and Carle (2012) reported that maximum ascorbic acid retention was observed when strawberries were thawed in a microwave oven.

Although overall there does not appear to be a consensus on the effect of different thawing rates on the quality of the thawed product, it appears that microwave thawing may offer some advantages with some foods.

12.2 Conventional and emerging thawing and tempering approaches

The main conduction-based thawing methods rely on air, water or steam condensation under vacuum.

12.2.1 Air assisted thawing

Air thawing systems transfer heat to the frozen material by conduction through the static air boundary layer at the product surface; the rate of heat transfer is a function of the difference in temperature between the product and the air, the air velocity, the relative humidity (RH), and the shape and the surface area of the product to be thawed. Air systems are very flexible and may be used to thaw any size of product from a whole beef carcass to individual strawberries.

12.2.1.1 Still air thawing—natural convection

Thin blocks (<10 cm) of meat or fish, or layers of fruit or vegetables, can be thawed overnight at room temperature and, provided the surface of the product does not become too dry, the thawed product can be perfectly acceptable. Air temperatures should not be greater than 15°C.

For thicker materials, natural convection based thawing is not recommended, since thawing times extend to days, rather than hours, and the surface layers may become warm and spoil long before the center is thawed. Still air thawing is practicable only on a small scale, because considerable space is required, the process is uncontrolled and the time taken is often too long to fit in with processing cycles. The sole advantage is that little or no equipment is required.

12.2.1.2 Moving air thawing—forced convection

The majority of commercial thawing systems use moving air as the thawing medium. Not only does the increased surface heat transfer (h value) caused by the moving air result in faster thawing, but it also produces allows for much better control than using natural convection. Control of relative humidity is important with unwrapped products to reduce surface desiccation and increase the rate of heat transfer to the product, 85–95% RH being recommended for meat (Bailey, James, Kitchell, & Hudson, 1974).

With 250 g slabs of meat, Zagrodzki, Niedzielski, and Kulogawska (1977) found that weight loss was a function of temperature, velocity, and relative humidity. In all cases, increasing the air temperature or decreasing the air velocity produced a decrease in percentage weight loss at 85–88% RH or an increase in weight gain at 95–98% RH. Changes ranged from a 2.5% weight loss at a temperature of 5°C, an air velocity of 5 m/s, a relative humidity of 85–88% RH, to a 0.51% weight gain at 25°C, 1 m/s, 95–98% RH. With thin products, the use of air impingement has been advocated and is claimed to be four times faster than still air on a 2 cm thick food product (Anderson & Singh, 2006).

12.2.1.3 Two-stage air thawing

Two-stage air thawing has often been proposed as a means of shortening the thawing process (James & James, 2002; Pham, 2014). In the first stage, a high air temperature is maintained until the surface reaches a predetermined set temperature, thus ensuring a rapid initial input of energy. The air temperature is then reduced rapidly and maintained below 10°C until the end of the thawing process. Heat flows from the hotter surface regions to the center of the frozen foodstuff, lowering the surface temperature to that of the ambient air. Since this temperature is below 10°C, and the overall thawing time is short, total bacteria growth is small.

The Danish Meat Research Institute developed a two-stage thawing system using almost saturated air between 35 and 60°C, followed by air between 5 and 10°C after the surface temperature of the product has reached 30 to 35°C (Anon, 1974). They claimed that for larger pieces of meat the first stage normally took 1 to 1.5 h, the second 15 to 20 h, reducing the time compared to single stage still air thawing systems that required 2–3 days or more. They claimed that weight loss was low and drip loss minimal.

12.2.2 Water-assisted thawing

The mechanism of heat transfer in water is similar to that in air, but because the heat transfer coefficients obtained are considerably larger, the thawing times of thinner cuts are effectively reduced (Akhtar, Khan, & Faiz, 2013; Archer et al., 2008; Bailey et al., 1974). However, there are practical problems that limit the use of water thawing systems; boxed or packaged goods (unless shrink-wrapped or vacuum-packed) must be removed from their containers before they can be water thawed, composite blocks of boned-out pieces or individual fish break up and disperse in the thawing tank, and handling difficulties preclude the use of large cuts such as carcasses. Furthermore, for larger cuts the heat transfer resistance is dominated by the slow conduction within the product.

12.2.3 Vacuum–heat-assisted thawing

A vacuum-heat thawing system operates by transferring the heat of condensing steam under vacuum to the frozen product (Archer et al., 2008; Bailey et al., 1974; James & Bailey, 1980). Theoretically, a condensing vapor in the presence of a minimum amount of a noncondensable gas can achieve a surface film heat transfer coefficient far higher than that achieved in water thawing. The principle of operation is that when steam is generated under vacuum, the vapor temperature will correspond to its equivalent vapor pressure. For example, if the vapor pressure is maintained at 1106 N/m^2, steam will be generated at 15°C. The steam will condense onto any cooler surface such as a frozen product. The benefits of latent heat transfer can be obtained without the problems of cooking that would occur at atmospheric pressure.

With thin materials, thawing cycles are very rapid, enabling high daily throughputs to be achieved. The advantage of a high heat transfer coefficient becomes less marked as material thickness increases, and beef quarters or 25 kg meat blocks require thawing times permitting no more than one cycle per day. Under these conditions, the economics of the system and the largest capacity unit available (10–12 tonnes) severely restrict its application.

Vacuum tumble thawing systems, in which the frozen product is continuously tumbled while steam under vacuum condenses on the exposed surfaces of the food, are a recent development. Very fast thawing is claimed to be obtained with small bulk-frozen individual products. However, delicate products such as fish fillets cannot be thawed in such systems.

12.2.4 Acoustic pressure-assisted thawing

The use of acoustic energy to thaw frozen foodstuffs was initially investigated in the 1950s (Brody & Antenevich, 1959). However, it was worked in the 1980s that showed more acoustic energy could be absorbed by frozen foods when an acoustic frequency in the relaxation frequency range of ice crystals in the food was applied (Kissam, Nelson, Ngao, & Hunter, 1981). Their experiments showed that, when 1500 Hz acoustic energy at 60 W was applied to 12.7 kg blocks of cod, 91 mm thick, it required 71% less time to thaw than with water immersion only. Further work by Miles, Morley, and Randell (1999), demonstrated that high power ultrasound could be used to thaw meat and fish at frequencies around 500 kHz, which conformed to the relaxation mechanism. Using frequencies and intensities around 500 kHz and 0.5 W/cm^2, surface heating was minimized, and beef, pork, and cod samples were thawed to a depth of 7.6 cm within about 2.5 h. It would therefore seem that acoustic thawing could be a promising technology if developed commercially.

12.2.5 Air-assisted tempering

All large-scale industrial tempering is carried out in air, and studies by Brown and James (2006) demonstrated the problems of air tempering. In air at a temperature of 3°C and an air velocity of 5 m/s, the center of 150 mm thick meat blocks rose to temperatures of −3°C after 22.5 h. Under these conditions, the surface layers of the meat would have spent many hours in a thawed condition that would be detrimental to both drip and optimal processing. Using lower temperatures avoids thawing and at the same time produces an optimum temperature difference for subsequent processing. However, tempering times are substantially extended. For example, times to −3°C using air at −1°C and 5 m/s were 37.5 h for 50% visual lean and 44.5 h for 100%. In general, frozen product on pallets is stacked in a temperature-controlled room and left for the time needed for the tempering requirements to be achieved (Table 12.1). In many cases, the environment is changed during the

Table 12.1 Tempering configurations, initial and final product and environmental temperatures and tempering times for cartons of frozen meat (approximately 27 kg in weight, 15 cm thick) in industrial systems

Factory	Configuration	Temperature (°C)			Tempering
		Initial	Final	Ambient	Time (d)
A	Palletized	−14 to −12	−3	−8 then −3 to −1	5 then 7
	Palletized	−10	−3	−8 then −3 to −1	1 then 6
	Palletized	−15	−2.5	−8 then −3 to −1	3.5 then 6
	Palletized	−18	−3	−8 then −3 to −1	6 then 5
B	Palletized	−18	−15 to −6	0	0.9
	Racked	−18	−4.5	14 then 0	0.2 then 0.7
C	Palletized	−20 to −18	−10 to −5	0	1.0
D	Palletized	−20	−6	−5 to −3	6 to 7
E	Palletized	−27 to −23	−8 to −5.5	−3 to −1	0.9
	Racked	−27 to −23	−5 to −4	−3 to −1	0.9
F	Palletized	−18	−8 to −3.5	−4	1.0
	On trolleys	−18	−5 to −3.5	−4	0.8

Source: James, S.J. (1986). Microwave thawing and tempering of frozen meat. In *ENCE heating and processing 1–3000 MHz, Cambridge, 7.4.*

process. Tempering times are often very long, up to 12 days. Times of less than 1 day can only be achieved if the following process can handle product at a much lower temperature.

In a few cases, pallets are split up at the start of the process and the frozen material is placed on racks or trolleys. This reduces the path for heat penetration and substantially speeds up the tempering process. However, splitting up pallets require considerable manpower and substantially increase the space required.

12.3 Electromagnetic-assisted thawing and tempering approaches

In all of the methods described above, the rate of thawing is a function of the transfer of heat from the thawing medium to the surface of the foodstuff, and the conduction of this heat into the center of the material (James & James, 2010). In theory, electrical systems should overcome these problems because heat is generated within the material and the limitations of thermal conductivity are circumvented. In such systems, the kinetic energy imparted to molecules by the action of an oscillating electromagnetic field is dissipated by inelastic collisions with surrounding molecules and this energy appears as heat. Thus, electromagnetic radiation can be used to heat foodstuffs.

Three regions of the electromagnetic spectrum have been used for such heating: resistive 50 Hz; radio frequency 3 to 300 MHz; and microwave 900 to 3000 MHz (James et al., 1983: James & James, 2002).

12.3.1 Resistive-assisted thawing and tempering

Placing it between two electrodes and applying a low voltage at normal mains frequency can heat a frozen foodstuff. As the electric current flows through the material, it becomes warm; this process is referred to as ohmic heating (Knirsch, Dos Santos, De Oliveira Soares, & Penna, 2010). Electrical contacts are required and product structure must be uniform and homogeneous, otherwise the current will take the path of least resistance, resulting in uneven temperatures and potentially runaway heating (Roberts, Balaban, Zimmerman, & Luzuriaga, 1998). Frozen food at a low temperature does not readily conduct electricity, but as it becomes warmer its electrical resistance falls and a larger current can flow, leading to more pronounced heat generation within the product (Roberts et al., 1998).

Bozkurt and Içier (2010) stated that "there is no commercial ohmic thawing system for meat products according to the best of authors' knowledge." They carried out laboratory scale resistive thawing studies on small ($2.5 \times 2.5 \times 5\ cm^3$ to $5 \times 5 \times 5\ cm^3$) lean beef samples under different voltage gradients. Thawing times (from −18 to +10°C) in the center ranged from 580 to 650 seconds in the resistive systems compared with up to 4000 seconds in air at 25°C.

12.3.2 High-voltage electric field-assisted thawing

Ohtsuki (1991) patented a process for the rapid thawing of food using electrons produced by a high-voltage electrostatic field. Laboratory studies on small ($2 \times 4 \times 4\ cm^3$) cubes of frozen pork tenderloin reported reductions of 0.5−1 log CFU/g during thawing (He, Liu, Nirasawa, Zheng, & Liu, 2013). Thawing time (from −20 to 0°C) was reduced when the applied voltage exceeded 6 kV with the application of 10 kV reducing the time from 64 to 40 minutes at a thawing temperature of 20°C. The actual origin of the heating effect was poorly explained but claimed to be similar to that of electrohydrodynamics described by Chen and Barthakur (1991) where the air ions in a strong electric field enhance heat transfer. In further studies (He, Liu, Tatsumi, Nirasawa, & Liu, 2014), it was claimed that the energy consumption of a high-voltage electric field (HVEF) thawing system was substantially less than water or microwave thawing. However, in the study the main source of heat used to thaw the meat came from the ambient air and the consequences of providing this in a commercial situation were not considered.

Hsieh, Ch, Wj, Huang, and Ko (2010) reported that the application of HVEF substantially reduced the time to temper small (190 g, 2.8 cm thick) samples of boneless chicken thigh from −16 to −5°C from 60 to 40 minutes. Thawing frozen beef and eggs were accomplished in 25–30% of the time using HVEF compared with conventional methods using the same temperature (Orlowska et al., 2014). Previous studies have indicated that HVEF could reduce the time for thawing frozen pork and chicken to 2/3 of the time spent in still air (He et al., 2013; Hsieh et al., 2010). Increasing the voltage and decreasing electrode distance in the HVEF system was found to reduce the thawing time of frozen pork tenderloin meat (He et al., 2014). Mousakhani-Ganjeh, Hamdami, and Soltanizadeh (2015) showed that thawing under HVEF significantly improves thawing rate and total volatile binding nitrogen of the frozen fish cubes. The highest thawing rate was 1.78 times greater than that obtained for the control samples. However, color, texture, and protein solubility of fish samples declined as a result of thawing under HVEF. Increasing the applied voltage decreased protein solubility and affected the hardness, gumminess, cohesiveness, and chewiness of the fish samples as compared to the control. In further work, Mousakhani-Ganjeh, Hamdami, and Soltanizadeh (2016) reported similar energy savings to those already mentioned by He et al. (2014).

12.3.3 Radio frequency-assisted thawing and tempering

During radio frequency thawing, heat is generated in the frozen foodstuff because of dielectric losses when a product is subjected to an alternating electric field. In an idealized case of radio frequency heating, the foodstuff, consisting of a regular slab of homogeneous material at a uniform temperature, is placed between parallel electrodes and no heat is exchanged with its surroundings. When an alternating electromotive force (emf) is applied through the electrodes, the resulting field in the slab is uniform, so the energy and the resultant temperature rise is identical in all parts of the food (Sanders, 1966).

In practice, this situation rarely applies. Foodstuffs are not generally in the shape of perfect parallelepipeds; frozen meat, for example, consists of at least two components, i.e., fat and lean. During loading, frozen meats pick up heat from the surroundings, the surface temperature rises, and the dielectric system is not presented with the uniform temperature distribution required for even heating.

By using a conveyorized system to keep the product moving past the electrodes and/or surrounding the material by water, commercial systems have been developed for blocks of oily fish and white fish (Jason & Sanders, 1962). Successful thawing of 13 cm thick meat, and 14 cm thick offal, blocks have also been reported (Sanders, 1966) but the temperature range at the end of thawing (44 min) was stated to be −2 to 19°C and −2 to 4°C, respectively, and the product may not have been fully thawed.

To overcome runaway heating with slabs of frozen pork bellies, studies by Satchell and Doty (1951) reported on coating the electrodes with lard, placing the bellies in oil, water, and saline baths, and wrapping the meat in cheesecloth soaked in saline solution. Only the last treatment was successful, however, even that was not deemed practical.

An industrial system was installed to thaw frozen blocks of boned-out poultry in the late 1980s (Anon, 1992). The continuous plant had a throughput of 450 kg per hour with four 12.5 kW generators operating at a frequency of 27.12 MHz

Table 12.2 **Commercial reported tempering times (Anon, 2013)**

Product	Configuration	Temperature (°C)		Time (min)
		Starting	Final	
Beef trimmings (7% fat)	20 kg block	−18	−3 to 0	25
Beef trimmings (15% fat)	20 kg block	−18	−4 to 0	25
Pork shoulder	20 kg block	−12	−4 to −1	18
Pork (20% fat)	16 kg vacuum bag	−12	−4 to 0	18
Beef	25 kg block	−20	−8 to −6	20
Beef	25 kg block	−20	−4 to −2	35
Whole chicken	10 kg in cardboard box	−18	−2 to 0	30
Chicken MDM 3 mm	12.5 kg block in plastic bag	−18	−2 to −1	10
Chicken trimmings	10 kg block	−18	−3 to 0	40
Boneless shoulder pork	27 kg block	−18	−4 to −1	40

providing the heating. It was claimed that the process reduced thawing time from 3 to 4 days to under 2 hours and cut weight loss from 7 to 0.5%.

Pizza, Pedrielli, Busetto, Bocchi, and Spinelli (1997) investigated RF thawing of pork muscles ($35 \times 10 \times 11\ cm^3$) using RF heating at a nominal power of 4 kW and a frequency of 27.5 MHz, measuring temperatures at three locations. While center point temperatures averaged 4°C, they found that near surface temperatures averaged approximately +13°C. When thawing 4 kg blocks of lean meat, Farag, Lyng, Morgan, and Cronin (2011) found the optimum RF conditions were 35 minutes of RF heating delivered in a noncontinuous fashion (20 minutes on, 10 minutes off, and followed by 15 min on) at 400 W, which gave a mean temperature of 0.2°C (SD 1.8). By comparison, conventional thawing was achieved in 50 hours and 20 minutes, which represented an 85-fold difference in thawing time. Comparable uniformity of temperature distribution was obtained by each method. The target range of −1 to +5°C could not be achieved with 50:50 lean/fat blocks or 100% fat.

Commercial systems claim to be able to temper a wide range of frozen products to final temperatures between −8 and 0°C in less than 40 minutes (Table 12.2).

12.3.4 Microwave-assisted thawing

Microwave thawing utilizes electromagnetic waves directed at the product through waveguides without the use of conductors or electrodes. Whilst the heating of frozen food by microwave energy is potentially a very fast method of thawing, its application is constrained by thermal instability. At its worst, parts of the food may be cooked whilst the rest is substantially frozen. This arises because the absorption by frozen food of electromagnetic radiation in this frequency range increases as the temperature rises, this dependence being especially large at about −5°C, increasing as the initial freezing point is approached. If for any reason during the treatment a region of the material is slightly hotter than its surroundings, proportionately more energy will be absorbed within that region and the original difference in enthalpy will be increased. As the enthalpy increases so does the absorption, and the unevenness of heating gets more pronounced at an ever-increasing rate. Below the initial freezing point, the temperature increase is held in check by thermal inertia, since for a given energy input the temperature rise is inversely proportional to the thermal capacity. If the microwave treatment is continued after the hot spot has reached its initial freezing point, the temperature rises at a very high rate. Lowering the power density to allow thermal conduction to even out the enthalpy distribution through the food can reduce such runaway heating. Orders of magnitude calculations have been made of the times involved for thermal stability and showed that processing times for standard blocks need to be over 8 hours (Jason, 1974a, 1974b; Ohlsson, 1984).

Since the main instability tends to occur at the surface, attempts have been made to cool the surface during thawing using air or liquid nitrogen (Bialod, Jolion, & Legoff, 1978; Les Micro-Ondes Industrielles, undated). Although experimentally successful, the systems were not economically viable.

A hybrid microwave/vacuum thawing system in which boiling of surface water at low temperature was used to cool the surface was developed in the 1980s (James, 1984). The system consisted of a cylindrical vacuum chamber approximately 1 meter in diameter and 1 meter long (Fig. 12.1). The chamber could be evacuated to absolute pressures as low as 10 mbar by a water ring pump in series with a rotary pump. Microwaves at a frequency of 915 MHz were introduced into the chamber via two waveguides positioned near the top at the front and rear of the plant. The microwaves were produced by a 25 kW generator, though power had to be limited to 2.5 kW to avoid arcing problems in the chamber. A large circular twisted disc was rotated within the chamber during thawing to produce a more even microwave field.

Figure 12.1 A hybrid microwave/vacuum thawing system.

In trials chilled meat (85% visual lean) was obtained from a commercial abattoir and made up into either whole blocks with dimensions of $61 \times 40 \times 15\ cm^3$ or half blocks $30 \times 40 \times 15\ cm^3$. The blocks were then wrapped in polyethylene film and placed separately in solid fiberboard cartons before freezing and storage at −20°C. After thawing, maximum surface temperatures were measured using infrared thermometry. A single-point hypodermic probe was then used to find the minimum temperature within the block. The block was then weighed and placed in an insulated box for 2 hours to determine the average temperature after thawing.

After thawing, the average meat temperature was 9.4°C (ranging from −1 to 23.3°C between blocks) and the maximum surface temperature measured on any block was 28.4°C (Table 12.2). Weight losses averaged to 7.6%, which appeared large, but weight loss in commercial systems was stated to range from 2 to 10%. The overall energy efficiency of the plant was 49% with 24% of the energy being absorbed by the structure of the chamber itself.

Virtanen, Goedeken, and Tong (1997) combined microwave energy and cold air with different ambient temperatures to reduce thawing time and avoid runaway heating during microwave-assisted thawing. The microwave power was cycled on and off using two-temperature control schemes to maintain a predetermined temperature gradient based on hot and cold points. Thawing time was reduced by as much as a factor of seven compared to convective thawing at ambient temperature when appropriate conditions were used (Table 12.3).

Yagi and Shibata (2002) applied for a patent for a similar process that used microwaves under vacuum to thaw foods. They claim that the rate of sublimation at the surface, together with a relatively low microwave power, resulted in even

Table 12.3 Initial weight and percentage weight loss from meat, thawing time, microwave power supplied, and maximum, minimum and average temperature after thawing

Run no.	Weight		Thawing time	Power	Temperature		
	Initial (kg)	Loss (%)	(h)	(kW)	Minimum (°C)	Maximum (°C)	Average (°C)
1	25.6	9.0	2.0	2.5	3.0	24.3	8.0
2	25.5	7.9	2.0	2.5	2.6	24.0	8.1
3	25.6	7.3	2.0	2.5	−2.0	12.7	4.0
4	24.5	10.3	2.0	2.5	21.2	26.7	23.3
5	33.2	5.9	2.0	2.5	−2.0	22.0	4.8
6	27.9	7.1	2.0	2.5	−0.2	25.0	6.2
7	11.9	6.5	1.0	2.5	4.6	14.9	12.2
8	11.9	6.2	1.0	2.5	7.0	17.5	13.0
9	12.3	7.0	1.0	2.5	−4.2	3.2	−1.0
10	11.5	9.4	1.0	2.5	15.5	28.4	20.5
11	11.9	7.5	1.0	2.5	−2.4	15.2	4.3

Table 12.4 Thawing results for claimed process

Product	Starting temperature (°C)	Microwave power (kW)	Final temperature (°C)		Thawing time (min)	Weight loss (%)
			Center	Surface		
3 × 10 kg blocks pork	−40	1.8	−1.9	−1.9	34.25	0.6
4 × 2 kg blocks tuna	−55	0.7	−1.5	−1.8	27.8	0.8
3 × 2 kg blocks beef	−40	0.6	−2.0	−1.1	24.5	

Source: Adapted from Yagi, S., & Shibata, K. (2002). *Microwave defrosting under reduced pressure*. US Patent Application 20020195447.

product temperatures and low weight losses during the process. The temperatures claimed to be achieved at the end of thawing processes lasting between 24.5 and 34.25 minutes ranged from −1.1 to −2.0°C (Table 12.4), and were very even and close to the final thawing point of the products.

A novel approach was adopted by Fathi, Lauf, and Mcmillan (2003) for thawing fruit juices or other foods that would flow in the thawed state. Their patent described an apparatus that included a microwave energy source; a microwave applicator that defines a cavity for applying microwave energy from the microwave source to a material to be thawed; and a shielded region that is shielded from the microwave source. The shielded region was physically joined to the cavity so that thawed material could flow from the cavity into the shielded region. Despite a widespread belief to the contrary, microwave-thawing systems have not been commercially successful. However, microwave-tempering systems have found successful niche applications in the meat and dairy industry.

12.3.5 Microwave-assisted tempering

Despite the widespread industrial use of tempered meat, there is little published process design data for meat tempering operations, with the exception of commercial claims for microwave processing units.

James and Crow (1986) provide some data on the use of microwaves for tempering meat blocks. The batch unit investigated was manufactured by the Raytheon Company and would accept five meat cartons in a single layer on a pallet that was pushed over rollers manually into the microwave chamber. Microwave power was provided by a 30 kW magnetron (variable down to 20 kW) operating at 896 MHz, from which the microwaves entered the chamber via waveguides situated at the top and bottom. Rotating metal discs were positioned above and below the product to provide a more uniform microwave field and the loaded pallet was subjected to cyclic lateral movements of 90 mm during treatment.

It is clear from Table 12.5 that blocks processed directly from frozen storage can be acceptably tempered in a batch microwave unit to a mean temperature of approximately −3°C (range −5 to 0°C) with no hot spots. Tempering times varied from 3.5 to 5.0 minutes with block types (1) and (2), and at least two combinations of microwave power and processing time yielded acceptable results. The frozen cartons of flank that contained a higher percentage of fat caused more runaway heating problems, and lower microwave power (20 kW) applied for 6.5 minutes was required to achieve the desired results.

Table 12.5 Effects of block types, weights of frozen meat and initial meat temperature on final meat temperature and condition of meat blocks tempered in a batch microwave unit

Block type (weight)	Initial temperature (°C)	Microwave power (kW)	Process time (min)	Final temperature (°C) (meat condition)
1 25−30 kg	−15	28	3.5	−7 to −2
			4.0	−4 to −2 (soft corners)
			4.5	−4 to −1 (soft surfaces)
			4.5	−3 to −2 (locally +2)
			4.75	−4 to −2 (surfaces +2)
		25	5.0	−3 to −2 (surfaces 0)
			5.0	−4 to 0
	−8 deep, surface −2	30	3.0	−5 to −2
		25	5.5	−3 to −2 (corners +3)
2 27−37 kg	−15	30	5.0	−5 to +2
			4.0	−5 to −4 (soft spots)
			4.5	−4 to −3 (soft spots)
		28	5.0	−4 to −2 (surface <2)
	−8 deep, surface −1	30	6.0	−4 to +27 (v. variable)
			6.0	−6 to +25 (v. variable)
			5.0	−5 to +6 (spot at +54)
3 25−27 kg	−15	30	4.0	−6 deep, −1 surface
			5.0	−5 deep, 0 surface
			6.0	−5 deep, some cooking
			6.0	−5 deep, some cooking
		20	6.5	−4 to −2, uniform
	−8 deep, surface −1	20	7.0	−5 deep, hot spots
			5.0	−5 uniform
		30	5.0	−5 deep, hot spots

Source: Adapted from James, S. J., & Crow, N. (1986). Thawing and tempering: Industrial practice. *Meat Thawing! Tempering and Product Quality, IFR-BL: Subject Day.*

In general, microwave tempering of blocks, which had been allowed to warm up in factory ambient temperatures for 8 hours, was unsatisfactory. Surface temperatures, especially at the corners of the meat blocks, rose to unacceptable levels and there was substantial drip loss from thawed surfaces.

These results indicate that it would be difficult if not impossible to produce a uniform power/time combination for all types of "standard 27 kg blocks." For optimal tempering, trials have to be carried out to determine the correct power and time setting for each type of block. Blocks sorted into batches of similar type should be processed directly from frozen storage under the predetermined conditions.

Successful tempering can be achieved in minutes using a microwave system, compared with the 1 to 14 days required in industrial air tempering systems. Continuous conveyorized microwave-tempering systems using either a single 60 kW magnetron or two 40 kW magnetrons can temper 2 to 2.5 tonnes per hour depending upon fat content.

In large-throughput operations the continuous microwave-tempering plant provides considerable flexibility, in that changes in raw material requirements, for the posttempering processes, can be accommodated in minutes. Using air tempering systems, at least 1 day and up to 8 days are required to accommodate equivalent changes.

Many advertisements for microwave systems claim higher product yield due to reductions in evaporative and drip loss during tempering. Since the majority, if not all, of conventional plants temper material in a wrapped form, evaporative losses are insignificant, while substantial periods at air temperatures above 0°C would be required before thawing of surface tissues occurred and drip became apparent.

12.3.6 Modeling of microwave-assisted thawing and tempering

The thawing times of various foods in conventional thawing systems can be predicted using finite difference techniques with good accuracy. Models have been developed and verified for the thawing of meat blocks, beef quarters, pork legs, etc., under a wide range of conditions (James & James, 2002).

However, it is much more difficult to model microwave thawing accurately. Ayappa (1997) in his review of microwave heating stated that early models of microwave heating used Lambert's law to describe the microwave power absorption. In the 1990s, models for transport processes were developed with the microwave power derived from Maxwell's equations. The modeling has revealed some interesting effects. When modeling the thawing of "Tylose" slabs (a meat simulant), Basak and Ayappa (1997) predicted that resonance, during which the microwave power absorption is high, causes a 2 cm thick slab to thaw more quickly than a 1 cm slab. Basak (2003) modeled the thawing of slabs of ice and found that thawing could start at the exposed and unexposed face of a slab when only one face was exposed to microwaves. However, when two faces were exposed, thawing could also commence in the center of the slab.

Taher and Farid (2001) predicted that it is possible to thaw meat under controlled conditions such that its surface temperature never exceeds 10°C.

They claimed that thawing time would be less than one-fifth of that required in conventional thawing. The results were verified under laboratory conditions by installing a control system that switched off the magnetron when the surface temperature reached 10°C. The control cycled the magnetron on and off to maintain the surface temperature at approximately 10°C.

Chamchong and Datta (1999) modeled food thawing in a microwave oven. They predicted that power cycling has an almost identical effect as continuous power at the reduced level of the average cycled power. As power level increases, the surface flux increases by the same fraction. At higher power levels, however, the outside thaws relatively faster. A "shield" develops due to a much reduced microwave penetration depth at the surface. Thus thawing time at higher power levels is reduced considerably. Temperature increases initially are nonuniform since the surface is heated at a faster rate than the interior.

Campañone and Zaritsky (2010) reported that predictions of their three-dimensional model were within 10% of experimental results for thawing of frozen meat. They used it to predict temperature and water concentration profiles under different thawing conditions in meat products and to simulate the effect of a fat layer located at the surface of the meat piece on temperature profiles. The model was also used to predict the advantages, in terms of temperature uniformity, of cooling the surface of the meat using refrigerated air during microwave thawing.

12.3.7 Current commercial microwave-assisted thawing and tempering systems

A wide range of batch and continuous commercial microwave systems for thawing and tempering have been developed since the 1970s. Many of the early systems were sold as thawing plants, i.e., heating frozen food to a temperature where no ice was present in the product. However, most systems could and can only successfully temper the food, and current systems are sold as microwave-tempering systems.

In the late 1980s it was believed that approximately 200 industrial microwave-tempering systems had been installed (Decareau & Peterson, 1986). Microwave drying and cooking installations numbered 30 and 16, respectively; so tempering was by far the most successful application of microwaves in food processing. The systems were used to temper meat, fish, butter, and berries. At that time there were 17 fully operational microwave systems for meat in the United Kingdom, 12 of which were small batch units and the remainder continuous tempering tunnels (James, 1986). A number of the batch systems were used either as one stage of a hybrid microwave/conduction system or to augment large conduction systems by fast tempering of small batches of urgently required material. No data have been located on commercial microwave use for thawing and tempering in the last 25 years.

A list of the companies that have in the past or are currently believed to manufacture commercial microwave-tempering systems is given in Table 12.6. Industrial systems can be batch or continuous and range in size from small systems that can

Table 12.6 Companies that currently or have in the past produced commercial microwave-tempering systems

Company	Country	Batch (B) or continuous (C)	Frequency (MHz)	Power (kW)	Contact
AMT (Advanced Manufacturing Technologies)	Australia	C	915 ± 10	10 to 75	davidm@AMTmicrowave.com
ANSA Technology	Malaysia	C			enquiry@ansatechno.com
APV	United Kingdom	B & C	896		
DanTech UK Ltd	United Kingdom	B & C	896 or 915	75 to 300	sales@dantechuk.com
Industrial microwave systems	United Kingdom	B&C	915	0 TO 30	sales@imwaves.com
Hermann Berstorff GmbH	Germany	C			
Cober Electronics Inc.	United States	B & C	915	0 to 100	sales@cober.com
Ferrite Inc.	United State	B & C	896 or 915	0 to 75	www.ferriteinc.com
LMI (Les Micro-ondes Industrielles)	France	C	2450	30 to 90	
Max Ibustrial Microwave	China	C	915 or 2450	0 to 40	info@maxindustrialmicrowave.com
Microdry Inc.	United State	C		0 to 150	engineering@microdry.com
Raytheon	United State	B & C			
Selo	The Netherlands	C			info@selo.com
Thermex-Thermatron Inc.	United State	B & C	915	0 to 200	sales@thermex-thermatron.com
Sairem	France	B & C	915	10 to 60	commercial@sairem.com

Table 12.7 Claimed hourly tempering capacity of MIP 12 for beef at −18°C with 100 kW generator to different final temperatures

Final temperature (°C)	Hourly tempering capacity (kg/h)	
	90% lean	50% lean
−7	6690	7940
−6	5670	7090
−4	4820	6180
−3	3640	5045

Source: Adapted from Ferrite Incorporated.

process one 25 kg block of food at a time for catering operations, to large continuous tunnels processing up to 8 tonnes per hour.

Typical of a small system is the MIP 4 (Ferrite, Inc., New Hampshire, USA), which has a 0.9 × 1.3 m footprint and is powered by a 40 kW magnetron operating at a frequency of 915 MHz. The manufacturer claims that it will temper 680 kg of raw frozen product per hour to a temperature from −2 to −1°C. A single 25 kg block of frozen meat can be tempered in 65 second. At the other end of the scale, the same company produces the MIP 12 that they claim will temper up to 7,700 kg of frozen product per hour. Up to four 75 kW generators can be coupled to each 2.5 m long tunnel. However, the throughput is very dependent on the composition of the product and the final temperature achieved after tempering (Table 12.7). A similar system from Industrial Microwave Systems Ltd (Buckinghamshire, UK) claims to temper 25 kg blocks of beef, chicken, pork fillet, fish fillets, fruit, and butter up to 100 mm high in 6 minutes. The product is tempered from −18 to −3°C and a 30 kW system is reported to process 8,000 kg per hour.

Typically, companies make a number of claims about their systems. For example, Thermex-Thermatron Inc. claims that they are able to provide systems in a large variety of shapes and sizes according to commercial requirements. Each module can apply up to 75 kW microwave power at a frequency of 915 ± 15 MHz to the product being heated, and two or more modules can be installed on the same conveyor. The systems are fabricated out of stainless steel for food applications. They claim that with a uniform load distribution within the oven, a multimode cavity applicator develops a uniform heat distribution in the entire oven. The belt material and configuration are selected based on the nature of the product being heated. Each end of the conveyor is provided with a special vestibule to suppress any microwave energy leakage into the environment. To assure uniform heat distribution in a large variety of load configurations, each oven section is provided with a waveguide splitter with dual microwave feed points and mode stirrers.

APV Baker installed probably the largest microwave installation in Europe to temper butter (Anon, 1992). There were three microwave tunnels each equipped with 2 × 60 kW generators operating at a frequency of 896 MHz and capable of tempering 7 tonnes of butter per hour. Each block can be tempered in 6 minutes so that the average tempering time for a 225-tonne batch has been reduced from 3–4 days to 1 day.

12.4 Conclusions and future trends

Due to the problems of runaway heating it is not thought that any current commercial microwave systems can deliver a controlled industrial thawing process. In such a process, the entire product would be above 0°C but none above 7°C at the end of the process. The product would then be suitable for further manual processing or sale without any appreciable increase in pathogen levels. If the product were to be heat processed after thawing, then these restrictions would not necessarily apply and microwave thawing could be practical.

Microwave-tempering systems have been in industrial use since the 1980s. They have many proven advantages over conventional tempering systems, especially in terms of thawing time, space requirements, and increased flexibility. However, their penetration in the industry is still limited due to their high initial cost and increased manpower and energy requirements.

Developments in computational modeling and control are progressing, which may eventually lead to more versatile microwave systems that can truly thaw products.

References

Akhtar, S., Khan, M. S., & Faiz, F. (2013). Effect of thawing on frozen meat quality: A comprehensive review. *Pakistan Journal of Food Science*, *23*(4), 198–211.

Anderson, B. A., & Singh, R. P. (2006). Modeling the thawing of frozen foods using air impingement technology. *International Journal of Refrigeration*, *29*, 284–304.

Anon. (1974). *A method of thawing deep-frozen meat products*. UK Patent Specification 1,355,164(A) - 1974-06-05.

Anon. (1992). Thawing frozen poultry. In: *Dielectric heating for industrial processes, UIE Tour Atlantique, Cedex 06, Paris* (pp. 110–113).

Anon. (2008). *Thawing and tempering*. WFLO Commodity Storage Manual.

Anon (2013). Pork, beef, poultry rapid defrosting with radio frequency. *Engineering Review*, *4*(June Stalam).

Archer, M., Edmonds, M., & George, M. (2008). *Seafood thawing*. Technical report, Seafish Authority.

Ayappa, K. G. (1997). Modelling transport processes during microwave heating: A review. *Reviews in Chemical Engineering*, *13*(2), 1–69.

Bailey, C., James, S. J., Kitchell, A. G., & Hudson, W. R. (1974). Air-, water- and vacuum-thawing of frozen pork legs. *Journal of the Science of Food and Agriculture*, *25*, 81–97.

Basak, T. (2003). Analysis of resonances during microwave thawing of slabs. *International Journal of Heat and Mass Transfer*, *46*(22), 4279–4301.

Basak, T., & Ayappa, K. G. (1997). Analysis of microwave thawing of slabs with effective heat capacity method. *AIChE Journal*, *43*(7), 1662–1674.

Baush, E. R., Caren, P. L., Stephen, R. L., & Tuemy, J. M. (1980). *Meat tempering practices and analysis of meat tempering systems at army installations*. Technical Report Natick/TR-80/030.

Bialod, D., Jolion, M., & Legoff, R. (1978). Microwave thawing of food products using associated surface cooling. *Journal of Microwave Power*, *13*(3), 269–274.

Bozkurt, H., & Içier, F. (2010). Ohmic thawing of frozen beef cuts. *Journal of Food Process Engineering*, *35*, 16–36.

Brody, A. L., & Antenevich, J. N. (1959). Ultrasonic defrosting of frozen foods. *Food Technology*, *13*, 109–110.

Brown, T., & James, S. J. (2006). The effect of air temperature, velocity and visual lean (VL) composition on the tempering times of frozen boneless beef blocks. *Meat Science*, *73*, 545–552.

Campañone, L. A., & Zaritsky, N. E. (2010). Mathematical modeling and simulation of microwave thawing of large solid foods under different operating conditions. *Food Bioprocess Technology*, *3*, 813–825.

Chamchong, M., & Datta, A. K. (1999). Thawing of foods in a microwave oven: I. Effect of power levels and power cycling. *Journal of Microwave Power and Electromagnetic Energy*, *34*(1), 9–21.

Chen, Y. H., & Barthakur, N. N. (1991). Potato slab dehydration by air ions from corona discharge. *International Journal of Biometeorology*, *35*, 67–70.

Decareau, R. V., & Peterson, R. A. (1986). *Current state of microwave processing. Microwave – processing and engineering* (pp. 22–23). Chichester, UK: Ellis Horwood.

Farag, K. W., Lyng, J. G., Morgan, D. J., & Cronin, D. A. (2011). A comparison of conventional and radio frequency thawing of beef meats: Effects on product temperature distribution. *Food Bioprocess Technology*, *4*, 1128–1136.

Fathi, Z., Lauf, R. J., & Mcmillan, A.D. (2003). *Microwave thawing apparatus and method.* US Patent application 20030192884.

Fik, M., & Macura, R. (2001). Quality changes during frozen storage and thawing of mixed bread. *Nahrung Food*, *45*(2), 138–142.

Haugland, A. (2002). *Industrial thawing of fish – to improve quality, yield and capacity. Ph.D. thesis*. Trondheim: Norwegian University of Science and Technology.

He, X., Liu, R., Nirasawa, S., Zheng, D., & Liu, H. (2013). Effect of high voltage electrostatic field treatment on thawing characteristics and post-thawing quality of frozen pork tenderloin meat. *Journal of Food Engeering*, *115*(2), 245–250.

He, X., Liu, R., Tatsumi, E., Nirasawa, S., & Liu, H. (2014). Factors affecting the thawing characteristics and energy consumption of frozen pork tenderloin meat using high-voltage electrostatic field. *Innovative Food Science and Emerging Technologies*, *22*, 110–115.

Hergenreder, J., Hosch, J. J., Varnold, K. A., Haack, A. L., Senaratne, L. S., Pokharel, S., … Calkins, C. R. (2011). The effects of freezing and thawing rates on tenderness and sensory quality of beef subprimals. *Journal of Animal Science*, *91*, 483–490.

Holzwarth, M., Korhummel, S., & Carle, R. (2012). Evaluation of the effects of differing freezing and thawing methods on color, polyphenol and ascorbic acid retention in strawberries (Fragaria x ananassa Dutch). *Food Research International*, *48*(1), 241–248.

Hsieh, C. W., Ch, L. A. I., Wj, H. O., Huang, S. C., & Ko, W. C. (2010). Effect of thawing and cold storage on frozen chicken thigh meat quality by high-voltage electrostatic field. *Journal of Food Science*, *75*(4), M193–M197.

James, S. J. (1984). Thawing meat blocks using microwaves under vacuum. In: *Proceedings of the 30th european meeting of meat research workers, Bristol, Paper 2:6* (pp. 61–62).

James, S.J. (1986). Microwave thawing and tempering of frozen meat. In: *ENCE heating and processing 1–3000 MHz, Cambridge, 7.4.*

James, S. J., & Bailey, C. (1980). Air and vacuum thawing of unwrapped boneless meat blocks. In: *The proceedings of the institute of refrigeration 76, session 1979–1980* (pp. 44–51).

James, S. J., & Bailey, C. (1983). The theory and practice of food thawing. In P. Zeuthen, et al. (Eds.), *Thermal processing and the quality of foods* (1st ed., pp. 566–578). London & New York: Elsevier Applied Science Publishers.

James, S. J., & Crow, N. (1986). Thawing and tempering: Industrial practice. *Meat Thawing/ Tempering and Product Quality, IFR-BL: Subject Day.*

James, S. J., & James, C. (2002). *Thawing and tempering. Meat refrigeration* (pp. 159–190). Cambridge: Woodhead Publishing.

James, S. J., & James, C. (2010). Freezing/thawing. Chpt 5, 105–124. In F. Toldrá (Ed.), *Handbook of meat processing*. Wiley-Blackwell.

James, S. J., Nair, C., & Bailey, C. (1984). The effects of freezing and thawing rates on the drip loss from frozen beef. In P. Zeuthen, et al. (Eds.),*Proc. COST '91 Seminar, Athens. In Thermal Processing and the Quality of Foods* (1st ed., pp. 807–813). London and New York: Elsevier Applied Science Publishers.

Jason, A. C. (1974a). *Thawing frozen fish, torry advisory note*. Ministry of Agriculture, Fisheries and Food, Torry Research Station.

Jason, A. C. (1974b). Microwave thawing – Some basic considerations. In: *Meat freezing; why and how? Meat research institute symposium no. 3, 43.31.*

Jason, A. C., & Sanders, H. R. (1962). Dielectric thawing of fish. *Food Technology*, *16*, 101–107.

Kissam, A. D., Nelson, R. W., Ngao, R., & Hunter, P. (1981). Water thawing of fish using low frequency acoustics. *Journal of Food Science*, *47*(1), 71–75.

Knirsch, M. C., Dos Santos, C. A., De Oliveira Soares, A. A. M., & Penna, T. C. V. (2010). Ohmic heating – A review. *Trends in Food Science & Technology*, *21*(9), 436–441.

Les Micro-Ondes Industrielles. (undated). *L.M.I. Gigatrons in the fish industry*. L.M.I. Epone, France.

Linares, C. P., Saavedra, F. F., Serrano, A. B., Les, P. A. Z., & Sosa, A. R. T. (2005). Effects of freezing temperature and defrosting method on pork quality characteristics. *Journal of Animal and Veterinary Advances*, *4*(12), 976–979.

Merts, I. (1999). *A review of methods for thawing meat*. Meat Industry Research Institute of New Zealand (MIRINZ).

Miles, C. A., Morley, M. J., & Randell, M. (1999). High power ultrasonic thawing of frozen foods. *Journal of Food Engineering*, *39*(2), 151–159.

Mousakhani-Ganjeh, A., Hamdami, N., & Soltanizadeh, N. (2015). Impact of high voltage electric field thawing on the quality of frozen tuna fish (*Thunnus albacares*). *Journal of Food Engineering*, *156*, 39–44.

Mousakhani-Ganjeh, A., Hamdami, N., & Soltanizadeh, N. (2016). Thawing of frozen tuna fish (*Thunnus albacares*) using still air method combined with a high voltage electrostatic field. *Journal of Food Engineering*, *169*, 149–154.

Ohlsson, T. (1984). Low power microwave thawing of animal foods. In P. Zeuthen, et al. (Eds.), *Proc. COST '91 Seminar, Athens. In Thermal Processing and the Quality of Foods* (1st ed., pp. 579–584). London and New York: Elsevier Applied Science Publishers.

Ohtsuki, T. (1991). Process for thawing foodstuffs. United States Patent, Patent Number: 5034236.

Orlowska, M., Lebail, A., & Havet, M. (2014). Electrofreezing. In Ramaswamy, et al. (Eds.), *Ohmic Heating in Food Processing* (pp. 423–440). Boca Raton London New York: CRC press.

Pham, Q. T. (2014). *Thawing. Encyclopedia of Meat Sciences* (3nd edn). Elsevier.

Pizza, A., Pedrielli, R., Busetto, M., Bocchi, M., & Spinelli, R. (1997). Use of radiofrequencies in the meat processing industry. Effects on the quality characteristics of meat and cooked meat products. *Industria Conserve*, *72*, 122–133.

Riihonen, L., & Linko, P. (1990). Effect of thawing on the quality of frozen mechanically deboned meat. *Journal of Agricultural Science in Finland*, *62*(5), 407–415.

Roberts, J. S., Balaban, M. O., Zimmerman, R., & Luzuriaga, D. (1998). Design and testing of a prototype ohmic thawing unit. *Computers and Electronics in Agriculture*, *19*(2), 211–222.

Sanders, H. R. (1966). Dielectric thawing of meat and meat products. *Journal of Food Technology*, *1*(3), 183–192.

Satchell, F. E., & Doty, D. M. (1951). High-frequency dielectric heating for defrosting frozen pork bellies. *Bulletin of the American Meat Institute Foundation*, *12*, 95.

Taher, B. J., & Farid, M. M. (2001). Cyclic microwave thawing of frozen meat: Experimental and theoretical investigation. *Chemical Engineering and Processing*, *40* (4), 379–389.

Virtanen, A. J., Goedeken, D. L., & Tong, C. H. (1997). Microwave assisted thawing of model frozen foods using feed-back temperature control and surface cooling. *Journal of Food Science*, *62*(1), 150–154.

Xin, W., Rui, H., & Jin-Hong, Z. (2015). Evaluation of the effects of different thawing methods on the texture, colour and ascorbic acid retention of frozen hami melon. *Journal of Food Science and Technology*, *50*(2), 116–1122.

Yagi, S., & Shibata, K. (2002). *Microwave defrosting under reduced pressure*. US Patent Application 20020195447.

Zagrodzki, S., Niedzielski, Z., & Kulogawska, A. (1977). Meat thawing under controlled conditions. *Acta Alimentaria Polonica*, *3*(21), 1, 11.

Packaging for microwave foods

13

R. Schiffmann
R.F. Schiffmann Associates, Inc., New York, NY, United States

13.1 Introduction

Microwave food product development is possibly the most challenging task the food technologist faces. Much of the reason for this surrounds those peculiarities of microwave heating that are so appealing to consumers: the speed of heating and cool oven operation, i.e., that the air inside the oven remains cool while the food becomes hot. There are also unusual temperature profiles, in large part due to the dielectric properties of the food, but also influenced by the size and shape of both the food and the container. As a result, the selection of the proper packaging system is more important with microwavable foods than with most other products. It controls how quickly and uniformly a food heats. In some cases, the package may actually contribute to the heating by providing a heated surface for browning and crisping, or a high-heat steam atmosphere for moisture retention and faster heating.

Microwave containers may be broken into two broad categories: passive and active containers. Passive containers serve to hold the food without contributing any heat to the product, rather like a coffee cup. On the other hand, active space containers actually interact with the microwaves to provide a source of heat. A good example is the susceptor trays used for crisping the bottom of microwavable pizza. Therefore, the first thing the product developer must decide is what role the package will have in relation to the heating phenomenon—i.e., will it be passive or active? The passive container would be for general reheating of simple food systems such as vegetables or entrees that do not require additional browning or crisping. The active container affects the heating, perhaps by providing surface heat, or directing the microwave energy to a specific place. Active containers are made of interactive materials such as susceptors for crisping pizza crusts or foil shielding to prevent a food item from becoming too hot, or to direct the microwave energy to specific areas to assist heat distribution. Adopting these two broad definitions, this chapter will examine the technical considerations that should be followed for the selection of the proper package design.

The Microwave Processing of Foods. DOI: http://dx.doi.org/10.1016/B978-0-08-100528-6.00013-9

13.2 Factors affecting temperature distribution in microwaved foods

Before describing the requirements for various containers, it is important to recognize the factors influencing the distribution of temperature in microwaved foods (Schiffmann and Sacharow, 1992, p. 104). These are:

- Microwave power absorption rate
- The shape of the food, its thickness, and density
- The weight of the food
- The thermal properties of the food—specific heat capacity and thermal conductivity
- The dielectric properties of the food
- The starting temperature of the food—refrigerated, room temperature or frozen
- Nature of the food–single or multicomponent food
- Covering, stirring, elevation of the product, use of rest periods during or after heating.

Many of these factors are directly related to the container. In those cases in which the heated food is not stirred, the thermal profile is often worse than that encountered when using conventional heating methods. However, the use of a proper container, often employing a cover or lid, can yield comparable results from the microwave oven, except when the food is frozen. In the latter case, the food's temperature distribution will be significantly inferior to that from a conventional oven. This is due to the very large difference in dielectric loss between frozen and nonfrozen items, the latter often being several thousand times more lossy, i.e., more microwave absorptive, than the former (see Table 13.1).

Another characteristic of microwave heating is that it may produce highly nonuniform thermal profiles in foods. Typical of this is the overheating that may occur in the center of baby food in small jars, while the surface remains much cooler. The cool ambient temperature in the microwave oven may produce other peculiar thermal profiles typified by the low surface temperature of foods that inhibits browning and crisping. The surface is never the hottest place in a microwaved food unless some auxiliary heat form is used, such as hot air or a susceptor.

Table 13.1 Typical dielectric properties (permittivities, loss factors, and penetration depths) for various food materials

Food	Permittivity	Loss factor	Penetration depth (cm)
Water (25°C)[a]	78.2	12.3	1.4
Ice (−12°C)[b]	3.2	0.003	1158
Beef (cooked, 30°C)[c]	27.2	9.6	1.1
Ham (precooked, 20°C)[d]	42.9	22.8	0.6
Mashed potato (30°)[e]	72.5	24.0	0.7

[a]Von Hippel (1954), measured at 3.0 GHz.
[b]Buffler (1993), measured at 2.45 GHz.
[c]Kent (1987), measured at 2.45 GHz.
[d]Kent (1987), measured at 2.8 GHz.
[e]Kent (1987), measured at 2.2 GHz.

13.3 Passive containers

It was only in the latter half of the 1970s that food containers other than aluminum or glass were used for microwavable foods. Today, plastic and paperboard containers dominate as containers for these products, while there is still some use of glass, but very little for aluminum. The fundamental characteristic of passive containers is that they are made of materials that will not interact with the microwave oven, i.e., will not be heated by the microwaves. Therefore, they are considered to be microwave transparent. Their principal function is to contain the food during storage, distribution, and heating. They may also have removable or self-venting covers.

The major considerations when selecting the type of container material are, as described by Schiffmann (1990):

- Must be microwave compatible—will not heat excessively, or prevent effective microwave heating
- Must be thermally compatible with the food—that is, it should not melt, distort, or otherwise affected by the hot food
- May need to withstand conventional oven temperatures—these are "dual oven use" containers
- Should provide shelf life properties commensurate with the food and its use
- Should not ignite, arc, or smoke during normal use
- Should not affect the flavor or color of the food
- Should not be discolored by the food
- Should not allow the migration of any materials into the food that might adulterate it
- Should be safe and easy to handle
- Should be grease resistant
- Should provide a moisture and oxygen barrier where required
- Sealability
- Cost.

The product developer must also be concerned with the following issues, described in detail below:

- The maximum temperatures achieved in the food and their effect upon the container
- The general shape and size of the container
- The effects of the mass and distribution of multicomponent products
- The materials of construction
- Covering and other handling issues.

13.3.1 Temperature

The penetration depth of the microwaves into the food primarily controls the temperature distribution within microwave-heated foods. This is defined by the dielectric loss properties of the food and the wavelength of the microwave energy in a simplified equation as:

$$D_p = \frac{\lambda_0 \sqrt{\varepsilon'}}{2f\varepsilon''}$$

where:

D_p = penetration depth, the depth at which ~63% of the energy has been absorbed, i.e., ~37% of energy remains.
λ_0 = free space wavelength; 12.2 cm at 2.45 GHz
ε' = relative dielectric constant or permittivity
ε'' = dielectric loss factor

Microwaves penetrate only short distances into most foods and so heating may be distributed in unusual ways. Typical penetration depths are shown in Table 13.1. The large discrepancy in penetration depth between water and ice usually causes large variations in temperature within a frozen microwaved food. This can be improved or lessened by the container, as discussed below.

There are several critical temperature issues that affect the selection of the container:

- The temperature of the food in contact with the container
- The temperature that the container itself reaches
- The effect the temperature of the container has on the temperature of the food.
- The effect of the ambient temperature in a conventional oven.

All four are interrelated in a complex manner.

13.3.1.1 Food contact temperature

Most often, since it is the point of entry of the microwave energy, the highest temperature will be found at or near the interface between food and the container, although this may vary considerably and is influenced by the shape and material of the container. If the container is reasonably transparent to microwaves, the container wall temperature will be a result of the food temperature and the cooling influence of the air and the oven floor. Since most microwaved food temperatures are controlled by water, the temperatures at the food/container interface will usually not exceed 100°C. However, exceptions do occur. For example, if the food contains large amounts of fat, such as on the surface of soups or in sauces or gravies, or a large amount of sugar as in preserves or syrups, or very high solids, then temperatures may become very high, 150–200°C or even higher, sometimes in local areas. For example, the fat on a soup will migrate to the edge of the container, remaining there due to interfacial tension, where it can be heated by the microwaves to localized high temperatures. This can cause the container to melt or scorch.

Another effect sometimes seen is splattering of food onto a container sidewall where it may be dehydrated and carbonized when microwaved, thus reaching temperatures high enough to melt most thermoplastics. Such dehydration may also occur when a partially emptied container is stored for one or more days. This problem can often be prevented by using container-wall draft angles which promote the flow of such splashing back into the food mass. It is important that tests be run to eliminate the problem.

13.3.1.2 Microwave heating of the container

In some cases the container itself may heat and this is usually not desirable. One situation occurs when the dielectric loss of the container material is high enough for microwave energy coupling. Two examples of this are thermoset polyesters in which the fillers may be lossy, or soda-lime glass that is also lossy. The latter should not be used as a microwave container as it will not only heat but is likely to break. Another situation occurs with thermoplastics such as polypropylene. As these materials heat to their heat distortion temperatures or softening points, they suddenly change from being nonlossy (microwave transparent) to highly lossy (microwave absorptive) and may exhibit thermal runaway, heating as fast or faster than the food, as their dielectric loss characteristics increase with temperature. In the example cited above, of the fat in a microwave-heated soup, the polypropylene container at the soup/container interface may reach temperatures above the approximately 115°C softening temperature at which the polypropylene absorbs microwave energy; the microwave will heat the container very quickly and nearly uncontrollably, melting a hole in the container. This is a more common occurrence than one might imagine.

Another situation in which the container must have high thermal stability is when it is for dual-use, i.e., microwave and conventional oven heating. The latter may require an oven temperature of 175°C or higher, so the container must be able to withstand very high temperatures.

13.3.2 Shape and size

The selection of the shape and size of a container, and its material of construction, is the most important decision the developer will make.

13.3.2.1 Shape

Some shapes are more amenable to good microwave performance than others. Some general rules to be followed are:

- *Simplicity of design.* The shape should be as regular as possible: oval or round, for example. Narrowing or restricting the body of a cylinder can cause the contents to erupt or explode out of it. A "lobster-shaped" container may seem a good way of marketing a microwavable version of this crustacean, but the peculiar shapes, with broad and narrow areas, will cause serious temperature nonuniformity problems.
- *Container geometry.* The most common nonmicrowave tray shape is rectangular. However, this is not the best shape for a microwavable tray. The thermal profiles usually found in rectangular trays are that the corners are hottest, the center is coolest, and temperatures near the side walls are slightly less than at the corners. The range of temperature from hottest to coolest is similar to that seen in conventional ovens when the initial food temperature is ambient , approximately 25°C, to refrigerated, i.e. <7°C. However, with frozen foods the difference from corners to center may exceed 50°C (see Figs. 13.1 and 13.2; Schiffmann and Sacharow, 1992, p. 106). If corners are unavoidable, they should be rounded as much as possible. Sidewalls should have generous draft angles commensurate

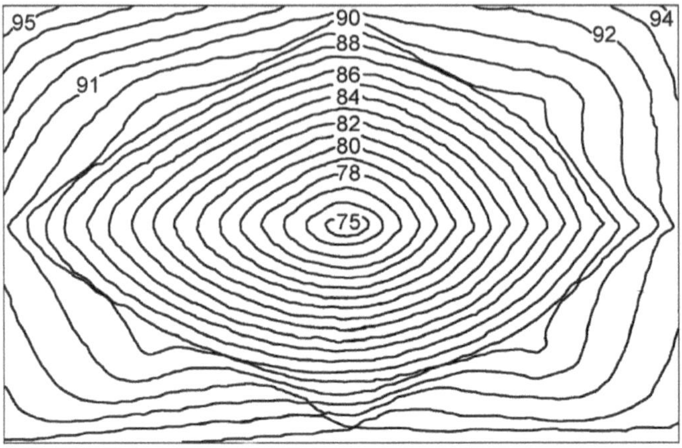

Figure 13.1 Temperature distribution in a typical food microwave heated in a rectangular container (14.0 × 17.8 cm). Note the highest temperatures (°C) are located at the corners and the coolest in the center.
Source: Ministry of Agriculture, Fisheries and Food: UK (Schiffmann, R. F., & Sacharow, S. (1992). *Microwave packaging*. Leatherhead, UK: Pira International, p. 106).

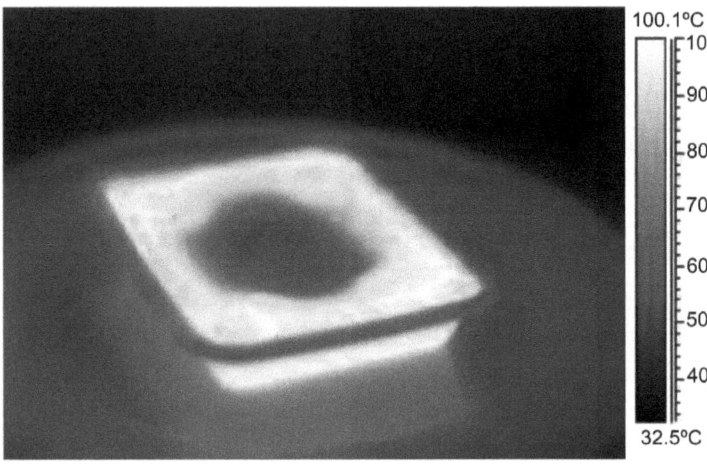

Figure 13.2 Thermal image of a tray of mashed potato taken after 5 min heating in a 750 watt turntable microwave oven (Budd, Hill, & Hooper, 1993).

with stacking, but not be so shallow as to allow a thin food profile around the edges that could overheat. A generous bottom radius is also of benefit, i.e., the bottom center should be slightly bowed or raised. This thins the volume above it that is usually the slowest to heat. It also reduces heat loss to the oven floor.

Small cylindrical containers, such as baby food jars, should be avoided since very high internal temperatures may be produced, which can cause injury when

Figure 13.3 Shelf-stable, single-serve tub product. Note the four vent holes in the lid.

consumed or may erupt due to the generation of high internal pressure. Larger cylindrical containers such as glass jars may be acceptable but can be a problem if there is a shoulder in the jar resulting in a neck opening significantly narrower than the majority internal diameter since this, too, may lead to excessively high internal pressure and eruption. There may also be a wide variation in internal temperature.

Tub-shaped containers, used for single-serve room temperature products, overcome many of the eruption hazards by using a rounded, sloping wall construction in conjunction with a steam vented cap, as seen in Fig. 13.3. The fairly large internal diameters also reduce intense internal spot heating.

13.3.2.2 Size

The size of containers is controlled by the food product's volume and the need to heat the product thoroughly without overheating any portion. For this reason, the net weight of most microwavable food products is in the range from 250 to 300 g. Although some higher weights are available, as described below, they require other means to improve temperature uniformity. The physical size of the most useful containers is on the order of 15 cm × 20 cm, with depths commonly of 2.5 to 5 cm.

13.3.3 Multicomponent containers

Multicomponent meals containing a protein (meat, poultry, or fish), starch, and vegetable are usually sold as frozen meals in the United States and the Australasia market, but primarily as refrigerated versions in the United Kingdom and continental Europe. The frozen meals are especially challenging due to the sometimes very diverse microwave heating properties of the various components. The formulator must try to balance the mass as it is somewhat lossy, i.e. microwave absorptive and may break and dielectric properties of the various components as much as possible, so their final temperatures are similar. The products are usually available in plastic or paperboard containers, often round or oval, although they may also be rectangular. Physically dividing the container into two or three compartments is a good way to keep the components separate during transportation. In that case, the microwave energy sees the separate compartments as separate trays, rather than parts of a single tray. This makes it easier to achieve more uniform heating since the shape, mass, specific heat capacity, and dielectric loss of each component may be individually controlled. In cases where the various components are arranged about the circumference of the dish and one meal component, such as ham, heats faster than the others, it is possible to achieve better results by placing the ham in the center of the dish and surrounding it with the other meal components which act as microwave shields. There have also been numerous uses made of foil shielding to achieve similar effects, as will be described below.

13.3.3.1 Covering

Covering these trays is useful to achieve more uniform heating. Recently, a variety of tray/cover systems have been supplied that contain various self-venting technologies including valves and or other means to release the steam pressure built up during microwaving. The steam itself adds to the uniformity of heating, as well as shortening the heating time and maintaining the moisture level in the food product. Products in these containers may be microwave cooked from raw or processed components.

Three main technologies are used for steam venting (self-venting) packages:

- *Valves*. These are generally described as "one-way valves." These vent the steam through a valve molded into the lid or wrap-film. An example is the "Wipf" valve made in Switzerland. A one-way valve, it allows the venting of gases evolved during storage as well as steam during cooking. It is claimed by CNF, the Swiss group that invented and market products using this valve under the labels "Dream Steam" and "Steamy Wonder," that valves offer an exact method of controlling and releasing the steam pressure, and thereby, controlling cooking better than nonvalve procedures (Keller, 1999).
- *Nonvalve systems*
  - *Holes*. These can be micropores or larger holes or slits, usually covered with a barrier film, adhesive film tape, or adhesive label to be removed by the consumer or opened by the internal steam pressure that allows the steam to vent; or the film lid may shrink, thereby exposing the vent hole.
  - *Cohesive-seal or seal-break systems*. These are various types that use different means to create a steam vent, such as a weak point at a tray/lid seal, or a material or device that is somehow microwave interactive and "melts" a hole or vent for steam release. Several systems, including Vacsys (Global Sealing Technologies, Ltd) are adhesive based. The Vacsys technology is based upon three major elements: special microwave absorptive inks; a gap in lamination adhesive; and slits in sealant film. The ink creates a hermetic seal and helps control venting. The gap allows passage of the steam from inside the package to come in contact with the special ink. The slit in the film allows the steam to enter between the lamination so it can travel through the adhesive gap until it reaches the special ink for venting (Hagino, 1998).

13.3.4 Covers

Covering food during microwaving has been shown by this author and others to improve the cooking performance (Schiffmann and Sacharow, 1992, p. 118). It improves the uniformity of temperature distribution, retains moisture, and shortens cooking time. Recent packaging innovations have included self-venting valves and other structures in or as part of the cover, as was described in detail earlier. These venting systems, in effect, create something similar to a steam pressure cooker within a microwavable container. They are being used extensively in Europe and Japan and are the subject of about 200 patents. The need for venting is necessary where covers are sealed to the container for protection during shipping and storage. If they were not vented in some way, the container cover could burst causing a mess in the oven and even injuring the user. Therefore, users are advised either to slit, pierce or, partially or completely, remove the cover, prior to microwaving. Self-venting covers add an element of safety by eliminating the need for the consumer's actions, as well as providing product benefits. The materials used for tray covers have been described earlier.

A different cover-vent design is used for the microwavable plastic tub products. These multilayer-barrier cylindrical tub-shaped containers are used for products such as macaroni cheese, lasagna, and chili. Because they contain large particulates in viscous sauces or gravies, there is a tendency for them to erupt during microwaving. In order to prevent the contents from erupting out of the container, the flexible polypropylene lid, seen in Fig. 13.2, contains large vent holes and is fitted over the tub, removing a sealed aluminum closure.

13.4 Packaging materials

When selecting the proper container material for use with a microwavable food, the guidelines described earlier should be followed (Schiffmann & Sacharow, 1992, p. 111):

Keeping those guidelines in mind, the following discussion looks at how well these factors are met by the various materials currently on the market.

13.4.1 Glass

Soda lime glass should never be used for microwavable products as it is somewhat lossy, can become hot and may break. Borosilicate glass jars are microwave transparent, but should be used sparingly for microwavable foods since there are significant drawbacks: the high cost of glass, its breakability, high shipping weight, and tendency to become hot due to heat transfer from the contents. As to microwavability, there are additional drawbacks:

- Their usual cylindrical shape causes focused interior heating. In large jars, such as those used for spaghetti sauce, there will be a ring of intense heat near the glass, while the center is significantly cooler. Temperatures near the top will also be high. Small jars, such as those used for baby food, may have intense hidden hot spots in the center, at or near 100°C, while the surface of the food and the glass may be only warm, creating a serious hazard (Schiffmann, 1988). For this reason, it is best if small jars of baby food or any other food are not heated in microwave ovens. If they are, the contents must be thoroughly stirred and tested prior to use. Stirring is advised for all foods heated in glass jars, or any microwave heated food for that matter, as are very conservative heating instructions.

- Under some conditions, glass jars may break during microwaving. This is especially true if there is a surface flaw that may occur during shipping or handling. If the interior becomes very hot while the outer surface is cool, due to the oven's cool ambient temperature, there can be a large enough difference in thermal expansion to cause the glass to crack.
- The usual cylindrical geometry of glass and the restriction at the neck is likely to cause "bumping" or eruption of the contents, especially if there are large particulates present.

13.4.2 Boil-in-bag and stand-up pouches

These flexible pouches have been used for many years, especially for frozen vegetables and side dishes. Originally heated by immersion in boiling water, they were easily adapted for microwave oven use, but with some problems. Since the melting point of the pouch must be above that of boiling water, they are usually made of coextruded plastics or laminations containing layers of medium density polypropylene (MDPE). It is important that these pouches be vented when used in the microwave oven or they will burst. Also, handling of the hot pouch can be difficult. The consumer can simply cut or pierce the pouch but the contents may spill out, so they should be placed on a microwavable plate.

Today, many self-venting pouches are available and in use globally, using some of the same technology described above for steam vented packaging.

13.4.2.1 Stand-up pouches

While boil-in-bag pouches are usually placed horizontally inside the microwave oven, stand-up pouches are heated vertically. They also have all their graphics on the pouch, whereas a boil-in-bag pouch is usually enclosed in an overwrap carton that contains the graphics and heating directions. Stand-up pouches are constructed of multiple layers of thermoplastics and other materials to provide structural support and barrier properties. Since aluminum cannot be used as a barrier in a microwavable pouch, other barriers must be used. A unique pouch construction employing a silicon oxide barrier is used for a wild rice product by Fall River Mills in California (Lingle, 2003). The structure is 48-ga PET/48-ga PET copolymer with a silicon oxide coating/60-ga biaxially oriented nylon/2.4-mil cast polypropylene. The outer PET is reverse gravure printed in eight colors. It may be retorted.

An often unsuspected problem may arise when microwaving a high fat content food such as a soup or a meat-in-sauce. If the fat separates and floats to the food's surface it will attach to the pouch's interior wall by interfacial tension. When microwaved, this interfacial fat can become exceptionally hot and melt through the pouch.

It is, therefore, essential that the product formulation prevents fat separation in these products. It is also important that the stand-up pouch be properly vented to prevent excess pressure build-up and eruption. While many self-vented pouches are available commercially, there are still many that require that the pouch be opened manually, by cutting the seal, prior to microwaving.

13.4.3 Paper and paperboard

A large number of different paperboard containers, folded or press-molded, are employed for microwavable foods in the United States. They are coated with various polymers for different food applications, the paperboard provides structural rigidity, while the coating provides chemical resistance and sealability. The choice of coating, and hence the choice of container, largely hinges on the maximum temperature likely to be reached by the food during microwaving. Other considerations are grease resistance, extent of oxygen and moisture barrier, sealability, dual-ovenability, and price.

There are three different paperboard tray types based upon their polymer coating. These coatings on the paperboard result from either extrusion coating of a molten polymer resin, adhesive or extrusion lamination of a previously fabricated polymer film, or roll coating of a polymer solution.

- Low-density polyethylene (LDPE) coated paperboard: since the polymer has a low melting point, this is only for the lightest microwave use, such as microwaving frozen vegetables. Food temperature should not exceed that of boiling water. LDPE also has poor grease resistance and is not suitable for conventional oven use, i.e., it is not dual-ovenable.
- High-density polyethylene (HDPE): has better temperature and grease resistance, but its higher cost and difficult sealability have blocked its adoption for food packaging.
- Polypropylene (PP): has a higher melting point and better grease resistance than polyethylene. However, it is still not suitable for dual-ovenable use.
- Polyethylene therephthalate (PET): has a use temperature of up to 205°C, so these trays are ideal when dual-ovenability is required. Such high temperatures may also occur during microwaving, especially at the surface interface between the food and the tray. They are also grease resistant and have modest gas barrier characteristics. PET coated trays, pressed and folded, have become the package of choice for many microwavable food products in the United States.
- Covers: either polymer coated paperboard or polymer films may be heat sealed to the trays. Commonly, both heat-sealable oriented PP and PET films are employed and their characteristics are shown in Table 13.2 (Bohrer and Brown, 2001, p. 402). These covers may also be steam-vented in the manner described above.

13.4.4 Polymer trays

The use of polymer trays was first encountered in the early 1980s when thermoset polyester trays were used for various frozen meals. Many consumers saved these trays for reuse and soon found themselves with more trays than they needed and stopped buying the products. High tray cost also led to their disuse. Replacing them

Table 13.2 Properties of microwavable paperboard trays

Package type	Materials	Maximum service temperature	Oxygen barrier	Moisture barrier	Grease resistance	Sealability	Dual ovenable?	Comments
Tray								
Tray	PET coated paperboard	205	Poor	Fair	Good	Fair	Yes	Pressed or folded trays
Tray	PP coated paperboard	125–135	Poor	Fair	Good	Good	No	Pressed or folded trays
Tray, sleeve, support, overwrap	LDPE coated paper or paperboard	95	Poor	Fair	Good	Fair	Yes	Light duty reheating
Overwrap								
Overwrap	Heat sealable oriented PET film	220	Poor	Good	Good	Good	No	Needs venting during reheat, barrier improved by coatings
Overwrap	Heat sealable oriented PP film	110	Fair	Fair	Good	Good	No	Needs venting during reheat, barrier improved by coatings

Source: From Bohrer, T. H., & Brown, R. K. (2001). Packaging techniques for microwavable foods. In A. K. Datta & R. C. Anantheswaran (Eds.), *Handbook of microwave technology for food applications (p. 402)*. New York: Marcel Dekker.

are trays made from two of the three polymers noted above: polypropylene (PP) and polyethylene terephtalate (PET). Since there is no paperboard to offer structural rigidity, low-density polyethylene (LDPE) is not useful for microwavable trays since it begins to distort at about 75°C.

13.4.5 Polypropylene

These trays are useful only for high-water containing products where temperatures are unlikely to go above 100°C. This means that they are not suitable for use with such high-fat products as soups, sauces, or gravies, or high-sugar products such as syrups and glazes. However, the cost of these trays is low and they are optically clear, so they are used for many microwavable products.

13.4.6 Polystyrene

Foamed polystyrene (PS) is popular for its thermal insulating properties and is used for take-out coffee cups and food trays. However, its thermal stability (softens or melts at <100°C) is too low for it to be usable as a tray material for microwavable processed foods.

13.4.7 Crystallized polyethylene terephtalate

Whereas amorphous PET begins to soften at temperatures over 63°C, trays made from crystallized polyethylene terephthalate (CPET) are stable up to 230°C, and thus are dual-ovenable (Robertson, 1993). They are manufactured by thermoforming an extruded PET sheet containing an added nucleating agent. Thermoforming is done in a hot mold where the tray is held long enough for the crystalline structure to develop. It is then transferred to a second mold for cooling (Proffit, 1991). Since they are not optically clear, these trays are usually pigmented, black being the color of choice for most products. These characteristics are summarized in Table 13.3 (Bohrer & Brown, 2001, p. 402).

13.4.8 Multilayer PP and PET

Containers made from these materials may have three or more layers to provide rigidity, moisture and oxygen barriers, and come in a variety of shapes and sizes. They may be hot-filled or retorted at 70 to 120°C and have room temperature shelf lives in the order of 24 months. The favored barrier options are EVOH (ethylene vinyl alcohol) or Nylon. Graphics are supplied using adhesive labels, shrink sleeves, silk screen printing, etc. In some cases, a foam-structured sleeve is used to provide some heat insulation, such as on the tub-shaped products described earlier.

Table 13.3 Properties of microwavable polymer trays

Package style	Materials	Maximum service temperature	Oxygen barrier	Moisture barrier	Grease resistance	Sealability	Dual ovenable?	Comments
Trays								
Thermoformed tray	CPET	220	Good	Fair	Good	Fair	Yes	Generally pigmented white or black
Thermoformed tray	PP	110	Poor	Good	Good	Good	No	Brittleness an issue at freezer temperatures
Thermoformed trays, clamshells	PS and foamed	80	Good	Fair	Good	Poor	No	Light duty reheating only
Thermoformed tray	LDPE	75	Poor	Good	Fair	Good	No	Light duty reheating only

Source: From Bohrer, T. H., & Brown, R. K. (2001). Packaging techniques for microwavable foods. In A. K. Datta & R. C. Anantheswaran (Eds.), *Handbook of microwave technology for food applications (p. 402)*. New York: Marcel Dekker.

13.4.9 Aluminum

This material has always been controversial for microwave oven use. However, it has several distinct advantages which, over the years, led many packaging and food processors to consider its use for microwavable foods. On the plus side is aluminum's cost, versatility in molding different shapes and sizes, excellent heat transfer characteristics when freezing, shelf life properties and recyclability. However, it has an image among consumers and most food processors that it is not suitable for microwave oven use. There are two major concerns—potential for damage to the magnetron and arcing. Numerous tests by this author and others have shown these concerns to be largely unjustified. Coated aluminum trays were marketed for microwave use for a period, the coating disguising the aluminum and forming a partial electromagnetic barrier to control arcing. However, this failed to overcome both consumer misperceptions and the real problems when used as foil packaging. Since foil is reflective, microwaves can only enter through the open top, causing heating profiles that are very different from those in transparent trays. Since heating though the sides and top are not possible, foil trays must be rather shallow, more in line with the penetration depth of the food therein, and probably should not exceed 2.5 cm deep. Two other phenomena are encountered with foil trays:

- Unusual thermal profiles, especially with rectangular shapes. These often have more exaggerated hot or cold spots than are seen with transparent trays. This results from the foil tray as a resonant structure and the creation of a standing wave pattern within the tray. Changing to oval or round containers can alleviate, but not eliminate, this problem.
- The lack of heating close to the foil/food interface. The foil suppresses the electric field close to its surface so there is little heating except by conductive heat transfer from the food zones beyond this. This can be particularly problematical with frozen foods.

13.5 Active containers

Without the provision of auxiliary heating means within a microwave oven, it is generally not possible to crisp or brown foods, unless special packaging techniques are employed. While combination ("combi") ovens are popular in Europe, where the main voltage is in the order of 240 volts, they represent less than 5% of annual sales in the United States where the ordinary household supply is only 120 volts and 15 amperes in the kitchen, making it nearly impossible to operate the magnetron plus an auxiliary heater simultaneously. Microwave-only oven cavities have ambient temperatures at or only slightly above room temperature, which is unsatisfactory to achieve the high temperatures required for browning and crisping, whose requirements are described below. To meet these needs, various active packaging technologies have been successfully developed.

13.5.1 Browning and crisping

Both are high-temperature phenomena but they differ in very specific ways.

13.5.1.1 Browning

This is a surface color change caused by the complex organic Maillard reaction, and is associated with such things as the browned crust on bread and pastry. The Maillard reaction depends upon four things:

- The presence of the Maillard reactants: an amino acid such as glycine, and a reducing sugar, such as fructose.
- A surface temperature of 100°C or more. Below this temperature the reaction proceeds very slowly, if at all.
- A water activity of approximately 0.8: above this, the reactants are too dilute; below it they are too concentrated. The reaction requires water to proceed.
- A neutral or slightly alkaline pH, approximately 7 and above.

Since the air in the microwave oven is cold, it is difficult to achieve the high surface temperature required. The cold temperature also causes surface condensation of water being pumped out of the food. This, combined with the increasingly high humidity in the oven during its operation, raises the water activity close to 1.0. So, at least two of the requirements for Maillard browning are difficult to achieve by ordinary means. While browning coatings, both liquid and powders, have been used to create a pseudobrowning effect, microwave susceptors are used in many products.

13.5.1.2 Crisping

Crisping requires even higher temperatures, 190°C and above, in order to dehydrate the surface and achieve the required water activity of approximately 0.1. Since microwaves heat the interior of the food, thereby raising the internal vapor pressure, a phenomenon called water-pumping occurs which actively forces water to the surface where it condenses, saturating it and making crisping impossible without additional means.

13.5.2 Characteristics of active containers

Active containers are defined as packages containing elements that interact with the microwave field to create conditions other than simply holding or containing the food while it is being microwaved. There are three basic classes of active containers:

- Shields: these prevent the microwaves from reaching all or part of the food.
- Susceptors: these become hot and transfer heat, primarily by conduction, to the surface of the food being microwaved.
- Field modifiers and patterned susceptors: these improve or focus the microwave energy distribution.

Steam-venting containers, as described earlier in this chapter, may also be considered to be active containers since the contained steam pressure positively affects the cooking performance.

13.5.2.1 Shields

Shielding is achieved by incorporating aluminum foil (typically 0.02–0.03 mm thick), as part of the container. For example, for a frozen meal in a multicompartment tray, it is possible to place ice cream in a totally shielded compartment of the tray—surrounded by aluminum foil—and keep it frozen while the rest of the frozen meal is heated by microwaves. It is also possible to place aluminum foil around portions of the tray to prevent overheating of the corners, sides, or edges of the food.

While the use of shields was originally accomplished by adhering aluminum patches to the surface of trays, a company in the Netherlands, Shieltronics, B.V., has developed a patented procedure that embeds the aluminum within the polyethylene or other plastic tray. In this way, the amount of microwave energy applied to a food component or components will be optimized to provide the optimal temperature for that component. For example, it would be possible to heat a protein, such as fish, with a vegetable, such as potato, and a salad at the same time, with the salad surrounded by the aluminum so it remains cold, while the microwaves are heating the other components.

13.5.2.2 Susceptors

The purpose of susceptors in various tray types is to provide the high temperature surfaces required for browning and crisping. Susceptors have been in commercial use since 1975 and, except for microwave popcorn bags, are used to achieve the localized effects of browning and crisping. Three classes of susceptor have been developed: pigment/binder coatings, chemical susceptors, and metalized conductive coatings. Only the latter are in extensive use for microwavable packaging. Vacuum-deposited aluminum is the most common metalized susceptor material in general use, although sputter-coated stainless steel has seen some commercial applications. The four basic layers consist of the following (see Fig. 13.4):

1. A polymer film heating surface (often 0.012 mm of heat-set, biaxially orientated PET) onto which is deposited:
2. a thin metal layer (usually vacuum deposited aluminum at 50–250 ohms per square) with a thickness measured in angstrom units.
3. An adhesive to bond the film to:
4. a substrate (usually paper or paperboard) that provides structural stability.

Vacuum metalization creates microscopically tiny islands of aluminum separated by microscopically tiny distances. As the microwave field oscillates, a current flows across these aluminum islands with the gaps providing a resistance to the flow, thereby causing resistive (I^2R) heating of the susceptor film. In order to be effective the metal thickness must be of the order of 30–60 angstroms, with a surface resistivity of approximately 0.01 mhos. Fig. 13.5 shows the absorption of microwave power as a function of surface resistivity (Andreasen, 1986). A comprehensive theoretical discussion of these structures was published in *Microwave World* (Habeger, 1997).

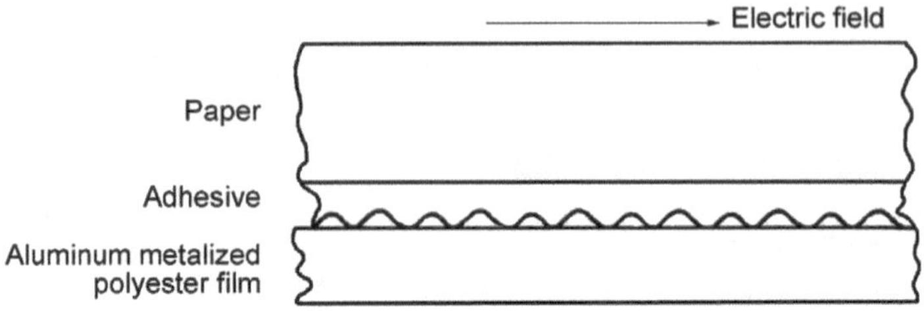

Figure 13.4 Construction of a typical microwave susceptor.
Source: From Schiffmann, R. F., & Sacharow, S. (1992). *Microwave packaging (p. 123)*. Leatherhead, UK: Pira International.

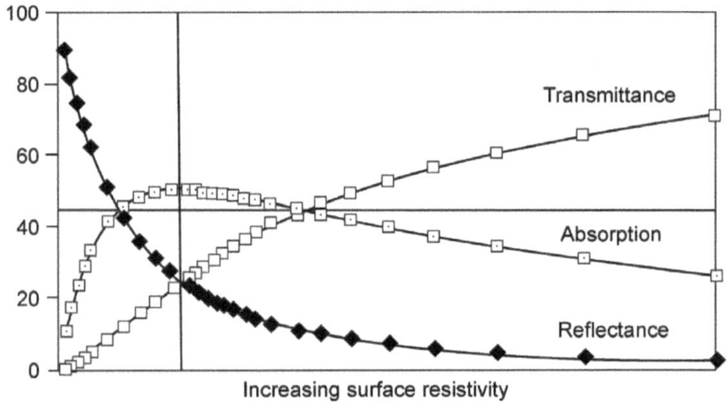

Figure 13.5 Percent microwave power absorption, reflection, and transmittance as a function of surface resistivity (Andreasen, 1986).

The food is usually in direct contact with the heating surface which may have a release coating. The metalized side of the susceptor is separated from the food surface by the PET film, which also protects the metal from chemical or physical damage. While the mass of the thin conductive coating in a susceptor is extremely small by comparison to that of the food, its lossiness is much greater than that of any food. Consequently it absorbs microwaves well despite being in intimate contact with the food, thereby providing high surface temperature. However, when the film temperature reaches approximately 190°C the oriented PET film cracks or crazes and becomes discontinuous, and there is a sharp, irreversible drop in lossiness and the susceptor's temperature falls. This mechanism provides a temperature safety limit. This is especially important for areas in which there is no food contact to act as a heat sink, otherwise the susceptor could cause the paper to burn. Such films can be used only once.

For effective use, the susceptor must be in intimate contact with the surface of the food, otherwise there is a dramatic fall-off of temperature between the susceptor and the food surface. Most susceptors are designed to provide such contact and may

Figure 13.6 The unique susceptor sleeve used with the Hot Pockets filled pastry. Note the large moisture-release vent hole that enhances the crisping efficiency.

even be folded to provide good contact for products such as egg rolls. A unique susceptor sleeve is provided for the Hot Pockets brand of stuffed sandwiches (Chef America). The sleeve also has two large vent holes that allow water vapor to escape which would otherwise prevent effective surface heating of the pastry shell (Pawlowski & Brown, 1988; see Fig. 13.6).

Another problem encountered with susceptors is uniformity of heating. Since these products are usually of large surface area, they exhibit hot and cold spots partially due to the oven's properties and also because of unevenness of deposition of the aluminum film. This is one reason for not making the susceptor dimensions too large—a diameter of 20 cm is a useful limit.

Susceptors can also be quite sensitive to location in the vertical plane. Raising them from the floor of the oven can improve susceptor performance simply by reducing heat loss to the floor. However, once raised, small changes in height can have major effects on efficiency of energy deposition in the susceptor.

One method for preventing overheating of the susceptor in areas at which it is not in good contact with the food is to use a patterned or fused susceptor as described in a US patent (Walters et al., 1996). Here, a regular pattern of Xs that do not have heating capability will generate current concentrating paths when imposed on a susceptor film. The pattern forces a concentration of currents between the aligned points of the Xs and generates a series of sites at which the breakdown occurs if heat transfer is poor (Bohrer & Brown, 2001, p. 416). A grid pattern of lines of active microwave heating function serves a similar role, and both are in commercial use.

Matching the active heating area to the shape of the product and varying the heating rate applied to different areas of the product may provide better browning and crisping performance than a single overall susceptor design. Patterned susceptors are useful for achieving different heating distributions. A new technology, Quilt Wave, manufactured by Graphic Packaging International, Inc. (Golden, CO, USA), "quilts" when exposed to microwave radiation, i.e., it expands into air-filled susceptor cells, which promote better surface contact. This better contact leads to more uniform surface heating, at the same time providing insulation to prevent heat loss to the environment, thereby maximizing the heat flux to the food surface. It has been engineered to run on conventional form—fill—seal equipment to form pouches and is FDA (USA Food and Drug Administration), CFIA (Canadian Food Inspection Agency), and EU (European Union) compliant for reconstituting frozen and chilled foods in microwave and conventional ovens.

Another problem of susceptors is their low thermal mass. If a large amount of water vapor is produced from the food it will inhibit the heating and effective use of the susceptor. For this reason, paperboard-backed, rather than paper-backed susceptors, often have superior browning and crisping performance.

13.5.2.3 Popcorn bags

The largest volume of metalized susceptors used in the United States is for the production of approximately one billion popcorn bags annually. This has nothing to do with browning or crisping. In this application, the susceptor film is bonded directly to the middle third of the interior of the down-side of the bag. The susceptor's function is to melt rapidly the hydrogenated soy bean oil (which otherwise couples microwave energy poorly) in which the popcorn kernels are mixed. Once melted, the microwave coupling to the oil is improved and this enhances the popping performance as measured by the volume of popped kernels and the number of nonpopped kernels remaining following popping. The use of susceptor lined popcorn bags has become standard for this industry. There have been a few successful nonpopcorn applications of susceptor bags. In the United States, this includes the heating and puffing of pork rinds, resulting in very large volume expansion and creating a crunchy edible foam-like structure, often spice-coated.

13.5.2.4 Other susceptor designs

Susceptors have also been bonded to either face of corrugated paperboard. When applied to the fluted side of the corrugate, the food contacts only the tips of the susceptor but the channels between them provide a convenient means for the escape of water vapor. When the susceptor is applied to the flat side, the food, such as pizza, is placed thereupon, while the fluted nonsusceptor side provides insulation to prevent heat loss to the floor of the oven. See Fig. 13.7.

13.5.2.5 Field modification

Field modification packages for microwavable foods redirect the energy of the microwave field in a predictable fashion to optimize the microwave heating performance of foods, overcoming two major drawbacks of microwave cooking with

Figure 13.7 Two types of fluted or corrugated susceptor. On the left, the fluted surface is coated with the metalized susceptor and provides channels for evolved moisture to escape. On the right, the metalized susceptor is on the flat top surface—the fluted corrugations insulate the susceptor from the floor.

respect to heating rates. The first is the tendency for overdone edges and underdone centers of food items. The second is providing different heating rates to different components of multicomponent meals.

The pioneering work in this area was done by The Aluminum Company of Canada and resulted in patented technology and a line of trays and lids named Microwave Match (Keefer, 1987, 1989). The lids were designed to affect the distribution of microwave energy in the container so as to cause more uniform heating. Some browning of the top surface was also demonstrated. The lid structure was not in contact with the surface but remote from it and contained a multitude of pieces of foil bonded to the lid. The pieces of foil took various geometric shapes such as a group of squares with rounded corners, or of arcs of different length, and so on. Both the shape of the foil and spacing were critical and selected for specific food items. The plastic dome lid was mated to an aluminum tray, in the original designs, or plastic or paperboard thereafter. This concept had limited success and is no longer in use. Since then, a great deal of research and development has been done by others, leading to a large number of patents and some commercial success.

There are three main uses of field modifiers: shields (described earlier), field distribution, and field intensification. These may be used singly or in combination to achieve the desired results. They have also been combined with susceptors, where the former redirects the microwave energy to these heating elements.

13.5.2.6 Theoretical basis for field modification

Electrical conductors, such as aluminum, of sufficient thickness prevent the transmission of microwave energy, thereby acting as reflective shields. However, when properly designed in strips and geometric patterns, the aluminum conductor can act as a *field intensifier* where it locally intensifies the microwave energy; or as a *field distributor* by transmitting the microwave energy to areas in the food that would otherwise not receive enough microwave energy. The value of such containers is shown in the following examples:

- Foods of large dimension: conventional pizzas can be quite large, 12 inches (30 cm) or more in diameter. However, microwave pizza is likely to be only 8 inches (20 cm) or less, even on a susceptor tray. The reason is that the pizza is "self-shadowing." That is, it prevents the microwave energy from reaching the center of the crust. In a microwave oven, the energy approaches the food load from all sides. However, the energy distribution is not equal. Edges and corners are heated from the side, top and bottom, but the center of a dimensionally large food such as pizza is microwave heated from only the top. While energy may be reflected from the floor, the outer portions of the pizza absorb microwave energy and effectively block it from reaching the bottom center of the pie, i.e, self-shadowing. The situation is even worse in foods of considerable depth, as shown in Fig. 13.8. It is possible to improve this condition by raising the pizza, or other large food item, an inch or more from the floor or turntable. But another solution is to use a field modifier placed under the pizza that causes more microwave energy to reach the bottom center. An example of such a field distributor is shown in Fig. 13.9. This is the MicroRite combined field distributor and susceptor tray manufactured by Graphic Packaging International, Inc. In this case, it is combined with the microwave susceptor that provides the actual heat needed to crisp the crust.

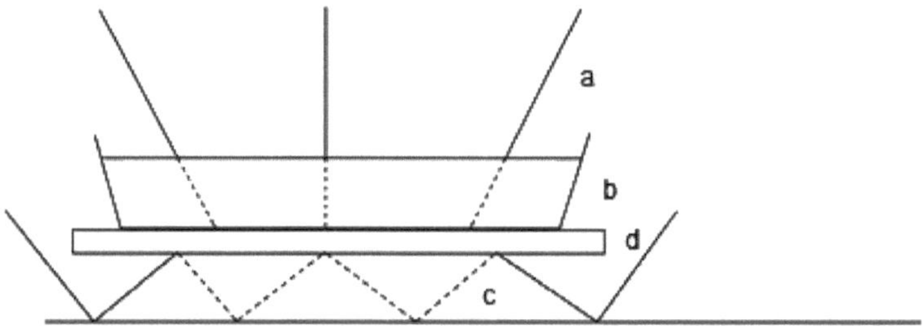

Figure 13.8 Self-shadowing reduces or prevents microwave energy (a) reaching the bottom center of large food items (b). The energy is absorbed and some attenuated as it passes through the food, thereby reducing or eliminating the energy that can be reflected from the oven floor (c). Note: (d) represents the turntable.

Figure 13.9 The MicroRite field distributor and susceptor tray for microwave pizza. *Source*: Courtesy of Graphic Packaging International.

- Another problem arises when the food load is very large, for example, a 1-kilo (2.2 lb) lasagna. If heated in a passive container, the temperature distribution is very nonuniform. The large mass requires an especially long time to heat thoroughly, particularly if the lasagna is frozen. Edges and corners not only will overheat but may char and dry out while the center remains relatively cold. By shielding the outer walls, energy deposition and, therefore, temperature rise is significantly reduced in those zones. The addition of a resonant loops structure to the bottom of the tray increases energy coupling (*field intensification*) and energy transfer from the tray bottom, thereby reducing the overall cooking time. Fig. 13.10 shows the energy distribution in lasagna cooked with and without using this technology, the graphs being provided by Graphic Packaging International, which manufactures these containers.
- A third problem addressed by these field-modifying structures is improving the heating of multicomponent meals. Since each meal component may have significantly different dielectric properties, as well as different specific heat capacities, masses, and shapes, temperature uniformity is difficult to achieve. A typical example is shown in Fig. 13.11

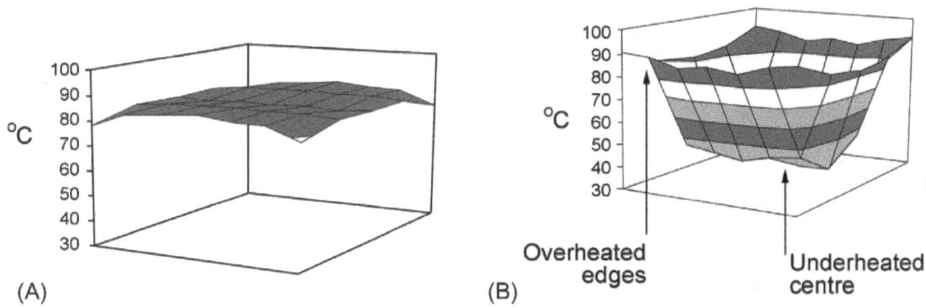

Figure 13.10 Temperature distributions in microwaved lasagna with (A) and (B) without field modification.
Source: Courtesy of Graphic Packaging International, Inc.

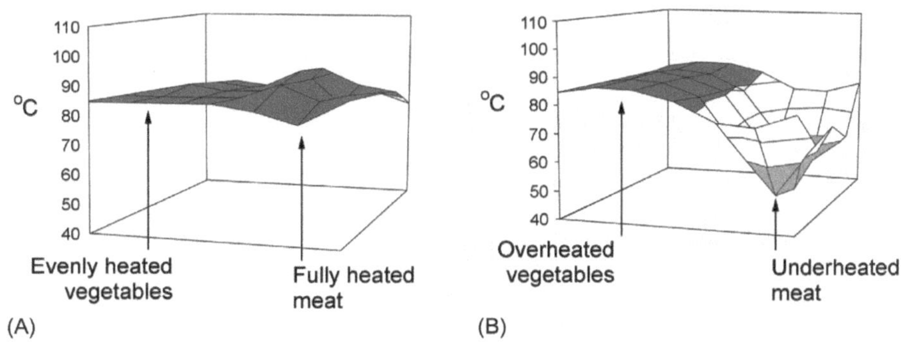

Figure 13.11 Temperature distributions in a multicomponent meal (A) with and (B) without field modification.
Source: Courtesy of Graphic Packaging International, Inc.

Figure 13.12 An example of a MicroRite container to provide more uniform product temperature as illustrated in Fig. 13.11.
Source: Courtesy of Graphic Packaging International, Inc.

showing overheated vegetables and underheated meat. By combining shielding on the sides of the tray and resonant loops under the meat (Fig. 13.12), it is possible to improve this situation dramatically.

The design and placement of all field modifiers must be tailored to specific food combinations, shapes, sizes, and temperature storage conditions. They are usually made by demetalization techniques employing chemical etching. However, adhesive bonding of the aluminum strips onto the fiber substrate can produce simple designs. A third manufacturing option is rotary die-cutting. In all these cases, the conductive metal is aluminum, usually at the thickness of common foil, 6–7 μm, at which it is both dimensionally and electrically stable.

Field modifiers may also be combined with metalized susceptors, enhancing the heating of the latter in shadowed areas in products such as fruit pies and pot pies.

- *Steam-assisted packaging*. In the microwaving of foods, one of the strongest phenomena is that the ambient temperature inside the oven is at or only slightly above room temperature and the relative humidity is low. This causes the surface of the food to be cooler than the interior. This also leads to more nonuniform heating than if the food were covered. Further, during the heating cycle, moisture within the food is actively pumped toward the surface, and condenses there due to the cool surface temperature, making crisping difficult and even impossible in most cases. Since the surrounding atmosphere is low in humidity, this "pumped" water vapor and other volatiles escape from the food, leading to toughness, loss of flavor, and other deleterious effects. This condition can be improved by covering the food with a microwave-transparent material. However, this requires that the cover can be vented so that water vapor or steam does not result in injury to the consumer caused by eruption or steam burns, when removing the cover or lid. It is also possible that a well-sealed cover may burst during microwaving, resulting in the contents being sprayed about the oven cavity and door.

This has led to the creation of numerous sealed steam-packages for products ranging from fresh or partially cooked vegetables, to meats, poultry, and fish. The steam may be vented in a controlled manner during the microwave heating cycle by the use of valves incorporated into the package, most often in the lid. This allows a significant steam atmosphere to be created surrounding the food without the dangers described above. Since steam has a high energy and, therefore, heating capacity, it provides a significant amount of heat, as well as moisture, to the surface of the food. The combination allows microwaves to cook or heat the inside of the food, while the steam heats the surface. This provides more uniform heating or cooking and significantly reduced moisture loss. Faster heating or cooking also results in better flavor, color, texture, and nutrient retention. How well they perform is largely a function of how well the valve systems perform. There are numerous valve-types including simple notches or holes. There are probably 200 to 300 patents for self-venting packages and related materials, mostly from Japan.

13.6 Microwave pasteurization of ready meals

Ready meals, i.e., microwavable refrigerated meals, are becoming more and more popular globally due to the perceived freshness of meals in the refrigerated section in the supermarkets, compared to those in the frozen section. While chilled meals have found mainly acceptance in countries where short transportation lines allow fast delivery to the supermarkets (e.g., in the United Kingdom), through unique microwave pasteurization technology, it is possible to extend the shelf life of these products in refrigerated conditions for as much as 30 days or more at 8°C without preservatives. This extended shelf life allows for longer distribution channels, while still having a decent refrigerated shelf life once it reaches the supermarkets. One such process, the MicVac process (Micvac AB, Moelndal, Sweden), is a successful commercial means for achieving such results (Reynolds, 2011). This process involves filling a thermoformed tray with raw food ingredients, such as meat, poultry, onions, celery, etc., as well as some partially precooked items, such as potatoes or rice. A flexible film lid is heat-sealed into place and then a special MicVac valve is applied. The filled and lidded trays are then transported through a microwave tunnel, also supplied by MicVac, during which the contents are cooked for 4 to 10 minutes, depending on the pasteurization requirements. During processing, the MicVac valve opens to release the internal pressure, and as soon as the pasteurization process is complete, the trays enter a spiral cooler and the MicVac valve closes, causing a partial vacuum inside the container during cooling. This results in finished, cooked, and pasteurized ready-to-eat meals with the extended shelf life described above.

In the United States, both a microwave pasteurization (MAPS: microwave-assisted pasteurization system) and sterilization (MATS: microwave-assisted thermal sterilization) system is being commercialized by 915 Labs. These processes were initially developed and patented by the Washington State University. They utilize microwave energy being applied to sealed food packs as they are transported, under water, through a chamber (which is pressurized in the case of MATS (Lingle, 2016)) while being exposed to 915 MHz microwave energy. Several of its products have been approved by the USFDA (Lingle, 2003; see also Chapter 10: Microwave-assisted pasteurization and sterilization—commercial perspective in this book).

13.7 Future trends

Perhaps the greatest innovations to assist in the design and development of new packaging systems have been computer-assisted design and 3D printing. Computers began having significant impact on the design and manufacturing of packaging with the introduction of CAD/CAM systems. By the late 1970s to early 1980s, it was possible for a packaging design team to create digitized design blueprints and send them directly to equipment that was capable of cutting the tools necessary for manufacturing of these packages. Now it was possible to accomplish in-house in a matter of a few days what might have taken weeks and months prior to this. The next innovation-leap occurred with the introduction of 3D printing. Now, the packaging designer was suddenly able to draw an idea for a new packaging form on a computer and send that design directly to a 3D printer, which, in a short time of minutes to hours, would produce a perfect representation of the idea. Prior to this, designers often had to send new packaging ideas either to in-house or outside groups to create mockups of the desired container. The incredible increase in speed and the reduction of costs made it possible to test ideas and whittle them down to the most successful ones in a very short time.

Both of these technologies have contributed and continue to contribute to the design and development of new packaging forms and allow the testing and the introduction new materials. When we consider how few years it has been since the introduction of both computer design technologies and 3D printing, we are bound to look forward to major innovations that have probably not even been thought of today.

There is increased interest in environmentally friendly packaging, that will be recyclable as well as producing less greenhouse gases. For example, investigations are underway to find alternate materials for the production of molded trays, such as Bagasse, a fiber from sugarcane that has very good oil resistance and can be used for molded tray containers, but requires a polymer liner for moisture resistance. Another example is wood pulp in Thailand which is currently a waste product that until now was burned or used for cattle feed, but could again be used in molded containers with very little or no greenhouse gas effect.

Whatever container the product developer settles upon, it cannot replace good food technology and formulation. One should never start by first picking the container and then trying to develop the product to match the container. Nor should one fully develop a product and then search for the proper container. The best and most successful technique is to bring packaging technology into the product's development early, once the basic form and characteristics of the product have been settled upon. Then the right container can be matched to optimize the product to achieve the greatest consumer satisfaction.

References

Andreasen, M. (1986). *New technologies to improve susceptor efficiencies*. Santa Rosa, CA: Microwave Packages, Optical Coating Laboratory, Inc.

Bohrer, T. H., & Brown, R. K. (2001). Packaging Techniques for Microwavable Foods. In A. K. Datta, & R. C. Anantheswaran (Eds.), *Handbook of microwave technology for food applications* (pp. 402–416). New York: Marcel Dekker.

Budd, C. J., Hill, A. D. C., & Hooper, G., (1993). Modeling microwave cooking: Theory and experiment. http://people.bath.ac.uk/mascjb/new2d.pdf.

Buffler, C. R. (1993). *Microwave cooking and processing*. New York, NY: Van Nostrand Rheinhold.

Habeger, C. C. (1997). Microwave interactive thin films. *Microwave World*, *18*(2), 8–22.

Hagino, J. (1998). *Wrapping device*. US Patent No. 5,839,832.

Keefer, R. M. (1987). *Microwave heating package and method*. US Patent No. 4,656,325.

Keefer, R. M. (1989). *Container for microwave heating including means for modifying microwave heating distribution and method of using same*. US Patent No. 4,818,568.

Keller, K. (1999). *Ventable microwave-safe food package*. European Patent No. PCT/ CH98/ 00546 (US Patent No. 6,607,764, 2003).

Kent, M. (1987). *Electrical and dielectric properties of food materials*. UK: Sience and Teechnology Publishers.

Lingle, R. (2003). Pouches stand up to retorting. *Packaging World*, 51–52, July.

Lingle, R. (2016). Food packaging identified for new MATS processing. *Packaging Digest*.

Pawlowski, T. D., & Brown, R. K. (1988). *Sleeve for crisping and browning of foods in a microwave oven and package and method utilizing same*. US Patent No. 4,775,771.

Proffit, P. J. G. (1991). Packaging of heat preserved foods in plastic containers. In J. A. G. Rees & J. Bettison (Eds.), *Processing and packaging of heat preserved foods*. Glaskow and London: Blackie and Son, Chapter 7.

Reynolds, P. (2011). Microwave pasteurization puts ready meals where consumers want them. *Packaging World*.

Robertson, G. (1993). *Food packaging* (p. 121). New York: Marcel Dekker.

Schiffmann, R. F. (1988). Hazards of microwave heating. *Microwave foods*, *88*, The Packaging Group.

Schiffmann, R. F. (1990). Microwave foods: Basic design considerations. *TAPP1 Journal*, *73*(3), 209–212.

Schiffmann, R. F., & Sacharow, S. (1992). *Microwave packaging*. Leatherhead, UK: Pira International, pp. 104, 106, 111, 118, 123.

Von Hippel, A. (1954). *Dielectric materials and properties*. Cambridge, MA: The MIT Press.

Walters, G. et al. (1996). *Multilayer fused microwave conductive structure*. US Patent No. 5,530,231.

Further reading

Lorence, M. W., & Pesheck, P. S. (2009). *Development of packaging and products for use in microwave ovens*. Cambridge, UK: Woodhead Publishing.

Han, J. (2014). *Innovations in food packaging* (2nd ed.). Acacemic Press.

14 The heating performance of domestic microwave ovens

S.J. James, C. James and G. Purnell
Food Refrigeration and Process Engineering Research Centre, Grimsby, United Kingdom

14.1 Introduction

Since the 1980s, the market for chilled and frozen food has grown rapidly. This is especially true for convenience products that are typically reheated rather than fully cooked in the home or at a catering establishment. The majority of these products are designed for rapid reheating in a domestic, or commercial, microwave oven. During processing, the manufacturers of these products aim to cook the products to a temperature that would ensure the reduction of pathogenic bacteria to a safe level. However, it is possible that the required reduction has not been achieved during processing or the product can become contaminated during distribution or in the home. Thus, microwave reheating must ensure that the product is heated to a safe temperature appropriate to the product.

While many food dishes are perfectly fine after reheating, others are not, particularly reheated cooked meats which often have a particular off-flavor, commonly called "warmed-over flavor" (WOF) (Tims & Watts, 1958). Thus, food manufacturers in recent years have sought to improve eating quality by introducing Not-Ready-To-Eat (NRTE) or Ready-To-Cook (RTC) products that have received minimal processing and rely on the consumer to fully cook the product (Schiffmann, 2014). This has led to renewed concerns over the performance of microwave ovens and uniformity of microwave cooking.

In particular, there has been concern over the growth of *Listeria monocytogenes* at the storage temperatures commonly encountered in retail and domestic storage of chilled foods (James & Evans, 1989). Walker, Bows, Richardson, and Banks (1991) showed that *L. monocytogenes* could survive after microwave heating. Recommended reheating temperatures in Europe have been based on the thermal resistance of *L. monocytogenes*, with a recommendation to reheat to a minimum temperature of 70°C for at least 2 min or an equivalent temperature/time condition, e.g., 75°C for 30 s, 65°C for 10 min, etc. (European Chilled Food Federation, 1996). In the United States, USDA guidelines are to heat to a minimum internal temperature from 71 to 77°C for meat and poultry products and 63°C for fish and seafood products (Schiffmann, 2014). Interestingly, studies on the microwave cooking of raw chickens (Farber, D'aoust, Diotte, Sewell, & Daley, 1998) showed that *Listeria* spp. survived after microwave cooking even though the temperatures measured were above 87°C.

The Microwave Processing of Foods. DOI: http://dx.doi.org/10.1016/B978-0-08-100528-6.00014-0

A small change in heating time in a microwave oven has a large effect on the survival of *Salmonella* spp. Levre and Valentini (1998) found that in inoculated meat loaves *Salmonella* survived after 3.5 min heating followed by a 2 min holding time but was not found when the heating time was increased to 4 min, while Tassinari and Landgraf (1997) found that *Salmonella typhimurium* survived in baby food, mashed potato, and stroganoff after microwave heating.

There have been a number of recorded occurrences of food poisoning associated with microwave heating. In one case (Gessner & Beller, 1994) it was related to microwave reheating of cooked pork slices. In another (Evans, Parry, & Ribiero, 1995) five individuals had food poisoning after consuming microwave-heated cooked rice that contained *Salmonella enteritidis* phage type 4 (PT4), while in the third (Bates & Spencer, 1995) *Salmonella* in microwave-poached eggs was believed to be the source. The most significant outbreak associated with microwave heating resulted in 401 cases of salmonella infection after the consumption of improperly cooked frozen NRTE pot pies (Anon, 2008). Investigations determined that 77% of patients who ate these pies cooked them in a microwave oven. Consumer confusion regarding microwaving instructions and the variability in microwave wattage was identified as a possible reason for a failure to cook the product properly.

At the same time as ensuring microbial safety, it is important that the food is not overheated so that the eating quality is unacceptable. This second criterion is far more difficult to define and is a function of time, temperature, and the food being heated. Some obvious limits can be set. If parts of the food rise to 100°C or above for any length of time, the food will dry out and ultimately burn. With some foods other aspects of the eating quality, i.e., taste, texture, smell, will be unacceptable well before this can occur. If the time/temperature treatment is too mild, then the food may appear undercooked. Investigations carried out since the late 1980s have revealed considerable variability in the ability of different models and types of domestic microwave oven to reheat chilled convenience meals. They have also revealed a large degree of nonrepeatability during heating. Some of the reasons for the variability and nonrepeatability have been identified whilst others are less defined. The development of a protocol to produce microwave reheating instructions for ready meals needs to take into account all the factors that are currently undefined as well as those that can be quantified.

14.2 Factors affecting food heating: microwave power output

The rate of temperature rise at a point in a food depends on the power (rate of energy input) at that point and the thermal properties of the food at that point. A microwave oven differs from a conventional oven because the presence or absence of a food and its dielectric properties change the amount of microwave power delivered.

Since 1990, most microwave oven manufacturers have adopted a single standard for measuring the power output for microwave ovens. The current standard (IEC, 2010) uses a 1 kg water load and the test is carried out using a cold oven in defined ambient conditions. A standard borosilicate glass dish is filled with 1 kg of water at 10°C, the dish placed at the center of the base of the oven and the time required to heat the water to 20°C determined. Since the amount of heat required to heat water by 10°C is known, the average power (W) delivered by the oven to the load can be calculated.

While microwave oven manufacturers should be using the IEC procedure to determine their ovens power, independent studies have shown that the advertised microwave power may not be the same as that measured using the IEC procedure. A recent study by Schiffmann (2014) showed that all eight microwave ovens used in an evaluation of microwave cooking had significantly lower "IEC" power than that declared by the manufacturers. Advertised microwave powers ranged from 1100 to 1200 W but measured IEC powers ranged from 842 to 966 W, 70.9 to 87.8% of the manufacturer's declared power. In the authors' own experience, the advertised microwave power seldom agrees with the measured IEC power and shows the importance of measuring the IEC power in any test program/study involving the use of a microwave oven.

James et al. (1994) used five ovens with different characteristics, including whether they had the facility to operate in a combination mode, i.e., with heated air (Table 14.1), to look at the effect of size of load, repeated use, and mains voltage supply on power output.

14.2.1 Size of load in oven

Since convenience meals commonly range in weight from 0.2 to 0.4 kg, the amount of power delivered to 0.42 and 0.25 kg water loads was measured and compared to the standard 1 kg load.

The power output into the different-sized water loads is shown in Fig. 14.1 together with the standard deviation of five replicated tests. In all ovens there was a reduction in power output as the size of the water load was reduced. In three of the

Table 14.1 Characteristics of ovens used

Oven code	Cavity size cu ft (m^3)	Type of turntable	Recommended food position	Combination
A	1.3 (0.037)	Metal	Metal rack	Yes
B	1.0 (0.028)	Glass	Center	No
C	1.45 (0.041)	Glass	Plate off-center	No
D	0.6 (0.017)	Glass	Plastic rack	No
E	1.15 (0.033)	None	Offset	No

Source: Adapted from James, S. J., Foster, A. M., Phillips, I. C., Wilkins, T. J., Swain, M. J., & Burfoot, D. (1994). *Effects on microwave power output of size of load, continuous (intermittent) use, position of load, and variations in mains supply voltage*. Microwave science series no. 11. London: MAFF Publications.

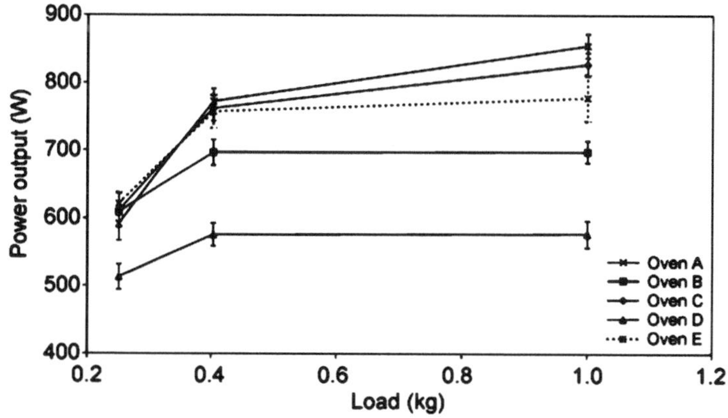

Figure 14.1 Power output into different sized water loads. Error bars ± 1 standard deviation.
Source: Adapted from James, S. J., Foster, A. M., Phillips, I. C., Wilkins, T. J., Swain, M. J., & Burfoot, D. (1994). *Effects on microwave power output of size of load, continuous (intermittent) use, position of load, and variations in mains supply voltage*. Microwave science series no. 11. London: MAFF Publications.

five ovens reducing the load from 1.0 to 0.42 kg produced a small, <20 W, change in power output. In the other two, the reduction was far greater. The power loss when the load was reduced from 1.0 to 0.25 kg was substantial, ranging from 64 to 263 W. In percentage terms, the reduction in power output when the load size was reduced from 1.0 to 0.42 kg ranged from 0.1 to 9.7%, and when the load was reduced from 1.0 to 0.25 kg it ranged from 11.1 to 30.8%.

Experimental work reported by Persch and Schubert (1995) on two different microwave ovens identified a similar trend in the reduction of power output as the volume of water load was reduced. Small amounts of microwave energy were absorbed by the small loads, increasing to a maximum for volumes greater than 1 liter. A recent work by Rother (2010) could explain this trend by a model of hit probabilities.

14.2.2 Repeated use of oven

IEC power determinations are carried out in cold ovens, i.e., ovens that have not been used for at least 6 h. In domestic and commercial situations many occasions occur where an oven may be used sequentially and the oven and its components will heat up.

Studies have shown that the output power of a microwave oven can slowly decrease during continuous, or repeated, use (Swain, Ferron, Coelho, & Swain, 2006; Cullen, 2010). This drop may be as much as 20%. Cullen (2010) measured an average wattage drop of between 8 and 22% in eight ovens after a 10 min preheat.

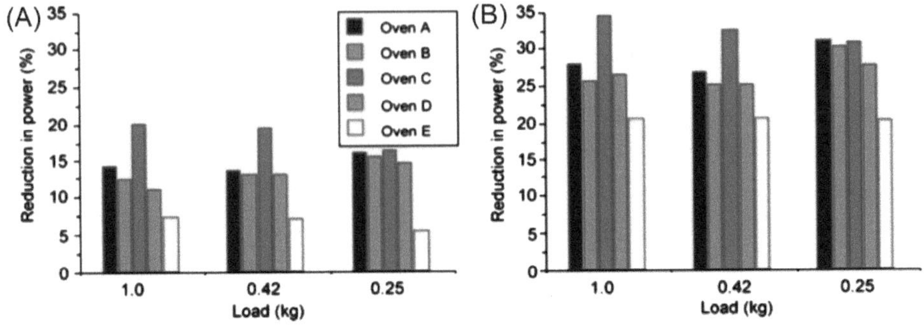

Figure 14.2 Percentage (A) apparent and (B) real reduction in power output after 0.5 h use. *Source*: Adapted from James, S. J., Foster, A. M., Phillips, I. C., Wilkins, T. J., Swain, M. J., & Burfoot, D. (1994). *Effects on microwave power output of size of load, continuous (intermittent) use, position of load, and variations in mains supply voltage*. Microwave science series no. 11. London: MAFF Publications.

To measure the effect of warming up on the power delivered into water loads, James et al. (1994) operated ovens from cold and subjected them to repeated water load power determinations over a period of 0.5 to 1.5 h. Tests were carried out using 1.0, 0.42, and 0.25 kg loads and each complete trial was replicated five times.

In the five ovens, the power delivered to a water load decreased as the oven became warm. The magnitude of the reduction over 0.5 h ranged from 35 to 167 W, and in percentage terms from 6 to 20% (Fig. 14.2A). However, the measurements underestimated the reduction in microwave power because a proportion of the heat gained by the food was by conventional heat transfer. Heat was transferred to the water by conduction and radiation from the warm walls and convection from the warm air. Separate experiments carried out without the microwave source being switched on showed that conventional heat input accounted for between 67 and 118 W of the warm power. The true reduction in microwave power output due to oven warming was, therefore, between 20 and 35% of the cold power (Fig. 14.2B).

Not all studies have shown a prewarming to have a significant effect on measured IEC power. A study by Schiffmann (2014) of eight ovens showed that power dropped only between 4.8 and 0.1% in three of the ovens, while the power increased from 1.0 to 3.4% in five of the other ovens. The prewarming step "consisted of heating one-liter of water at full power, in a microwavable container, for 10 min, removing the water and allowing the oven to cool down with the oven's door open for at least 15 min or until the turntable was no hotter than 27°C."

14.2.3 Variation in mains supply voltage

Microwave ovens are designed to operate at a specified supply voltage. In the United Kingdom until 1995 the electricity supply industry was contracted to provide its consumers with an electricity supply of 240 V, which could deviate by

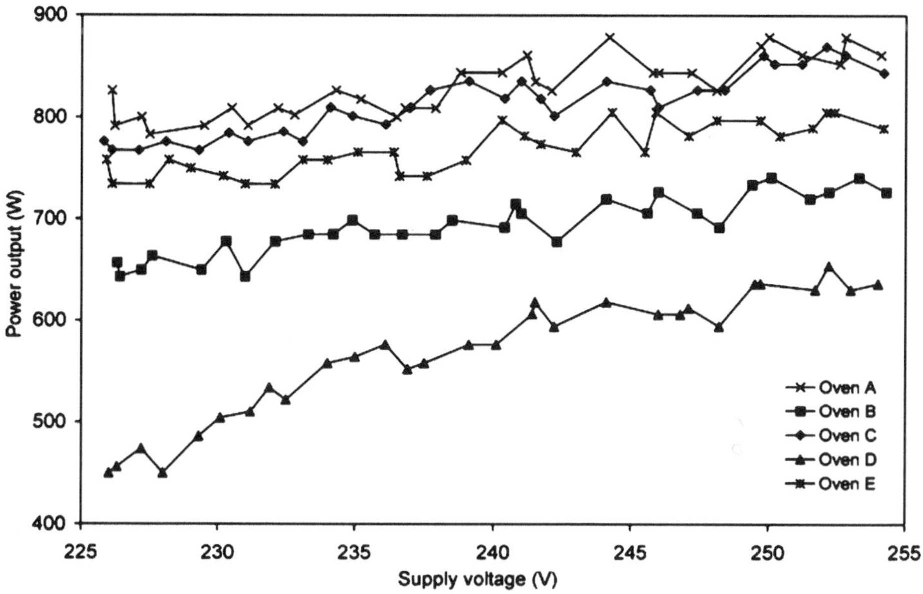

Figure 14.3 Power delivered into a 1 kg water load at different mains supply voltages. *Source*: Adapted from James, S. J., Foster, A. M., Phillips, I. C., Wilkins, T. J., Swain, M. J., & Burfoot, D. (1994). *Effects on microwave power output of size of load, continuous (intermittent) use, position of load, and variations in mains supply voltage*. Microwave science series no. 11. London: MAFF Publications.

not more than ± 6% of this value. James et al. (1994) found that there was a substantial reduction in power delivered to a 1.0 kg water load as the voltage input to the oven was reduced through the permitted range (Fig. 14.3). The magnitude of the change varied considerably between ovens. In four of the five ovens the change was less than ± 8% of the output at 240 V. In the fifth oven the total change over the ± 6% voltage range was almost 36%.

14.3 Factors affecting food heating: reheating performance

Studies were carried out by Burfoot et al. (1990) using 102 individual ovens from 70 different models to look at variability between and within models and the effect of oven features on reheating performance, while studies by Swain, Foster, Phillips, and James (1994, 1995) looked at the effect of initial product temperature, continuous use of the oven, and variability in commercial meals on reheating.

In the Burfoot study, a standard tray of 420 g of mashed potato at 5°C was placed at the geometric center of the turntable (carousel), or floor of the oven where no turntable was present. Each oven was operated at its power setting closest to 650 W and the time of heating was determined such that the energy dissipated into the product at the stated power rating and the heating time equalled 240 kJ. After heating, the tray was placed into a polystyrene mould and a "temperature hedgehog" inserted into the potato. The "hedgehog" consisted of nine probes made from 2 mm diameter wooden rods with four thin wire (0.076 mm diameter) chromel-alumel thermocouples located along each rod. The temperatures in the mashed potato were monitored at 4 s intervals for 2 min after the probes were placed in the potato. Data on the highest temperature achieved at each location were analyzed.

The mean temperature of the mashed potato ranged from 92.8 to 98.9°C. However, 32 of the 36 positions were close to the periphery of the pack and would bias the average temperature of the potato. The maximum temperature of the 1080 measured in the mashed potato, in any of the ovens was always within ± 1.1°C of 100°C, which indicates that localized boiling occurred in all the ovens. The minimum temperature of the 1080 measured in the mashed potato, in any of the ovens ranged from 44.4 to 95.6°C.

14.3.1 Uniformity of reheating

There are considerable differences in the uniformity of reheating in different microwave ovens. Fig. 14.4A and B shows examples of temperatures measured at each position on sensors within mashed potato in ovens that showed the lowest and highest degree of nonuniformity of heating. In most but not all of the ovens, the lowest temperature was located on the sensor that passed through the geometric center of the potato.

The difference between the highest and lowest temperatures measured in each oven, which is the strictest test of nonuniformity, ranged from approximately 5 to 57°C (Fig. 14.5). Temperature differences of <10°C were measured in 39 of the ovens, <20°C in 49, and <30°C in 67 ovens from the 102 individual ovens tested. The respective figures for numbers of models were 22, 30, and 43 of the 70 models of microwave oven tested. However, because the highest temperature was always 100 ± 1.1°C, the minimum temperature achieved is a good measure of nonuniformity. The mean of the lowest temperatures measured in the five trials with each oven is used in the following comparisons.

14.3.2 Repeatability within ovens

In the Burfoot study (Burfoot et al., 1990), tests using mashed potato were repeated five times with each oven. The standard deviation on the mean of the five lowest temperatures achieved with each oven varied from 0.6°C to 8.0°C. Reheating in 63 of the 102 individual ovens (representing 70 different models) was very repeatable with standard deviations of approximately 2°C or less. Nine ovens produced standard deviations were greater than 4°C, and in four of these, standard deviations were greater than 6°C. Respective figures for models were 44 ≤2°C, five >4°C, and one model of the 70 different models had a standard deviation over

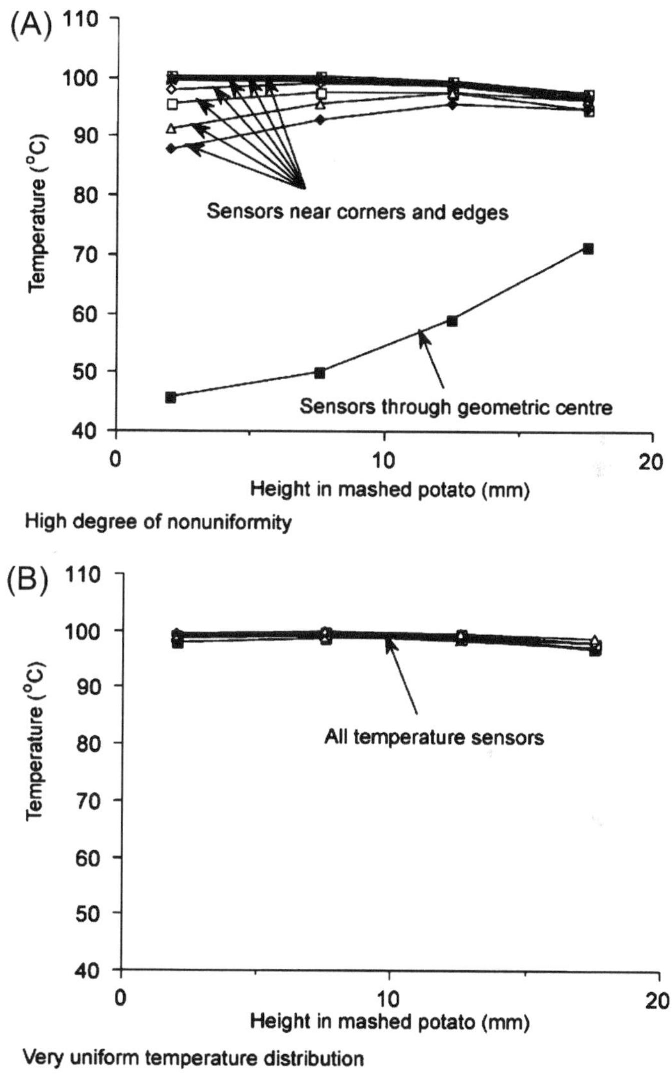

Figure 14.4 Temperature at different heights in mashed potato after heating in ovens that produced (A) a high or (B) a low degree of nonuniformity of temperature distribution. *Source*: Adapted from Burfoot, D., James, S. J., Foster, A. M., Self, K. P., Wilkins, T. J., & Phillips, I. (1990). Uniformity of reheating in domestic microwave ovens. In R. W. Field & J. A. Howell (Eds.), *Process engineering in the food industry: 2 Convenience foods and quality assurance* (pp. 33–44). Elsevier Science.

6°C. Forty of the 41 ovens that achieved a lowest temperature of ≥90°C also produced standard deviations of <2°C. Standard deviations at high temperatures are lower because water in the mashed potato boils at approximately 100°C, producing a natural limit to the temperature that can be achieved during the heating process.

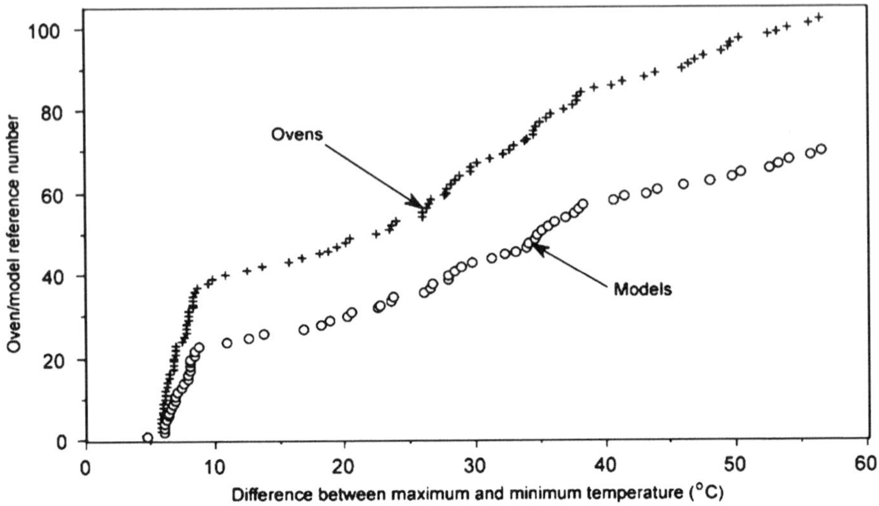

Figure 14.5 Maximum difference between measured temperatures in microwave ovens.
Source: Adapted from Burfoot, D., James, S. J., Foster, A. M., Self, K. P., Wilkins, T. J., & Phillips, I. (1990). Uniformity of reheating in domestic microwave ovens. In R. W. Field & J. A. Howell (Eds.), *Process engineering in the food industry: 2 Convenience foods and quality assurance* (pp. 33–44). Elsevier Science.

There was no other apparent relationship between the variability in lowest temperatures and the magnitude of the temperature.

In an assessment of nineteen microwave ovens, Pitchai, Birla, Jones, and Subbiah (2012) found that the overall heating performance, such as mean heating rate and uniformity of heating, of ovens did not significantly vary between two replications that were performed 4 h apart. However, there were significant differences, between two replications, in the heating rate at various locations in the containers. The authors considered this to be particularly challenging for development and validation of microwave heat transfer models, as it is hard to produce consistent experimental microwave heating results for validation.

14.3.3 Repeatability within replicate models

In the Burfoot study (Burfoot et al., 1990), the differences between the mean of the lowest temperatures of five replicate ovens of each of four models that achieved temperatures >90°C (models A to D in Fig. 14.6) were less than 2°C and there were no significant differences between the replicates within a model type. In the other four models, higher differences (4 to 8°C) were measured between mean lowest temperatures from ovens of the same model. However, because the variations in reheating performance of these ovens were greater, only the highest and lowest means from models E, F, and G were statistically different. The maximum

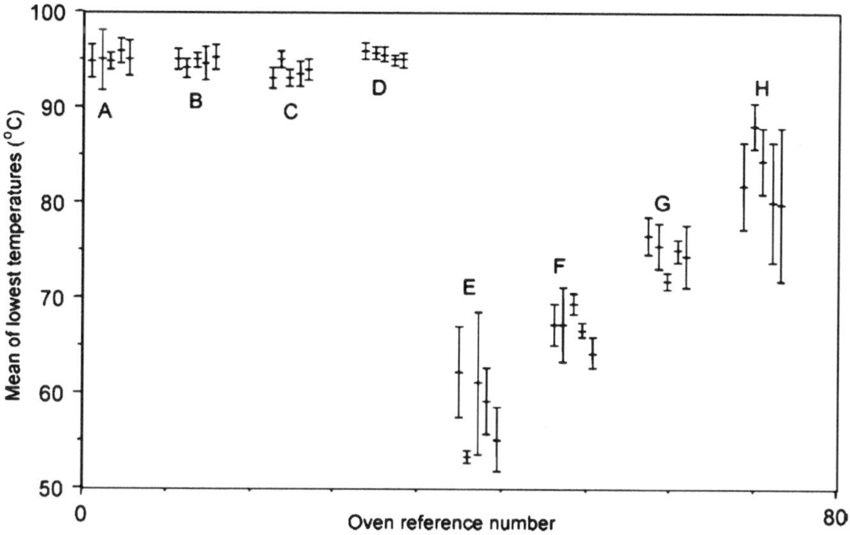

Figure 14.6 Mean of lowest temperatures in replicated models.

temperature difference in the means (8°C) represents a variation of approximately 10% in temperature rise.

14.3.4 Oven features

The Burfoot study (Burfoot et al., 1990) found that there was no statistically significant relationship between the maximum stated power or the power used in the tests of the microwave ovens and the lowest temperature. Cavity size had little effect on the lowest temperature (Fig. 14.7) with both very high (>90°C) and low (<60°C) temperatures being found at all cavity sizes from 0.017 to 0.034 m^3. Too few ovens with sizes outside this range were present to see if there was any trend in large or small cavities. There was no significant linear relationship between the mean of the lowest temperature and the oven power to cavity size ratio (Fig. 14.7) as the degree of fit ($R = 0.15$) was exceptionally poor.

In a study of 19 ovens, Pitchai et al. (2012) found that the mean heating rate (°C/min) increased with cavity size (particularly in ovens with a cavity size over 0.045 m^3). However, this effect was attributable to a positive correlation between the rated power and cavity size. Therefore, they concluded that rated power is a more important direct factor than cavity size for heating rate and nonuniform heating.

In the Burfoot study (Burfoot et al., 1990), a range of mean lowest temperatures from <50 to >95°C was measured in ovens with and without grill elements. The sample size was too small to see if there was any significant effect of the presence or absence of a grill element or of its type of construction. However, these

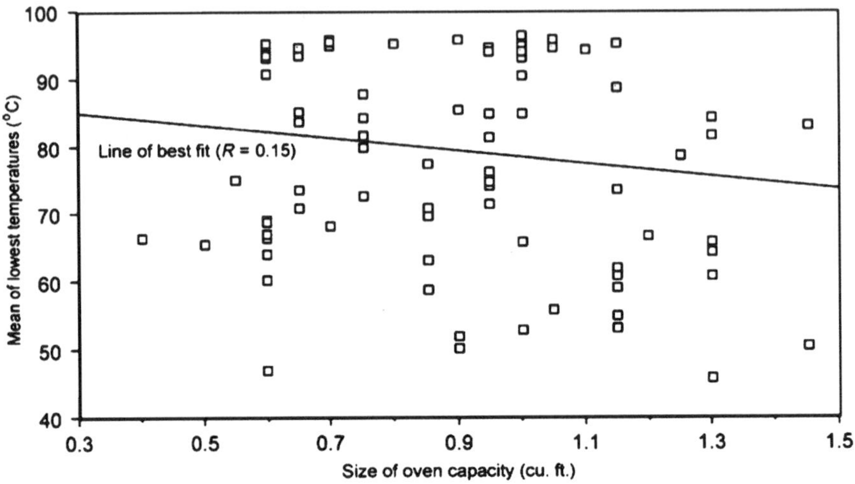

Figure 14.7 Mean of the lowest temperature measured in ovens of different cavity size. *Source*: Adapted from Burfoot, D., James, S. J., Foster, A. M., Self, K. P., Wilkins, T. J., & Phillips, I. (1990). Uniformity of reheating in domestic microwave ovens. In R. W. Field & J. A. Howell (Eds.), *Process engineering in the food industry: 2 Convenience foods and quality assurance* (pp. 33–44). Elsevier Science.

data show that it is possible to produce minimum temperatures above 90°C in a microwave oven with a grill.

It is well documented that rotating turntables (carousels) contribute to greater uniformity of product heating (Manickavasagan, Jayas, & Vadivambal, 2009; Oliveira & Franca, 2002). Geedipalli, Rakesh, and Datta (2007) found a 37 to 43% increase in uniformity of temperatures when employing the use of a turntable. However, they postulated that even greater temperature uniformity could be achieved if a system could be built that moves the food also from top to bottom (and vice versa) within the oven cavity. Since not all microwaves are equipped with turntables, cooking instructions often include manual product rotation.

In microwave ovens with turntables, the turntable may be metallic or nonmetallic. In the Burfoot study (Burfoot et al., 1990), no oven with a metal turntable achieved a mean lowest temperature above 80°C, and two of the ovens without turntables achieved temperatures above 80°C (Fig. 14.8). Fifty-four of the 77 ovens with nonmetallic turntables achieved mean lowest temperatures above 80°C, and 42 of these achieved temperatures above 90°C. Temperatures below 50°C were measured in two ovens, one with a nonmetallic and the other with a metallic turntable.

The waveguide from the magnetron of a domestic microwave oven may feed the cavity via the side or top. Studies indicate that the heating of small loads in an oven may be affected by the location of the input port of the magnetron (Pitchai, Brahma, Birla, Jones, & Subbiah, 2011). The heating rate of small loads (250 g water) increases as the distance of load from the turntable center increases for side port microwave ovens but not in top feed microwave ovens.

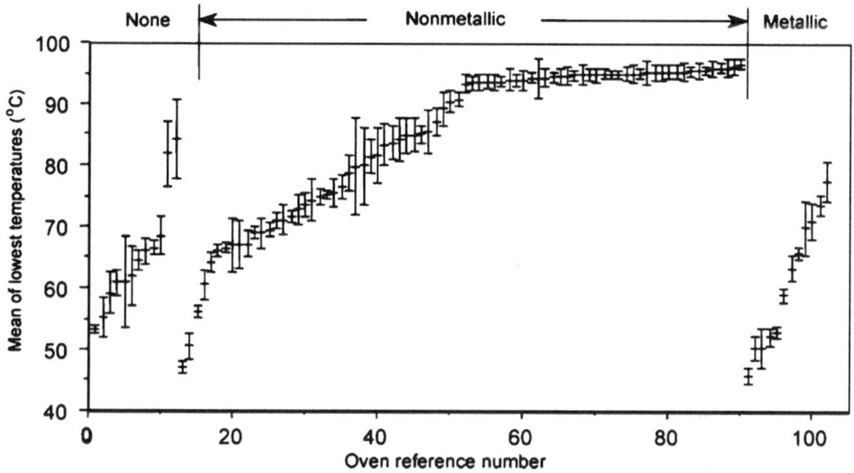

Figure 14.8 Mean of lowest temperature measured in ovens with different turntables. Error bars ± 1 standard deviation.
Source: Adapted from Burfoot, D., James, S. J., Foster, A. M., Self, K. P., Wilkins, T. J., & Phillips, I. (1990). Uniformity of reheating in domestic microwave ovens. In R. W. Field & J. A. Howell (Eds.), *Process engineering in the food industry: 2 Convenience foods and quality assurance* (pp. 33–44). Elsevier Science.

Most researchers have found an increase in heating uniformity when using an on/off cycle during heating (Fakhouri & Ramaswamy, 1993; Gunasekaran & Yang, 2007; Oliveira & Franca, 2002). The thawing setting on a microwave oven uses cycled power to reduce "runaway heating" problems, however Chamchong and Datta (1999) found no advantage to using cycled power levels for thawing.

14.3.5 Initial product temperature

To determine the effect of initial temperature, Swain et al. (1995) measured temperatures in trays of mashed potato after heating from −20, −5, 5, and 20°C. Reheating times used, with each oven/food temperature combination, were calculated from energy/mass data obtained from a summary of information printed on commercially available ready prepared meals. Energy/mass (heating time × power/mass) values of 958, 918, 571, and 515 kJ/kg were used when heating from − 20, − 5, 5, and 20°C, respectively.

The lowest temperatures measured in the potato after heating increased with the initial temperature of the potato (Table 14.2). As the lowest temperature in the potato increased, the temperature uniformity tended to improve. Only one oven produced a mean lowest temperature below 70°C and that only occurred after heating from −20°C.

Ovens that heated the frozen potato to above 90°C also performed well when heating the chilled potato. The smaller lower-powered ovens performed better than

Table 14.2 Mean and standard deviation (values in brackets) of lowest temperature measured after heating from different initial temperatures

Oven code	Lowest temperatures achieved after heating from initial temperatures of			
	−20°C	−5°C	5°C	20°C
1	61.0 (6.6)	73.2 (11.3)	84.8 (2.6)	90.9 (2.6)
2	93.1 (2.3)	90.8 (3.4)	93.6 (1.0)	94.2 (2.1)
3	93.0 (1.3)	91.4 (1.6)	93.3 (1.0)	94.8 (1.3)
4	75.3 (1.7)	81.5 (12.9)	84.4 (4.4)	89.3 (5.2)
5	72.9 (8.4)	80.8 (4.1)	87.2 (3.0)	89.9 (4.2)
Overall mean	79.1	83.5	88.7	91.8

Source: Adapted from Swain, M. J., Foster, A. M., Phillips, I. C., Wilkins, T. J., James, S. J., Radhakrishnan, C. D., et al. (1995). *Effect of initial temperature of food and continuous use of a microwave oven on the temperature of food after reheating*. Microwave science series no. 13. London: MAFF Publications.

other models when heating from frozen temperatures, which suggest that the use of low operating powers may be beneficial when reheating foods from frozen.

14.3.6 Repeated use of oven

Since the output power of domestic microwave ovens was known to fall with continuous use, Swain et al. (1995) investigated whether such a fall in power would affect the temperatures achieved in foods during reheating.

A polyethylene terephthalate (PET) tray containing mashed potato (0.42 kg) was heated in an oven operating at the setting closest to 650 W for a time sufficient to provide an apparent energy input of 240 kJ. Immediately after heating, a further tray of potato was heated and the process continued for 1.5 h. The lowest temperature achieved in each tray of potato was measured after the tray was removed from an oven. Two ovens, consistently over the 1.5 h period, produced mean lowest temperatures in the potato that exceeded 90°C and no significant fall in temperature with time. The mean lowest temperatures achieved in the potato in the other ovens fell from approximately 81 to between 71 and 76°C during prolonged use.

A further study (Swain, James, & Swain, 2008) was carried out in 16 different models of domestic microwave ovens using tests based on those of Swain, Spinassou, and Swain (2008). Tests were carried out in "hot" ovens that had been in use for 15 min prior to the heating test. All of the test loads had lower minimum measured temperatures (mean of three replicates) after heating in the hot ovens compared to when they were heated for the same time in the cold ovens (ovens that had not been used for 6 h). Minimum measured temperatures fell from a mean of 70.3°C in the cold ovens to mean of 61.3°C (range 50.3–67.6°C) in the hot ovens. In 12 out of the 16 ovens (75%) these reductions in minimum temperature were statistically significant.

14.3.7 Variability in foods

Swain et al. (1994) investigated the variability of reheating chilled ready meals using a regime that minimized the other factors known to influence reproducibility of power output or temperature distribution in domestic microwave ovens. Forty-four packs each of lasagne, chicken with pineapple and cashews (chicken meal), and vegetable rogan josh (vegetable meal), were bought from a local supermarket, and 44 packs of prepared mashed potato (mash) were stored for 22 to 30 h in a chillroom at 5 ± 1°C. A microwave oven was then preheated by using it to heat 1.5 kg of water for 30 min and each of the meals heated for a set time. When the meal was removed from the oven a prewarmed temperature "hedgehog" with 20 temperature sensors was inserted into the meal and the temperatures logged for 20 s. The whole procedure was repeated in three different microwave ovens.

The lowest mean standard deviations (20 temperature positions) were found in the mashed potato. They ranged from 2.9 to 3.9°C (Table 14.3) and were a factor of between 1.3 and 2.1 lower than those found in the meals. Based on the average standard deviation, no particular ready meal stood out as being any less or more reproducible than any other. Similar factors, 1.3- to 2.0-fold differences, were found in the mean range data (Table 14.3). There was no clear indication in these data of any consistent difference between the ovens.

In two of the ovens (Y and Z) the minimum temperature in the mashed potato packs after heating was always found in the same position. However, in the third oven (oven X) the minimum temperature was found in eight different thermocouple positions measured by the "hedgehog." The minimum temperature position was most variable in the chicken ready meal, ranging from eight different positions in oven Z to 14 in oven X. In the vegetable meal the minimum temperature position was equally variable (seven positions) in all three ovens. The lasagne minimum temperature positions ranged from four (oven Y), which was the least variable of all the meals (excluding mashed potato), to eight different positions (oven Z), which was as variable as the chicken meal in the same oven.

Table 14.3 Mean standard deviation (mean SD) and mean range (°C) at the 20 thermocouple positions

Oven code		Chicken	Lasagne	Vegetable	Mashed potato
X	Mean SD	5.3	5.2	6.1	3.9
	Mean range	22.7	22.7	26.6	17.8
Y	Mean SD	6.1	5.0	6.2	2.9
	Mean range	24.4	22.1	26.5	13.2
Z	Mean SD	6.6	6.0	6.2	3.6
	Mean range	26.2	26.0	25.9	16.0

Source: adapted from Swain, M. J., Foster, A. M., Phillips, I. C., & James, S. J. (1994). *Variation of uniformity of microwave reheating within commercial chilled convenience meals*. Microwave science series no. 10. London: MAFF Publications.

Table 14.4 Mean, standard deviation (SD), and range (°C) at the thermocouple position which has the lowest mean temperature (44 replicates)

Oven code		Chicken	Lasagne	Vegetable	Mashed potato
X	Mean	78.7	71.9	68.4	69.5
	Position	2	2	2	10
	SD	8.4	5.7	7.2	5.0
	Range	31.0	23.3	31.0	25.3
Y	Mean	69.8	72.5	60.9	63.1
	Position	9	10	10	9
	SD	5.3	6.8	5.9	2.7
	Range	23.7	36.3	25.7	12.9
Z	Mean	70.4	74.6	70.0	64.1
	Position	1	2	2	1
	SD	7.1	4.9	6.4	3.1
	Range	28.1	26.6	26.0	14.0

Source: Adapted from Swain, M. J., Foster, A. M., Phillips, I. C., & James, S. J. (1994). *Variation of uniformity of microwave reheating within commercial chilled convenience meals*. Microwave science series no. 10. London: MAFF Publications.

Data from the position that had the lowest temperature on average over the 44 replicates are shown in Table 14.4. At these positions, the standard deviations in the mashed potato ranged from 2.7 to 5.0 and were a factor of 1.1 to 2.5 lower than those found in the meals.

The 44 replicated temperatures at each lowest average temperature position for all oven/food combinations were allocated to 5°C temperature bands to produce the percentage temperature distributions shown in Figs 14.9, 14.10, and 14.11. In all the mashed potato cases, the temperature distributions tended to follow a bell-shaped normal distribution and were more compact than those for the ready meals. The chicken and vegetable based meals tended to have an uneven distribution, indicating less repeatability than lasagne at the lowest average temperature position.

14.3.8 Experimental validation of microwave heating

Researchers have explored various methods to measure temperatures during microwave-only and microwave combination heating (see also Chapter 15: Measuring temperature distributions during microwave processing). However, there are limitations associated with all of these techniques. Most methods can not measure temperature "on-line" during microwave heating and thus can only map the final temperature distribution.

Fiber-optic thermometry can measure temperatures during microwave heating, however there is a limit to the number of probes that can be used for measuring the temperature inside the food (Nott & Hall, 2005; Rakesh, Datta, Amin, & Hall, 2009).

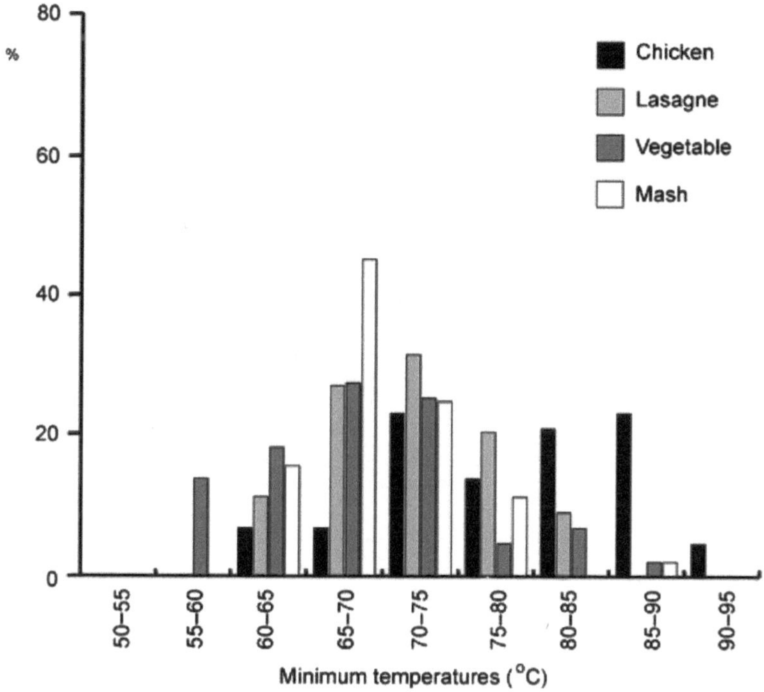

Figure 14.9 Distribution of replicated temperatures at mean lowest temperature position (44 replicates) after reheating of three meals and mashed potato in microwave oven X. *Source*: Adapted from Swain, M. J., Foster, A. M., Phillips, I. C., & James, S. J. (1994). *Variation of uniformity of microwave reheating within commercial chilled convenience meals*. Microwave science series no. 10. London: MAFF Publications.

Measuring in a microwave oven with a turntable during heating also presents some problems, but can be solved using a counterrotating turntable system such as the one developed by Phippen and Burfoot (1990).

The most common method of mapping temperature distributions is to measure after microwave heating using a "hedgehog" array of thermocouples (James, Swain, James, & Swain, 2002; Pitchai et al., 2012; Schiffmann, 2014; Swain et al., 1994). However this method is destructive; requires modification for the wide array of microwave containers that are now available with different shapes and sizes; and is difficult to implement for containers that have restricted access, e.g., a jar (Nott & Hall, 2005).

Infrared thermal imaging is a noninvasive and nondestructive technique that can be used both off- and on-line. However, this method is limited by the fact that it can measure temperature profiles only for the exposed surface. This is a problem since microwaves heat the food load volumetrically, and this leads to the formation of hot and cold spots inside the body of the food (Nott & Hall, 2005; Rakesh et al. 2009).

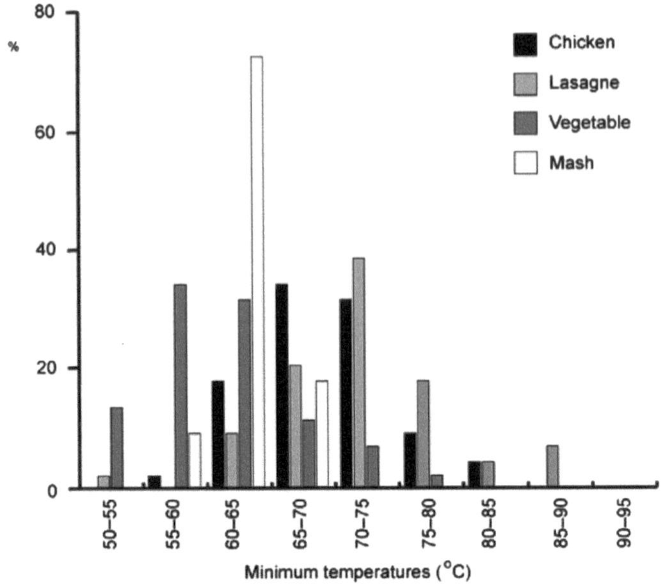

Figure 14.10 Distribution of replicated temperatures at mean lowest temperature position (44 replicates) after reheating of three meals and mashed potato in microwave oven Y. *Source*: Adapted from Swain, M. J., Foster, A. M., Phillips, I. C., & James, S. J. (1994). *Variation of uniformity of microwave reheating within commercial chilled convenience meals*. Microwave science series no. 10. London: MAFF Publications.

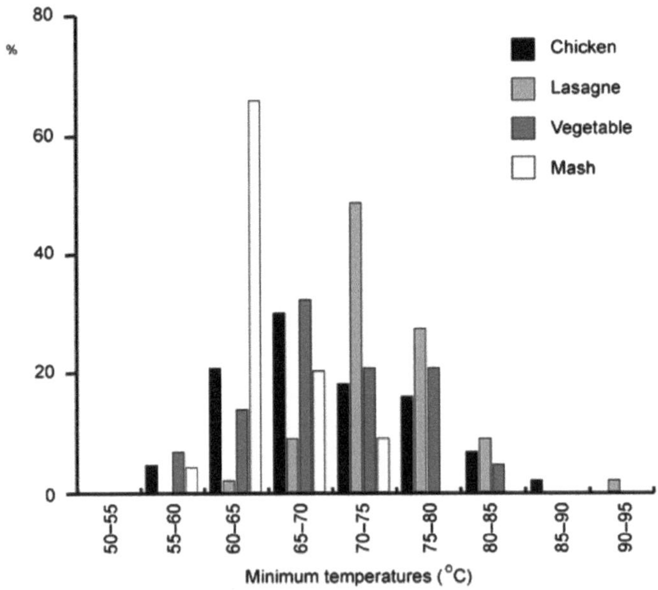

Figure 14.11 Distribution of replicated temperatures at mean lowest temperature position (44 replicates) after reheating of three meals and mashed potato in microwave oven Z. *Source*: adapted from Swain, M. J., Foster, A. M., Phillips, I. C., & James, S. J. (1994). *Variation of uniformity of microwave reheating within commercial chilled convenience meals*. Microwave science series no. 10. London: MAFF Publications.

Magnetic resonance imaging (MRI) has been used for temperature mapping in microwave heating studies (Bows, Patrick, Nott, & Hall, 2001; Knoerzer, Regier, & Schubert, 2006; Knoerzer, Regier, Hardy, Schubert, & Schuchmann, 2009; Nott & Hall, 2005; Rakesh et al. 2009). These studies have been purely experimental and/or for microwave-only heating.

14.4 Food simulant materials for testing oven heating performance

A wide range of potential food simulant materials have been suggested and used to mimic heating responses of foods heated in microwave fields. Historically, wet thermal fax paper has been used in demonstrating microwave heating nonuniformity in domestic ovens (Bradshaw, Delport, & Van Wyk, 1997). The problem with this approach was that it had to be used in an empty cavity. Presence of a food product will alter the electromagnetic field and therefore the severity of heating nonuniformity. Moreover, it does not provide quantitative assessment of heating uniformity.

Swain, Russell, Clarke, and Swain (2004) carried out an extensive literature search to identify potential food simulant materials that could be used to mimic heating responses of chilled foods heated in microwave fields (Table 14.5). Each prospective food simulant material was initially assessed against a set of criteria that included ease of manufacture, stability, and formability.

Knoerzer, Regier, Erle, Pardey, and Schubert (2004) developed a model food suitable for microwave heating and microwave vacuum drying experiments, which was composed of water, fructose, NaCl, $CuSO_4$, cellulose, agar-agar, and starch, which was primarily used to evaluate induced heating patterns in various microwave systems and validating numerical models predicting these patterns (Knoerzer, Regier, & Schubert, 2008). The $CuSO_4$ was added to the model food to reduce relaxation times in magnetic resonance imaging trials, yielding 3D temperature maps of model food cylinders with a diameter of ~30 mm and a height of 20–50 mm in less than 5 s. While other studies at the University of Nebraska, Lincoln have used gellan gum as a model food (Birla, Pitchai, Subbiah, & Jones, 2009, 2010) and whey protein gel (Chen et al., 2013), before falling back on mashed potato in subsequent studies (Chen et al., 2014). Whey protein gel has been used for validating microwave heat transfer models, as its properties match those of meat products (Tang et al., 2008; Wang et al., 2009).

In order to identify hot and cold spots during microwave heating a number of studies have used containers with compartments (James et al., 2002; IEC, 2010; Pitchai et al., 2012).

The IEC 60705 standard suggests the use of a square container (228 mm × 228 mm × 30 mm) made of a material transparent to microwaves for measuring the performance of microwave ovens (IEC, 2010). The square container consists of 25 compartments in which 1000 g of water is poured to measure the uniformity of heating. The compartment separators have a small hole positioned in the approximate center of each partition allowing water to flow from one compartment to

Table 14.5 Some of the materials proposed for simulants

Material name	Use
Bentonite (bentonite + glycerin)	Thermal sterilization in cans
EWSG/Gellan gum	Muscle phantom
EPG (ethylene glycol + polyethylene powder)	Fat phantom
Gelatine	Gelling agent
Sodium alginate	Gelling agent
Hydrophilic organic polymer (TX151)	Muscle phantom
Laponite	Synthetic clay
Carbopol	Ohmic heating test liquid
SIK gel (carrageenan, sugar, glycerol)	Food simulant
SIK liquid (sugar, glycerol)	Liquid food simulant
IEC batter (wheat flour, eggs, sugar)	Test batter
Bacto agar	Gelling agent
Polyvinyl alcohol + sodium borate solution	Crosslinked gel
Gellan gum	Gelling agent
Whey protein	Gelling agent
Starch solutions (soup analogues)	Soup simulant
Biogel (methylmethacrylate n-vinyl-2-pyrolidone)	Hydrophilic food simulant
Plasticine	Modeling material
PVA + Dimethylsulfoxide	Gel
Tylose (hydroxyethylmethylcellulose)	Meat thermal simulant
Carboxymethylcellulose	Cellulose gum
Glycol + cellulose	Cellulose based material
Glucomannan	Gelling agent (foam)
Mashed potato	Fast food test simulant
Polyacrylamide + polycrylic acid	Hydrogel

Source: Adapted from Swain, M. V. L., Russell, S. L., Clarke, R. & Swain, M. J. (2004). Development of food simulants for microwave oven testing. *International Journal of Food Science and Technology*, *39*(6), 623–630.

another. In this method, a microwave oven's uniformity of heating can be measured by noting the difference between the highest and lowest temperature increase of each of the 25 compartments over the mean temperature rise of all 25 compartments.

Pitchai et al. (2012) designed a test rig consisting of a round polypropylene container (300 mm diameter × 100 mm height) with 36 equal volume compartments. The design of the container ensured that the surface area (44 cm^2) of all compartments were the same.

14.5 Methodology for identifying cooking/reheating procedure

Developing microwave reheating instructions for a food product is not a trivial task. The food manufacturer or testing laboratory will only have a limited time, access to only a small range of ovens and a limited space on the packaging for the instructions. Until the mid-1990s the only information on an oven would be the power delivered into a 1 kg water load. Investigations carried out by George, Richardson, Hooper, and Burnett (1992) led to an alternative categorization in the United Kingdom that labeled ovens from A to E in five steps based on the power delivered to a 0.35 kg water load (O'Leary, 1993).

When George, Evans, Hooper, Campbell, and Dobie (1995) investigated the results of following manufacturers' heating instructions on chilled and frozen packs of lasagne, they reported that "the reheating instructions given on the food package failed to deliver a thermal process equivalent to 70°C for a period of two minutes." They developed six-stage microwave reheating instructions based on the 0.35 kg power test. These used longer reheating times, rest periods, a 90° turn, and retention of the film lid (pierced). Reheating times ranged from 6.25 to 12.75 min depending on the product and the oven used, resulting in total heating times from 10.25 to 16.75 min for an individual meal.

James (1993) suggested a 14-point plan that, if used to develop reheating guidelines for a convenience meal, would take into account most of the available technology and knowledge. He suggested that many of the points identified in the protocol would have already been covered during the development of the product being tested. In these cases, the whole process could be condensed and carried out in a few days. In an attempt to help ensure the highest standards of safety and quality for microwave reheated products, the UK Microwave Working Group (comprising representatives from oven and food manufacturers, the UK Ministry of Agriculture, Fisheries and Food, retailers, consultants, academics, trade, research, consumer, and standards organizations) developed guidelines (Richardson & Gordon, 1997) for food manufacturers and retailers that set out procedures necessary for the development and verification of reheating instructions for foods intended for use in microwave ovens. The guide also provides advice on the equipment and environment for controlled testing, power output measurement, the UK labeling scheme for ovens and foods, the symbols for use on microwave ovens and food packs, and references to the relevant European standards. It also includes guidance on the assessment of the impact of consumer abuse of the developed instructions in terms of product safety and quality.

As highlighted in the introduction, concern for the safe cooking of NRTE or RTC foods in microwaves in the United States has led to both the Grocery Manufacturers Association (www.gmaonline.org) and the American Frozen Food Institute (www.affi.org), e.g., drafting guidelines for their members producing NRTE foods, outlining standardized protocols for the validation of consumer cooking instructions. These guidelines also encourage food manufacturers to ensure that label text is clear, concise, and helpful to the consumer in achieving a fully cooked product, whether that is conducted in a conventional appliance, such as an oven or toaster oven, or in a microwave oven.

14.6 Determining the heating performance characteristics of microwave ovens

The only performance-related information provided to the consumer purchasing a microwave oven is the power output measured using a standard 1 kg water load, which provides little information about how well the oven actually heats such a food product as a chilled convenience meal. In the United Kingdom, a categorization letter (A to E) corresponding to the power output into a 0.35 kg water load is also stated. This goes some way toward informing the consumer how much power is available to heat a smaller food load, but does not give any indication about how well that power is distributed within the food. For example, two ovens having an identical power output may heat the same type of food load very differently; one oven may be prone to produce unacceptable "cold spots" whereas the other may achieve a more satisfactory uniform temperature distribution.

Swain, Swain, and Russell (1998) proposed a test procedure to determine specifically the heating performance of domestic microwave ovens when heating chilled convenience meals. The procedure uses standard test loads that simulate the heating and weight loss characteristics of chilled convenience food. It aimed to be reproducible, relate to consumer use, ensure that the test load attained specified minimum temperatures for food safety, provide a measure of consumer acceptability, and be equally applicable to all types of domestic microwave oven. In order to carry out tests that met all the stated requirements, it was necessary to replace foods having inherent biological variability with a heating test load that had reproducible behavior.

Tests on a range of chilled convenience meals available in the United Kingdom, using continuous fiber-optic temperature measurement during heating, indicated that the meals could be separated into two broad categories according to heating times, i.e., "slow" and "fast" response. Food simulants were developed to represent the two food categories. A good match between food simulant and meal was achieved for the main indicators, i.e., temperatures at central and intermediate locations and weight loss.

A food simulant (developed and tested by Swain et al., 2004) made from TX151 (Oil Center Research International, Lafayette, LA, USA), a hydrophilic polymer, in a TX151:NaCl:H_2O ratio of 22.2:0.7:77.1 by weight was the best match to a "slow" heating chilled convenience meal like lasagne. A simulant made up using Carbopol (BF Goodrich Performance Materials, Cleveland, OH, USA), a cross-linked polyacrylic acid copolymer, in a Carbopol:polyethylene:H_2O ratio of 10:10:80 by weight was the best match to a "fast" heating chilled food like rice. The manufacture of the simulants only required the use of standard test kitchen equipment (e.g., standard liquidizer, bowl mixer, refrigerator). Both food simulants have the advantage that their heating characteristics can be modified by changing the composition ratios to match other food products.

The simulants were placed into three different types of microwaveable food trays (rectangular, two compartment, and four compartment) to provide three alternative simulant/packaging combinations to represent the range available. The most commonly available meal was represented by the rectangular crystalline polyethylene terephthalate (CPET) tray (external dimensions $156 \times 111 \times 35$ mm) containing 0.35 kg of the TX151 based "slow" heating food simulant.

Fig. 14.12 shows a flowchart that outlines the sequence of events required to carry out the test procedure. The temperature distribution after heating from 5°C to a defined minimum temperature (72°C) was measured in the food simulant using a specially designed multipoint thermocouple "hedgehog" probe consisting of 39 fine hypodermic stainless steel sheathed type "T" thermocouples.

The weight of the food simulant was recorded before and after heating and used as an indicator of product deterioration. The greater the weight lost after heating to a minimum temperature of 72 °C, the greater the likelihood of irreversible degradation. A relatively small weight loss ($<5\%$) would indicate that the consumer would be unlikely to notice any quality degradation. With product weight loss above approximately 15%, the consumer would notice excessive shrinkage, with edges of the product dried out and in extreme cases evidence of burning or charring. The TX151 based simulant mimicked these quality deterioration effects.

The simulant load temperatures at 60 and 120 s post heating, weight loss, heating time, and power output data were entered onto a spreadsheet template that automatically analyzed the data. The tester could choose to view the detailed analysis, which evaluated such characteristics as evenness of heating; top to bottom, side to side, and front to back heating balance; edge overheating; and average rate of heating; or a simplified analysis based on a scoring system of oven performance. A "3D" plot representing the temperature distribution within the food simulant load after heating was also automatically produced. This test has subsequently been adopted for comparative consumer testing of domestic microwave ovens. The application of this test procedure applied to seven microwave ovens was described by Swain, Spinassou, et al. (2008).

James et al. (2002) stated that "many of the tests currently used to assess the reheating of a foodstuff in microwave ovens are either subjective, or use commercially produced chilled meals." Previous studies have clearly shown that there is considerable variability between individual packs of the same commercial product. Consequently, when using commercial packs, there is a considerable degree of uncertainty whether differences in performance between ovens are true differences, not artefacts caused by pack to pack variability. Additionally, it is difficult to compare current tests with those carried out in previous years due to changes in commercial production and in some cases the unavailability of some of the meals previously used.

James et al. (2002) developed a simple procedure for comparing the three most important reheating characteristics of a domestic microwave oven—its "true" power, temperature variability, and repeatability. The "true" power being absorbed by the food governs the reheating time. The temperature variations within the food after reheating control the safety and quality of the food. An oven's repeatability

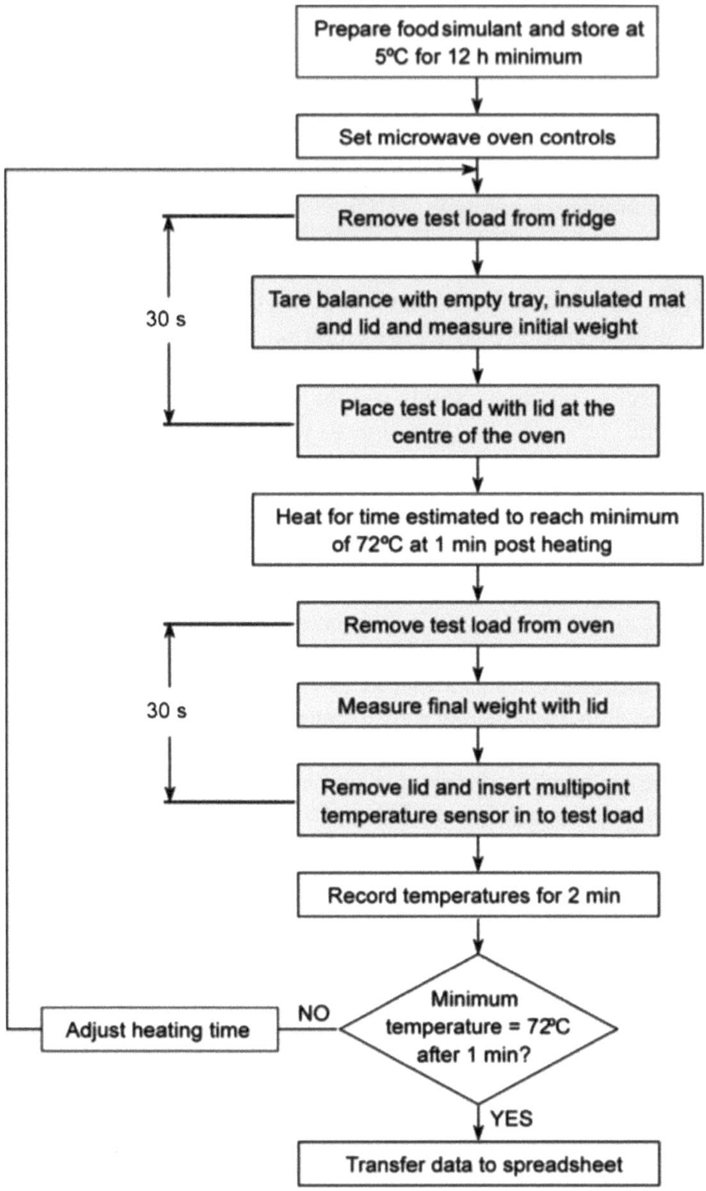

Figure 14.12 Flowchart of the sequence of events required to carry out the test procedure to determine the heating performance of microwave ovens.

shows its ability to produce the same performance on subsequent occasions and hence reliably produce an identical product.

The procedure required two simply prepared test materials made up from readily available chemically consistent raw materials. Samples were prepared using a simple protocol in the type of packs commonly used for commercial ready-meals. The only addition was an easily manufactured PTFE frame, which divided the pack into 12 compartments. Three identical tests, one with the liquid test material, the second with the solid, and the third with a combination, were required to perform the whole procedure. It was recommended that each be replicated a minimum of five times. One additional stage was carried out with the liquid test material.

The data produced in each replicated test was reduced to three numbers that were measures of the true power, variability, and repeatability of the oven. Further simple analysis, which weighted the relative importance of each factor to the consumer, produced a single value for the oven's relative reheating performance. Any simple spreadsheet program could reduce the 195 individual pieces of temperature data gathered in the total procedure to a measure of oven performance.

14.7 Conclusions and future trends

It is clear that many factors relating to the design and operation of a microwave oven and the product being heated affect the temperatures reached in the product, their uniformity, and their repeatability. Current standard methods of assessing microwave power provide limited information for the food manufacturers developing heating instructions for chilled and frozen foods.

Although reproducible methods, in addition to the standard methods of testing the performance of domestic microwave ovens, have been developed for specific products (e.g., chilled convenience meals), their uptake has been mostly limited to consumer test and academic research laboratories. Further methods of testing need to be developed (and greater use made of them) to provide reliable data to assess both the oven itself and the suitability of foods to ensure safe heating.

While modeling may have a role to play in the development of optimal microwavable food products, it will not replace the need for manufacturers to experimentally validate the safety of their cooking instructions, particularly manufacturers of NRTE foods.

References

Anon. (2008). Multistate outbreak of Salmonella infections associated with frozen pot pies – United States. *Morbidity and Mortality Weekly Report, 57*(47), 1277–1280.

Bates, C. J., & Spencer, R. C. (1995). Survival of Salmonella species in eggs poached using a microwave oven. *Journal of Hospital Infection, 29*(7), 121–127.

Birla, S. L., Pitchai, K., Subbiah, J., & Jones, D. (2009). Development of novel microwave cooking model for not-ready-to-eat foods. In *International microwave power institute's 43rd annual symposium*, Washington, DC, July 8–10.

Birla, S. L., Pitchai, K., Subbiah, J., & Jones, D. (2010). Effect of magnetron frequency on heating pattern in domestic oven. In *International microwave power institute's 44th annual symposium*, Denver, July 14–16.

Bows, J. R., Patrick, M. L., Nott, K. P., & Hall, L. D. (2001). Three-dimensional MRI mapping of minimum temperatures achieved in microwave and conventional food processing. *International Journal of Food Science & Technology*, *36*(3), 243–252.

Bradshaw, S., Delport, S., & Van Wyk, E. (1997). Qualitative measurement of heating uniformity in a multimode microwave cavity. *Journal of Microwave Power and Electromagnetic Energy*, *32*(2), 88–95.

Burfoot, D., James, S. J., Foster, A. M., Self, K. P., Wilkins, T. J., & Phillips, I. (1990). Uniformity of reheating in domestic microwave ovens. In R. W. Field & J. A. Howell (Eds.), *Process engineering in the food industry: 2 Convenience foods and quality assurance* (pp. 33–44). Elsevier Science.

Chamchong, M., & Datta, A. K. (1999). Thawing of foods in a microwave oven: I. Effect of power levels and power cycling. *Journal of Microwave Power and Electromagnetic Energy*, *34*, 8–21.

Chen, J., Pitchai, K., Birla, S., Gonzalez, R., Jones, D., & Subbiah, J. (2013). Temperature-dependent dielectric and thermal properties of whey protein gel and mashed potato. *Transactions of the ASABE*, *56*(6), 1457.

Chen, J., Pitchai, K., Birla, S., Negahban, M., Jones, D., & Subbiah, J. (2014). Heat and mass transport during microwave heating of mashed potato in domestic oven—model development, validation, and sensitivity analysis. *Journal of Food Science*, *79*(10), E1991–E2004.

Cullen, L. D. (2010). *Factors affecting heating of calzones in microwaves. MSc thesis.* Kansas State University, Manhattan, Kansas, USA.

European Chilled Food Federation (1996). *Guidelines for good hygienic practice in the manufacture of chilled foods* (p. 59). London: European Chilled Food Federation.

Evans, M. R., Parry, S. M., & Ribiero, C. D. (1995). Salmonella outbreak from microwave cooked food. *Epidemiology and Infection*, *115*(4), 227–238.

Fakhouri, M. O., & Ramaswamy, H. S. (1993). Temperature uniformity of microwave heated foods as influenced by product type and composition. *Food Research International*, *26*, 89–95.

Farber, J. M., D'aoust, J. Y., Diotte, M., Sewell, A., & Daley, E. (1998). Survival of *Listeria* spp. on raw whole chickens cooked in microwave ovens. *Journal of Food Protection*, *5*, 1465–1469.

Geedipalli, S. R. R., Rakesh, V., & Datta, A. K. (2007). Modeling the heating uniformity contributed by a rotating turntable in microwave ovens. *Journal of Food Engineering*, *82*, 359–368.

George, R. M., Evans, D. G., Hooper, G. I., Campbell, G. M., & Dobie, P. A. P. (1995). *Assessment and improvement of the manufacturers' reheating instructions for microwavable lasagne based upon the voluntary UK new microwave labelling scheme.* Microwave science series no. 14. London: MAFF Publications.

George, R. M., Richardson, P. S., Hooper, G. I., & Burnett, S. (1992). *Microwave reheating of ready meals: Characterisation and categorisation of domestic microwave ovens based upon a 350 g water load.* Microwave science series no. 6. London: MAFF Publications.

Gessner, B. D., & Beller, M. (1994). Protective effect of conventional cooking versus use of microwave ovens in an outbreak of salmonellosis. *American Journal of Epidemiology*, *139*(7), 903–909.

Gunasekaran, S., & Yang, H. W. (2007). Effect of experimental parameters on temperature distribution during continuous and pulsed microwave heating. *Journal of Food Engineering*, *78*, 1452–1456.

IEC (2010). *Household microwave ovens – Methods for measuring performance* (4th ed., International Electrotechnical Commission). Geneva: IEC Publications 60705:2010.

James, C., Swain, M. V. L., James, S. J., & Swain, M. J. (2002). Development of a methodology for assessing the heating performance of a domestic microwave oven. *International Journal of Food Science & Technology*, *37*(8), 879–892.

James, S. J. (1993). Factors affecting the microwave heating of chilled foods. *Food Science and Technology Today*, *7*(1), 28–36.

James, S. J., & Evans, J. A. (1989). Temperatures in the retail and domestic chill chain. Cost 91 bis. In: *The effects of processing and distribution on the quality and nutritive value of foods*. Gothenburg, Sweden.

James, S. J., Foster, A. M., Phillips, I. C., Wilkins, T. J., Swain, M. J., & Burfoot, D. (1994). *Effects on microwave power output of size of load, continuous (intermittent) use, position of load, and variations in mains supply voltage.* Microwave science series no. 11. London: MAFF Publications.

Knoerzer, K., Regier, M., & Schubert, H. (2006). Microwave heating: A new approach of simulation and validation. *Chemical Engineering & Technology*, *29*(7), 796–801.

Knoerzer, K., Regier, M., & Schubert, H. (2008). A computational model for calculating temperature distributions in microwave food applications. *Innovative Food Science & Emerging Technologies*, *9*(3), 374–384.

Knoerzer, K., Regier, M., Erle, R., Pardey, K. K., & Schubert, H. (2004). Development of a model food for microwave processing and the prediction of its physical properties. *Journal of Microwave Power and Electromagnetic Energy*, *39*(3 + 4), 67–78.

Knoerzer, K., Regier, M., Hardy, E. H., Schubert, H., & Schuchmann, H. P. (2009). Simultaneous microwave heating and three-dimensional MRI temperature mapping. *Innovative Food Science & Emerging Technologies*, *10*(4), 537–544.

Levre, E., & Valentini, P. (1998). Inactivation of Salmonella during microwave cooking. *Zentralblatt fu r Hygiene und Umweltmedizin*, *6*, 431–436.

Manickavasagan, A., Jayas, D. S., & Vadivambal, R. (2009). Non-uniform microwave heating of ready-to-eat chicken pies. *Canadian Biosystems Engineering*, *51*, 3–39.

Nott, K. P., & Hall, L. D. (2005). Validation and cross-comparison of MRI temperature mapping against fibre optic thermometry for microwave heating of foods. *International Journal of Food Science & Technology*, *40*(7), 723–730.

O'leary, M. (1993). UK voluntary labelling scheme for domestic microwave ovens and microwaveable foods. *Trends in Food Science and Technology*, *4*(3), 355–357.

Oliveira, M. E. C., & Franca, A. S. (2002). Microwave heating of foodstuffs. *Journal of Food Engineering*, *53*, 347–359.

Persch, C., & Schubert, H. (1995). Characterization of household microwave ovens by their efficiency and quality parameter. In *Proceedings of the 5th International Conference on Microwave and High Frequency Heating*, (p. S31). University of Cambridge.

Phippen, N., & Burfoot, D. (1990). Continuous measurement of temperatures in microwave ovens with turntables. *Laboratory Practice*, *39*, 9–11.

Pitchai, K., Birla, S. L., Jones, D., & Subbiah, J. (2012). Assessment of heating rate and non-uniform heating in domestic microwave ovens. *Journal of Microwave Power and Electromagnetic Energy*, *46*(4), 229–240.

Pitchai, K., Brahma, S., Birla, S., Jones, D., & Subbiah, J. (2011). Effect of location of small loads on heating rate and uniformity in domestic microwave ovens. In: *IMPI's 45th annual microwave power symposium*, New Orleans, USA (pp. 29–40).

Rakesh, V., Datta, A. K., Amin, M. H. G., & Hall, L. D. (2009). Heating uniformity and rates in a domestic microwave combination oven. *Journal of Food Process Engineering*, *32*(3), 398–424.

Richardson, P. S., & Gordon, K. H. (Eds.), (1997). *Guidelines on the verification of reheating instructions for microwaveable foods – Guideline no. 14*. Gloucestershire, UK: Campden and Chorleywood Food Research Association.

Rother, M. (2010). Mikrowellenunterstützte thermische Prozesse: Auslegung, Regelungsstrategien und Potentiale, Dissertation Uni Karlsruhe.

Schiffmann, R. F. (2014). Microwave ovens and food safety: Preparation of not- ready-to-eat products in standard and smart ovens. *Ingenierías*, *XVII*(64), 38–54.

Swain, M. J., Ferron, S., Coelho, A. I. P., & Swain, M. V. L. (2006). Effect of continuous (intermittent) use on the power output of domestic microwave ovens. *International Journal of Food Science & Technology*, *41*(6), 652–656.

Swain, M. J., Foster, A. M., Phillips, I. C., & James, S. J. (1994). Variation of uniformity of microwave reheating within commercial chilled convenience meals. Microwave science series no. 10. London: MAFF Publications.

Swain, M. J., Foster, A. M., Phillips, I. C., Wilkins, T. J., James, S. J., Radhakrishnan, C. D., et al. (1995). *Effect of initial temperature of food and continuous use of a microwave oven on the temperature of food after reheating*. Microwave science series no. 13. London: MAFF Publications.

Swain, M. J., James, S. J., & Swain, M. V. L. (2008). Effect of power output reduction of domestic microwave ovens after continuous (intermittent) use on food temperature after reheating. *Journal of Food Engineering*, *87*(1), 11–15.

Swain, M. J., Spinassou, A., & Swain, M. V. L. (2008). A test procedure to characterise the heating performance of domestic microwave ovens. *International Journal of Food Science & Technology*, *43*(1), 15–23.

Swain, M. J., Swain, M. V. L., & Russell, S. L. (1998). A new test procedure to determine the heating performance of domestic microwave ovens. *International Appliance Manufacturing*, *98*, 72–77.

Swain, M. V. L., Russell, S. L., Clarke, R., & Swain, M. J. (2004). Development of food simulants for microwave oven testing. *International Journal of Food Science and Technology*, *39*(6), 623–630.

Tang, Z., Mikhaylenko, G., Liu, F., Mah, J.-H., Pandit, R., Younce, F., et al. (2008). Microwave sterilization of sliced beef in gravy in 7 oz trays. *Journal of Food Engineering*, *89*(4), 375–383.

Tassinari, A. D., & Landgraf, M. (1997). Effect of microwave heating on survival of *Salmonella typhimurium* in artificially contaminated ready-to-eat foods. *Journal of Food Safety*, *10*, 239–248.

Tims, M. J., & Watts, B. M. (1958). Protection of cooked meats with phosphates. *Food Technology*, *12*, 240.

Walker, S. J., Bows, J. R., Richardson, P. S., & Banks, J. G. (1991). Listeria survival in chilled retail products: Effect of recommended microwave cooking. Microwave science series no. 1. London: MAFF Publications.

Wang, Y., Tang, J., Rasco, B., Wang, S., Alshami, A. A., & Kong, F. (2009). Using whey protein gel as a model food to study dielectric heating properties of salmon fillets. *LWT-Food Science and Technology*, *42*(6), 1174–1178.

Measuring temperature distributions during microwave processing

15

K. Knoerzer[1], *M. Regier*[2] *and H. Schubert*[3]
[1]CSIRO Agriculture and Food, Melbourne, VIC, Australia, [2]Trier University of Applied Sciences, Trier, Germany, [3]Karlsruhe Institute of Technology (KIT), Karlsruhe, Germany

15.1 Introduction

Microwave ovens are commonplace in households and are established there as devices of everyday use. While microwaves are used in the food industry, e.g., for pasteurization, it is still a technology that hasn't seen the uptake one would expect from such a versatile technology. Reasons for this fact are, e.g., the conservatism of the food industry, but also the high capital expenses that are involved in retrofitting an existing food plant (Decareau, 1985).

Microwave heating has been proposed as an alternative for traditional heating methods for several decades for a number of food manufacturing processes. Examples are (re)heating, baking, (pre)cooking, tempering of frozen food, blanching, pasteurization, sterilization, and dehydration.

For an extensive overview see Decareau (1985), Metaxas and Meredith (1983), Metaxas (1996), Roussy and Pearce (1995), Buffler (1993), Chandrasekaran, Ramanathan, and Basak (2013), and other chapters of this book. While their objectives differ, the aims are established in all applications through an increase of temperature.

Where the process has a preservation function, as in pasteurization, sterilization (see Chapter 10: Microwave-assisted pasteurization and sterilization—commercial perspective and Chapter 11: Microwaves for microbial inactivation—efficiency and inactivation kinetics), and drying (see Chapter 8: Microwave-assisted drying), it is critical to reach a minimum time–temperature treatment or a target water content, i.e., a minimum water activity in drying applications to assure microbial safety.

In other processes, temperatures should remain below thresholds to achieve a desired product quality (see Chapter 8: Microwave-assisted drying, Chapter 9: Microwave-assisted blanching, Chapter 10: Microwave-assisted pasteurization and sterilization—commercial perspective, and Chapter 13: Packaging for microwave foods) with a maximum preservation of color, flavor, etc.

In traditional heating and drying methods, heat transfer is mostly limited by heat conduction and so it is usually easy to identify locations of slowest heating or

The Microwave Processing of Foods. DOI: http://dx.doi.org/10.1016/B978-0-08-100528-6.00015-2

drying. This fact could lead to overheating near surface regions and, therefore, microwave heating, as a volumetric heating method, is proposed to improve product quality.

However, there are a number of microwave intrinsic factors that cause nonuniform heating, e.g., through focusing effects, corner and edge heating, inhomogeneous electromagnetic field distributions, or in particular the shape and the composition of the heated material (see Chapter 7: Microwave-assisted frying). As a consequence, heating patterns can differ significantly from one product to another and, therefore, in industrial microwave processes the quality and safety of the products can be compromised (Mudgett, 1986), and care needs to be taken to ensure sufficient heating uniformity (Tang, 2015). In addition to the conservatism of the food industry and the high capital expenses, it is largely for the reason of nonuniformity and associated issues with respect to food safety and quality, that microwave processes have not been adopted widely in the food industry.

To optimize microwave processes and their impact on food safety and quality, it is essential to understand the temperature distributions throughout the heated products.

15.2 Methods of measuring temperature distributions

To date, there are a number of methods to measure temperature distributions in products that are placed in electromagnetic fields; e.g., shielded thermocouples, fiber optic thermometers, infrared (IR) thermography, microwave radiometry, temperature indicators, e.g., liquid crystal foils, thermopaper, and model substances, and more recently a technique developed based on magnetic resonance imaging (MRI). Some of these methods are also discussed in Knoerzer, Murphy, Fresewinkel, Sanguansri, and Coventry (2012).

Many of these methods, however, have profound disadvantages, which will be discussed in the following sections.

15.2.1 Thermocouples

Without significant effort, normal thermocouples cannot be used in microwave devices, since the intense electromagnetic fields which cause the product heating also disturb the operation of these conventional temperature probes. Furthermore, these probes contain metallic wires and cause pronounced disturbance of the fields, leading to overheating and possible arcing and damage to the device, as well as to the product. Moreover, the method is invasive and only a poor spatial resolution is achievable.

15.2.2 Fiber optic probes

While the use of fiber optic sensors avoid the problem of an interaction with the electromagnetic field, the achievable spatial resolution, at least for conventional fiber optic probes, is still generally not sufficient, either. New developments, however, allow measurements along the fiber, which increases the information depth by providing one-dimensional temperature distributions across the insertion length of the fiber (Samiec, 2011). The fiber optic probes must be in direct contact with the product to be measured, i.e., this is also an invasive method.

15.2.3 Model substances

A different approach to assess temperature distributions in microwave applications is the use of model substances (or substances added to a gelled model food) that change their properties (e.g., color, pH value, structure, etc.) after reaching a certain temperature, or time/temperature combination. Published methods include the use of a color change of the model substance (Risman, Ohlsson & Lingnert, 1993), another one a coagulation effect of contained proteins (Wilhelm & Satterlee, 1971), while others report on inducing chemical reactions (e.g., Maillard reaction) and the produced chemical markers can be used to assess the temperature uniformity. Care has to be taken to match the other important physical properties of the model to the real product (e.g., dielectric and thermal properties). This is a rather outdated method, however, and is rarely used today.

15.2.4 Infrared thermography

Infrared thermography provides very good spatial resolution and is well-established for measuring surface temperatures. Even in real time with decent time resolution, infrared sensors can provide a reasonably accurate readout of the surface temperature of an object that is in the sensor's field of view. However, in common with conventional contact probes, infrared sensors are metallic and cannot be placed inside the electromagnetic fields of a microwave heating assembly. Furthermore, most measurements have to be done through a shielded window because the sensitive electronics inside the infrared sensors requires extensive electromagnetic shielding to allow accurate measurements in the presence of electromagnetic (EM) fields. Thus, the efficiency and accuracy of the temperature measurement, which in any case represents only the surface temperature of one part of the heated product, is reduced.

15.2.5 Microwave radiometry

The principle and, therefore, the advantages of microwave radiometry are similar to infrared thermography (Stephan & Pearce, 2004), except that the thermal energy detected and used as the basis for temperature estimation is in the microwave frequency range rather than the infrared region. As the penetration depth of radiation in the microwave frequency range is much higher than in the IR frequency range also subsurface temperatures can be observed. While penetration is in the order of several centimetres, i.e., temperatures can be sensed up to this depth in the product, the spatial resolution of this approach is much lower than that of IR due to the significantly longer wavelengths of microwaves (Gardner, Matousek, & Stone, 2016).

15.2.6 Temperature indicators

Temperature indicators, e.g., liquid crystal foils, thermal paper, or other chemical markers are also established methods for measuring temperature distributions (Feher, 1997; Grünewald & Rudolf, 1981; Zhang, Tang, Liu, Bohnet, & Tang, 2014). But these methods also provide mainly surface temperature mapping. A further disadvantage of thermopaper in particular is the fact that only information of temperature being above or below a certain temperature is accessible. With the exception of liquid crystals, most other temperature indicators undergo an irreversible change and transient temperature evolution cannot be mapped.

15.2.7 Magnetic resonance imaging

By using MRI as a tool for temperature measurement almost all the above mentioned problems can be avoided (Knoerzer, Regier, Hardy, Schubert, & Schuchmann, 2009; Knoerzer, Regier, & Schubert, 2008; Nott, Hall, Bows, Hale, & Patrick, 2000).

MRI enables noninvasive three-dimensional measurement of temperature distributions, offline, i.e., after heating, as well as online in purpose-designed microwave systems; the probe does not have to be in contact with the product, the product does not have to be sliced to get information about subsurface temperature measurements, any product containing water can be observed during or after microwave heating, and decent spatial resolutions can be achieved.

The major disadvantage of this technique is, however, that the costs for an MRI tomograph exceed the costs for all devices mentioned above by orders of magnitude.

15.3 Basic principles of different temperature mapping methods

15.3.1 Thermocouples

Thermocouples are fabricated from two electrical conductors made of two different metal alloys. The conductors are typically built into a cable having a heat-resistant sheath, often with an integral shield conductor. At one end of the cable, the two conductors are electrically shorted together. This end of the thermocouple, referred to as the hot junction, is thermally attached to the object to be measured. The other end, i.e., the cold junction, sometimes also referred to as reference junction, is connected to a measurement system. It should be noted that the "hot" junction, which is somewhat of a misnomer, may in fact be at a temperature lower than that of the reference junction if low temperatures are being measured.

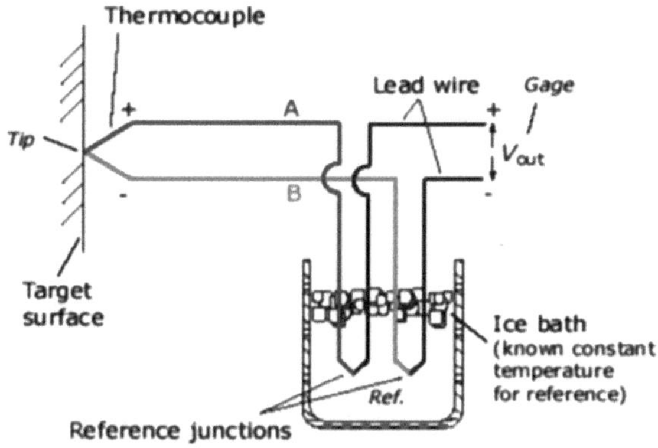

Figure 15.1 Sketch of thermocouple junction (efunda, 2016).

Thermocouples generate an open-circuit voltage, called the Seebeck effect or Seebeck voltage, discovered 1821 (Seebeck, 1821), that is proportional to the temperature difference between the hot and reference junctions:

$$\Delta V = S \cdot (T_{\text{hot}} - T_{\text{ref}}) \tag{15.1}$$

With:

ΔV: change in voltage
S: Seebeck coefficient (dependent on the thermocouple's metal composition)
T_{hot}: temperature at hot junction in K
T_{ref}: temperature at cold/reference junction in K

Since thermocouple voltage is a function of the temperature difference between junctions, it is necessary to know both voltage and reference junction temperature in order to determine the temperature at the hot junction. Consequently, a thermocouple measurement system must either measure the reference junction temperature or control it to maintain it at a fixed, known temperature. Fig. 15.1 shows the sketch of a thermocouple junction formed by joining two metallic alloys, A and B. Fig. 15.2 shows standard commercial thermocouples.

15.3.2 Fiber optic probes

The basis of the conventional fiber optic thermometry is the temperature-dependent photoluminescence of optically excited phosphors. The key element of fiber optic probes is the sensor consisting, e.g., of a small amount of manganese-activated magnesium fluorogermanate mounted at the tip of the optical fiber working in a temperature range from −200°C up to +450°C.

When the sensor material is excited with blue-violet light, the phosphor exhibits a deep red fluorescence. Fig. 15.3 shows the spectrum of this material. A pulse of blue-violet light to excite the phosphor to fluorescence is provided by a filtered

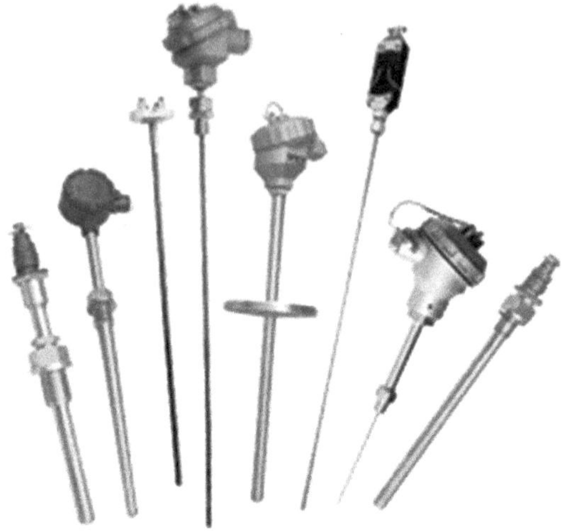

Figure 15.2 Standard industrial commercial thermocouples (GEM Elecrotech Corporation, 2016).

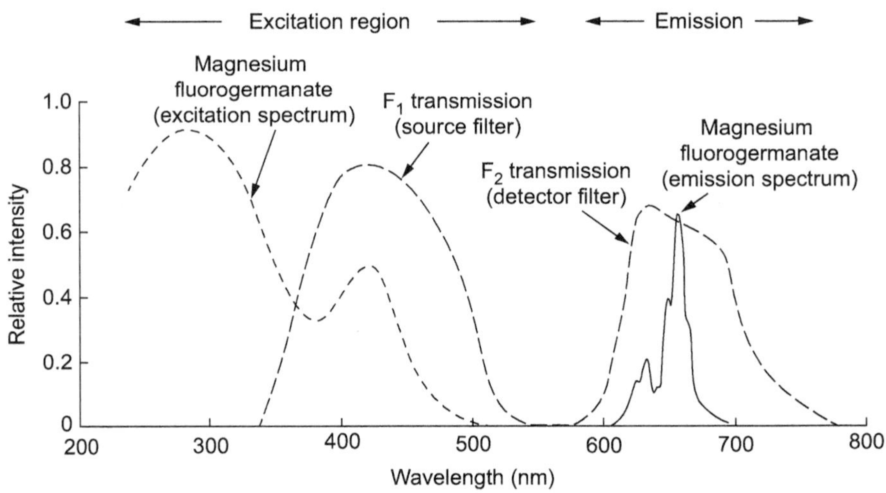

Figure 15.3 Room temperature spectrum of magnesium fluorogermanate. Dashed lines indicating passbands of source and detector filters (Luxtron Corporation, 1992).

xenon flash lamp. After the activation pulse, the intensity of fluorescent radiation decays exponentially. After a certain time t_1, the signal S_1 is measured and the time $t_1 + \tau$ is determined where the signal decayed to S_1/e. The actual phosphor temperature can be correlated with a digital look-up table (see Fig. 15.4).

The method for measurement of the decay time is shown schematically in Fig. 15.5. The sketch of a fiber optic device is shown in Fig. 15.6.

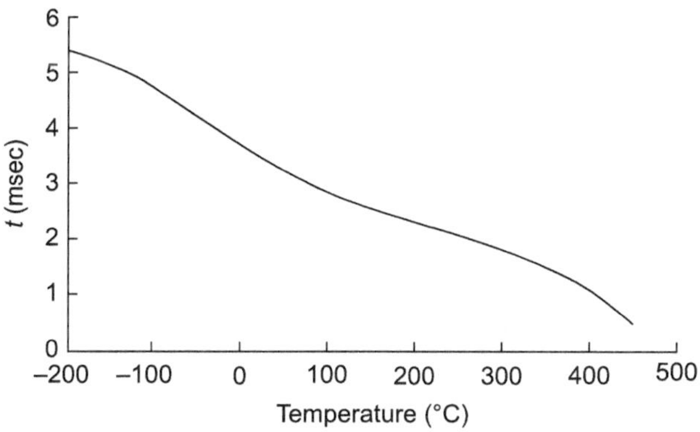

Figure 15.4 Emission intensity and fluorescent decay time of the magnesium fluorogermanate phosphor as a function of temperature (Luxtron Corporation, 1992).

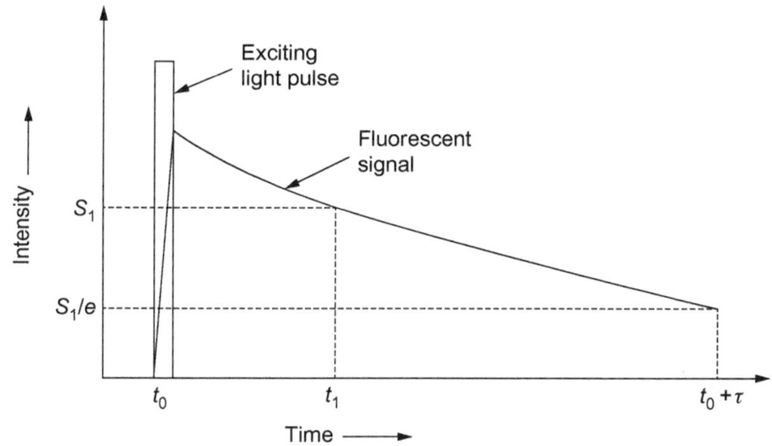

Figure 15.5 A schematic representation of the method used to extract the fluorescent decay time of the phosphor sensor (Luxtron Corporation, 1992).

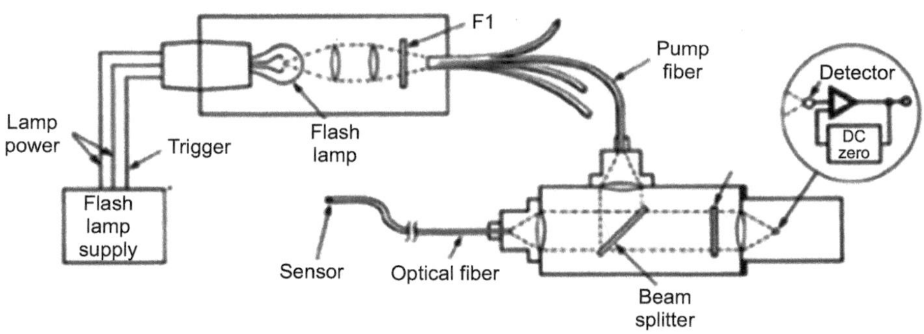

Figure 15.6 Cross-sectional view of Luxtron's model 755 (Luxtron Corporation, 1992).

Given that these conventional fiber optic probes only measure in a discrete location, the number of probes and in particular their placement is critical for accurate and relevant data acquisition. In scenarios such as microwave heating, where the local appearance of hot spots is generally not known (or may not be constant over the course of the process), the wrong placement of the fibers can lead to misinterpretations of the acquired data. Pseudocontinuous systems can utilize the reflected light from the entire fiber (by measuring the Raman or Brillouin components of the reflected light through optical time domain reflectometry), which is useful for very long fibers (in the km range) where they allow a spatial resolution in the meter range; however, this is not relevant for microwave applications, where temperature gradients develop over much shorter length scales.

Latest developments though allow the observation of the Rayleigh components of the reflected light (fiber imperfections along the fiber backscatter light) through coherent optical frequency domain reflectometry. In these systems, a continuous wave tunable laser light is coupled into an interferometer, where one interferometer arm is built by the fiber. By interfering the Rayleigh scattered light of the fiber with that of the fixed interferometer arm, the so-called beat frequency is dependent on the location along the fiber, where the light is scattered. The frequency increases (i.e., the wavelength decreases) with increasing distance from the source. According to Guemes, Fernandez-Lopez, and Soller (2010), shifts in the refractive index of the fiber, that are caused by a local temperature (or strain) change, lead to shifts in the local spectral frequency of the Rayleigh backscatter. Furthermore, accumulated changes in the refractive index along the fiber produce a time (of flight) shift of the Rayleigh backscatter return loss amplitude pattern. Thus, a cross correlation of the backscatter amplitude time-domain or frequency-domain data gives the spectral and temporal shifts, which correlate to the local temperature distribution. Commercial systems based on this technology can achieve spatial resolutions of 10 μm (Samiec, 2011), the resolution in temperature can achieve 0.1°C in the range of −50°C up to 300°C. Measurement rates are in the range of several Hertz.

15.3.3 Model substances

Model substances must represent the relevant properties of actual foods. For microwave heating scenarios, sometimes it is sufficient that only the dielectric properties (permittivity ε' and loss factor ε'') are representative for those of the food to be assessed, in a certain temperature interval and at a frequency of 2.45 GHz (Risman et al., 1993). But often, especially at lower microwave powers or longer heating times, it is also essential that the thermal conductivity and the heat capacity of the model food fit to the values of real food stuff.

The following factors can be temperature-dependent and have been used for temperature mapping (Risman et al., 1993):

- texture (or viscosity)
- optical transparency
- melting temperature

- coagulation or solidification
- color change (e.g., by changing of pH-value with added indicator)
- Maillard reaction products (Zhang et al., 2014)

Often gels are used, because they are easy to cut and mold to an exact shape. Color change over small temperature intervals of optically transparent gel helps in determination of hot or cold spot areas. Also melting or coagulation of model foods can indicate temperature changes (Risman et al., 1993). Another approach to assess temperature distributions is based on a temperature-induced chemical reaction in an indicator substance. For example, the US Army Natick Research Center has developed such a method to locate hot and cold spots in foods (Kim & Taub, 1993). Their method was based on an induced Maillard reaction in which specific temperature/time-dependent chemical compounds were produced. They used this, e.g., for assessing temperature uniformity in ohmic heating (Kim, Taub, Choi, & Prakash, 1996), radio-frequency processing (Wang, Lau, Tang, & Mao, 2004), and microwave-assisted thermal sterilization (Wang et al., 2009). However, to assess the reaction products, time-consuming high performance liquid chromatography (HPLC) had to be performed. In later studies, a computer vision method is reported, which was developed based on the formation of such reaction products, which allows rapid visualization of the heating pattern of the food (mainly surface, or subsurface when transparent gels were used) (Pandit, Tang, & Mikhaylenko, 2007; Zhang, Liu, Nindo, & Tang, 2013).

15.3.4 Infrared thermography

All objects (especially hot ones) emit thermal radiation that is in the visible and even the ultraviolet portion of the EM spectrum as well as the infrared, e.g., an incandescent light bulb or the sun.

Furthermore, every object at temperatures above Absolute Zero (0 K or −273.15°C) emits electromagnetic radiation, the so-called thermal radiation, for ambient temperatures, much in the infrared portion of the EM spectrum.

Reasons for this radiation are vibrating atoms and molecules behaving like Hertz' dipoles emitting electromagnetic waves. The frequency of the corpuscle vibration and, therefore, the intensity of the EM radiation increases with increasing temperature.

Planck's radiation law (Eqs. 15.2a and b) describes the radiation intensity at one discrete frequency,

$$I(\nu)d\nu d\Omega = \frac{2h\nu^3}{c^2} \cdot \frac{1}{e^{\left(\frac{h\nu}{kT}\right)} - 1} d\nu d\Omega \tag{15.2a}$$

$$I(\lambda)d\lambda d\Omega = \frac{2hc^2}{\lambda^2} \cdot \frac{1}{e^{\left(\frac{hc}{\lambda kT}\right)} - 1} d\lambda d\Omega \tag{15.2b}$$

With:

k: Boltzmann-constant
T: absolute temperature in K
h: Planck's constant
c: speed of light in m/s
λ: wavelength in m
ν: frequency in Hz
Ω: solid angle element

whereas Stefan-Boltzmann's law (Eq. 15.3) describes the total energy emitted by a black body integrated over all wavelengths per area. It is directly proportional to temperature to the power of 4:

$$I = \frac{2k^4T^4}{h^3c^2} \cdot \int_0^\infty \frac{x^3}{(e^x - 1)} dx = \frac{2k^4\pi^4}{15h^3c^2} T^4 = \sigma \cdot T^4 \quad \text{with} \quad x = \frac{hc}{k\lambda T} \tag{15.3}$$

With:

I: intensity of radiation in W/m^2
σ: Stefan-Boltzmann-constant

For real foods, Stefan-Boltzmann's law has to be extended with an efficiency factor ε (Eq. 15.4).

$$I = \varepsilon \cdot \sigma \cdot T^4 \tag{15.4}$$

With:

ε: emission coefficient, which has to be known for the studied material, but is close to 1 for many food substances

What is invisible to humans, particularly when only thermal infrared is present, can be "seen" by a thermal imager (see Fig. 15.7), or more precisely, a thermal imaging camera. A thermographic recording camera can be constructed in two different ways:

- First-generation commercial IR systems from the mid-1960s up to the early 1990s included a lens and moving mirrors or prisms that scan the image, and focus the heat rays on *one* infrared sensor. The scan moves over the subject area in a television-like pattern. These mechanical scanning devices increased the size of the system as well as their weight and power consumption. Also, because of this mechanical scanning system the thermal sensitivity can be compromised.
- From the beginning of the 1990s, the thermographic systems evolved and matured and a new detector technology, the so-called focal plane array (FPA), was introduced. This FPA is an IR detector that incorporates rows and columns of individual sensors eliminating the requirement for scanning systems. An FPA system usually enables resolutions of 160×120 or 320×240 pixels, although, more expensive cameras can achieve resolutions up to 1280×1024 pixels. Frame rates are in the order of several Hertz which gives decent time resolution for transient temperature mapping.

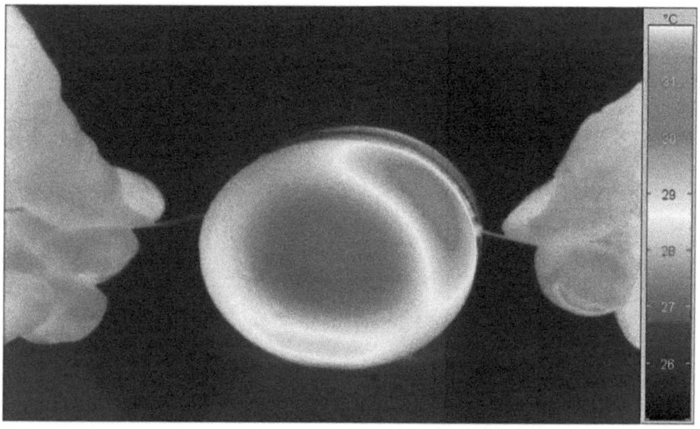

Figure 15.7 IR image of slicing of a heated model food cylinder.

The basis of both systems is the same: From the measured radiation pattern the actual temperatures can be calculated.

In general, commercial infrared cameras integrally detect a certain bandwidth of the radiation emitted by the observed object (approximately $1\ \mu m < \lambda \leq 6\ \mu m$) according Planck's radiation law (see Eqs. 15.2a and b) and the temperature is calculated from the calibration curve of the camera manufacturer.

15.3.5 Microwave radiometry

Microwave radiometry (MWR) allows to measure the power in the microwave region of the thermal radiation of heated food. Similarly to the IR thermography, albeit at lower spatial resolution due to the longer wavelengths of the microwaves (Gardner et al., 2016), it is possible to observe brightness temperatures of the sample. The brightness temperature is:

$$T_{B,i} = \frac{P_i}{k \cdot Df_i} = (1 - R_i) \cdot \frac{P_{\text{sample}}}{k \cdot Df_i} = (1 - R_i) \cdot T_{B,\text{sample}} \tag{15.5}$$

With:

$P_{\text{sample},i}$: thermal radiation power emitted by the sample in W
P_i: power received by the antenna in a bandwidth Df_i around a center frequency f_i in W
R_i: power reflection coefficient at the skin-antenna interface at f_i
k: Boltzmann's constant

According to the Rayleigh-Jeans law (an approximation of Planck's law (Eqs. 15.2a and b) for high frequencies), at microwave frequencies the thermal radiation intensity is proportional to the absolute temperature and, therefore:

$$T_{B,i} = \int_{afv} W_i(r) \cdot T(r) \cdot dv \tag{15.6}$$

With:

$T(r)$: absolute temperature in an incremental sample volume dv located at radius r in K
$W_i(r)$: radiometric weighing function
afv: antenna's field of view

Therefore, the measured brightness temperature is $T_{B,\text{meas},i} = T_{B,\text{sample},i} = (\int_{afv} W_i(r)\, T(r)\, dv)/(1 - R_i)$, and in general, is dependent upon:

- the actual temperature distribution throughout the samples viewed;
- the receiving antenna's weighing function, which in turn depends on its operating frequency, dimensions, the permittivity of the material with which it is filled, and the microwave attenuating properties of the samples;
- microwave reflections occurring at interfaces between different sample regions and between these regions and the measuring equipment.

Since the attenuation of microwaves in samples and the antenna characteristics are frequency dependent, the temperature–depth profile within the sample underneath the antenna can be found using a multifrequency radiometer and a stable solution to the inverse problem of retrieving the temperature–depth dependence from a set of $T_{B,\text{meas},i}$. Internal temperature at depths of up to 4 cm or more in samples may be obtained using multifrequency MWR (Hand et al., 2001).

15.3.6 Liquid crystal foils

Liquid crystal is a unique substance, which exists between the solid and the isotropic-liquid phase of some organic compounds, selectively scattering incident light. The theory behind its optical characteristics involves the behavior of the molecular structure. Each liquid crystal compound possesses a helical structure with a characteristic pitch length (Demus, Goodby, Gray, Spiess, & Vill, 1998). These pitch lengths are in the range of the wavelength of visible light, and can be altered by changing the external stimulus, typically, temperature or shear stress. Since the fundamental chemical structure is unaffected by the external stimulus, a thermochromic liquid crystal coating can respond repeatedly to the same physical changes; i.e., unlike most other thermal indicators, the change is fully reversible and, therefore, transient temperature changes can be monitored (Balasubramaniam & Sastry, 1995). There are two families of liquid crystal materials which have the helical structure, namely chiral-nematic and cholesteric.

The event temperature and clearing point temperature are used to describe the properties of liquid crystals. The lowest temperature where liquid crystals scatter visible light is called the event temperature. At a temperature below the liquid crystal's event temperature, the liquid crystal is in the solid state and appears transparent. At a temperature above the clearing point temperature, it will enter the pure liquid state and will revert back to being transparent. At the clearing point temperature, the helical pitch of the liquid crystal exceeds the wavelengths of visible light. The reflected color spectrum of liquid crystals varies continuously from the longer wavelengths (i.e., red) corresponding to event temperature to shorter wavelengths (i.e., blue) corresponding to

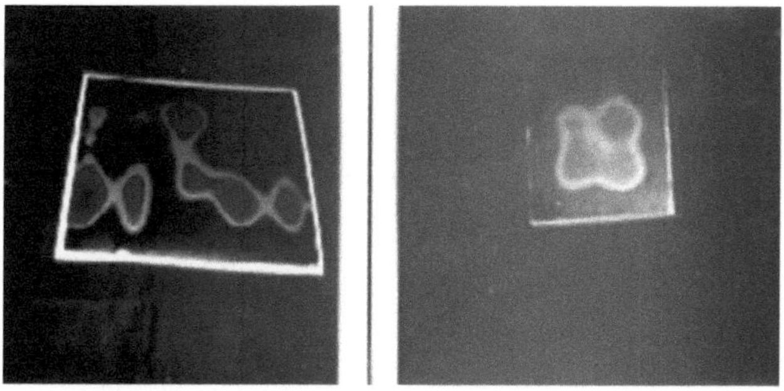

Figure 15.8 Temperature distribution at the bottom of an empty microwave oven after introducing power (left); surface temperature distribution of a microwave heated cube (right). Both determined by liquid crystal foils.

clearing point temperature (see Fig. 15.8). Liquid crystals transmit a significant amount of the incident light with virtually no modification. Generally, they are viewed against a nonreflecting, i.e., black background, preventing the transmitted light from getting reflected without adversely affecting the interpretation of selectively scattered light from the liquid crystals.

A color/temperature designator describes the response of a typical liquid crystal chemical make-up. This helps in selecting the liquid crystal composition (i.e., formulation) for a particular application. For example, "R35C5W" designates a formulation which signifies that the red start or event temperature of the liquid crystal is 35°C and the blue start temperature is 5°C above the red start temperature. This provides a crude estimate of the bandwidth of liquid crystals. The liquid crystals can be either narrow band or wide band formulations. The narrow band formulations have bandwidths below 1 or 2°C while wide band formulations have bandwidths between 5°C and 30°C (Panigrahi, 2000).

15.3.7 Thermal paper

Thermal paper is a special fine paper that is coated with a chemical that changes color when exposed to heat. The active slice of the thermal paper contains temperature-sensitive chemicals, the so-called leuco dyes and color developers; the dyes melt after reaching a certain temperature and change their appearance from transparent to usually black (sometimes blue or red coatings are used) through a reaction with the developer (see Fig. 15.9).

Generally, the thermal paper consists of three different slices (Wikipedia, The Free Encyclopedia, 2016):

- The thin base paper.
- The thermo-active slice with a thickness of approximately 5 μm, containing thermo-sensitive chemicals and additional ingredients, e.g., stabilizers against static electrical charge, durability, UV-resistance, and sensitizers that optimize colorization temperature and facilitate mixing of dye and developer.
- A protective top coating to reduce fading of the thermal image caused by UV light, water, oils, etc.

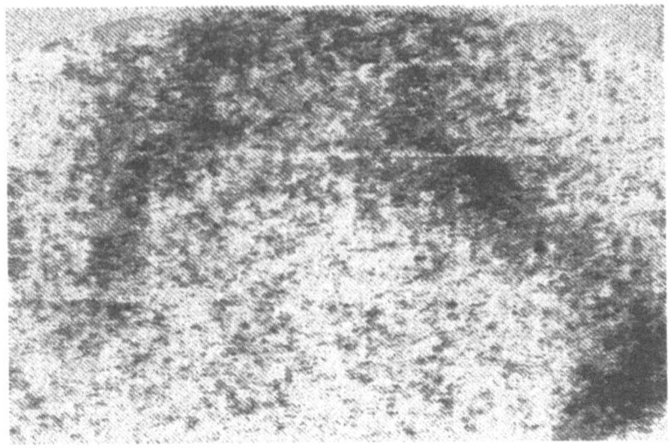

Figure 15.9 Typical appearance of a heated thermal paper (black regions are above the certain temperature, white regions below (Feher, 1997)).

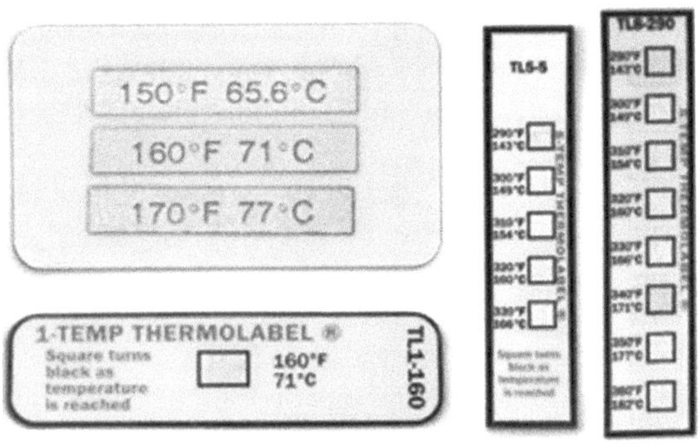

Figure 15.10 Commercial multitemperature thermal labels (DeltaTrak, 2016).

Commercial thermo-labels are formulated to react within a few seconds when the rated temperature is reached. Various single and multi-temperature labels are commercially available, with ratings ranging from 32°C to 260°C, see, e.g., Fig. 15.10 (DeltaTrak Europe, 2016).

As each section of a label reaches its rated temperature, that section responds with a sharply defined color change from white to black. While this is a cheap and reliable measurement approach to indicate whether a certain temperature has been reached during the process (an important indication for example in autoclaving), this is pretty much all the information the thermal paper can provide.

15.3.8 Magnetic resonance imaging

Many protons (like ^1H, the hydrogen proton) have a nucleus spin and, therefore, a magnetic moment μ (see Fig. 15.11).

Bringing the randomly distributed and oriented magnetic moments (see Fig. 15.12A) in a magnetic field B_0 (see Fig. 15.12B) causes an alignment in parallel or antiparallel

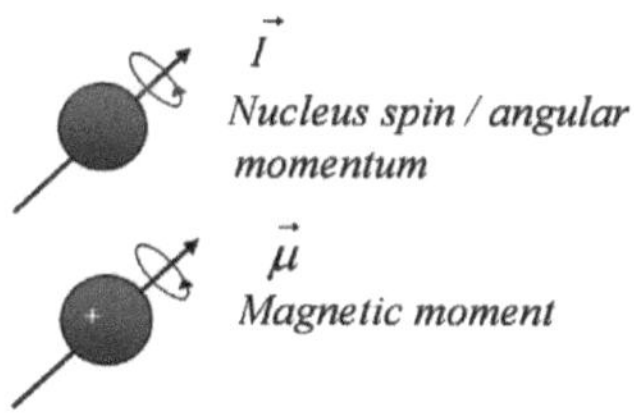

Figure 15.11 Nucleus spin and magnetic moment of the hydrogen proton.

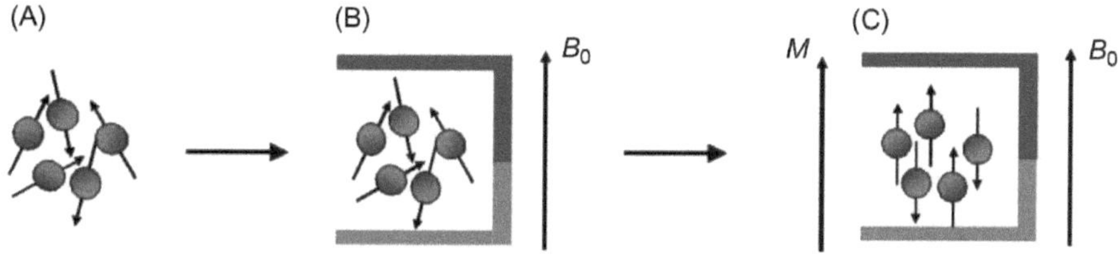

Figure 15.12 (A) Nonarranged magnetic moments; (B) brought into an external magnetic field B_0; (C) alignment of magnetic moments.

orientation according the external magnetic field and as a result the magnetization M (see Fig. 15.12C).

$$\vec{\mu} = \gamma \cdot \vec{I} \tag{15.7}$$

$$\vec{M} = \sum \vec{\mu} \tag{15.8}$$

With:

μ: magnetic moment
γ: gyromagnetic relation (constant for ^1H)
I: angular momentum
M: magnetization

When a radio frequency (RF) pulse is applied by a surrounding radio frequency coil, the alignment of the magnetization M can be inverted into the XY-plane and thereby brought into phase (see Fig. 15.13).

After switching in the XY-plane, the external magnetic field causes a precession of the magnetization M with a specific frequency, referred to as Lamor frequency:

$$\omega_L = \gamma \cdot B_0 \tag{15.9}$$

With:

ω_L: Lamor frequency in s^{-1}
B_0: external magnetic field in T

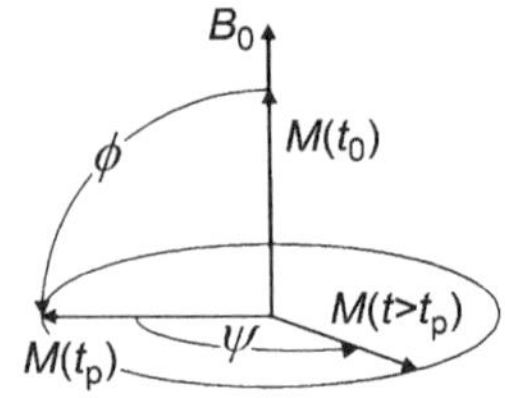

Figure 15.13 Sketch of inverting the magnetization M in the XY-plane and bringing into phase.

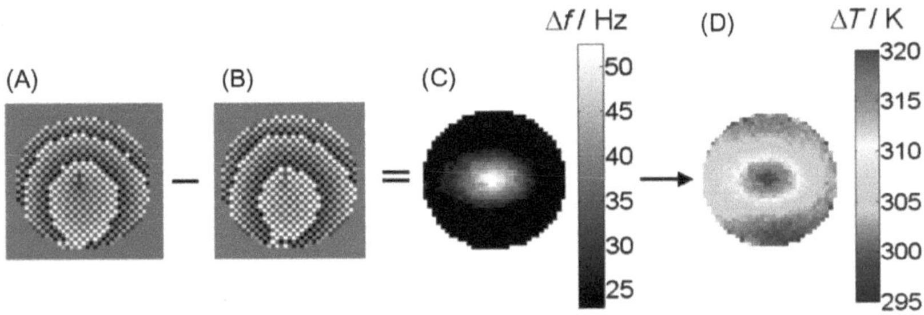

Figure 15.14 (A) MRI phase image of a model food at known initial temperature (T_1 = const); (B) phase image of heated sample; (C) phase difference distribution (C = A − B); (D) temperature distribution (calculated from phase difference with knowledge of a chemical shift of $\Delta f/\Delta T = 0.01$ ppm/°C).

This precession induces an AC voltage in the RF coil, the measurable MR-signal which is directly proportional to the spin density, and therefore the proton density and thus the water content.

Once the RF pulse is switched off, the phase relationship will decay with a characteristic time constant (referred to as T_2) and their alignment will relax with a different time constant T_1; both those parameters are temperature-dependent. A further parameter measurable by MRI and temperature dependent is the self-diffusion coefficient; however, this self-diffusion coefficient and the time constants also show a strong dependence on the composition (and thus the viscosity) of the solution in the food.

Another parameter, almost independent of the composition of the solution in the food, and measurable by MRI is referred to as the chemical shift. The method of temperature mapping using this parameter is based on the temperature dependence of the water proton chemical shift: Increasing temperature leads to breaking of hydrogen bonds and therefore to a stronger shielding of the hydrogen proton, due to a reduced hydrogen–oxygen-distance, by oxygen-located electron clouds. As a result, the precession frequency decreases and therefore the phase (spin angle) of the processed signal ($\Delta f/\Delta T = 0.01$ ppm/°C; Hindman, 1966). From a measured phase difference between the sample with known initial temperature and the heated sample, the actual temperature distribution can be calculated (see Fig. 15.14).

Due to the universality and great advantages of temperature mapping using MRI (already mentioned in Section 15.2), experiments and results of MRI studies will be presented in the following section.

15.4 Experiments and results of MRI study of microwave-induced heating pattern

15.4.1 Experimental procedures

The experiments were conducted in a Bruker Biospin/Oxford Instruments Super-Wide-Bore cryomagnet with a maximum sample diameter of 64 mm and a magnetic flux of 4.7 T (see Fig. 15.15).

For measuring phase distribution, and therefore temperature, a gradient echo pulse sequence (Kimmich, 1997) was utilized. A variable number of slices (from 10 up to 50 slices) perpendicular to the *z*-axis with a thickness of 1 mm, a two-dimensional resolution of 64 by 64 pixels, and therefore a three-dimensional

Figure 15.15 Tomograph used for temperature mapping experiments.

resolution of 1 mm^3 were measured. The overall measuring time was below 13 s with an accuracy of approximately ± 2 K.

The obtained MRI-data (binary 3D-matrices) was exported and analyzed with purpose-developed functions in Matlab 6.5.

15.4.1.1 External microwave heating

Model food samples with a height of 30 mm and a diameter of 54 mm (see Fig. 15.17; Knoerzer, Regier, Erle, Pardey, & Schubert, 2004) were heated in a microwave oven (Panasonic Pro II NE 2740; continuous power supply) with varying heating time and microwave power.

Before and after microwave heating, the samples were brought into the MRI tomograph for measuring phase distributions.

To validate the temperature distribution measured by MRI, three single slices were compared qualitatively and quantitatively with two-dimensional temperature distributions measured using infrared thermography (Mitsubishi IR-M700).

15.4.1.2 Internal microwave heating

To avoid temperature equalization between the heating procedure and measuring temperature distribution in the tomograph, a microwave device was developed together with GIGATHERM AG (Hermann, 2004; Knoerzer et al., 2009) which enabled the introduction of microwave power directly into the magnet (see Fig. 15.16).

With this device it was possible to measure three-dimensional temperature distributions during microwave heating. Heating experiments with different shapes of the model food and different real foods were performed and temperature changes

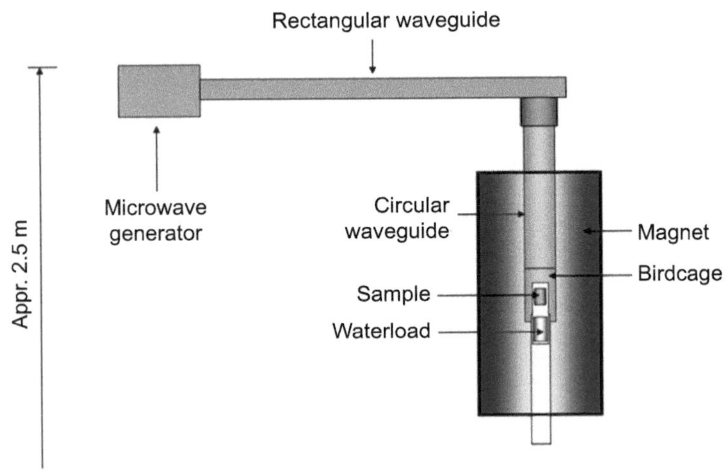

Figure 15.16 Sketch of the device for introducing microwaves into the MRI tomograph.

at discrete points, as measured by the MRI approach, were validated using fiber optic thermography.

15.4.2 Results and discussion

15.4.2.1 External microwave heating

As explained above, heating experiments were conducted externally in a microwave oven with continuous power supply. Fig. 15.17 shows the qualitative comparison of temperature distributions measured by MRI and infrared thermography.

As can be seen, the MRI data qualitatively agree well with the data obtained from infrared thermography. However, a quantitative comparison showed a strong deviation of temperatures measured by the two methods (see Fig. 15.18). This strong deviation of the expected agreement ($T_{MRI} = T_{IR}$) can be explained by temperature equalization and pronounced surface cooling after slicing of the model food for IR measurements.

15.4.2.2 Internal microwave heating

Measuring microwave heating inline makes the observation of temperature changes at discrete spots of the heated food possible. Fig. 15.19A shows the three-dimensional temperature distribution of a heated model food sphere at a certain time ($d_{modelfood} = 50$ mm, $P_{microwave} = 27$ W). Fig. 15.19B shows the transient temperature profile at the selected point (Fig. 15.19A) measured by MRI (solid line) and fiber optic thermography (dotted line). A good agreement between the two methods was observed.

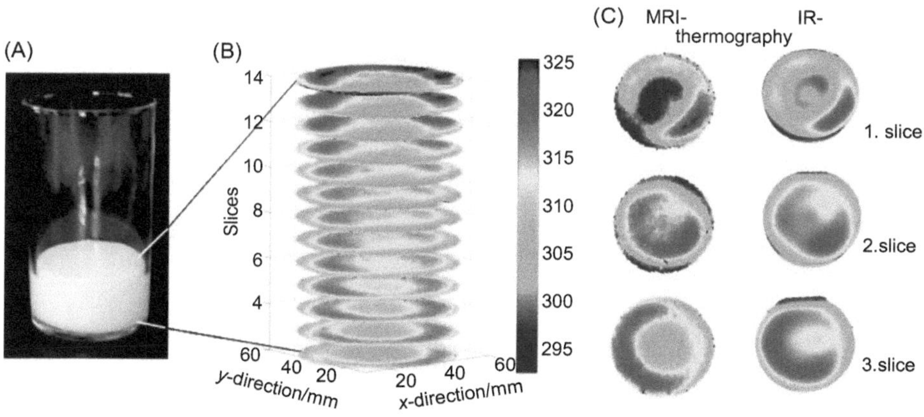

Figure 15.17 (A) Microwave heated model food sample in a glass. (B) MRI determined temperature distribution of heated model food (15 slices, slice thickness 1 mm, slice distance 1 mm). (C) Comparison of MRI and IR thermography of three selected slices.

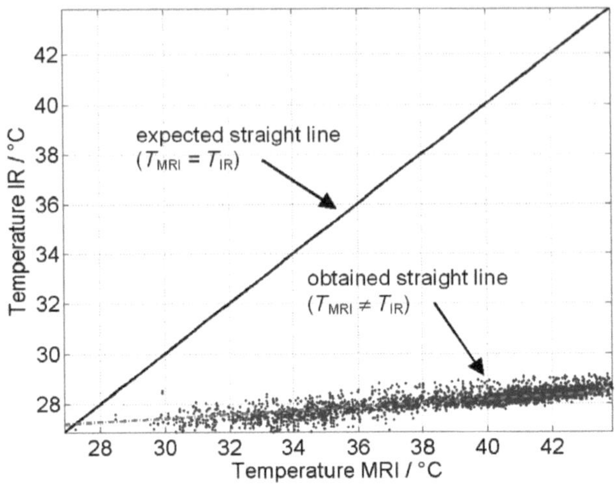

Figure 15.18 Quantitative comparison of MRI and IR thermography.

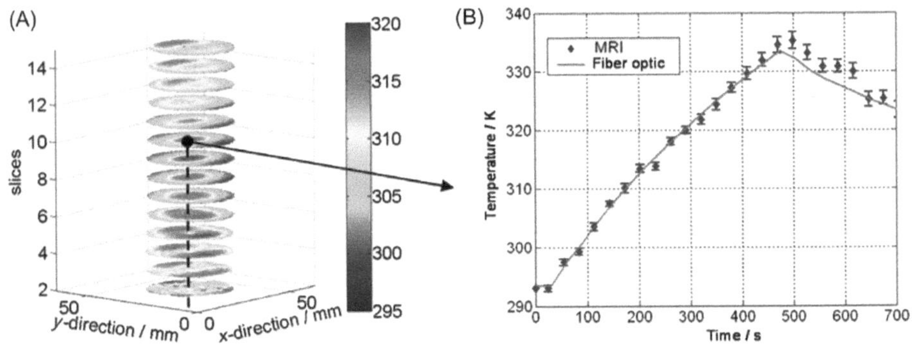

Figure 15.19 (A) MRI determined temperature distribution in heated model food cylinder; point of location of fiber optic sensor is indicated. (B) Comparison of temperature profiles measured by MRI and fiber optic thermography.

15.5 Summary

In this chapter, different methods for measuring temperatures during or after microwave processing, their advantages and disadvantages, and their fundamental principles have been presented and discussed.

It was shown, that the use of thermocouples, likely the cheapest method for temperature mapping, is not feasible in a microwave field because of the metal components; furthermore, the achievable spatial resolution is poor.

While the first problem of thermocouples can be overcome by using fiber optic probes, the achievable spatial resolution is often not sufficient, either.

A different approach for temperature mapping mentioned in this chapter was the use of model substances and chemical markers that change their properties or undergo a chemical reaction during heating. Care has to be taken to match the physical properties of the model to the real food in order to get accurate predictions of the temperature distributions induced during microwave heating.

For determining surface temperatures three different methods were presented in this chapter: infrared thermography, liquid crystal foils, and thermal paper.

However, even though these methods are established for surface temperature mapping, especially in microwave processing as a volumetric heating method, it is mostly not sufficient, measuring only surface temperatures.

For measuring temperatures noninvasively inside the product, two methods were presented. Multifrequency microwave radiometry allows to get information about subsurface temperatures; however, the achievable spatial resolution is usually insufficient for the often complex heating patterns induced by microwave heating. The method of temperature mapping using MRI, however, was shown to circumvent almost all the problems inherent to the other discussed methods, and it is possible (even inline during heating) to measure three-dimensional temperature distributions noninvasively at high spatial and temporal resolution.

All the methods discussed in this chapter can give useful insights into the temperature distribution induced by microwave heating; however, only the method of MRI can measure temperature distributions within the product at a sufficient resolution.

The disadvantage of MRI is that it is a very complex and expensive technique, which makes it currently only suitable for research.

References

Balasubramaniam, V. M., & Sastry, S. K. (1995). Use of liquid crystals as temperature sensors in food processing research. *Journal of Food Engineering*, *26*(2), 219–230.

Buffler, C. R. (1993). *Microwave cooking and processing: Engineering fundamentals for the food scientist*. New York, NY: AVI Book.

Chandrasekaran, S., Ramanathan, S., & Basak, T. (2013). Microwave food processing – A review. *Food Research International*, *52*(1), 243–261.

Decareau, R. V. (1985). *Microwaves in the food processing industry*. Orlando, FL: Academic Press Inc.

DeltaTrak Europe. (2016). *Thermal labels*. >http://deltatrakeurope.be/index.php/yt-products/labels> Accessed 13.03.16.

Demus, D., Goodby, J., Gray, G. W., Spiess, H.-W., & Vill, V. (1998). *Handbook of liquid crystals – Vol. 2A: Low molecular weight liquid crystals I*. Weinheim: Wiley – VCH.

efunda. (2016). *Thermocouples: Theory*. <http://www.efunda.com/designstandards/sensors/thermocouples/thmcple_theory.cfm> Accessed 13.03.16.

Feher, L. E. (1997). Simulationsrechnungen zur verfahrenstechnischen Anwendung von Millimeterwellen für die industrielle Materialprozesstechnik, Wissenschaftliche Berichte FZ KA 5885, Karlsruhe, Forschungszentrum Karlsruhe.

Gardner, B., Matousek, P., & Stone, N. (2016). Temperature spatially offset Raman spectroscopy (T-SORS): Subsurface chemically specific measurement of temperature in turbid media using anti-stokes spatially offset Raman spectroscopy. *Analytical Chemistry*, *88*(1), 832–837.

GEM Elecrotech Corporation. (2016). *Thermocouple types*. <http://gemelectrotech.com/thermocoupletypes.html> Accessed 13.03.16.

Grünewald, T. H., & Rudolf, M. (1981). Messung der Temperatur und der Temperaturverteilung um Mikrowellenfeld. *Zeitschrift für die Lebensmittelwirtschaft*, *32*(3), 85–88.

Guemes, A., Fernandez-Lopez, A., & Soller, B. (2010). Optical fiber distributed sensing – Physical principles and applications. *Structural Health Monitoring*, *9*(3), 233–245.

Hand, J. W., Van Leeuwen, G. M. J., Mizushina, S., Van de Kamer, J. B., Maruyama, K., Sugiura, T., ... Edwards, A. D. (2001). Monitoring of deep brain temperature in infants using multi-frequency microwave radiometry and thermal modelling. *Physics in Medicine and Biology*, *46*(7), 1885–1903.

Hermann, A. (2004). *Aufbau und Inbetriebnahme einer Vorrichtung zur Einkopplung von Mikrowellen in einen NMR-Tomographen, Diplomarbeit*. Universität Karlsruhe (TH).

Hindman, J. C. (1966). Proton resonance shift of water in gas and liquid states. *Journal of Chemical Physics*, *44*, 4582–4592.

Kim, H. J., & Taub, I. A. (1993). Intrinsic chemical markers for aseptic processing of particulate foods. *Food Technology*, *47*(1), 91–97.

Kim, H. J., Taub, I. A., Choi, Y. M., & Prakash, A. (1996). Principles and applications of chemical markers of sterility in high-temperature-short-time processing of particulate foods. In T. C. Lee, & H. J. Kim (Eds.), *Chemical markers for processed and stored foods* (pp. 54–69). Washington, DC: American Chemical Society.

Kimmich, R. (1997). *NMR-tomography-diffusometry-relaxometry*. Berlin, Germany: Springer.

Knoerzer, K., Murphy, A. B., Fresewinkel, M., Sanguansri, P., & Coventry, J. (2012). Evaluation of methods for determining food surface temperature in the presence of low-pressure coo plasma. *Innovative Food Science & Emerging Technologies*, *15*, 23–30.

Knoerzer, K., Regier, M., Erle, U., Pardey, K. K., & Schubert, H. (2004). Development of a model food for microwave processing and the prediction of its physical properties. *Journal of Microwave Power and Electromagnetic Energy*, *39*(3 + 4), 67–78.

Knoerzer, K., Regier, M., Hardy, E. H., Schubert, H., & Schuchmann, H. P. (2009). Simultaneous microwave heating and three-dimensional MRI temperature mapping. *Innovative Food Science & Emerging Technologies*, *10*(4), 537–544.

Knoerzer, K., Regier, M., & Schubert, H. (2008). A computational model for calculating temperature distributions in microwave food applications. *Innovative Food Science & Emerging Technologies*, *9*(3), 374–384.

Luxtron Corporation (1992). Fluoroptic temperature sensing. *The Operator's Guide, Model, 755*.

Metaxas, A. C. (1996). *Foundations of electroheat*. Chichester, UK: John Wiley & Sons.

Metaxas, A. C., & Meredith, R. J. (1983). *Industrial microwave heating*. London, UK: Polu Pelegrinus Ltd.

Mudgett, R. E. (1986). Dielectric properties of foods. In R. E. Mudgett, & R. V. Decareau (Eds.), *Microwaves in the Food Processing Industry* (pp. 15–56). Orlando, FL: Academic Press Inc.

Nott, K. P., Hall, L. D., Bows, J. R., Hale, M., & Patrick, M. L. (2000). Three-dimensional MRI mapping of microwave induced heating patterns. *International Journal of Food Science and Technology*, *34*, 305–315.

Pandit, R. B., Tang, J., & Mikhaylenko, G. (2007). A computer vision method to locate cold spot in foods during microwave sterilization processes. *Pattern Recognition*, *40*(12), 3667–3676.

Panigrahi, P. K. (2000). *Colourful world of liquid crystals.* <http://www.iitk.ac.in/infocell/Archive/dirmar2/techno_crystals.html> Accessed 13.03.16.

Risman, P. O., Ohlsson, T., & Lingnert, H. (1993). Model substances and their use in microwave heating studies. *SIK Report, 588*, (pp. 1–10).

Roussy, G., & Pearce, J. A. (1995). *Foundations and industrial applications of microwaves and radio frequency fields.* Chichester, UK: John Wiley & Sons.

Samiec, D. (2011). Verteilte faseroptische Temperatur- und Dehnungsmessung mit sehr hoher Ortsauflösung. *Photonik, 6*, 34–37.

Seebeck, Th. J. (1821). Magnetische Polarisation der Metalle und Erze durch Temperaturdifferenz. Abhandlung der Akademischen Wissenschaften 1823. Berlin, Germany.

Stephan, K. D., & Pearce, J. A. (2004). Continuous non-contact remote temperature sensing during microwave heating with microwave radiometry. In: *38th annual microwave symposium, Toronto, Canada, July 2004*.

Tang, J. (2015). Unlocking potentials of microwaves for food safety and quality. *Journal of Food Science, 80*, 1776–1793.

Wang, Y., Lau, M. H., Tang, J., & Mao, R. (2004). Kinetics of chemical marker M-1 formation in whey protein gels for developing sterilization processes based on dielectric heating. *Journal of Food Engineering, 64*(1), 111–118.

Wang, Y., Tang, J., Rasco, B., Wang, S., Alshami, A. A., & Kong, F. (2009). Using whey protein gel as a model food to study the dielectric heating properties of salmon (*Oncorhynchus gorbuscha*) fillets. *LWT-Food Science and Technology, 42*, 1174–1178.

Wikipedia, The Free Encyclopedia. (2016). *Thermal paper.* <https://en.wikipedia.org/w/index.php?title = Thermal_paper&oldid = 704913731> Accessed 13.03.16.

Wilhelm, M. S., & Satterlee, L. D. (1971). A 3-dimensional method for mapping microwave ovens. *Microwave Energy Application Newsletter, 4*(5), 3.

Zhang, W., Liu, F., Nindo, C., & Tang, J. (2013). Physical properties of egg whites and whole eggs relevant to microwave pasteurization. *Journal of Food Engineering, 118*, 62–69.

Zhang, W., Tang, J., Liu, F., Bohnet, S., & Tang, Z. (2014). Chemical marker M2 (4-hydroxy-5-methyl-3(2H)-furanone) formation in egg white gel model for heating pattern determination of microwave-assisted pasteurization processing. *Journal of Food Engineering, 125*, 69–76.

16 Microwave plant requirements and process control for advanced applications

M. Ozcelik[1] *and P.-A. Püschner*[2]
[1]Ankara University, Ankara, Turkey, [2]Püschner GmbH & Co KG, Schwanewede, Germany

16.1 Basic microwave design

As microwaves offer volume heating (see Chapter 1: Introducing microwave processing of food: fundamentals of the technology) and rapid heat transfer, microwave technology can be used for debottlenecking and new products with better properties.

Compared to conventional heat transfer based on heat conductivity, microwaves enable rapid volume/3D heating in the penetration depth. If product dimensions are smaller than the penetration depth of microwaves, core temperatures are generally higher than the surface temperatures in comparison with a conventionally heated and dried product (see Fig. 16.1). Therefore, this so-called 3D or volume heating has the following advantages:

1. A temperature gradient directed toward the surface, i.e., inside temperatures are higher than those outside giving rise to a higher partial pressure that drives the evaporating liquid to the surface.
2. Consequently, the superficial layer does not dry out completely and surfaces remain permeable.
3. The liquid evaporating inside the product is emitted through the pore structure of the solid material's macrocapillary system, resulting in a high drying velocity.
4. The heating of water and most organic solvents occurs selectively—due to the greater dielectric losses of water as compared to the product to be dried.
5. Swift and thorough drying of moist products with low thermal conductivity.

In vacuum drying, the boiling point is reduced in accordance with the vacuum level. If the boiling point is below the triple point, water is sublimated (transfer from the solid stage directly to the steam phase).

Fig. 16.2 shows a basic industrial design of a microwave system. The magnetron, in which microwave energy is generated, is installed on a waveguide launcher. The generated electromagnetic energy is guided inside the waveguide structure and coupled into the cavity where the product is heat-treated and dried.

Electromagnetic energy is reflected within the structure. The amount of reflected energy varies according to the dielectric properties of the product, the dimensions of the cavity, and its matching toward the microwave source. In order to achieve best possible efficiency, the amount of reflected energy can be

The Microwave Processing of Foods. DOI: http://dx.doi.org/10.1016/B978-0-08-100528-6.00016-4

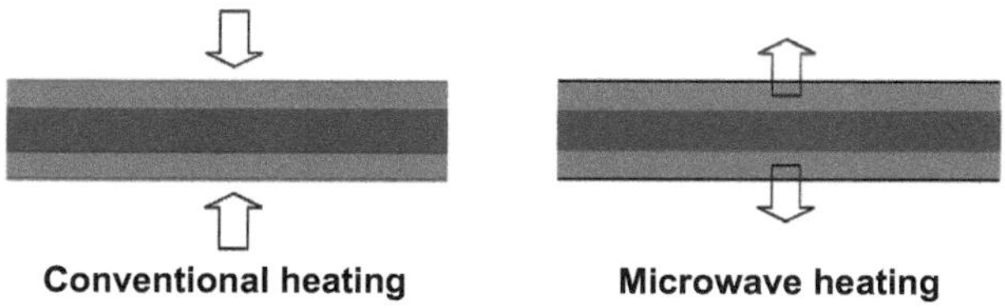

Figure 16.1 Comparison of conventional and microwave heat transfer (colors show temperature levels: blue (dark gray in print versions) cold; orange (light gray in print versions), warm; and red (mid-gray in print versions), hot).

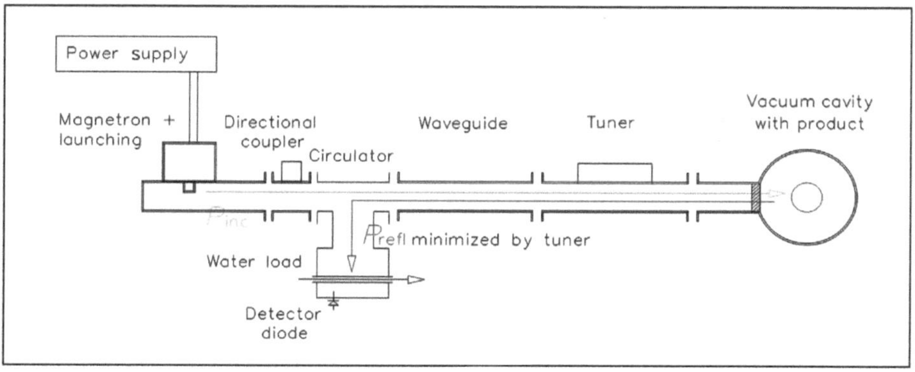

Figure 16.2 Schematic design of an industrial microwave plant according to Good Manufacturing Practice (GMP) for industrial microwaves.

minimized using a tuner in order to better match the microwave system to the load conditions prevailing inside the cavity.

The installed circulator is working as a load protection for the magnetron. Depending on the quality of the used circulator, the reflected energy is mainly absorbed by the water load at the circulator. This is needed for enabling the magnetron to work within its specifications and generate its nominal set microwave power provided by the power supply which drives the magnetron.

Directional couplers measure the real forward and reflected microwave power. Their position for installation inside the waveguide system may vary. The measurement of the reflected power can also be performed in the water load of the circulator. This setup is a must for a reliable and reproducible industrial microwave system which can be used for GMP processes. After proper matching, the reflected microwave energy should be lower than 10% of the forward microwave power. The measurement of forward and reflected power allows to determine the microwave power thermally absorbed by the product.

16.1.1 Field distribution

In most cases, a homogeneous field distribution is the key for proper microwave heating and/or drying processes (see Chapter 8: Microwave-assisted drying). In

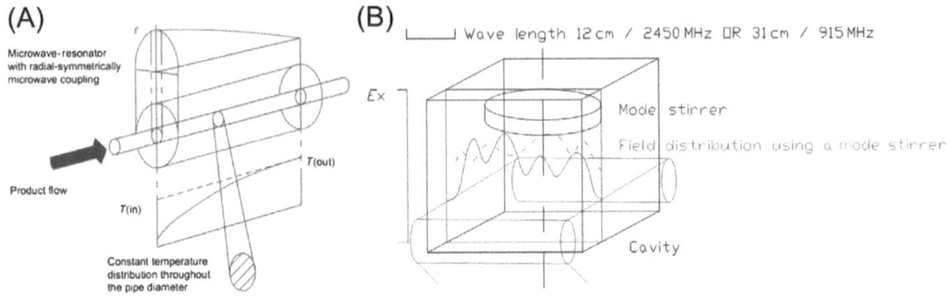

Figure 16.3 Field distribution of a ground mode applicator for heating viscous food products (A-left) and a multimode cavity inside a continuous dryer (B-right).

general, we distinguish between a ground mode applicator type and a multimode chamber oven applicator type. The ground mode applicator type has a better and defined energy distribution. More recent examples of microwave pasteurization and sterilization of liquids are using a ground mode applicator type (see Fig. 16.3A), but most of the drying chambers and, especially vacuum chambers, are based on multimode chambers. Compared to ground mode applicators, multimode chamber dimensions are considerably greater than the microwave wave length. The free space wave length for 2450 MHz is in the order of 12 cm and chamber dimensions can be in the order of 0.5 to 1.5 m. The coupled microwave energy is creating microwave field distributions with a great number of minima and maxima called modes (see Fig. 16.3B). Optimization of the field distribution for these multimode chambers is challenging and it is impossible to achieve a totally constant field distribution inside the cavity. Compared to the ground mode applicator with a defined field distribution, the condition for heat transfer is almost stable for different dielectric materials. The multimode chamber field distribution will change significantly due to dielectric material changes, which occur during drying phases. The higher the number of modes, the more difficult it becomes to control the mode distribution. If these modes are generated by multimagnetron systems, cross-talking between the magnetrons will occur and may reduce efficiency, if the magnetrons are not protected by circulators (Metaxas & Meredith, 1988).

In the concept of atmospheric drying and heating using batch or continuous microwave dryers, the position of the microwave coupling and the use of mode stirrers to change the multimode field distribution can result in more uniform heating within the residence time inside the dryer and therefore create more evenly dried products (Pueschner, 2005).

In a microwave vacuum batch system, drums are mainly used to mix and tumble the product during drying. Fig. 16.4 shows a generator setup of a vacuum vessel where an open plastic drum is installed to take the product load dosed from the top by a butterfly valve and a hopper. During microwave drying, the open plastic drum is turned clock- and anticlockwise, e.g., $\pm 90^\circ$ to mix and tumble the product during drying. After drying, the product is discharged through the bottom butterfly valve. Microwave energy is applied through the microwave systems from each side.

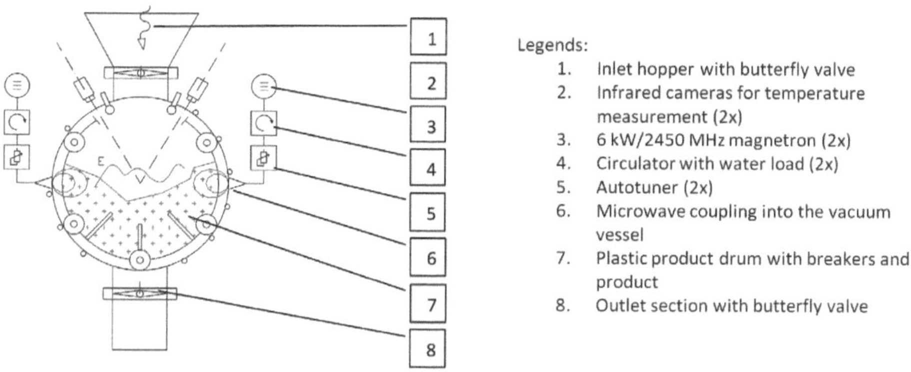

Figure 16.4 Setup of a multimode batch vacuum drum dryer using two microwave systems.

Infrared cameras are measuring the bulk temperature of the product. In the main drying phase (see Fig. 16.6) where the amount of moisture contained in the product is high, most microwave energy is absorbed directly inside the product in front of the coupling systems as shown in Fig. 16.4. The more water is evaporated, the deeper the microwave can penetrate into the product. When the product is getting drier, not all microwave energy is completely absorbed while the microwaves are traveling and passing the product volume. This remaining microwave energy will be reflected at the metal walls and will create standing wave patterns within the vacuum chamber, where areas with high and low field densities are created. Therefore, during the entire drying process, mixing and tumbling of the product is essential to ensure even drying results at the end.

16.1.2 Process control

The aim of industrial microwave heating/drying processes is to obtain products of better quality thanks to a faster process at lower costs as compared to conventional drying systems. During microwave drying, several properties such as mass, volume, moisture content, temperature, texture, and color of the material may change significantly. In microwave drying of foods, the complex dielectric permittivity of water in liquid stage is high

$$\varepsilon = 70 - j\,10.$$

Mainly the imaginary part, the so-called loss factor is responsible for dielectric heating.

As the amount of water is removed during the drying phase, such dielectric losses of the product itself become more determining for the process, when most of the water is removed during liquid stage drying (see Fig. 16.5, left).

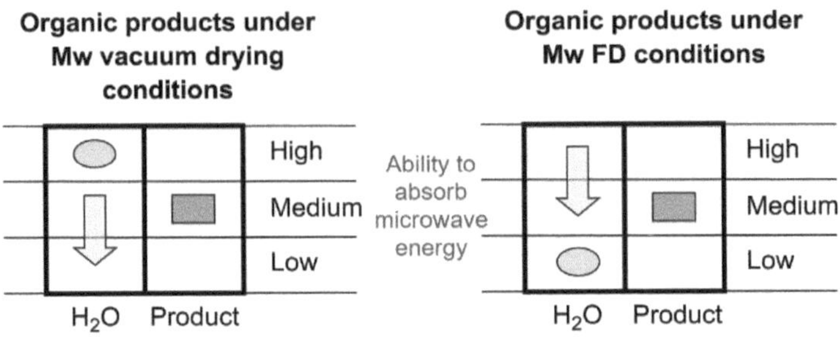

Figure 16.5 Ability to absorb microwave energy from water (left column) and organic products (right column) for products in liquid stage drying (+20°C, left) and under freeze drying conditions (−20°C, right).

Another case is shown in Fig. 16.5 (right) under freeze-drying conditions. When the wet product is frozen, the complex dielectric permittivity of water is approximately

$$\varepsilon = 3 - j\,0.003$$

This means that the water is almost "transparent" for microwaves, and dielectric heating is mainly driven by the dielectric properties of the organic product itself. Therefore, also the penetration depth of microwaves is considerably larger under freeze-drying conditions.

The challenging points of microwave heating and vacuum drying are:

- Rapid heat transfer with limited uniformity of field distribution (risk of hot spots);
- High energy densities in end drying can damage the products and cause plasma discharges;
- End point determination of the drying process has to be safe in order to avoid overdrying and overheating of the product.

The process control has to ensure that the abovementioned points be properly controlled for each specific food drying applications under microwave vacuum drying or under microwave freeze-drying conditions.

In the main drying phase according to Fig. 16.6, maximum microwave energy is applied in order to run a fast and efficient drying process. When the product contains a sufficient amount of water, the boiling point, which depends upon the applied vacuum level, controls the product temperature (e.g., approx. 100°C at normal atmospheric drying and 20°C at 20 mbar in vacuum drying).

When the product is almost dried at approximately 10−15% residual moisture content on a wet basis, the microwave energy density has to be reduced as the amount of water is unable to keep temperatures low around the boiling point. This is the most critical point to find the right settings to reduce the amount of microwave power for end drying in order not to exceed critical temperatures where product properties might be damaged. Furthermore, it is crucial to find the right setting to stop the drying process at a defined point at the end.

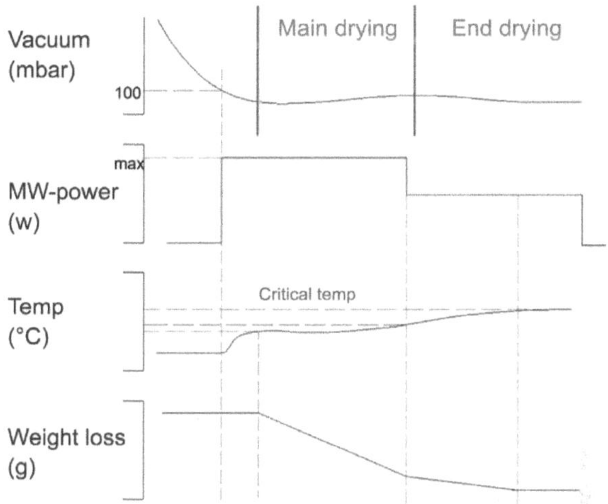

Figure 16.6 Simplified microwave vacuum drying diagram.

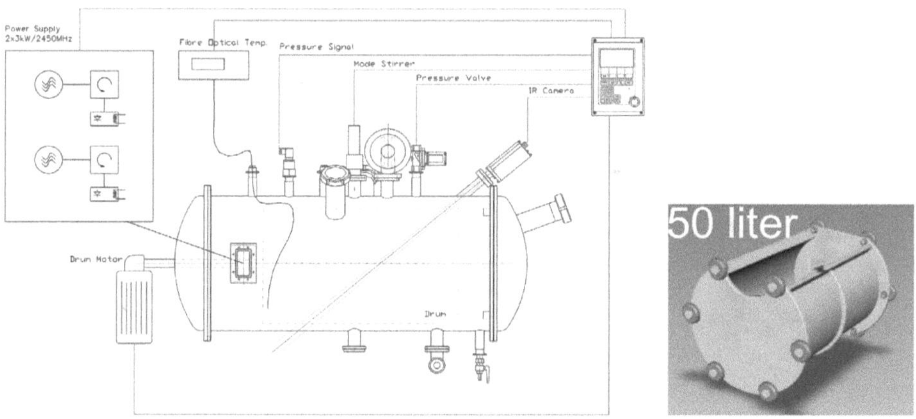

Figure 16.7 Simplified block diagram of a 2 × 3 kW/2450 MHz microwave vacuum drum dryer with a 200 L vacuum vessel and a 50 L plastic drum.

The process control system (PLC) of the microwave dryer has to be capable to do this job in a proper way. The settings for different recipes must be stored inside the PLC to deal with different dryer settings for the various products.

The drying diagram in Fig. 16.6 was performed inside a microwave vacuum drum dryer as described in the block diagram shown in Fig. 16.7.

After activation of the vacuum pump, the vacuum chamber is evacuated. At a pressure level of 100 mbar, the microwave power is switched on at a max value suitable for the product to maintain fast drying without exceeding the breakdown voltage. The product temperature measured by the infrared camera increases to the boiling point which depends on the actual vacuum level (e.g., 20 mbar at about 20°C). In this main drying phase, the weight loss from water evaporation is high. At a certain point, the dielectric properties of the product itself exert a dominating influence when the moisture content of the product decreases in the range of 5–15% MC wet basis. As a result, the product temperature rises, even when the vacuum level is low. At this point, the microwave power is reduced without exceeding critical temperatures above 50–60°C at which the product properties will be stressed. At a certain product temperature and a certain microwave energy uptake, the drying process is stopped. Before breaking the vacuum, the product can be kept under low vacuum conditions to achieve a homogeneous moisture distribution for small and larger particles inside the product bulk.

For end point determination, the following conditions can be used by the PLC to determine the final drying point:

1. Product temperature in the final drying phase. If the microwave power is switched off and the products still present a reasonably high water content, the product temperature will significantly fall back again toward the boiling point. If the product temperature stays at the same level, this indicates that the water content of the product is rather low.
2. Reflected microwave energy increases toward the end of the drying phase.
3. Due to the lower steam rate, the vacuum pressure drops at the end of the drying phase.

16.1.3 Process safety and relevant standards

Microwaves have a great number of applications used in many different sectors. Microwaves are similarly used in the food and pharmaceutical industries. In the food industry, microwaves are used for drying, cooking, tempering, pasteurization, and sterilization (Giese, 1992).

The necessity of food safety is indisputably due to health, social, and economic considerations. Food safety is a priority when manufacturing food products. Safe food products are those which are free from any physical, chemical, and biological risks throughout their shelf life. The term food safety, on the other hand, refers to the entirety of all preventative measures taken against possible physical, chemical, and biological hazards. The notion of GMP was proposed by the FDA (Food and Drug Administration) to properly describe any unhygienic conditions. The GMP is a cornerstone of the contemporary methods of food safety systems. Contamination is to be understood as the existence of any unwanted material in the food product. Potential threads that may cause food safety hazards are determined by examining the contamination sources (Fig. 16.8).

Process safety is a key in GMP processes in the food and pharmaceutical industry. It includes:

- Process validation in research and development, process development, up-scale strategies and proof-of-principles, etc.;
- User requirements for machinery to be used by food producers;
- Special requirement for food safety;
- Risk/hazard analyses;

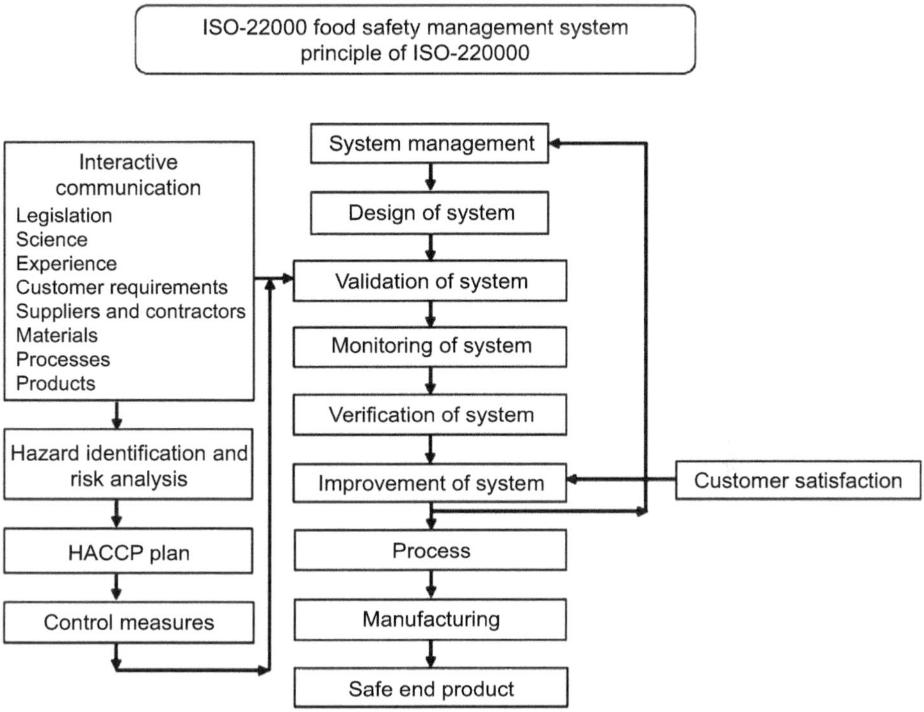

Figure 16.8 ISO 22000 flow chart (Jones, 2007).

- Observance of all relevant standards (*International Organization for Standardization* (ISO), International Electrotechnical Commission (IEC), Food and Drug Administration (FDA), etc.).

Food safety management systems are: Hazard Analysis and Critical Control Points (HACCP), Good Agricultural Practice (GAP), Good Manufacturing Practice (GMP), Good Hygienic Practice (GHP), Good Laboratory Practice (GLP), Good Distribution Practice (GDP), Good Trade Practice (GTP), Good Veterinary Practice (GVP) (Motarjemi and Lelieved, 2014).

The food industry is required to have an integrated food safety assurance system which consists of implementing the codes of good practices, applying the HACCP system activities, and so forth.

Food manufacturers have to comply with regulatory food safety standards and ensure that their products do not violate their norms.

The HACCP system consists of the following principles (Ricetti, 2013):

- Perform hazard analyses;
- Identify critical control points;
- Establish critical limits for critical control points;
- Establish monitoring requirements for critical control points;
- Establish corrective actions;
- Establish procedures for ensuring the HACCP system to work as intended;
- Establish record keeping procedures.

Almost all food producers are focused on minimally processed foods, in order to meet customers' expectations with regard to fresh-like high-quality convenience foods. These products are subjected only to mild treatment and nonthermal processing and additive-free materials. This means microbial growth must be controlled and limited below harmful thresholds and, at the same time, the nutritional and organoleptic properties of food must be preserved (Dogan Halkman, Yücel, & Halkman, 2014).

Many studies have been conducted to compare the retention of vitamins in many vegetables, fruit, and other foodstuffs subject to conventional and microwave treatment. In literature, it is often shown that vitamin retention in microwave food is equivalent to or better than conventional foodstuffs (Decareau, 1992). According to the latter, nutritional values of microwave applied food would be comparable with conventional applications.

The challenge is to use processes which can produce bacteria-free or reduced products while retaining the natural flavor, odor, and texture. Nonthermal technologies tested in food processing are high hydrostatic pressure, pulsed electric fields, ultrasound, cold plasma, intense light pulses, oscillating magnetic field, ultraviolet light, microwave radiation, irradiation, dense phase carbon dioxide, and shock waves. This technology can be used for decontamination, pasteurization, and, in some cases, sterilization.

The effect of microwave radiation is thermal and nonthermal. Temperature increase in the product during microwave heating is accepted as the destruction of microorganisms is mainly due to exposure to thermal effects. Most nonthermal effects result from a proposed direct interaction of the electric field with specific molecules in the reaction medium and the electrostatic polar effect. In many research shows there are remarkable conflicts between the studies of the nonthermal effects of microwave radiation. Microbial destruction in microwave is considered to be a result of thermal effects only (Dogan Halkman et al., 2014).

An efficient food safety program identifies hazards that are reasonably likely to occur and eliminates them by changing the process design to a well-designed controllable process ensuring elimination or reduction of hazards to acceptable levels and an environment preventing recontamination. To eliminate the hazards from the product, cleaning is one of the most critical steps in the production process—cleaning of the contact surfaces of manufacturing, process, and support equipment. Often, Cleaning-In-Place systems (CIP) are used to minimize or eliminate contamination of batches produced in the process equipment. The CIP-installation is a system consisting of various components including tanks, vessels, pumps, heat exchangers, valves, piping, and instrumentation, as well as control/monitoring systems. CIP systems are utilized to clean equipment, process installations, and associated piping in a reproducible and effective way within the system. The system must be able to reproducibly achieve the appropriate parameter levels to effectively clean the process equipment. The cleaning process for each product may have different parameters to optimize the cleaning operation (Holah, 2014).

The CIP equipment must be of hygienic design to eliminate or minimize contamination between cleaning cycles. CIP vessels and piping should allow full drainage and be able to be stored in a dry state between uses. When designed properly, the CIP system should be essentially self-cleaning. The overall process equipment subject to CIP has to conform to the latest 3A Sanitary Standards, ASME BPE, and/or equivalent international standards (Hasting, 2008). All CIP solution delivery devices (spray balls, spray nozzles, etc.) should be designed to ensure complete internal coverage and contact with the various cleaning solutions and rinses in the CIP procedure.

As far as the example of industrial vacuum drum dryers as shown in Figs. 16.7 and 16.14 is concerned, all materials in contact with the food product, e.g., the drum, have to be food grade and approved by the local regulatory authority. Furthermore, these parts have to be manufactured with traceability certificates. All steel parts in contact with dust and evaporated water, such as the vacuum vessel, piping, etc., have to be manufactured in high quality steel, mostly stainless steel grade 316 L/1.4404. The inner surfaces must have a surface roughness of $R_a < 1.6\,\mu m$ (R_a is the arithmetic average of the roughness profile). All weld seams have to conform to standards such as the American Society for Testing and Materials (ASTM: A269-94a, A269-95, etc.). Even if manufacturers issue cleaning recommendations, users will develop their own cleaning procedures in their production environments. Design points critical in respect of cleaning have to be identified during the design phase in accordance with process objectives (Holah, 2014).

Furthermore, there are specific standards that are relevant for industrial microwave installations. In terms of safety, the *IEC 60519-1/-6 (2011)* in *"Electroheat Installations – Part 6: Specifications for Safety in industrial Microwave Heating Equipment"* have to be complied with. Besides other requirements, the allowed microwave leakage has to be below $<5\,mW/cm^2$ at a distance of 5 cm from all touchable surfaces (see Chapter 1: Introducing microwave processing of food: fundamentals of the technology) of the industrial microwave equipment; furthermore, inspection doors, which can be opened without the help of special tools, have to be interlocked by two independent safety switches and controlled by a safety relay using two independent circuit breakers.

The IEC standard *IEC 61307 (2011) "Industrial Microwave Heating Installations – Test Methods for Determination of Power Output"* is relevant for the design of an industrial microwave heating system. Calorimetric test conditions have to be measured in order to define the thermal power of the microwave system.

16.2 Industrial application

16.2.1 Examples of vacuum applications in liquid stage

16.2.1.1 Example 1: leaves of basil

Herbs such as mint, thyme, basil, parsley, peppermint, and cilantro to name only a few, are regarded as high market value products. The majority of herbs are sold in dehydrated form, which is traditionally obtained through hot air drying at temperatures in excess of 60°C. Drying of herbs at 60°C or higher temperatures induces damage in the form of discoloration, loss of flavor volatiles, and collapsed cellular structures (Therdthai & Zhou, 2009). Such undesirable outcomes of the hot air drying process can be avoided by using microwave vacuum drying in the dehydration process of herbs.

Mint leaves subjected to hot air and microwave vacuum drying exhibit substantial differences in terms of drying time, color, structure, and rehydration behavior. For example, while the hot air drying process of a batch takes about 90 min at 60°C, it only takes less than 20 minutes using 3 kW of microwave power at a pressure of 50 mbar to obtain the same level of residual moisture in the microwave vacuum drying of an identical batch. Hot air-dried mint leaves exhibit a dark brown color as a result of chlorophyll degradation induced by high temperature and oxygen, whereas microwave vacuum-dried leaves exhibit an appealing light-green/yellow color. Scanning electron micrographs of hot air-dried mint leaves reveal a dense microstructure, which is in sharp contrast to the porous microstructure in the microwave vacuum-dried leaves. Due to the porous microstructure, the microwave vacuum-dried mint leaves have higher rehydration rates compared to the hot air-dried leaves (Therdthai & Zhou, 2009).

Fig. 16.9 shows the basil leaves before and after microwave vacuum drying in a 2×3 kW/2450 MHz microwave vacuum drum dryer *μWaveVac0650* (Fig. 16.7) (Pueschner GmbH & Co KG, Bremen, Germany; installed in Turkey). As can be seen, the green color is pleasantly retained. A recipe programmed in the PLC (Programmable Logic Controller) of the microwave plant is controlling the process as shown in Fig. 16.10.

As shown in Fig. 16.10, the set value for the vacuum is 40 mbar using a 250 qm/h liquid seal water ring pump. Microwave power was started at 3 kW. The reflected microwave power was detected by the PLC to be too high and the autotuner minimized the reflected microwave energy after some minutes down to below 2%. Due to the high thermal energy within the basil, the generated steam caused the vacuum level to rise up to 50 mbar as the vacuum pump had not been powerful enough to maintain the set value of 40 mbar. Product temperatures are increasing up to around 50°C. Microwave power was reduced by the absorbed microwave energy of 0.5 kWh. The product drum speed was reduced and the product drum turning angle was increased from 60° to 85°.

Figure 16.9 Basil leaves in a 50 L drum before and after drying (initial mass = 1 kg—after drying approx 100 g).

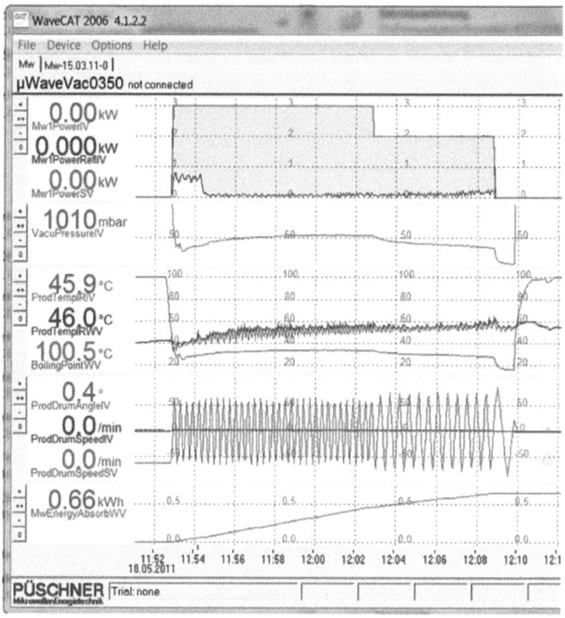

Figure 16.10 Drying of basil within 17 min using a microwave vacuum drum dryer.

This lab process shown in the preceding example is scaled for a larger industrial process using a 30 kW/915 MHz *μWaveVac30100* (Pueschner GmbH & Co KG, Bremen, Germany, installed in France).

The drum is loaded with 10 kg of fresh basil and is capable of drying this batch in 40–45 min. Following the drying process, the drum is turned by 180° to remove the dried product through the bottom butterfly valve. Then, the drum is rotated back into its initial position and fresh herbs are loaded through the feeding system into the drum. Also, a CIP system is installed to run cleaning cycles when the product is changed from, e.g., basil to, e.g., coriander. The system is operating in a three-shift production schedule with 5.000 operation hours per year (Table 16.1)

The above table shows production data of the industrial 30 kW/915 MHz microwave vacuum drum dryer, which has an annual capacity of 8 tons dried product. The concept can be scaled up further by installing more production units in parallel, enabling factories to increase their capacity by installing additional units at a later point in time.

16.2.1.2 Example 2: carrot chips

Carrots are known to contain the highest content of β-carotene (Béttega et al. 2013; Lin, Durance, & Scaman, 1998)—an organic compound which is thought to reduce the risk of cancer and age-related degenerations (Suvarnakuta et al., 2005)—among vegetables of human diet. They are also rich in other vitamins, minerals, and fibers and also very low in fat content. Due to the fact that carrots offer an abundance of

Table 16.1 Main plant and drying specifications as well as consumption rates of the microwave vacuum dryer in Fig. 16.11

Batch volume in product drum	500	L
Batch initial weight of product	10	kg
Removed water (90% down to <5% MC) 85%	8.8	kg
Drying time for basil	0.75	h
Capacity dried product based on 1 h drying time	1.60	kg/h
Max MW power	30	kW/915 MHz
Average MW power for drying	25	kW
Electrical supply power of machine	45	kVA
Cooling water at 20–25°C	60	L/min
Cooling water power by chiller	12	kW

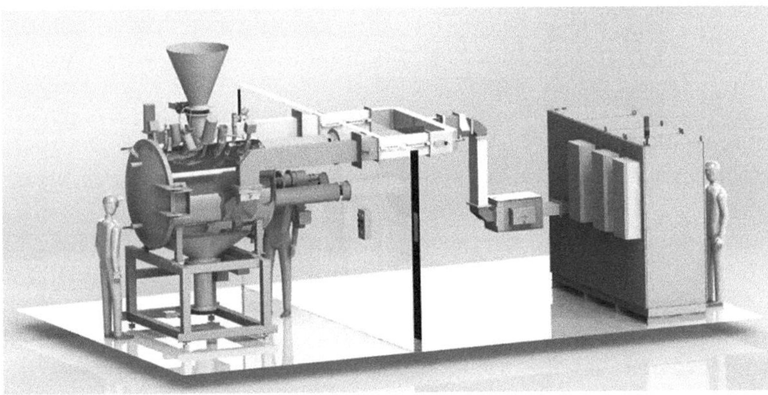

Figure 16.11 30 kW/915 MHz vacuum drum dryer with automated feeding and discharging system.

health benefits, dried carrots would be an excellent candidate for nutritious food products (e.g., instant powdered soups, snacks, cakes, etc.) provided that the nutrient composition is retained after drying.

Dehydration of carrots can be achieved gently via microwave vacuum drying which provides fast drying at low temperatures. In order to cut down the drying time, samples are prepared in slices, chips, or in shredded form as opposed to cylinders or cubes. In contrast to hot air-dried carrots, the microwave vacuum-dried carrot chips have improved rehydration behavior, highly retained nutritional value, "puffed" texture, and an appealing color (Lin et al., 1998). Regarding the nutrient retention, low-temperature microwave vacuum-drying process helps preserving the vitamin C which is unstable when subjected to heat. The high content of vitamin C in dried carrot also ensures the retention of other vitamins which generally exhibit better thermal resistance.

Figure 16.12 12–24 kW/2450 MHz microwave vacuum drum dryer.

However, the vacuum level happens to be an important factor that affects the retention of β-carotene. There is usually an optimum vacuum level at which the β-carotene retention is at its maximum (Béttega et al. 2013; Suvarnakuta et al., 2005). Therefore, the vacuum level should be controlled accurately in microwave vacuum drying of carrots.

In the following example, the carrot samples were microwave vacuum dried in a 12 kW/2450 MHz *μWaveVac1290* (Fig. 16.12) (Pueschner GmbH & Co KG, Bremen, Germany, installed in Hungary).

The initial weight of the samples was 25 kg, with an initial moisture content of 89% on wet basis. Drying time to achieve a final moisture content of 4–7% was approximately 3 h. The absorbed microwave energy was (12.8 + 2.7) 15.5 kWh and the outlet weight was 2.8 kg. The drying process is shown in Fig. 16.13. The microwave power was started with 7 kW at 30 mbar and was reduced to 6 kW as the vacuum increased to 50 mbar. As the reflected microwave energy and temperatures were increased to above 45°C, the microwave power was reduced for a short time and then stopped.

The carrots were cut into slices and filled into a 300 L plastic drum (see Fig. 16.14). The variation of dried slices is shown in Fig. 16.15.

Compared with the example of basil, carrots are mechanically more stable and higher product quantities (mass) can be loaded into a smaller drum (Table 16.2).

16.2.2 Freeze-drying application

Simply speaking, a microwave freeze dryer is a conventional freeze dryer which, however, additionally permits the introduction of microwaves into the drying chamber (Duan, Zhang, Li, & Mujumdar, 2008). The whole drying process is performed under vacuum environment in which water is removed through sublimation,

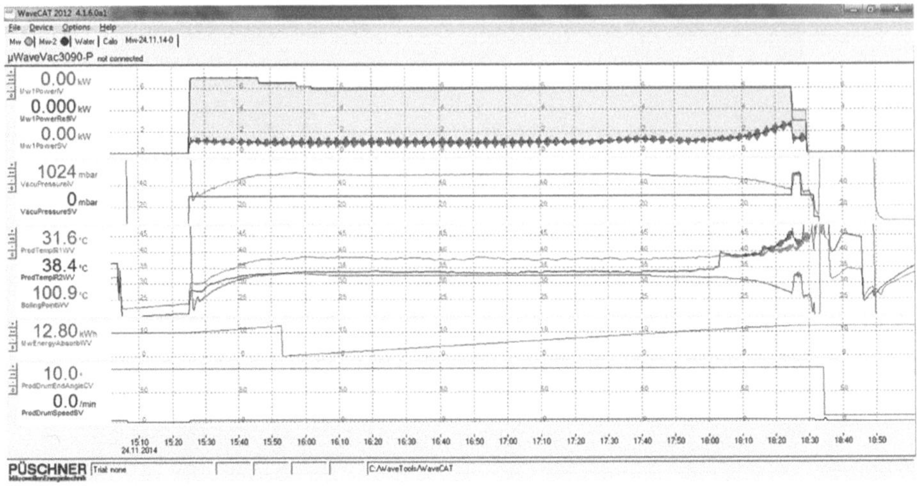

Figure 16.13 Monitoring of drying parameters (microwave power, pressure, temperatures, drum movement) utilized for microwave vacuum drying of carrots.

Figure 16.14 300 L drum with initial load of 25 kg of carrots.

Figure 16.15 Dried carrot chips.

Table 16.2 Main plant and drying specifications as well as energy consumptions (Fig. 16.15)

Batch volume in product drum	300	L
Batch initial weight of product	25	kg
Removed water (87% down to <5% MC)	22.2	kg
Drying time for carrots	3	h
Capacity dried product based on 1 h drying time	0.93	kg/h
Max MW power	7	kW/2450 MHz
Average MW power for drying	5.5	kW
Electrical supply power of machine	10	kVA
Cooling water at 20–25°C	30	l/min
Cooling water power by chiller	4	kW

Figure 16.16 Frozen ice cube with boiling water by microwaves.

i.e., direct phase change from solid to gaseous state. Therefore, the quality of microwave freeze-dried products is comparable to conventionally freeze-dried produces. The main advantage is that the difficulties of heat transmission found in freeze-drying are reduced to a great extent by volumetric microwave heating. As water in the liquid phase presents significantly high dielectric losses, dielectric losses under microwave freeze conditions are very low. Therefore, the microwave energy is mainly absorbed by the organic molecules. To understand this important difference between liquid and frozen stage, Fig. 16.16 shows an ice cube with a hole. The hole was filled up with liquid water. Then, the ice block with water was placed into a microwave chamber oven. After a short while, the water started to boil and the ice remained cold.

For that reason, sublimation occurs within the food material without being affected by the dry zone. Water in the frozen condition below −10°C to −15°C presents significantly smaller dielectric losses than the product itself. Therefore, also the penetration depth is far greater compared to the liquid state and, the electromagnetic energy can still be converted into thermal energy due to the dielectric losses of the organic product. As a result, microwave technology offers a great opportunity to increase the drying rate in freeze-drying. Experiments and numerical predictions show that the drying rate was significantly increased and the drying cost reduced thanks to microwave heating (Duan, Zhang, Mujumdar, & Wang, 2010; Wu, Tao, Chen, & Deng, 2004).

By application of microwave freeze-drying, the freeze-drying process can be significantly shortened by virtue of the following unique advantages:

- Microwave volumetric heating deep inside the product bed as water is almost transparent due to its frozen condition; the penetration depth is mainly driven by the dielectric losses of the organic product itself;
- Rapid energy dissipation throughout the material;
- More efficient drying in the falling rate period and total energy savings (Duan et al. 2010; Feng & Tang, 1998; Torringa, Esveld, & Scheewe, 2001).

In conventional freeze-drying, the heat gradient is working from outside to inside (i.e., the closer to the surface, the higher is the temperature). The limitation in transferring heat gives rise to a long drying time, especially when water is sublimated in fruit or vegetables and when the cellular product increasingly becomes a thermally insulating material. The greatest advantage in applying microwaves is that of the energy being generated throughout the product in the penetration depth.

As water is frozen, the penetration depth of microwaves increases and microwaves can penetrate deep into the vegetable bulk and whole fruit. Total volumetric heating can be ensured under such condition, and sublimation is not only occurring in thin layers as it is the case in conventional freeze-drying. Instead, sublimation occurs deep inside the volume of vegetables and fruit. This has a tremendous impact on the increase of the drying rate. Due to this fact, the drying phase of most of the microwave freeze-drying processes is within the range of 2–6 h based on an initial moisture content of 70–98% to the range of around 5% MC on wet basis, compared to 10–40 h in conventional freeze-drying.

16.2.3 Challenges

The risk of plasma discharge in low pressure environments has already been mentioned before. Microwave freeze-drying application should work below 1 mbar to ensure a sufficient safety margin to the minimum of the breakdown voltage shown in Fig. 16.17. The requirement for an advantageous microwave design is essential for transferring reasonable microwave energy densities without generating plasma. Plasma discharges can happen when the electric field intensity in the vacuum chamber is above a threshold value of the electric breakdown value (see Fig. 16.17). Ionization of residual gas molecules present in the vacuum chamber leads to plasma generation (Sanborn & MacDonald, 1950), which may cause excessive temperatures and therefore burns on the product surface. The occurrence of this

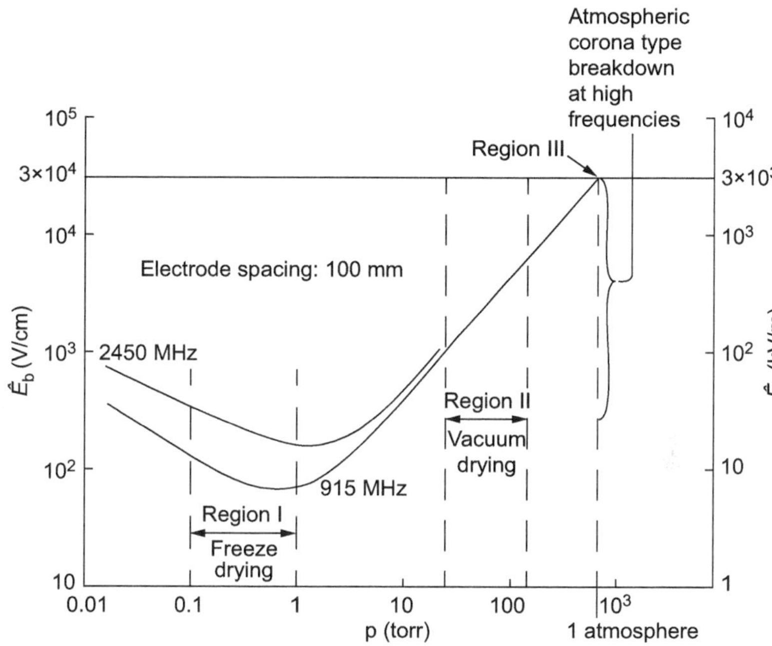

Figure 16.17 Electric field strength versus pressure profile showing critical values leading to plasma formation (Metaxas & Meredith, 1988).

phenomenon is responsible for immense energy losses and excessive heating of the dry zone of the material by which the final product will suffer severe damages (Arsem & Ma, 1985; Gould & Kenyon, 1971). Furthermore, it is important to design a chamber presenting a minimum local electromagnetic field strength (Duan et al., 2010) to avoid plasma discharges. Hot white plasma may occur mainly at coupling due to high electromagnetic energy density. Cold blue/purple plasma discharge can also occur in the product bed. Especially grape and apple skin can create plasma conditions as these skins have a certain electric conductivity. Dust particles can ignite the plasma, too.

However, this limitation is partly overcome in more recent microwave freeze-dryer designs. Low ripple microwave power supplies with excellent matching to the product load into the microwave vacuum system are of essential importance. By shaping the coupling and the load situation, the microwave vacuum system can be optimized. As the breakdown voltage of 2450 MHz has its minimum at around 1 mbar, vacuum freeze-drying can be reduced further to 0.2–0.5 mbar. This is more cost-intensive from the vacuum system side and even creates a considerably higher steam volume. On the other hand, however, the process re-attains more stability as the breakdown voltage is about 2–3 factors higher than at its minimum.

The typical duration of conventional freeze-drying is in the range of 10 to 40 h, depending on the type and quantity of vegetables or fruit to be dried. In the final drying stages, when the residual moisture content cannot follow the sublimation temperature against the heat transfer by heated shelves, the product temperature will typically rise above 0°C. As a result, product temperatures increase further and may negatively affect the product quality at around 50 to 60°C. When the end drying period is long, the product is exposed to these elevated temperatures for hours.

Moreover, thermal conductivity of the product becomes very poor and heat transfer into the product core will take a long time. This gets more pronounced with increasing product size which, in particular, presents a problem with whole fruit, such as strawberries, raspberries, and blueberries, among other fruit and vegetables. Due to excessive heating in this drying stage, volatiles and other product components, such as flavors, are affected and evaporated or thermally degraded. Therefore, product textures remain excellent with good rehydration properties, but will, however, present significant losses of vitamins and flavors. As microwave freeze-drying is in the range of a few hours, thermal exposure to temperatures above 20°C is far shorter and volatiles are retained and better preserved (Rahm, Ambros, & Kulozik, 2015).

As shown in Fig. 16.18, a microwave process will start after the product has been exposed to vacuum levels at around 1 mbar. The product will follow the sublimation temperature depending upon the vacuum level (1 mbar $T_{subl} = -20°C$). Then, the microwave power is switched on and the product is exposed to microwave energy. As a result, temperatures will increase from −20 to −15°C. Sublimation is taking place constantly until the moisture content reaches approximately 15% MC (wet base). The mechanical stability of the product is maintained by the moisture level. After that, product temperatures rise toward 0°C. At around 10% MC (wet base), the 0°C level is passed. The mechanical stability of the product will drop. The minimum mechanical stability is determined by the height of the product bed inside the drum. The product is not allowed to be subjected to mechanical damages. For example, strawberries are extremely sensitive to excessive mechanical stress. As a result, strawberries will change to a cone shape and thereby lose most of their outer texture. In this critical phase, microwave power can be reduced until reaching the target moisture content of 5% wet base. In this stage, product temperatures can reach approx 40–60°C for a short time.

16.2.3.1 Example 3: raspberries

Raspberries are an important commercial fruit crop, widely grown in all temperate regions worldwide. In this example, 60 kg of raspberries were microwave freeze-dried in an 800 L drum in about 6.5 h.

60 kg of raspberries are loaded into an open 800 L plastic drum and are placed into the 12 kW/2450 MHz microwave freeze-dryer *μWaveVac12120fd* (Pueschner GmbH & Co KG, Bremen, Germany, installed in Chile) (Table 16.3). Vacuum is applied and when reaching approximately 0.2 mbar, the microwave power is started. The open plastic drum is moved forward and backward, moving and mixing the product inside the drum. After some cycles, the drum is stopped

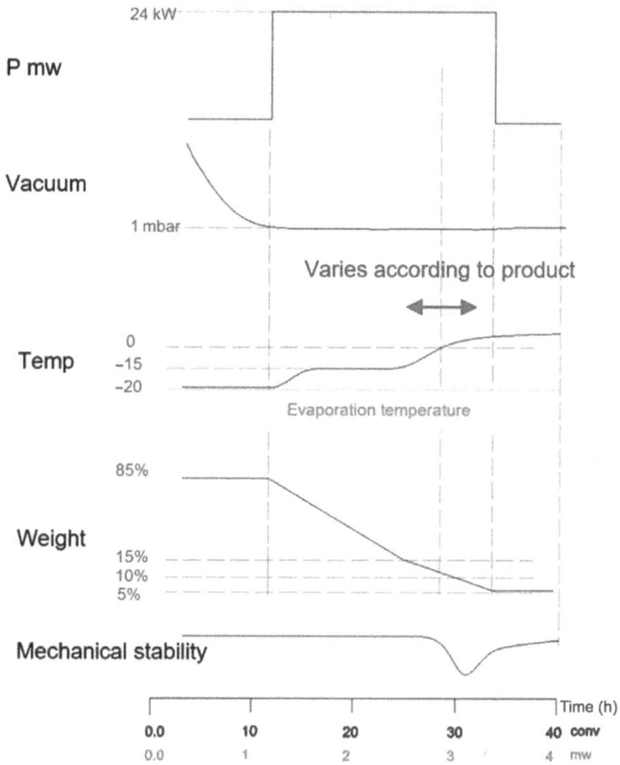

Figure 16.18 Typical drying profile in a conventional freeze-drying process (blue timescale (top timescale in print versions)) and assisted by microwaves (red time line (bottom time line in print versions)).

Table 16.3 Main plant and drying specifications as well as energy consumption (Fig. 16.22)

Batch volume in product drum	800	L
Batch initial weight of product	60	kg
Removed water (77% down to <6% MC)	50	kg
Drying time for raspberries	6.5	h
Capacity dried product based on 1 h drying time	1.5	kg/h
Max microwave power	12	kW/2450 MHz
Average MW power for drying	8–10	kW
Electrical energy microwave & vacuum system for 6.5 h	83	kWh
Electrical Energy of chiller (double stage compressor type at −65°C) for 6.5 h	58	kWh
Cooling water at 20–25°C	30	L/min
Cooling water power by chiller	8	kW

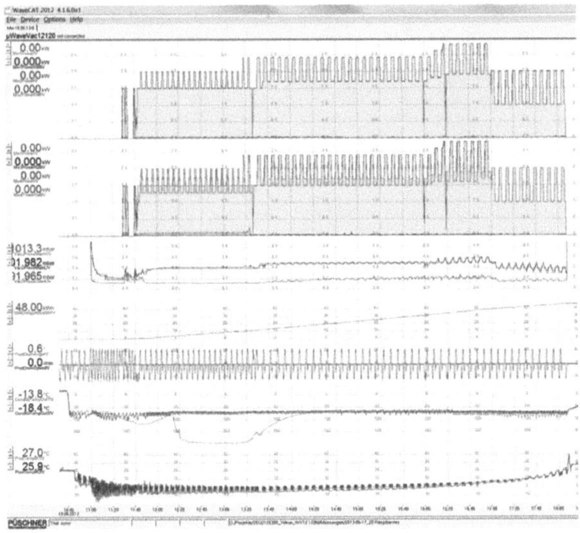

Figure 16.19 Monitoring of the drying process parameters for 60 kg initial load of raspberries.

and the microwave power levels are increased from 4 × 2.0 to 2.5 kW (upper two graphs in Fig. 16.19). When the product is moved, the microwave power levels are decreased to 4 × 1.4 to 1.6 kW. Due to the steam volume in the vacuum vessel, the pressure level is increasing to 0.4 to 0.5 mbar (middle graph in Fig. 16.19). Due to the huge steam volume, the pressure at the condenser still remains lower at approximately 0.2 mbar. As product temperatures measured by the infrared cameras are moving through the 0°C line, the microwave power is increased to 4 × 2.7 kW fast mechanical stabilization of the berries (see Fig. 16.20 lower graph). When product temperatures have reached a value of 40–45°C, the process will be stopped. The capacity of the *μWaveVac12120fd* is approximately 7.5 t dried product based on 5.000 h operation hours per year (Fig. 16.21).

A comparison between conventional and microwave-assisted freeze-drying is difficult and can only be performed in the framework of specific commercial projects installed in the market. In installations operating since 2010, microwave freeze-dryers have been installed in the range below 2 kg/h dried products to which predominantly some 10–12 kW of microwave energy is applied. Some general data from these projects indicate that drying periods of MW systems are 3–5 times faster whilst the running cost is approximately 50%. As electric energy prices vary considerably worldwide, these figures may be different from one country to the other. The investment cost is in the range of 20–50% less in comparison to conventional freeze-dryers. But the main argument brought forward in favor of microwave freeze-dryers is the better quality output due to a reduced drying time.

Brix	13%
Initial weight	60 kg (77% MC)
End weight / MC / AW	10 kg / 5.6% / 0.37
Drying time	6.5 h
Max temp	45°C
Applied Mw energy	48 kWh

Figure 16.20 Main drying parameters (left), single raspberries (middle), and 60 kg initial load of raspberries in the 800 L drum (right).

Figure 16.21 12 kW/2450 MHz microwave drum freeze dryer.

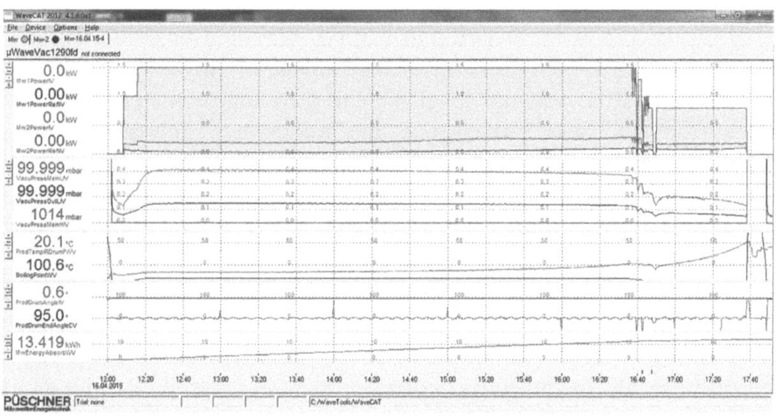

Figure 16.22 Monitoring of the drying process parameters for 20 kg initial load of blueberries.

Blueberries d6-8mm	9 brix
Initial weight	20 kg
End weight	3.3 kg
Drying time	5.5 h
Max temperature	50°C
Applied Mw energy	13.5 kWh

Figure 16.23 Main drying parameters (left), single blueberries cut after drying (middle), and initial load of 20 kg of blueberries in the 300 L drum (right).

16.2.3.2 Example 4: blueberries

Blueberries are producing special interest due to their high amount of antioxidants, in particular, anthocyanins. This antioxidant is known to contribute to scavenging of free radicals produced by metabolic processes. Health effects to human metabolism have been explored and studied in comparatively recent research over the past few years only (Jian & Monica Giusti, 2010). Therefore, blueberries are regarded as health-promoting fruit. To retain and preserve aroma, vitamins and especially anthocyanins, the final drying temperature is critical and has to be kept below 50°C. Periods of thermal exposure to temperatures between 30–50°C have to be kept as short as possible.

Blueberries have a hard skin which will not allow evaporated water to pass. The surface has to be pretreated to become porous in order to avoid excessive interior pressure levels. As ripe high bush blueberries generally have a glucose concentration in the range of 10°Bx to 15°Bx, the vacuum level has to be kept even lower to compensate for the increase in sublimating point for these highly concentrated glucose solutions.

20 kg of blueberries were loaded into a 300 L open plastic drum (see Fig. 16.23 right) and exposed to a vacuum level of 0.15 mbar. Microwave power was started with 2 × 1.5 kW and the vacuum level increased to approximately 0.4 mbar. After about 4.5 h, the 0°C was passed and the microwave power was reduced to 2 × 0.8 kW until 50°C product temperature was reached and measured by the infrared cameras. After that the process was stopped and the final moisture content (wet basis) was approximately <5% MC (see Fig. 16.23). The total energy absorbed and used in the process was 13.5 kWh.

Drying trials were similarly carried out at the TUM (Technical University of Munich) Institute of Food Process Engineering and Dairy Technology at Weihenstephan on a small 1 kW/2450 MHz lab microwave freeze-dryer, a conventional freeze-dryer, and a convectively dried product in order to measure anthocyanin retention. By means of the RP-HPLC, separation and simultaneous determination of anthocyanins is possible (Berente, 2004). To determine the content of individual anthocyanins, cyanidin-3-glucoside was used as a reference and the anthocyanins

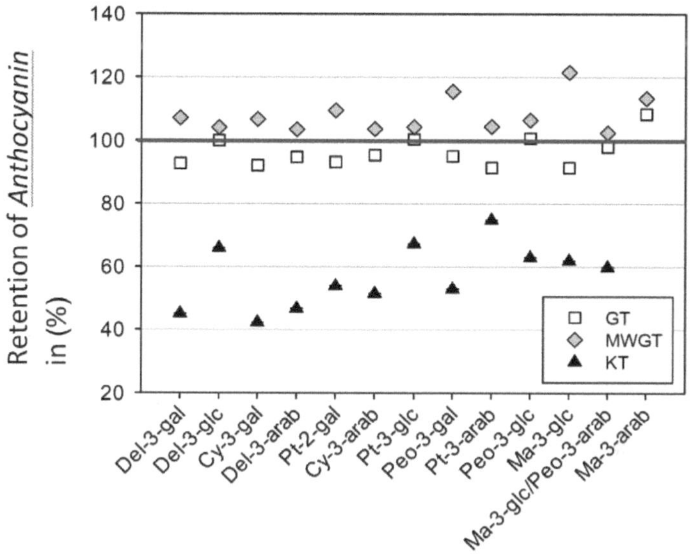

Figure 16.24 Comparison of anthocyanin retention in convectively dried blueberries (KT), conventionally freeze-dried (GT), and microwave freeze-dried (MWGT) as well as fresh blueberries (red reference line (reference line in bold in print versions)) (Rahm et al., 2015).

using RP chromatography determined (see Fig. 16.24) (Rahm et al., 2015). Retention values of more than 100% may be due to an enhanced extractability.

16.2.4 Combined freeze-drying and liquid stage applications

In general, toward the end of the freeze-drying process, the temperature will pass the 0°C line and increase further, where the remaining water will change from its solid to the liquid state. The longer the freeze-drying process can keep the product temperature at values between −5 and −10°C, the better the textures and overall quality of the final product. In addition, temperatures of above 50°C should be generally avoided. There are certain products, e.g., potato chips or pumpkin whose temperature profile during the process has to pass the starch gelatinization point to develop the right taste and flavor profile for a pleasant snack product. For these products, it would be appropriate that the product in the freeze-dryer be permitted to be heated up at a certain point in order to initiate melting of the remaining water by increasing the vacuum to approximately 20 mbar. This will enable transmission and conversion of an increased amount of microwave energy. Furthermore, when reaching a certain moisture content at which the product is still soft, the product can also be puffed, i.e., expanded by means of water evaporation and increased by the internal pressure. Also at higher temperatures Maillard reactions can develop additional tastes (e.g., apple slices). Therefore, combined processes will allow development of new product properties.

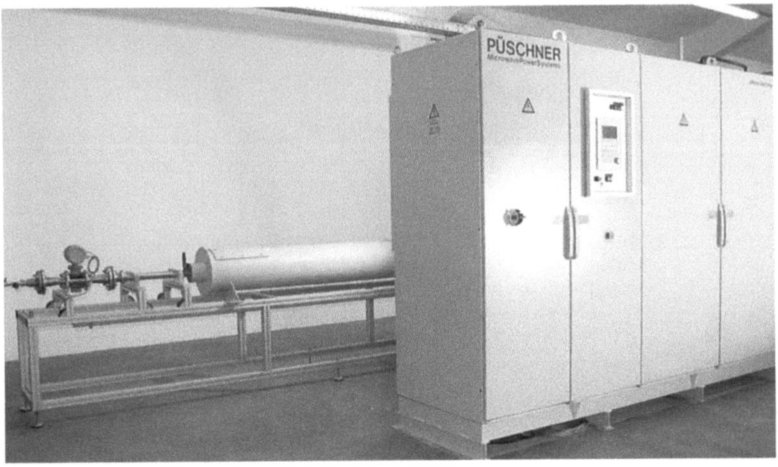

Figure 16.25 100 kW/915 MHz microwave-flow heater for pasteurization of fruit pulps (*μWaveFlow10040*, Pueschner GmbH & Co KG, Bremen, Germany, installed in Turkey).

16.2.5 Microwave flow heater for pasteurization and sterilization

Microwave flow heaters are used in various industrial applications:

- Pasteurization of liquid and viscous products in the food industry;
- Sterilization of liquid products on the food/pharmaceutical sector;
- Drying of casings in sausage production;
- Heating of minced meat mixtures after the extruder for flashing;
- Preheating of viscous products before spray dryers or vacuum dryers;

Pasteurization of fruit pulps, pastes, and other viscous mixtures processed in the food industry requires throughput rates of several tons per hour with uniform product heating to temperatures of 70–80°C. The microwave flow heater in Fig. 16.25 can handle a throughput of 1–2 tons per hour for heating/pasteurization/sterilization. The design is capable to operate at pressure levels of maximum 30 bar at a maximum temperature of 300°C.

The microwave applicator principle is described in Fig. 16.3. The applicator is based on a ground mode resonator which minimizes issues associated with the formation of hot spots and temperature nonuniformity in general. Due to radially symmetric coupling of the microwave power into the applicator, a defined field distribution is established. In the central axis of the cylindrical applicator a straight product tube of microwave transparent material is installed through which the product is pumped. Since microwave energy is absorbed immediately throughout the product volume, heating periods prove to be extremely short.

This technology is used at Boehringer Ingelheim for virus inactivation, where protein solutions are heat treated. The heating time from 20°C up to 140°C is achieved within a range of 20–100 ms with a throughput rate of 50–75 L/h. After

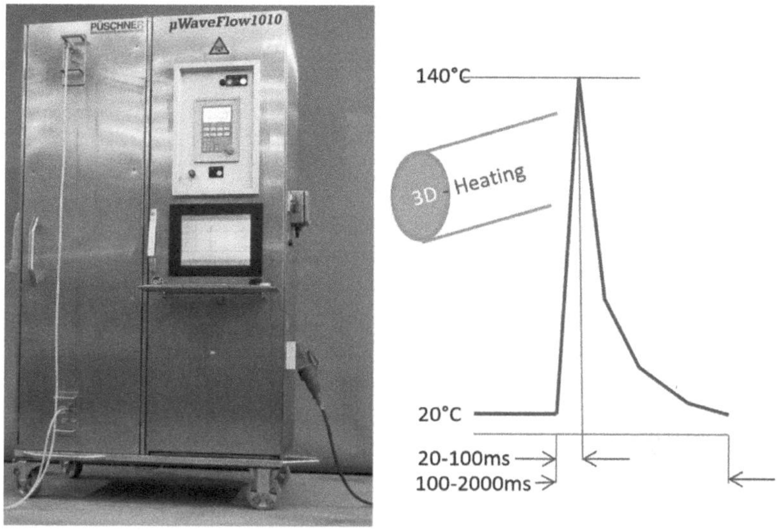

Figure 16.26 Mobile 10 kW/2450 MHz flow heater for virus inactivation (*μWaveFlow1010*, Pueschner GmbH & Co KG, Germany, installed in Germany).

Table 16.4 Pasteurization of pork blood using a 30 kW/915 MHz *μWaveFlow3040*

Temp in (°C)	Temp out Mw (°C)	Mw (kW)	Flow (L/h)	Residence time (s)	Total germrate	*Staphylocuccus aureus* bacterium	*Escherichia coli* bactierium
Initial					650.000	140	360
20	68.7	25	450	5.9	3.800	5	<1
26	81.2	28	370	7.1	<10	<10	<10

ultrafast heating, the protein product flow is cooled down by means of a high-performance heat exchanger. The sterilization step is selective for virus inactivation in sensitive protein solutions, which are not stressed in quality because of the ultrafast heating up and cooling time (Feuser et al., 2006; Fig. 16.26).

Recently, ultrahigh-temperature treatment of milk concentrates with high dry matter has gained significant interest in the dairy industry as conventional heat exchangers are limited in their performance in respect of heat transfer and may also cause fouling on heated surfaces. The objective in this process is to inactivate bacterial spores in a highly efficient way without causing thermal degradation to the product components. These systems can, for instance, be used prior to a spray drying process. Other applications of microwave inactivation, e.g., pasteurization of blood, which is very sensitive due to its low coagulation temperature has also been successfully proven using a 30 kW/915 MHz *μWaveFlow3040* (see Table 16.4).

16.2.6 Summary

The concept of utilizing microwave technology in industrial food applications is based on better qualities, higher energy efficiency, smaller equipment footprints, as well as lower environmental impact through reduced energy and water consumption. An appropriate process control is the key to successful implementation, robust processes, and the required process outcomes. As engineering and process understanding were developed further for microwave processes, the design of microwave plants for advanced applications was enhanced throughout the past 10 years. Also PLC systems have become by far more efficient and the availability of high quality microwave systems enables advanced applications like microwave freeze-drying and others. On top of this, the microwave plants have to observe food safety standards like GMP and FDA. These projects developed from lab equipment via pilot schemes to production. Only food companies with Research and Development understanding and patience for longer lasting development periods will at the end benefit from these results.

With the fast microwave vacuum and microwave freeze-drying technology, more volatiles and nutrients can be retained and, other quality attributes, such as color and texture, can be maintained in a better way in comparison to conventional drying or freeze-drying. Especially the application of microwave freeze-drying in industrial processing is of increasing interest, particularly because of the reduction of running and capital costs as well as drying time. Such unique volumetric heating provides benefits which are not achievable by other, more conventional methods.

16.2.7 Prospects

In the 21st century, consumers will become more aware of the importance of highly nutritious, natural products. For these premium products, there is a niche market where consumers are willing to pay higher prices for high value food. Microwave technology can play an important role in this field.

Also, microwave-assisted preservation (vacuum and freeze-drying, pasteurization) can be very attractive for decentralized processing, e.g., for farmers who want to avoid paying fees for toll processing in centralized facilities, or merely sell their products at low prices to the major food manufacturers (i.e., increase their revenue), and to establish their own product brands in the market place.

Following successful proof of principle in the lab, global food manufacturers especially are still somewhat reluctant to introduce this technology on a large scale to their production as only a few companies specializing in microwave installations can offer this technology. This technology is available at commercial scale for food manufacturers willing to bear the challenge of new food processing methods. Since small and medium size food manufacturers are more innovative, have shorter decision-making chains, and are more willing to take risks involved in innovative technologies, they will most likely be those daring to establish novel high value products based on microwave technology in niche markets.

References

Arsem, H. B., & Ma, Y. H. (1985). Aerosol formation in the microwave freeze-drying of beef. *Biotechnology Progress*, *1*, 104–110.

Berente, B. (2004). HPLC-Analyse von Anthocyanen im Rotwein und Klassifizierung deutscher Rotweine mittels multivariater statistischer Methoden. Dissertation. Chemisch-Geowissenschaftliche Fakultät. Friedrich-Schiller-Universität Jena.

Bettega, R., Rosa, J. G., Correa, R. G., & Freire, J. T. (2013). Comparison of carrot (Daucus carota) drying in microwave and in vacuum microwave. *Brazilian Journal of Chemical Engineering*, *31*(02), 403–412.

Decareau, R. V. (1992). *State of the microwave-power industry*, 65–66.

Dogan Halkman, H. B., Yücel, P. K., & Halkman, A. K. (2014) Non-thermal processing: Microwave. *Encyclopedia of food microbiology* (2nd ed., pp. 962–965).

Duan, X., Zhang, M., Li, X., & Mujumdar, A. S. (2008). Ultrasonically enhanced osmotic pretreatment of sea cucumber prior to microwave freeze drying. *Drying Technology*, *26*(4), 420–426.

Duan, X., Zhang, M., Mujumdar, A. S., & Wang, R. (2010). Trends in microwave-assisted freeze drying of foods. *Drying Technology*, *28*, 444–453.

Feng, H., & Tang, J. (1998). Microwave finish drying of diced apples in spouted bed. *Journal of Food Science*, *63*(4), 679–683.

Feuser, J., Heinz, C., Woog, S., Ambrosius, D., Püschner, P., & Graute, R. (2006) *Recovery of biological products XII" (RXII) Litchfield, Arizona, USA, (April 2–7, 2006).*

Giese, J. (1992). Advances in microwave food processing. *Food Technology*, *49*(9), 118–123.

Gould, J. W., & Kenyon, E. M. (1971). Gas discharge and electric field strength in microwave freeze-drying. *Journal of Microwave Power*, *6*(2), 151–167.

Hasting, A. P. M. (2008). Designing for cleanability. In A. Y. Tamime (Ed.), *Cleaning-in-place: Dairy, food and beverage operations* (pp. 81–106). Oxford: Blackwell Publishing.

Holah, J. (2014). *Food safety management, chapter 24. Hygiene in food processing and manufactoring*. ISBN 978-0-12-381504-0.

Jian, He, & Monica Giusti, M. (2010). Anthocyanins: Natural colorants with health-promoting properties. *Annual review of food science and technology, vol. 1*, 163–187, Volume publication date April.

Jones, P. (2007). *ISO 22000 Series*. Johannesburg: ISO Workshop Conformity Assessment.

Lin, T. M., Durance, T. D., & Scaman, C. H. (1998). Microwave characterization of vacuum, air and freeze dried carrot slices. *Food Research International*, *31*(2), 111.

Metaxas, A. C., & Meredith, R. (1988). *Industrial microwave heating*. London, UK: Peter Peregrinus Ltd, ISBN:0-906048893.

Motarjemi, Y., & Lelieveld, H. (2014). *Food safety management – A practical guide for the food industry*. ISBN:978-0-12-381504-0.

Pueschner, P. (2005). Improving microwave process control. In: Schubert, H. and Regier, M. *The microwave processing of foods*. ISBN 978-1-85573-964-2.

Rahm, K., Ambros, S., Kulozik, U. (2015). Can microwave-freeze drying substitute the conventional freeze drying process for the preservation of high quality fruit products? In: *International conference EFFoST international conference, 10–12 November 2015, Athens, Greece.*

Suvarnakuta, P., Devahastin, S., & Mujumdar, A. S. (2005). Drying Kinetics and B-carotene degradation in carrot undergoing different drying processes. *Journal of Food Science*, *70*(8).

Riccetti, S. (2013). *Equipment reliability through maintenance engineering*. Taylor&Francis Group.

Sanborn, C. B., & MacDonald, A. D. (1950). *Limits for the diffusion theory of high-frequency gas discharge breakdown*. Technical Report No 132, Massachusetts Institute of Technology, Research Laboratory of Electronics.

Therdthai, N., & Zhou, W. (2009). Characterization of microwavevacuum drying and hot air drying of mint leaves (*Menthacordifolia* Opis ex Fresen). *Journal of Food Engineering*, *91*, 482–489.

Torringa, E., Esveld, E., & Scheewe, I. (2001). Osmotic dehydration as a pre-treatment before combined microwave-hot-air drying of mushrooms. *Journal of Food Engineering*, *49*, 185–191.

Wu, H. W., Tao, Z., Chen, G. H., & Deng, H. W. (2004). Conjugate heat and mass transfer process within porous media with dielectric cores in microwave freeze drying. *Chemical Engineering and Science*, *59*(14), 2921–2928.

Further reading

Ahmed, J., Ramaswamy, H. S., & Raghavan, G. S. V. (2008). Dielectric properties of soybean protein isolate dispersions as a function of concentration, temperature and pH. *LWT-Food Science Technology*, *41*, 71–78, 77.

Ang, T. K., Ford, J. D., & Pei, D. C. T. (1978). A research note: Optimal modes of operation for microwave freeze drying of food. *Journal of Food Science*, *43*, 517–526.

Annual Report of Ministry of Food Processing Industries (MOFPI), Government of India, 2011.

Burke, R. F., & Decareau, R. V. (1964). *Advances in food research*, *13*, 88.

Copson, D. A. (1962). *Microwave heating*. Westport, CN: AVI Publishing.

Decareau, R. V. (1985). *Pasteurization and sterilization. Microwaves in the food processing industry*. New York: Academic Press, Orlando, MA.

Erle, U., & Schubert, H. (2001). Combined osmotic and microwave–vacuum dehydration of apples and strawberries. *Journal of Food Engineering*, *49*, 193–199.

FAO (Food and Agriculture Organization), WHO (World Health Organization). (2006). *Food safety risk analysis: a guide for national food safety authorities*. Food and Nutrition Paper 87. <http://www.fao.org/docrep/012/a0822e/a0822e00.pdf> 20.10.14.

FDA (U.S. Food and Drug Administration), FSIS (Food Safety Inspection Service, USA). 2003. *Interpretative summary: Quantitative assessment of the relative risk to public health from listeria monocytogenes among selected categories of ready to eat meats*. <http://www.fda.gov/downloads/Food/FoodScienceResearch/UCM197329.pdf> 20.10.14.

Figiel, A. (2009). Drying kinectics and quality of vacuum–microwave dehydrated garlic cloves and slices. *Journal of Food Engineering*, *94*, 98.

Genin, N., & Rene, F. (1995). Analysis of the role of the glass transition in the methods of food preservation. *Journal of Food Engineering*, *26*, 391–408.

Hippel, V. (1954). London; London; Chapman & Hall Ltd *Dielectric materials and applications*. New York: John Wiley & Sons Inc.

Hu, X. P., & Mallikarjunan, P. (2005). Thermal and dielectric properties of shucked oysters. *LWT-Food Science Technology*, *38*, 489–494.

Hu, Q. G., Zhang, M., & Mujumdar, A. S. (2006). Drying of edamames by hot air and vacuum microwave combination. *Journal of Food Engineering*, *77*(4), 977–982.

Huang, L. L., Zhang, M., Mujumdar, A. S., Sun, D. F., Tan, G. W., & Shameel, T. (2009). Studies on decreasing energy consumption for a freeze-drying process of apple slices. *Drying Technology*, *27*(9), 938–946.

Kardum, J. P., Sander, A., & Skansi, D. (2001). Comparison of convective, vacuum, and microwave drying chlorpropamide. *Drying Technology*, *19*(1), 137–143.

Krokida, M. K., & Maroulis, Z. B. (1997). Effect of drying method on shrinkage and porosity. *Drying Technology*, *15*(10), 2441.

Krokida, M. K., & Maroulis, Z. B. (1999). Effect of microwave drying on some properties of dehydrated quality products. *Drying Technology*, *17*(3), 449.

Li, Z., Raghavan, G. S. V., & Wang, N. (2010). Carrot volatiles monitoring and control in microwave drying. *LWT-Food Science and Technology*, *43*, 291.

Lin, T. M., Timothy, D. D., & Scaman, C. H. (1999). Physical and sensory properties of vacuum microwave dehydrated shrimp. *Journal of Aquatic Food Product Technology*, *8*(4), 41–53.

Litvin, S., Mannheim, C. H., & Miltz, J. (1998). Dehydration of carrots by a combination of freeze drying, microwave heating and air or vacuum drying. *Journal of Food Engineering*, *36*(1), 103–111.

Mageean, P., Kristott, J. U., & Jones, S. A. (1991). Physical Properties of sugars and their solutions. *Scientific and Technical Surveys*, *172*.

Mujumdar, A. S. (2006). *Guide to industrial drying* (3rd ed.). Mumbai: Colour Publications Pvt. Ltd.

Mujumdar, A. S., & Menon, A. (1995). Drying of solids: Principles, classification, and selection of dryers. In A. S. Mujumdar (Ed.), *Handbook of industrial drying* (2nd ed., p. 1). New York: Marcel Dekker.

Ohlsson, T., & Bengtsson, N. (2001). Microwave technology and foods. In S. Talor (Ed.), *In advances in food and nutrition research* (pp. 65–140). New York: Academic Press.

Overbosh, P., & Blanchard, S. (2014). *Principles and systems for quality and food safety management*. Matro AG. Düsseldorf, Germany: Food Safety Management.

Ratti, C. (2001). Hot air and freeze-drying of high-value foods: A review. *Journal of Food Engineering*, *49*(4), 311–319.

Regier, M., & Schubert, H. (2005). Introducing microwave processing of food: principles and technologies. In H. Schubert, & M. Regier (Eds.), *Microwave processing of foods* (pp. 1–9). Woodland Publishing Ltd.

Schiffmann, R. F. (1992). Microwave processing in the U.S. food industry. *Food Technology*, *56*, 50–52.

Schiffmann, R. F. (2006). Microwave and dielectric drying. In A. S. Mujumdar (Ed.), *Guide to industrial drying* (3rd ed., pp. 286–292). Mumbai: Colour Publications Pvt. Ltd.

Schleged, W. (1992). Commercial pasteurization and sterilization of food products using microwave technology. *Food Technology*, *12*, 62–63.

Strumillo, C., & Adamiec, J. (1996). Energy and quality aspects of food drying. *Drying Technology*, *14*(2), 423–448.

Sunderland, J. E. (1980). Microwave freeze drying. *Journal of Food Process Engineering*, *4*, 195–212.

Sunjka, P. S., Rennie, T. J., Beaudry, C., & Raghavan, G. S. V. (2004). Microwave—convective and microwave—vacuum drying of cranberries: A comparative study. *Drying Technology, 22*(5), 1217—1231.

Tang, J. (2005). Dielectric properties of foods. In H. Schubert, & M. Regier (Eds.), *Microwave processing of foods* (pp. 22—24). Woodland Publishing Ltd.

Tao, Z., Wu, H., Chen, G., & Deng, H. (2005). Numerical simulation of conjugate heat and mass transfer process within cylindrical porous media with cylindrical dielectric cores in microwave freeze-drying. *International Journal of Heat and Mass Transfer, 48*, 561—572.

Vega-Mercado, H., Gongora-Nieto, M. M., & Barbosa-Canovas, G. V. (2001). Advances in dehydration of foods. *Journal of Food Engineering, 49*, 271—289.

Viet, L. (1980). *Microwave freeze drying method and apparatus*. US Patent No 4,204,336.

Xu, Y. Y., Zhang, M., & Tu, D. Y. (2005). A two-stage convective air and vacuum freeze-drying technique for bamboo shoots. *International Journal of Food Science and Technology, 40*(6), 589—595.

Zhang, M., Jiang, H., & Lim, R. (2010). Recent developments in microwave-assisted drying of vegetables, fruits, and aquatic products — Drying kinetics and quality considerations. *Drying Technology, 28*, 1307—1316.

Zhang, H., & Datta, A. K. (2003). Microwave power absorption in single-and multiple-item foods. *Transactions of the Institution of Chemical Engineers, 81*, 257—266.

Zhang, M., Tang, J. M., & Mujumdar, A. S. (2006). Trends in microwave-related drying of fruits and vegetables. *Trends in Food Science & Technology, 17*(10), 524—534.

International microwave standards

IEC 60519-1. (2011). *Safety in electroheat installations — Part 1: General requirements.*

IEC 60519-6. (2011). *Electroheat installations — Part 6: Specifications for safety in industrial microwave heating equipment.*

IEC 61307. (2011). *Industrial microwave heating installations — Test methods for determination of power output.*

Improving the heating uniformity in microwave processing

17

B. Wäppling Raaholt and S. Isaksson
SP Food and Bioscience, RISE Research Institutes of Sweden, Gothenburg, Sweden

17.1 Introduction

In this chapter, an overview is given, focusing on methods for making microwave heating distribution as uniform as possible, with examples given for some selected applications. Since the space is limited, this overview cannot be complete; instead, some important theoretical remarks will be given where methods for improving the uniformity in microwave heat distribution are exemplified or referred to. This chapter provides a starting point; the interested reader can find more information on the details in the references.

The problem of nonuniform heating is described, exemplified, and discussed. References are also given to previous works. Moreover, some selected methods and tools are mentioned which are helpful when designing food products intended for microwave heating. The techniques referred to in this chapter are possible to use for improving microwave heating uniformity in many applications, including but not limited to, e.g., different applications of industrial microwave processing, reheating and cooking in household microwave ovens, among others. Finally, a section on future trends is given, followed by sources of further information and advice.

17.2 Heat distribution and uniformity in microwave processing

Microwaves are defined as electromagnetic waves with frequencies in the range between 300 MHz and 300 GHz, which correspond to wavelengths (in vacuum or air) between 1 m and 1 mm. For food processing applications, however, the frequencies are in practice limited to ISM (Industrial, Scientific, and Medical) bands, of which 2450 ± 50 MHz is the most common one and applies almost worldwide. There are however regional variations; in, e.g., the United States, another recognized ISM

The Microwave Processing of Foods. DOI: http://dx.doi.org/10.1016/B978-0-08-100528-6.00017-6

frequency is 915 ± 15 MHz, whereas in European countries[1] the latter frequency is not defined as an ISM band.

Over the years, microwave heating of foods has sometimes been associated with uneven heating, due to the "hot and cold spots" which may be present in the food product after microwave processing. The temperature distribution in microwave-heated foods is determined by the thermophysical properties of the food product as well as by the distribution of the dissipated microwave power. The microwave heating distribution in the food is, in turn, determined by the electric field inside the microwave cavity or applicator, by the dielectric properties of the food, and by the microwave frequency. The electromagnetic field pattern can be controlled by the oven design (i.e., size and shape of the cavity or applicator, including the design of the waveguide system), but heating uniformity is also influenced by composition, position, and geometry of food, as well as package during heating. This will be dealt with further in subsection 17.2.1 and for more detail see Wäppling Raaholt (2015).

More on the principles of microwave heating of foods can be found in the literature and in other chapters of this book. The subject has also been described in detail by several authors, e.g., Bengtsson and Risman (1971), Metaxas and Meredith (1983), Buffler (1993), Ohlsson and Bengtsson (2001), Wäppling Raaholt, Holtz, and Isaksson (2013), and Wäppling Raaholt (2015).

17.2.1 Factors influencing microwave heating uniformity

Several interacting factors influence the heating result in a microwave oven. Factors which affect the heating uniformity are, among others, the design of the waveguide or applicator system and of the cavity, but also the geometry and size of the food material, as well as its dielectric and thermal properties. Additionally, the package composition and geometry will influence the heating result, as will the spacing between neighboring products if several items are heated simultaneously. Furthermore, not only electromagnetic field pattern, and heat conduction, but also water transport in the food may influence the overall temperature distribution. The microwave oven mode properties, related to cavity volume modes and trapped surface waves (Risman, 1994), play a dominant role for heating from above and below, respectively. Trapped surface waves may occur over (or in) a dielectric slab when the angle of incidence from the dielectric medium to the air–dielectric interface is greater than the critical angle θ_c, defined by,

$$\sin(\theta_c) = 1/\sqrt{\varepsilon}. \tag{17.1}$$

[1] In the United Kingdom, 896 MHz is a recognized frequency.

Total internal reflection will occur when the angle of incidence is greater than or equal to the critical angle, with the resulting field outside the dielectric decaying exponentially away from the interface.

For multicomponent ready meals, additional factors like the geometry, size, and placement of the various components, as well as the distance between neighboring components will also be important for the heating result. More on this subject is discussed in Section 17.4.1.

17.2.2 Heating phenomena which influence uniformity

Examples of phenomena related to microwave heating uniformity are, among others, *edge overheating* caused by strong electric fields parallel to an edge of a food load, *runaway heating* in, e.g., frozen foods, and *standing wave patterns* within the food itself.

17.2.2.1 The runaway heating phenomenon

Thermal runaway when heating foods refers to a significant change of food temperature as a result of heating, which corresponds to a rapid increase in temperature.

One example is frozen foods, where most of the water content is found as ice crystals within the food item. Pure ice has a very low dielectric loss factor as compared to thawed foods. However, for selected freezing temperatures approximately one tenth of the water remains unfrozen as a strong salt solution inside the food, which contributes to the absorption of microwave energy in frozen foods (Ohlsson & Bengtsson, 2001). Since the first thawed regions have much higher loss factors, they will heat very rapidly in comparison to still frozen regions (Buffler, 1993). Since the loss factors between frozen and nonfrozen areas differ considerably in a temperature range close to 0°C, thermal runaway heating may occur when thawing frozen blocks of foods.

Another example is nonwater-containing materials, where very high temperatures may be achieved during heating. On the other hand, for water-containing foods the achieved temperatures will be only slightly higher than the boiling point of water, since in that case the temperature increase is rapid but limited.

17.2.2.2 Edge and corner overheating

When heating or thawing frozen foods, edges and corners thaw first, due to the so-called edge (or corner) overheating effect. See Fig. 17.1. Edge overheating is caused by scattering phenomena. It can be explained by the fact that at the boundary between the food item and the surrounding air, one of the boundary conditions when solving the electromagnetic problem leads to the continuity of the parallel component of the electric field. Edge overheating is strongly influenced by the polarization and incident angle of the incident field, the angle and curvature of the edges, the presence of other scatterers close to the edge, and the permittivity

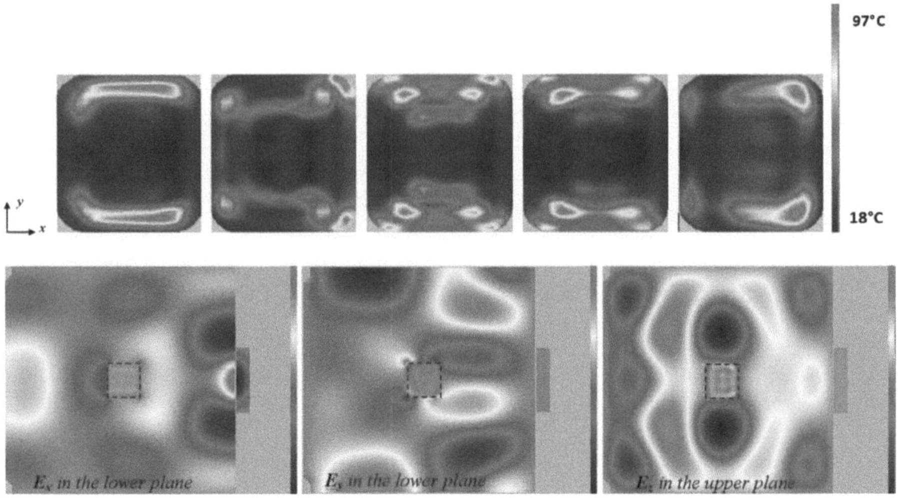

Figure 17.1 The upper part of the figure illustrates simulated temperature distribution at microwave heating of one rectangular food block (50 mm × 60 mm × 60 mm) with rounded corners and top edges in a model microwave oven with a sinusoidal waveform excitation (2.45 GHz) and rectangular TE_{01} excitation field (Wäppling Raaholt & Janestad, 2003). From left to right: heating patterns in lower plane (5 mm from the lower surface of the block), middle plane (at 30 mm height), and upper planes (10 mm and 5 mm, respectively, as measured from the top surface). The corners and vertical edges are overheated. So are the horizontal edges in the x direction at the upper surface, as well as the horizontal edges in the y direction at the lower surface. The lower part of the figure indicates the strong electric field components in parallel with the edges, and at the corners (normalized values). The contours of the load are marked with a dotted black line. The electric field components, E_x and E_y in the lower plane and E_y in the upper plane (5 mm from the upper surface), are shown. The simulations were performed using the Quickwave-3D software (QWED Sp. Z.o.o., Warsaw, Poland).

of the heated food materials (Sundberg, 1998a). For food items, where loss mechanisms due to ionic conduction as well as dielectric relaxation are present, the effective loss factor $\varepsilon''_{\text{effective}}$, where both types of losses are included, is often used (Metaxas, 1996).[2] The relative effective loss factor is then described by:

$$\varepsilon''_{\text{effective}} = \frac{\sigma}{\omega\varepsilon_0} + \varepsilon'' \tag{17.2}$$

where σ is the electrical conductivity (S/m), ω is the angular frequency (rad/s), ε_0 is the absolute permittivity of free space, and ε'' is the loss factor. The dissipated

[2] For applications related to microwave heating of foods, the loss factor refers to two types of losses into heat: the dipole relaxation and the ion conductivity. At other frequencies, however, other mechanisms may be more dominant (Hasted, 1973).

time-averaged power density p_{diss} (W/m^3) in typical food products is proportional to the square of the electric field,

$$p_{\text{diss}} = \frac{1}{2}\text{Re}\left\{\vec{E}\cdot\vec{J}^*\right\} = \frac{1}{2}\omega\varepsilon_0\varepsilon''_{\text{effective}}\left|\vec{E}\right|^2 \tag{17.3}$$

where $\varepsilon''_{\text{effective}}$ is the relative effective loss factor of the food material, $\left|\vec{E}\right|$ is the amplitude (peak value) of the electric field intensity (V/m), and $\vec{J}^*$ is the complex conjugate of the electric current density (A/m^2). In Eq. (17.2), it has been assumed that the relative complex permeability μ is equal to 1, which is true for foodstuffs since they are not magnetic materials.

At the edges of the food item, microwaves approach the food from two directions. Furthermore, electric fields of two polarizations are parallel to a surface at the boundary. The continuity boundary condition of the parallel electric fields explains the resulting concentration of the energy distribution to the sharp edges. This is one of the dominating heating uniformity problems in rectangularly shaped foods. Analogously, for corners, the corresponding will be the case for electric fields of three polarizations, resulting in an even more pronounced heating at the corners. In some cases, more than 15% of the microwave energy which is absorbed by the food sample may be lost by the edge overheating phenomenon (Risman, 1992). An investigation of the edge overheating effect for high-permittivity dielectrics is found in Sundberg (1998a). Furthermore, resonances between food items in the near vicinity of each other are illustrated in Fig. 17.2.

17.2.2.3 Center overheating

For foods with convex surfaces, e.g., spheres and cylinders, the power distribution is concentrated to the geometrical center for certain diameters (Ohlsson & Risman, 1978), due to refraction and reflection phenomena. This center overheating is influenced by different factors, e.g., geometry and size of the food components, as well as the complex permittivity of the food. In Fig. 17.3, the phenomenon of center overheating is exemplified.

17.2.2.4 Standing wave patterns in microwave-heated foods

There are several different types of standing wave phenomena in foods during microwave heating (Ryynänen, Risman, & Ohlsson, 2004). Standing waves may appear *between* plane upper and lower surfaces, *at* larger surfaces, *at edges* of larger surfaces (which in turn is partly due to the edge overheating effect, partly due to surface wave phenomena), and finally *within* certain thick loads. The latter phenomenon is due to internal resonances, i.e., standing waves.[3]

The standing wave patterns of absorption and reflection, i.e., internal "hot and cold spots" can be quantified by modeling in simplified scenarios, as described by

[3] It should be noted that external resonance phenomena, like the exploding egg effect, are not a standing wave phenomenon.

Figure 17.2 The heating profiles in food loads located close to each other, after heating in a model microwave oven with a sinusoidal waveform excitation (2.45 GHz), and rectangular TE_{01} excitation field (Wäppling Raaholt & Janestad, 2003). Resonances between cylindrically shaped food items in the near vicinity of each other are illustrated in the uppermost figures.
The lower part of the figure: the corresponding case for several rectangular blocks (nine blocks close to each other; the leftmost figure shows the result for the middle horizontal plane, while the second figure from the left shows heating profiles 5 mm from the upper surface of the same blocks). The two rightmost figures illustrate that the heating pattern in nine blocks will show similarities to the corresponding result for a single large block, having the same volume as the packs of nine ones.

Ryynänen et al. (2004) by using an extension of the transverse resonance method (Harrington, 1961) in order to include the behavior of all TM and TE waveguide modes propagating in the vertical direction.

There are also other kinds of phenomena occurring when heating foods, e.g., those which are related to different penetration depths[4] for different kinds of food materials (see Section 17.4.1, where this is described for multicomponent foods).

17.3 Heating effects related to uniformity

New microwave products have been developed on existing and private labels. In several cases, good results have been achieved, but some food products do not perform well from a quality point of view in microwave cooking, reheating,

[4] For a definition of the penetration depth in this context, the reader is referred to Buffler (1993).

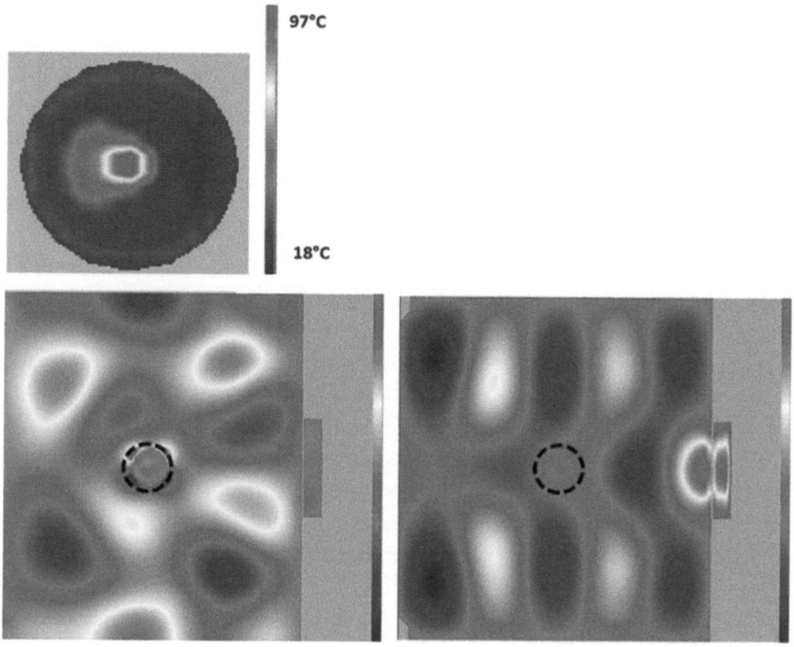

Figure 17.3 Center overheating in cylindrically shaped meatloaf, placed with the circular bottom area at the turntable in a microwave oven. The load has thawed during heating (relative permittivity 52-j20 at 20°C). The radius of the cylinder was set to 20 mm, and the height of the cylinder was selected to be much larger than the radius, so that the heating in the center of the cylinder at half its height would not be affected too much by heating effects at the edges. The upper figure shows the simulated heating profile in the cylinder at the middle horizontal plane (total height of cylinder = 160 mm). The next two figures illustrate the simulated electric field pattern (normalized values) in the oven cavity and in the load, with the cylinder placed in the middle, for the *y* and the *z* electric field components, in the middle horizontal plane. The cylindrical load is marked with a dotted black line.

or thawing. Quality variations connected with heating uniformity are related to differences in temperature and moisture distribution in microwave-heated foods, as compared to conventional cooking. Higher moisture levels in the surface areas of microwave-heated foods result in less development of desirable "cooked food" flavors and to variations in product appearance, which may lower sensory quality.

Ingredients especially designed for modification of microwave heating characteristics have been suggested to partly solve the problems encountered. Suggestions of such ingredients are, e.g., flavors and salts; other examples are sweeteners and starches (Grimwood, 1989; Katt, 1991; Pszczola, 2003; Sobel, Vasisht, & Gaonkar, 2014). Additional examples are soy, where Serventi, Sachleben, and Vodovotz (2011) suggested addition of soy to improve textural properties of microwaved par-baked products. For breaded products, Kang and Cheng (2014) suggested addition of microcapsules with a high dielectric constant and low specific heat capacity to a battered layer in order to create a higher temperature in the crust than in the interior of fish nuggets. The idea was to prevent migration of water vapor in fish nuggets to the crust during microwave heating. Mizrahi et al. (2012) suggest mechanisms that describe typical thermal characteristics of microwave heating that affect the texture. Use of emulsifiers and fibers is suggested to decrease the toughness of microwave-reheated bread (e.g., Miller & Hoseney, 1997).

Water and salt content are major factors in affecting the dielectric properties of a material. These in turn determine to what extent and how a material will absorb microwave energy and convert it into heat, how much the microwave wavelength will be reduced in the material, as well as how strong the reflection and refraction phenomena will be.

Even if microwave heating of foods may result in noneven temperatures, it should be noted that food nutritionists agree on the nutritional advantages of microwave cooking, e.g., in terms of improved retention of vitamins and minerals in vegetables (Leon Garcia, 2013; Ohlsson, 1989; Ohlsson & Åström, 1982; Yahyaei, Ghavami, Gharachorloo, Larijani, & Mazhari, 2014).

Uniform heating is important also from a microbiological perspective. A subject of previous publicity, particularly in the United Kingdom in the 1990s, was concern about insufficient inactivation of pathogenic bacteria in microwave-heated foods. The risk that sufficiently high temperatures are not reached everywhere in the foods during microwave processing was discussed (e.g., Walker, Bows, Richardson, & Banks, 1991 and Chapter 11: Microwaves for microbial inactivation—efficiency and inactivation kinetics of this book). Survival of *Listeria monocytogenes* during microwave heating has been evaluated by several researchers, among others Sheen, Huang, and Sommers (2012) and Cole, Holyoak, and Cole (1991). When a temperature of 70°C is reached and maintained for at least 2 min throughout a food product, there is a substantial reduction in the numbers of *L. monocytogenes* (Walker et al., 1991). However, such risks with microwave heating do not differ from those of traditional heating. Both types of heating methods require that foods are cooked throughout to the desired temperatures in order to inactivate food-poisoning bacteria that may be present in the food (Ohlsson, 1991). Efforts toward solving the problem by improvement of the manufacturer's reheating instructions are reported, e.g., by (George et al., 1995), where an investigation into the reheating of lasagne meals, a multicomponent product by that time known to be difficult to heat uniformly in a microwave oven, is described. Reheating instructions on the product label were, where necessary, further developed to allow a minimum of 70°C for 2 min to be achieved in the samples of commercially available lasagne. The new instructions incorporated longer heating times, rest periods, a 90° turn, and retention of a pierced film lid throughout reheating.

Much of the discussion on safety and quality of foods is characterized by limited understanding of the qualitative and quantitative importance of factors related to heating uniformity. Solutions have been found to many of the previous concerns since the early 1990s, thanks to improved heating uniformity of ovens, improved cooking and reheating instructions, and novel methods for designing food products in a way which gives as high level of temperature uniformity as possible after microwave heating. Today, methods are available which can serve as valuable tools in the design of microwavable foods, with the aim to improve uniformity in heating patterns (Wäppling Raaholt, 2000; Wäppling Raaholt, Galt, & Ohlsson, 1999; Wäppling Raaholt, Ohlsson, & Risman, 2001). More on this can be found in Sections 17.4.1 and 17.6.

17.4 The role of heating uniformity in some applications

17.4.1 Household and institutional microwave heating

Today's methods for optimizing microwave heating uniformity may often help eliminating previous problems with nonuniform heating patterns (Wäppling Raaholt et al., 1999, 2001; Wäppling Raaholt, 2000). As a first step, the factors of relevance for the present application are sorted out (e.g., food and oven geometry parameters, dielectric properties, etc.), using a fractional factorial scheme applied on modeling of microwave heating of the food product in question. Next, a suitable selection of the relevant factors is studied further, and the optimal settings of these factors are found by optimizing the heating uniformity with respect to the selected factors. It is possible to influence the heating distribution favorably, and in many cases it can be controlled toward the desired heating uniformity. At the same time, the heating times can often be significantly reduced. By optimizing the design of the food product, based on the mentioned methods, it is possible to avoid uneven temperatures. It should be pointed out that, for a successful result, electromagnetic modeling and mathematical optimization methods should be used as tools *in combination with* knowledge on microwave engineering and food science. Analogously, the method for optimizing heating uniformity can also be used as an aid in oven design (Wäppling Raaholt & Risman, 2000).

Microwave heating of multicomponent foods in a multimode oven, such as household ovens with a ceiling (mode) stirrer, may often be problematic due to nonuniform energy absorption. There are several reasons behind this. As in other microwave heating applications, the oven design will of course influence the electromagnetic field distribution and thus the temperature distribution in the food item (Buffler, 1993; Wäppling Raaholt, 2000, 2015). Furthermore, the different dielectric and thermophysical properties of the components will affect the heating result, as will microwave reflections caused by the boundaries between different components (Mudgett, 1986; Wäppling Raaholt, 2015). The latter phenomenon is a result of the fact that each component will differ in microwave penetration depth d_p, due to the different relative permittivities between components. Furthermore, the microwave reflections at the interfaces between air and food material will depend on permittivity for each component, frequency, as well as on the microwave oven mode properties. As described earlier (Section 17.2.2), several different *selective heating phenomena* like edge overheating, center overheating, and runaway heating will also play a role.

Methods for evaluating temperature distribution and heating uniformity have been developed, e.g., for the IEC standard works (IEC Standard 60705, 2014) which define the main performance characteristics of household microwave ovens, of interest to the user, and also specifies methods for measuring these characteristics. Such measurements must, however, always be related to the performance in heating of actual food products (Ohlsson, 1981).

Heating uniformity in microwave-heated foods of different geometry have been studied experimentally using temperature measurements (e.g., fiber-optic probes, temperature indicators, microwave radiometry, infrared thermography, and magnetic resonance imaging; see Chapter 15: Measuring temperature distributions during microwave processing), as well as numerical modeling of the electromagnetic fields and the resulting temperature distributions in the food (see Section 17.5; and Chapter 18: Simulation of microwave processes). In a study of composition of a layered product, a hamburger, and its microwave heating uniformity, results by Ryynänen et al. (2004) suggest that uneven heating is difficult to improve by recipe modifications, and the resulting changes in dielectric properties alone.

17.4.2 Industrial microwave heating

The phenomenon of uneven temperatures in foods has been described for industrial applications by several authors. Here follows a selection of some examples on industrial microwave heating were heating uniformity is discussed. In some of these examples, solutions for improving the heating uniformity are suggested. For a more complete picture of industrial microwave heating applications, the reader is referred to Wäppling Raaholt (2015), Ohlsson and Bengtsson (2001), and Metaxas and Meredith (1983).

In multiapplicator tunnel ovens, the dominating causes of nonuniform heating are load diffraction phenomena in combination with the applicator heating pattern (Sundberg, Risman, & Kildal, 1995). Among possible industrial applications of such ovens are: drying of wood (Antti, 1997), drying of various foods (Funebo & Ohlsson, 1997; Malafronte et al., 2012), as well as pasteurization and sterilization of foods in tubular systems (Wäppling Raaholt, Isaksson, & Hamberg, 2016) and conveyorized systems (Sundberg et al., 1996) (see also Chapter 10: Microwave-assisted pasteurization and sterilization—commercial perspective). Tubular systems, designed as described by Wäppling Raaholt et al. (2016), offer a more uniform heating than conventional in-flow heating systems, mainly due to the volumetric microwave heating and the rapid come-up times to reach target temperature. Since in conveyorized systems, as described by Sundberg et al. (1996), the packaged ready meals on the conveyor belt are moving under (or above) the applicator, its momentaneous overall microwave efficiency will often vary significantly, which sometimes results in an even more pronounced heating nonuniformity. On the other hand, irregular fields in the feed direction will be evened out by the continuous movement of the food on the conveyor, similar to the function of the rotating turntable in a household microwave oven. In addition, the nonlinear behavior of magnetron sources complicates the overall efficiency problem. If circulators are used, this problem could however be reduced. Sundberg (1998b) studied the analysis and design of industrial microwave tunnel ovens, and suggested a quantification of heating uniformity in multiapplicator tunnel ovens (Sundberg, Kildal, & Ohlsson, 1998). More details are found in Section 17.6.

An industrial unit operating under "steady state conditions," may give better load matching and larger power efficiency. In general, industrial microwave applications often show a more even field and temperature distribution than domestic microwave ovens (Ohlsson & Bengtsson, 2001). Single-mode applicators can be useful in some applications, for instance to establish locally high field strengths. In some cases, they may also be designed to give an unusually uniform field inside the food material. However, for a specific food and a specific application, such applicators typically have to be carefully tailored and small changes in food properties or other conditions are likely to affect the electromagnetic field distribution considerably.

Risman and Ohlsson (1975) describe a tubular TM_{020} applicator, which was designed for continuous microwave processing of pumpable liquid or semiliquid foods. In this kind of center-heating applicator, the radial heating profile can be adapted to match the food flow relatively well (Ohlsson, 1993), which typically would also have a maximum in the center of the pipe. This matching between heating and flow profile is done by the choice of an appropriate tube diameter. Inspired by the work of Risman and Ohlsson (1975), together with a patent by Risman (1997), Isaksson, Bondeson, and Ohlsson (2002a) used numerical modeling and simulation to design a system for continuous microwave heating of pumpable foods, with the aim to make the design permit larger load radii with as small an axial field variation as possible also at 2450 MHz, by superposition of two single modes. The design of the feed system, as well as of the circular cavities, are both important for a successful heating result. In order to get the desired heating pattern associated with each specific mode, a proper design is essential, especially for applications with microwave heating of high loss loads. This concept of tubular microwave heating is studied further with regard to heating uniformity in Wäppling Raaholt et al. (2016).

17.5 Modeling and simulation of microwave processes as a tool for improving heating uniformity

17.5.1 Background

Modeling plays an important role as a tool to give an increased knowledge and understanding of microwave heating of foods. This knowledge may be used to optimize the process with respect to certain quality aspects, such as increased heating uniformity, which in turn is also related to improved quality from a microbiological as well as a sensory and nutritional point of view.

The mechanisms behind the general behavior in an empty cavity are described in several textbooks (e.g., Kok & Boon, 1992). However, for the case of microwave heating of a food item, the situation generally becomes much more complex. An analytical or exact solution of Maxwell's equations is unfortunately extremely difficult or impossible for all but the simplest product geometries and cavity configurations.

Hence, the predominating approach today is to use numerical modeling and simulation; a survey of computational methods for electromagnetics and microwaves is given by, e.g., Booton (1992) and Rylander, Ingelsten, and Bondeson (2013).

Modeling of microwave processes involves solving for the electromagnetic fields, described by Maxwell's equations (see Maxwell, 1954), coupled to the heat transfer problem.

RISE (Research Institutes of Sweden) and its unit SP Food and Bioscience[5] have a long tradition in modeling of microwave heating, with work starting already in the 1970s (Ohlsson & Bengtsson, 1971; Ohlsson & Risman, 1978). With a development of modern computers toward increasing processing capabilities, numerical modeling and simulation of electromagnetic fields in the ovens and foods during heating has become a valuable tool when designing microwave heating units as well as microwavable food products and packages. For microwave processing of foods, the electromagnetic problem can be solved numerically by several methods. The most common ones are: the finite difference time domain FDTD method (Gwarek, 1988; Lau & Sheppard, 1986; Sundberg et al., 1996; Taflove, 1980; Yee, 1966), the finite element method (FEM) (Dibben & Metaxas, 1994), and the Method of Moments (MM) (Yamashita, 1990). Furthermore, hybrid methods like the stable FEM-FDTD method (Rylander & Bondeson, 2000) have been developed. Several commercial software packages are today available, which have offered increased opportunities to use modeling as a tool to find the optimal solutions for microwave processing, with regard to design of food products but also oven design (Nablaworks, 2016; Rylander et al., 2013).

By coupling electromagnetic models to heat transfer models, it is possible to predict the temperature distribution within the processed foods. As was described by Wäppling Raaholt (2000) and Fu and Metaxas (1994) included coupling of the electromagnetic to the thermal process using the method of lines (Fu & Metaxas, 1993). Ma et al. (1995) suggested a combined electromagnetic and thermal FDTD model, however with time steps in the two algorithms of very different order of magnitude. Torres and Jecko (1997) suggested the use of a time-scaled form of the heat transfer equation in order to overcome this difficulty. In both papers, temperature-dependent thermal properties were included in the model, though only to a limited extent due to the complexity of the numerical model. However, these and other authors limited their study to "pure" microwave heating (without convection) and held the thermal properties of the microwave-heated dielectrics constant. Zhang and Datta (2000) also studied a microwave heating process, using a finite element (FE) frequency domain software for the microwave part, and a separate FE software for the heat conduction part. Temperature dependence of dielectric properties and weak

[5] Formerly SIK—the Swedish Institute for Food and Biotechnology.

convection were taken into account. However, they assumed constant thermal properties and did not include any surface evaporation. Wäppling Raaholt (2000) and Wäppling Raaholt et al. (2001, 2002) modeled and validated combined microwave and forced air heating of foods in a microwave combination oven, using the FDTD method, taking into account the temperature dependence of the thermophysical properties, convective heat transfer at the boundaries, and evaporation in the product.

Other developments are modeling of microwave defrosting of foods in three dimensions (Wäppling Raaholt & Janestad, 2003), where the strong enthalpy and temperature dependence of dielectric and thermal properties were regarded.

17.5.2 Modeling as a tool in product development

Knowledge in coupled electromagnetic and heat transfer modeling (Wäppling Raaholt, 2015; Wäppling Raaholt & Janestad, 2003; Wäppling Raaholt et al., 2001) can, in combination with knowledge and experience in food science and microwave engineering, give enhanced opportunities to improve the design of microwavable food products. This has been demonstrated by Wäppling Raaholt et al. (1999), Wäppling Raaholt (2000), and Wäppling Raaholt et al. (2001). Since then, several projects where these methods were used have been performed successfully. Fig. 17.4 illustrates the predicted power distribution as a result of numerical modeling of microwave heating of a compact ready meal.

17.6 Techniques for improving heating uniformity

For several types of ovens, the introduction of a mode stirrer will assist for a selection of the possible modes that can exist in the cavity. These are correlated to the rotational speed of the mode stirrer. Alternatively, or as a combination, movement of the food through the field on a rotating turntable at the bottom of the cavity makes it possible for temperature differences to be partly leveled out, since the food is then subjected to different field patterns (also the fields in the oven are affected accordingly). Even if rotating turntables and mode stirrers may to some extent result in more uniform heating patterns, the need for a proper design of the food and package (as well as of the relevant oven parameters) remains.

As mentioned in the previous section, modeling of microwave processes may serve as a tool to learn more about a process, and for improving the heating uniformity by finding the optimal values of relevant food and package parameters, or alternatively relevant oven parameters. In many practical situations it can also be used to decrease the number of experiments required to find the best design of a food product. The previously empirically based product development of microwave foods has at least in part of the industry segment been replaced by numerical modeling, computer simulations, and optimization (Wäppling Raaholt, 2015).

For analysis of industrial microwave ovens, Sundberg et al. (1998) suggested the introduction of the concept of cost functions in order to classify the performance of a specific oven design. The cost functions assign a figure of merit to the field

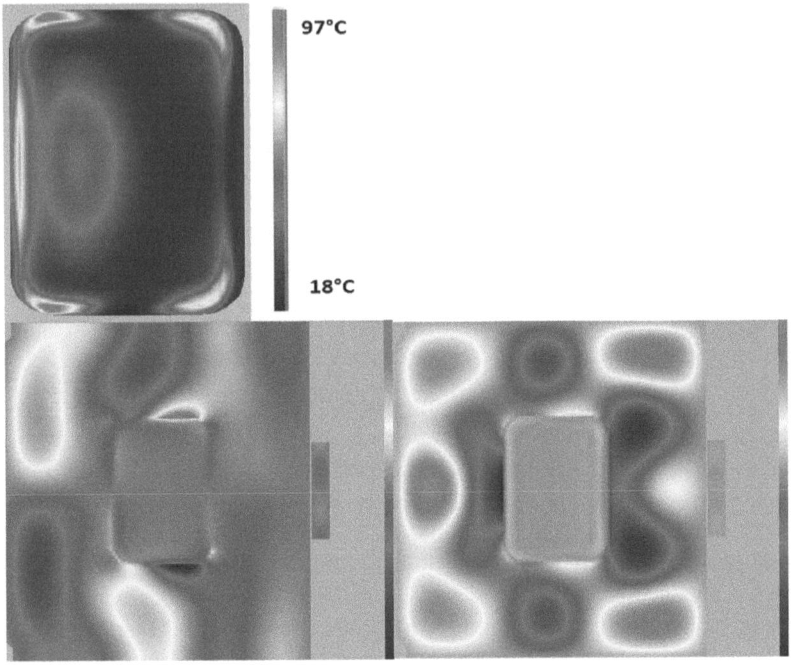

Figure 17.4 The predicted microwave heating distribution as determined by numerical modeling and simulation of microwave heating of a compact ready meal (the top part of the figure). In this example, the meal is a rectangularly shaped load with rounded corners. The top figure illustrates the result for the horizontal cross section located 5 mm below the upper surface of the load. At that plane, overheating occurs at the corners and at the left edge. The electric field components in the *y* and *z* directions in the cavity and load are illustrated in the lower part of the figure (normalized values). Model oven as described in Wäppling Raaholt and Janestad (2003), with a sinusoidal waveform excitation, and rectangular TE_{01} excitation field.

distribution within the oven, and serve as a measure of how close the field generated by the oven comes to the desired field distribution. Sundberg introduced two different cost functions, one that will give as uniform heating as possible in the load (Eq. 17.3), by minimizing the difference between the hottest and coldest part of the food, and one that will suppress the horizontal component of the electric field in order to minimize overheating of the edges of closely positioned food packages, transported on a conveyor belt (Eq. 17.4).

$$f_1(a,b,c,X) = \max\left|\vec{E}\right|^2 - \min\left|\vec{E}\right|^2 \tag{17.4}$$

$$f_2(a,b,c,X) = \frac{\sum |E_x|^2 + \sum |E_y|^2}{\sum |E_z|^2} \tag{17.5}$$

where the sums are taken over all points included in the set of data. Sundberg further suggests the use of "simulated annealing" (Press et al., 1992), as a suitable optimization method. The reason for this is the presence of many local minima, due to the strong variation of the field inside the cavity and within the food load.

In a patent by Risman (1997), a technique for exciting circular-cylindrical TM_{1n0} modes is described. Isaksson et al. (2002a) studied how this mode could be combined with a TM_{0n0} mode, with the aim to enable larger load radii. Isaksson (2002; Isaksson et al., 2002b) also presented a principle for enlarging the cavity length, if desired. The concept was also applied on the design of a microwave process where a meat product is continuously heated, as briefly described in Section 17.4.2. Later, this kind of process has been used in other applications, such as rapid tubular microwave sterilization of particulate soups (Wäppling Raaholt et al., 2016).

The penetration depth d_p is considerably higher (approximately 2.5 times larger) for microwave frequencies of 915 MHz than at 2450 MHz, under the assumption of similar permittivities at both frequencies. This has been used, e.g., in applications with continuous tubular microwave heating to overcome the disadvantage of limited penetration depth for the 2450 MHz (Isaksson, Bondeson, & Ohlsson, 2002b, Nykvist & Decareau, 1976). However, at lower frequencies, the temperature dependence of the complex permittivity becomes stronger (Ohlsson, Bengtsson, & Risman, 1974), which has to be taken into account in the electromagnetic modeling. This results in more time-consuming 3D simulations than in other cases. A simple and fast tool, based on a semianalytical method for solving the eigenmode problem and for estimation of the heat profile to get a quick view of the behavior of the process, was developed and suggested by Isaksson et al. (2002a). The system of equations for the electromagnetic axial fields is given by Jin (1993). Under the assumption of an axially uniform field, and no azimuthal variation of ε and μ, the modes become TE (transverse electric) and TM (transverse magnetic). For TM modes, this leads to the radial solution to Bessel's differential equation

$$\frac{1}{r}\frac{d}{dr}\left(r\frac{\varepsilon}{k_t^2}\frac{dE_z}{dr}\right) - \frac{\varepsilon m^2}{k_t^2 r^2}E_z + \varepsilon E_z = 0 \tag{17.6}$$

where r is the radial position, m is the angular mode index, and k_t is the transverse wave number. This approach has several advantages, especially in the estimation of the heat profile for scenarios with markedly strong temperature dependence on the dielectric properties.

Wäppling Raaholt et al. (1999), Wäppling Raaholt et al. (2001), and Wäppling Raaholt (2000) suggest a statistical criterion for a quantitative measure on the heating uniformity, namely, the standard deviation of the dissipated power distribution normalized by the mean value, for a representative plane of the food load. The plane is selected as described by Wäppling Raaholt et al. (1999). Practical examples on results for improving the heating uniformity of real food products are given in, e.g., Wäppling Raaholt et al. (1999, 2001) and Wäppling Raaholt (2000).

As described in the study by Wäppling Raaholt et al. (1999), an error norm to control the grid computational accuracy was defined, as given in (Eq. 17.7).

$$ERROR = \frac{\sqrt{\sum_{(i,j,k)\in M}(P^{\Delta x}(i,j,k) - P^{0.9\Delta x}(i,j,k))^2}}{\sum_{(i,j,k)\in M}\left|P^{\Delta x}(i,j,k)\right|} \leq \text{epsilon} \quad (17.7)$$

where M is a set of points in the food load, epsilon is a given error parameter, $P^{\Delta x}(i,j,k)$ and $P^{0.9\Delta x}(i,j,k)$ denote the power density at a point $((i+½)\Delta x,$ $(j+½)\Delta y,$ $(k+½)\Delta z)$ for grids with cell sides Δx and $0.9\Delta x$, respectively. A corresponding error norm to control the convergence of iterations was defined, as given in Eq. (17.8).

$$ERROR = \frac{\sqrt{\sum_{(i,j,k)\in M}(P^{n+1}(i,j,k) - P^{n}(i,j,k))^2}}{\sum_{(i,j,k)\in M}\left|P^{n+1}(i,j,k)\right|} \leq \text{epsilon} \quad (17.8)$$

Several examples on successful results on design of food products, intended for microwave heating, have been given since then. In Figs. 17.5–17.6, three such examples are given; more on this is found in Section 17.7. Furthermore, Figs. 17.7–17.8 illustrate the possibilities to use modeling as a tool for predicting the microwave heating pattern for two additional food products.

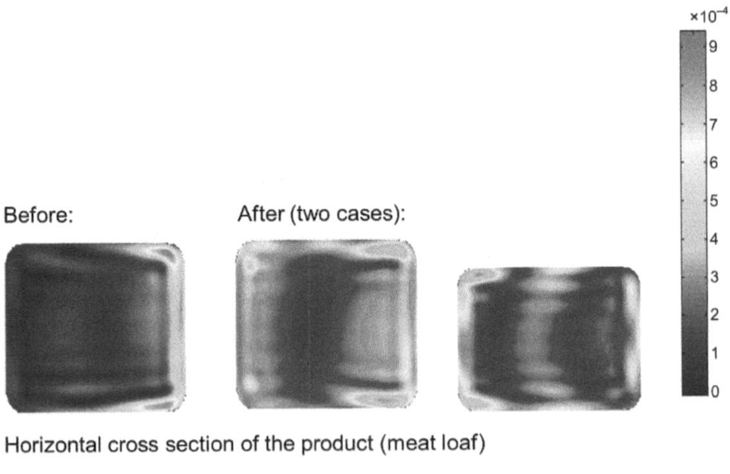

Figure 17.5 Simulated microwave power distribution in meatloaf, as a compact ready meal, before (left) and after (the two figures to the right) optimization (colorbar values in kW/cm^2). After optimization, the geometry and dimensions of the package were changed. In the optimized cases, the overheated and underheated areas are less pronounced.

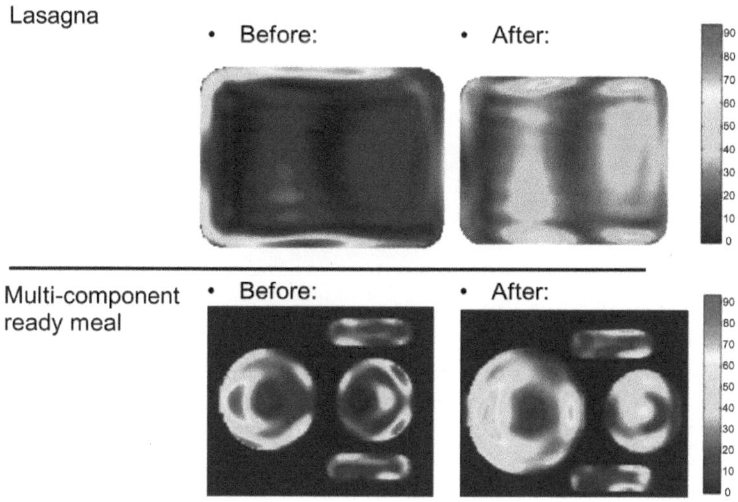

Figure 17.6 Predicted dissipated microwave power in lasagne and in a multicomponent ready-meal, before and after optimization (values in °C), before taking heat transfer into account. The multicomponent meal consists of a meat patty, mashed potatoes, and carrots. In both cases, the same volume of the meal is maintained after optimization. Geometries, dimensions, and placement of the individual components were changed. The optimized meals have a more even heating distribution.

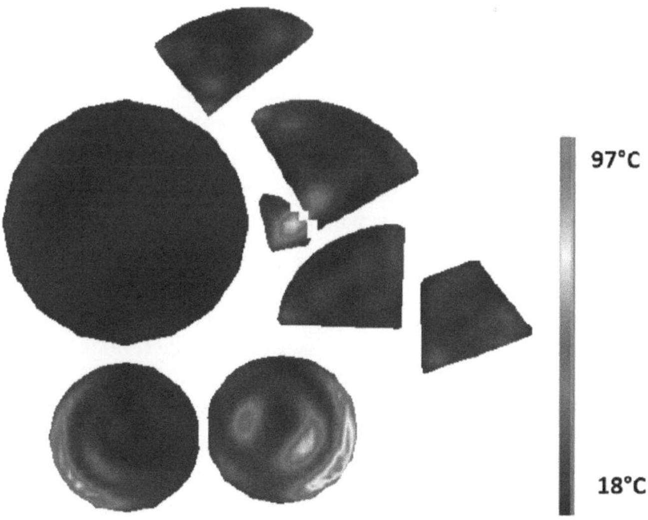

Figure 17.7 Predicted microwave heating distribution in a multicomponent ready meal. The meal is heated from frozen, and consists of mashed potato (the larger component), two meat patties (at the bottom), pieces of parsnip (upper mid part of figure), and Swedish turnip (middle right). The heating uniformity would be more even if the product were designed in a way which would allow a quicker thawing of several of the components and a more even heating distribution between the components.

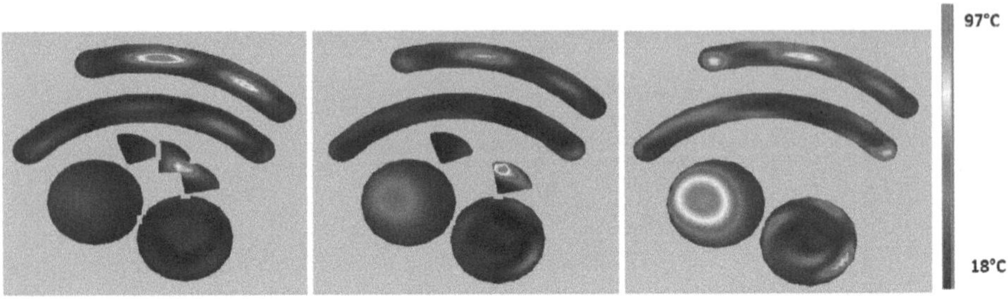

Figure 17.8 Predicted microwave heating pattern in a multicomponent ready meal, consisting of two sausages, two components of mashed potato, and pieces of blanched vegetables. From left to right: three different planes through the ready meal, lower, middle, and upper planes, located 5 mm from each other. The lower plane is located 5 mm above the bottom of the meal. The overheated areas and the nonuniformity between different components indicate the need for optimization of the product, similar to the solution for the previous case (see Fig. 17.7). Compare to Figs. 17.5–17.6.

17.7 Applications to particular foods/processes

Today, methods are available (Wäppling Raaholt, 2000; Wäppling Raaholt et al., 1999; Wäppling Raaholt et al., 2001), which give the opportunity to design microwavable food products with improved microwave heating characteristics. This is demonstrated for three types of ready-meals, meatloaf (Wäppling Raaholt et al., 1999), lasagne, and a multicomponent ready meal (Wäppling Raaholt et al., 2001), and illustrated in Figs. 17.5–17.6. In Figs. 17.7–17.8, the corresponding heating results for two additional food products are illustrated.

Among industrial applications designed to give uniform heating, a continuous microwave process of liver paste can be mentioned. Based on a technique for achieving uniform heating by combining selected modes, Isaksson et al. (2002a) use the concept for a microwave process where meat, which is transported in the axial direction of a circularly cylindrical tube geometry, is uniformly heated. In the process, two single-mode circularly cylindrical cavities are involved, each carrying a different mode (TM_{020} and TM_{120} mode). This combination of modes allows for uniform heating if designed appropriately. Isaksson et al. (2002b) analyzed the heating by a simplified one-dimensional approach, extended by introducing temperature-dependent dielectric properties. The validity of the one-dimensional results was evaluated by full-geometry finite difference time domain simulations (utilizing the software package Quickwave-3D (QWED Sp. Z o.o., Warsaw, Poland)), which showed good agreement. This approach allows for analysis also at lower frequencies, where the dielectric properties show stronger temperature dependence.

The study was demonstrated on microwave processing of liver paste at 2450 MHz. In Fig. 17.9, the experimental plant for tubular microwave heating of foods is shown. The same principle has also been applied on a pilot plant intended for sterilization of pumpable foods that have high viscosities or contain particulates that are large

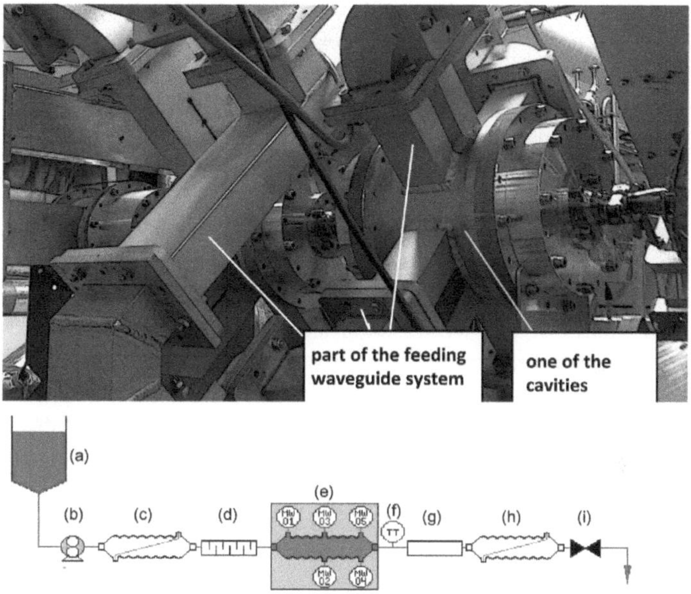

Figure 17.9 A pilot plant for continuous tubular microwave processing of pumpable foods at 2450 MHz, e.g., for pasteurization and sterilization of particulate foods: schematic view (top view) of a tubular microwave system, including: balance tank (a), product pump (b), preheater (c), static mixer (d), microwave heating unit (e), spatial temperature measurement unit (f), holding tube (g), cooler (h), and pressure regulator (i), and (above): cavities with parts of feeding waveguide system.

enough to delimit speed of heating in traditional heat exchanger-based processes (Wäppling Raaholt, 2015; Wäppling Raaholt et al., 2016).

17.8 Future trends

In the 1990s and 2000s, domestic microwave ovens started to become a standard kitchen appliance in developed countries (Wäppling Raaholt, 2015). In the United States, the number of households that owned a microwave oven increased from 1% in 1971 to approximately 25% in 1986 (Liegey, 2001), while 90% of the US households owned a microwave oven in 1997 (US Bureau of Labor Statistics) (Cox & Alm, 1997; Liegey, 2001). In 2008, as much as 96% of the US households owned at least one microwave oven (USDA, 2009). The amount of Australian families that owned a microwave oven was, according to a market research from 2008, 95%. In Canada, the number of households that owned a microwave oven grew from fewer than 5% of the households in 1979, to more than 88% by 1998 (Williams, 2000). In France, 76% of the households owned a microwave oven in 2014, which is an increase from the 40% of households that had a microwave oven in France in 1994 (Freedonia, 2009). The adoption of household microwave ovens was slower in developing countries, as expected. In India, for example, only 5% of the households in 2013 owned a microwave oven, while more than 31% owned refrigerators (Statista, 2015). Another example is Vietnam, where the corresponding increase in ownership of household microwave ovens for people with low income was from 7% in 2002 to 16% in 2008 (USDA, 2009). In South Africa, on the other hand, the ownership of microwave ovens among people with low income showed an almost two-fold increase from 20% in 2002 to 39% in 2008 (USDA, 2009).

The amount of households that own a microwave oven is steadily growing also in several other parts of the world, such as in Eastern Europe, where a growing middle class has shown an increased usage of household appliances, such as microwave ovens. Furthermore, there is a strong growth of the number of households that own a microwave oven in Africa, the Middle East, and in Latin America, partly due to rising per capita incomes, partly due to the increased number of households in these regions. According to a recent study, the increase in the number of microwave ovens in the households worldwide in 2013 was the largest among all product categories in the white goods segment of household appliances (Freedonia, 2006, 2009). This increasing trend is expected to continue in the years to come, with a significant, almost unexploited, market potential in numerous countries, particularly in developing regions.

The increasing knowledge on electromagnetic field distribution during microwave heating and the interactions between the factors which influence the heating results, will gradually improve oven performance. Furthermore, the ongoing international standardization of performance and testing methods will lead to improved heating performance.

Modeling based product development in combination with packaging development (including different materials, shielding, susceptors, geometry, etc., see also Chapter 13: Packaging for microwave foods) has increased rapidly in the food industry during the latest decades.

Furthermore, improvements in oven components, such as solid state components, will lower size, oven weight, and production cost. The number of combination ovens, combining microwave heating with convection or radiative heating is expected to grow continually, at least in Europe.

Industrial microwave processing has been considered very promising for decades, but practical application has previously developed slowly. There are several reasons for expecting an increasing growth in the future:

Today's microwave heating units have a reliability which is well comparable to that of alternative processing equipment, which makes microwave heating with its special advantages more attractive. Moreover, there is a trend in the food industry toward continuous processing lines and on-line process control. Thus, there will be an accelerating need for microwave heating as a unit operation for very rapid and in-depth heating. The advent of more sophisticated applicators, giving more uniform power distribution and improved process control will be a contributing factor; both for the tubular heating of pumpable liquids, and for the continuous heating of food packages on conveyor belts.

Tempering, pasteurization, drying, and snacks production will probably also in the future be the dominating applications, but with a steadily increasing number of applications.

With 915 MHz as a recognized ISM frequency (at least in North America), it will be important for industrial processing, mainly because of its advantages in penetration depth and generator power efficiency. However, properly designed 2450 MHz systems will also in the future be a strong alternative for many applications.

Further developed *modeling tools* will be *invaluable tools for modern product design of microwavable foods*. By combining today's optimization tools with knowledge on microwave engineering and food science, unique possibilities for optimizing and developing microwave food products are generated. These tools will also enable modeling of combination processes, where microwaves are combined with other heating techniques. The tools will *reduce the time from product development to market* for food products within the microwavable segment, with *increased possibilities to* meet the increased requirements from consumers and *respond to market needs.*

17.9 Sources of further information and advice

Among the research areas of interest at SP Food and Bioscience,[6] within microwave engineering is since many years improvement of heating uniformity in foods during microwave heating. Much of this work has previously been published. Parts of the present research will be submitted as papers to scientific journals, and presented at workshops and seminars internationally.

An introduction to the subject is found in Wäppling Raaholt (2015) and Bengtsson and Ohlsson (1974). For further information on dielectric properties and their consequences for heating uniformity, the reader is referred to Buffler and Stanford (1995) and Ohlsson and Bengtsson (2001). The subject of the principles for microwave heating is described by several authors, e.g., Bengtsson and Risman (1971), Ohlsson and Risman (1978), Ohlsson (1983), Metaxas and Meredith (1983); Walker (1987), Buffler (1993), Ohlsson and Bengtsson (2001), and Wäppling Raaholt (2015). For a review of the fundamental aspects on electromagnetics and numerical modeling and simulation, see e.g., Dibben (2001) and Rylander et al. (2013) and following Chapter 18, Simulation of microwave processes in this book.

References

Antti, L. (1997). *Microwave drying of wood components in a continuous conveyor system.* Technical Report 95-09638, NUTEK, the Swedish National Board for Industrial and Technical Development.

Bengtsson, N., & Risman, P. O. (1971). Dielectric properties of foods at 3 GHz as determined by a cavity perturbation technique. II. Measurements on food materials. *Journal of Microwave Power*, *6*(2), 107–123.

[6] Part of RISE Research Institutes of Sweden.

Bengtsson, N. E., & Ohlsson, T. (1974). Microwave heating in the food industry. *Proceedings of the IEEE, 62*(1), 44–55.

Booton, R. C. (1992). *Computational methods for electromagnetics and microwaves*. John Wiley and Sons Inc.

Buffler, C. R. (1993). *Microwave cooking and processing. Engineering fundamentals for the food scientist*. New York: AVI, van Nostrand Reinhold.

Buffler, C. R., & Stanford, M. (1995). The effects of dielectric and thermal properties on the microwave heating of foods. *Microwave World, 12*(4), 15–23 , 1991, and 16(1)

Cole, P. J., Holyoak, C. D., & Cole, M. B. (1991). Thermal inactivation of *Listeria monocytogenes* during a process simulating temperatures achieved during microwave heating. *Journal of Applied Bacteriology, 70*, 489–494.

Cox, W. M., & Alm, R. (1997). Time well spent: The declining real cost of living in America, 1997 Annual Report (Federal Reserve Bank of Dallas): 22 (see Exhibit 8), archived from the original (PDF) on 13 August 2006, http://www.dallasfed.org/assets/documents/fed/annual/1999/ar97.pdf (12.08.99).

Dibben, D. (2001). Electromagnetics: Fundamental aspects and numerical modelling; chapter 1. In A. K. Datta, & R. C Anantheswaran (Eds.), *Handbook of microwave technology for food applications*.

Dibben, D. C. & Metaxas, A. C. (1994). Time domain finite element analysis of multimode microwave applicators loaded with low and high loss materials. In: *Proc. int. conf. microwave and high frequency heating, 17–21 September. St John's College, Cambridge, UK* (A3.1–A3.4).

The Freedonia Group (2006). *World major household appliances: World industry study with forecasts to 2009 & 2014 (Study #2015)*. Cleveland, Ohio: The Freedonia Group. January 2006, Table VI-5: France cooking appliances supply & demand (million dollars); Jan. http://www.freedoniagroup.com/brochure/20xx/2015smwe.pdf (11.08.15).

The Freedonia Group. (2009). World major household appliances: World industry study with forecasts to 2013 & 2018 (Study #2588) http://www.freedoniagroup.com/brochure/25xx/2588embro.pdf, Dec. (13.08.15).

Fu, W. B., & Metaxas, A. C. (1993). MAXEMOL – A numerical scheme for the solution of Maxwell's equations using the Method of Lines. In: *Proceedings of microwave and high frequency heating, Göteborg, Sweden* (paper A5).

Fu, W. B., & Metaxas, A. C. (1994). Numerical prediction of three-dimensional temperature distributions inside a microwave oven using the method of lines. In: *Dig. 2nd int. conf. computation in electromagnetics, Nottingham* (pp. 255–258).

Funebo, T., & Ohlsson, T. (1997). Microwave-assisted dehydration of fruits and vegetables using applicators. In: *Proc. of the 6th Int. Conf. On Microwave and High Frequency Heating, Fermo, Italy* (pp. 229–232).

George, R. M., Evans, D. G., Hooper, G. I., Campbell, G. M., & Dobie, P. A. P. (1995). *Assessment and improvement of the manufacturers' reheating instructions for microwavable lasagne based upon the voluntary UK new microwave labelling scheme*. London: MAFF Publications.

Grimwood, N. (1989). The convenience culture – Producing the quality in microwave foods. *I. European Food Drink Review, 3*(Autumn), 11–12.

Gwarek, W. K. (1988). Analysis of arbitrarily shaped two-dimensional microwave circuits by finite-difference time-domain method. *IEEE Transaction on Microwave Theory Technology, 35*(4), 738–744.

Harrington, R. F. (1961). *Time-harmonic electromagnetic fields* (pp. 158–163). New York: Mc Graw Hill Book Co.

Hasted, J. B. (1973). *Aqueous dielectrics*. London: Chapman and Hall.
IEC Standard 60705: 4th Edition. (2014). *Household microwave ovens – Methods for measuring performance*. Geneva, Switzerland.
Isaksson, S. (2002). *Tubular microwave heating. Thesis for the Degree of Licentiate of Engineering*. Chalmers University of Technology.
Isaksson, S., Bondeson, A., & Ohlsson, T. (2002a). Computer-aided design of a continuous tubular microwave process, using the TM_{0n0} and TM_{1n0} modes. In S. Isaksson (Ed.), *Paper II in Tubular Microwave Heating*. Chalmers University of Technology, (Thesis for the Degree of Licentiate of Engineering).
Isaksson, S., Bondeson, A., & Ohlsson, T. (2002b). Tubular microwave heating of meat with temperature-dependent dielectric properties. In S. Isaksson (Ed.), *Paper III in Tubular Microwave Heating*. Chalmers University of Technology, (Thesis for the Degree of Licentiate of Engineering).
Jin, J. (1993). *The finite element method in electromagnetics*. New York: Wiley.
Kang, H. Y., & Chen, H. H. (2014). Preparation of thermally stable microcapsules with a chitosan–silica hybrid. *Journal of Food Science*, *79*, E1713–E1721. Available from http://dx.doi.org/10.1111/1750-3841.12544.
Katt, J. L. (1991). The effects on starches and sugars on microwave cooking. *Microwave World*, *12*(2), 19–23.
Kok, L. P., & Boon, E. (1992). *Microwave cookbook for microscopists. Art and science of visualization*. Coulomb Press Leyden.
Lau, R. W. M., & Sheppard, R. J. (1986). The modelling of biological systems in three dimensions using the time-domain finite-difference method: I. The implementation of the model. *The Institute of Physics*, *31*, 1247–1256.
Leon Garcia, M. M. (2013). *Impact of microwave processing on quality of high value shelf stable fruit products*. Open Access Theses. Paper 48. http://docs.lib.purdue.edu/open_access_theses/48.
Liegey, P. R. (2001). *Hedonic quality adjustment methods for microwave ovens in the U.S.* CPI, Bureau of Labor Statistics, United States Department of Labor. Oct, 16th http://www.bls.gov/cpi/cpimwo.htm (11.04. 15).
Ma, L., Paul, D.-L., Pothecary, N., Railton, C., Bows, J., Barratt, L., ... Simons, D. (1995). Experimental validation of a combined electromagnetic and thermal fdtd model of a microwave heating process. *IEEE Trans MTT*, *43*(11), 2565–2572.
Malafronte, L., Lamberti, G., Barba, A. A., Raaholt, B., Holtz, E., & Ahrné, L. (2012). Combined convective and microwave assisted drying: Experiments and modelling. *Journal of Food Engineering*, *112*(4), 304–312 , Oct
Maxwell, J. C. (1954). *A treatise on electricity and magnetism. Vol. 1–2, Unabrigated republication of the last third revised edition 1892*. New York: Dover Publication.
Metaxas, A. C. (1996). *Foundations of Electroheat – A unified approach*. Chichester, UK: John Wiley and Sons.
Metaxas, A. C., & Meredith, R. J. (1983). *Industrial microwave heating. IEE Power Engineering Series 4*. Piscataway NJ: Peter Peregrinus Ltd, 357 pp.
Miller, R. A., & Hoseney, R. C. (1997). Method to measure microwave-induced toughness of bread. *Journal of Food Science*, *62*(6), 1202–1204.
Mizrahi, S., et al. (2012). Mechanisms of objectionable textural changes by microwave reheating of foods: A review. *Journal of Food Science*, *77*, R57–R62. Available from http://dx.doi.org/10.1111/j.1750-3841.2011.02515.
Mudgett, R. E. (1986). Electrical properties of foods. Ch 7. In M. A. Rao, & S. S. V. Rizvi (Eds.), *Engineering properties of foods* (pp. 32–90). New York: Marcel Dekker Inc.

Nablaworks. (2016). *A list of commercial computational electromagnetics codes and their publishers*. http://www.nablaworks.com/CEM_Software.html (assessed 06.05.16).

Nykvist, W. E., & Decareau, R. V. (1976). Microwave meat roasting. *Journal of Microwave Power, 11*(1), 3–24.

Ohlsson, T. (1981). Methods for measuring temperature distribution in microwave ovens. *Microwave World, 2*(2), 14–17.

Ohlsson, T. (1983). Fundamentals of microwave cooking. *Microwave World, 2*(4), 4–9.

Ohlsson, T. (1989). Auswirkungen von Mikrowellenerwärmung auf den Närhwärt. In: *Internationales Seminar Lebensmittel und Verpackungen für Mikrowellenerwärmung* (4 p.).

Ohlsson, T. (1991). More uniformity in heat from microwave ovens. In: *Food technology international – Europe 1991, London 1991* (pp. 77–78, 80–81). SIK publication 552.

Ohlsson, T. (1993). In-flow microwave heating of pumpable foods. In: *Paper presented at international congress on food and engineering, Chiba, Japan, May 23–27* (7 p.).

Ohlsson, T., & Åström, A. (1982). Sensory and nutritional quality in microwave cooking. *Journal of Microwave Power, 17*(4), 320–321.

Ohlsson, T., & Bengtsson, N. (2001). Microwave technology and foods. *Advances in Food and Nutrition Research, 43*, 65–140.

Ohlsson, T., & Bengtsson, N. E. (1971). Microwave heating profiles in foods. A comparison between heating experiments and computer simulation. A research note. *Microwave Energy Applications Newletter, 4*(6), 3–8 , SIK publication 232

Ohlsson, T., Bengtsson, N. E., & Risman, P. O. (1974). The frequency and temperature dependence of dielectric food data as determined by a cavity perturbation technique. *Journal of Microwave Power, 9*(2).

Ohlsson, T., & Risman, P. O. (1978). Temperature distribution of microwave heating – Spheres and Cylinders. *Journal of Microwave Power, 13*(4), 303–310 , SIK publication 320

Press, W. H., Teukolsky, S. A., Vetterling, W. T., & Flannery, B. P. (1992). *Numerical recipes in C: the art of scientific computing*. New York: Cambridge University Press.

Pszczola, D. E. (2003). Delivery systems help send the right message. *Food Technology, 57* (4), 68–85.

Risman, P. O. (1992). Metal in the microwave oven. *Journal of Microwave Power and Electromagnetic Energy, 13*(1).

Risman, P. O. (1994). Confined modes between a lossy slab load and a metal plane as determined by a waveguide trough model. *Journal of Microwave Power and Electromagnetic Energy, 29*(3).

Risman, P. O. (1997). *Tubular microwave applicator*. US Patent No. 5,834,744.

Risman, P. O., & Ohlsson, T. (1975). Theory for and experiments with a TM_{02n} applicator. *Journal of Microwave Power, 10*(3), 271–280 , SIK Publication 286

Rylander, T., & Bondeson, A. (2000). Stable FEM-FDTD hybrid method for Maxwell's equations. *Computer Physics Communications, 25*, 75–82 , March

Rylander, T., Ingelsten, P., & Bondeson, A. (2013). *Computational electromagnetics . Texts in applied mathematics* (2nd ed.). New York: Springer Science and Business Media.

Ryynänen, S., Risman, P. O., & Ohlsson, T. (2004). Hamburger composition and microwave heating uniformity. *Journal of Food Science, 69*(7), 187–196 , Sep

Serventi, L., Sachleben, J., & Vodovotz, Y. (2011). Soy addition improves the texture of microwavable par-baked pocket-type flat doughs. *Journal of thermal analysis and calorimetry, 106*(1), 117.

Sheen, S., Huang, L., & Sommers, C. (2012). Survival of *Listeria monocytogenes*, *Escherichia coli* O157:H7, and *Salmonella* spp. on catfish fillets exposed to microwave heating in a continuous mode. *Journal of Food Science, 77*(8), E209–214 , Aug

Sobel, K. R., Vasisht, N., & Gaonkar, A. G. (2014). Parts I–III in microencapsulation in the food industry: A practical implementation guide. In A. G. Gaonkar, N. Vasisht, A. R. Khare, & R. Sobel (Eds.), *Technology & Engineering*. Academic Press, Elsevier.

Statista. (2015). *Household penetration rate of home appliances in India in 2013*. http://www.statista.com/statistics/370635/household-penetration-home-appliances-india/ (11.07.15).

Sundberg, M. (1998a). An investigation of the edge overheating effect for high-permittivity dielectrics. Paper F. In M. Sundberg (Ed.), *Analysis of industrial microwave ovens. Ph. D. thesis*. Department of Microwave Technology, Chalmers University of Technology.

Sundberg, M. (1998b). *Analysis of industrial microwave ovens. Ph. D. thesis*. Department of Microwave Technology, Chalmers University of Technology.

Sundberg, M., Kildal, P.-S., & Ohlsson, T. (1998). Moment method analysis of a microwave tunnel oven. *Journal of Microwave Power and Electromagnetic Energy*, *33*(1), 36–48.

Sundberg, M., Risman, P. O., & Kildal, P. -S. (1995). Quantification of heating uniformity in multi-applicator tunnel ovens. In: *Proc of the 5th int AMPERE conference on microwave and high frequency heating, Cambridge, UK, Sept.*

Sundberg, M., Risman, P. O., Kildal, P.-S., & Ohlsson, T. (1996). Analysis and design of industrial microwave ovens using the finite difference time domain method. *Journal of Microwave Power and Electromagnetic Energy*, *31*(3), 142–157.

Taflove, A. (1980). Application of the finite-difference time-domain method to sinusoidal steady-state electromagnetic-penetration problems. *IEEE Transactions on Electromagnetic Compatibility*, *EMC-22*(2), 191–202.

Torres, F., & Jecko, B. (1997). Complete FDTD analysis of microwave heating processes in frequency-dependent and temperature-dependent media. *IEEE Transactions on Microwave Theory and Techniques*, *45*(1), 108–117.

USDA (United States Department of Agriculture), (2009). *Ownership of household amenities among selected countries*. (online) http://www.ers.usda.gov/media/612721/householdamenities.xls (03.06.15).

Walker, J. (1987). The secret of a microwave oven's rapid cooking is disclosed. *Scientific American*, (2), 98–102.

Walker, S. J., Bows, J., Richardson, P., & Banks, J. G. (1991). *Listeria survival in chilled retail products: Effects of recommended microwave cooking*. MAFF, Food Safety Directorate, Ministry of Agriculture, Fisheries and Food, Microwave Science Series No 1.

Wäppling Raaholt, B. (2000). *Finite difference time domain analysis of heating of foods in household microwave ovens – Improving the heating uniformity of microwave heated foods. Thesis for the degree of Licentiate of Engineering*. Chalmers University of Technology, Also Paper 1 in.

Wäppling Raaholt, B. (2015). *Applications of industrial microwave heating*. Chalmers University of Technology, Dept for Signals and Systems, SP report 2015:38, ISSN 0284-5172, Nov.

Wäppling Raaholt, B., Galt, S., & Ohlsson, T. (1999). FDTD simulation of a microwave heating process: effects of food parameters. In: *Proceedings of the 7th international conference on microwave and high frequency heating, Valencia, Spain.*

Wäppling Raaholt, B., Holtz, E., & Isaksson, S. (2013). Application of Microwave Technology In Food Preservation And Processing. In S. Bhattacharya (Ed.), *Conventional and advanced food processing technologies*. Blackwell Publishing Ltd/John Wiley & Sons Ltd.

Wäppling Raaholt, B., Isaksson, S., & Hamberg, L. (2016). *Continuous tubular microwave heating of homogeneous foods: Evaluation of heating uniformity in pilot-scale at constant total microwave power*. Paper IV in SP report 2015:38, ISSN 0284-5172, Nov. 2015; submitted to Journal of Microwave Power, Oct. 2015, accepted for publication Jan, 2016. In print, March, 2016.

Wäppling Raaholt, B., & Janestad, H. (2003). *Modelling of microwave defrosting of foods using a coupled electromagnetic and heat transfer approach, technical report.* SIK's Internal.

Wäppling Raaholt, B., Ohlsson, T., & Risman, P. O. (2001). Microwave heating of ready meals – FDTD simulation tools for improving the heating uniformity. In: *Proceedings of the 8th international conference on microwave and high frequency heating, Bayreuth, Germany.*

Wäppling Raaholt, B., & Risman, P. O. (2000). FDTD simulation of a microwave heating process: Effects of oven parameters on heating uniformity. In: *Book of proceedings of the 3rd int. conf on predictive modelling, Leuven, Belgium, September* (pp. 271–273).

Wäppling Raaholt, B., Scheerlinck, N., Galt, S., Ohlsson, T., & Nicolaï, B. (2002). A combined electromagnetic and heat transfer model for heating of foods in microwave combination ovens. *Journal of Microwave Power and Electromagnetic Energy*, *37*(2), 97–111.

Williams, C. (2000). *Income and expenditures.* Canadian Social Trends — Catalogue No. 11-008 (Statistics Canada) (59): 7–12. (online) http://publications.gc.ca/Collection-R/Statcan/11-008-XIE/0030011-008-XIE.pdf, Winter 2000 (03.08.15).

Yahyaei, M., Ghavami, A., Gharachorloo, M., Larijani, K., & Mazhari, S. Z. (2014). Investigating the changes in β-carotene concentrations of carrot and sweet corn using different methods of heat treatments. *Journal of Food Biosciences and Technology*, *4*(2), 15–20 , Summer & Autumn

Yamashita, E. (1990). *Analysis methods for electromagnetic wave problems.* Boston: Artech House.

Yee, K. S. (1966). Numerical solution of initial boundary value problems involving Maxwell's equations in isotropic media. *IEEE Transaction on Antennas Propagation*, *AP-14*(4), 302–307.

Zhang, H., & Datta, A. K. (2000). Coupled electromagnetic and thermal modeling of microwave oven heating of foods. *Journal of Microwave Power and Electromagnetic Energy*, *35*(2), 71–85.

Further reading

Harrington, R. F. (1968) *Field computation by moment methods (Reprint Edition).* R.E. Krieger, Malabar, Fla., 1987; Original Edition.

Risman, P. O. (1991). Terminology and notation of microwave power and electromagnetic energy. *J. Microwave Power and Electromagnetic Energy*, *26*(4).

Risman, P. O. (1998). A microwave oven model: Examples of microwave heating computations. *Microwave World*, *19*(1), 20–23.

Silvester, P. P., & Ferrari, R. L. (1983). *Finite elements for electrical engineers.* New York: Cambridge University Press.

Simulation of microwave processes

18

S.L. Birla[1] *and K. Pitchai*[2]
[1]ConAgra Foods, Omaha, NE, United States, [2]University of Nebraska-Lincoln, Lincoln, NE, United States

18.1 Introduction

Microwave (MW) processing of materials is a technology that can provide the material processor with a powerful, and significantly different, tool to process materials that may not be amenable to conventional means of processing or to improve the performance characteristics of existing processes. However, due to the complexity of microwave interactions with materials, simply placing a sample in a microwave oven and expecting it to heat efficiently will seldom lead to success.

Failure to realize expected benefits from microwave processing is mainly a result of inadequate interaction among researchers, materials engineers, process designers, and microwave engineers. In most cases, the basic equipment (e.g., generators, applicators, power supplies) for microwave processing applications is commercially available. However, the methodology for system integration, including system design, special applicator design, rapid equipment prototyping, and process control, is inadequate. It must be recognized that foods cannot be heated efficiently and uniformly if simply placed in a microwave oven without consideration of specific microwave/materials interactions.

There are many factors affecting microwave heating uniformity including: dielectric properties (dielectric constant and dielectric loss factor), thermal properties (specific heat capacity and thermal conductivity), and physical properties (size, shape, density, and location in the package) of foods (Zhang & Datta, 2000). These factors affect the degree of food components interaction with microwave energy and are important to consider for optimizing food design to achieve uniform heating. Microwave food product development is very time-consuming and expensive. However, use of the computer simulation tool can accelerate the product development cycle and reduce the cost of production. Current methods for designing of foods for microwave heating are mostly based on trial and error. A computer simulation of microwave heating of foods could facilitate proper designing of foods to assure microbial safety.

An important element of microwave process development and system design is the capability to model electromagnetic interactions. The electromagnetic equations for calculating electric field distribution inside a domestic oven or an industrial plant cannot be solved solely by analytical methods. This situation requires to find a numerical method which can solve the complex electromagnetic equations in

The Microwave Processing of Foods. DOI: http://dx.doi.org/10.1016/B978-0-08-100528-6.00018-8

minimum computational time. The advanced numerical methods are a better option to solve those equations with better accuracy in solutions. An understanding of the variation of dielectric properties with temperature and processing state is crucial for simulations and process modeling. Numerical modeling can be used to optimize generator or applicator system design, established achievable processing windows, and conducted realistic process simulations for given dielectric properties, sample size, and desired processing conditions.

18.2 Review of literature

MW heating processes have been mathematically modeled using many numerical techniques. There have been many numerical methods developed and implemented in electromagnetic software packages to develop a predictive model for the microwave heating process. The most predominant numerical methods used for electromagnetic field equations are as follows: Finite-Difference-Time-Domain (FDTD), Finite Element Method (FEM), Finite Volume Method (FVM), Transmission Line Method (TLM), Method of Moments (MOM), Boundary Element Method (BEM), Asymptotic-Expansion Methods (AEM), and Partial Element Equivalent Method (PEEC). The two most common methods are: FDTD and FEM. The FDTD-based software are typically used for solving an electromagnetic field where frequency response is needed over the range of frequency band as it requires only reasonable time and computer resource for solving for a band of frequencies, whereas the FEM is used for solving single frequency response for computational domain adaptively meshed with effective matrix preconditioning (Celuch & Kopyt, 2009). Studies related to the modeling of the microwave heating process started with Ohlsson and Bengtsson (1971). They used the FDTD technique to model microwave heating in infinite slabs of salted ham and beef. Temperature profiles from numerical model prediction and experimental measurement compared favorably. In earlier studies, microwave heat transfer modeling was performed using analytical methods (Shou-Zheng & Han-Kui, 1988; Watanabe et al., 1978). With the advent of powerful computing technologies, solutions of microwave heating can be obtained by using different numerical methods such as FDTD method (Dinčov et al., 2004; Ma et al., 1995) and FEM (Zhang & Datta, 2000). Other studies have modeled microwave heating by solving the heat and mass transfer equation and assuming a source term with exponential decay (Lambert's law) instead of solving the set of Maxwell's equations for the electromagnetic field (Campanone & Zaritzky, 2005; Ni, Datta, & Parmeswar, 1999; Zhou, Puri, Anantheswaran, & Yeh, 1995). Due to variation in geometry and dielectric properties of foods, modeling consists of solving specific and limited problems in microwave heating. However, modeling can be a useful tool to study the effect of modeling parameters influencing microwave heating. Accurate modeling of microwave heating processes requires bringing the most effective methods for solving for both electromagnetism and thermodynamics (Kopyt & Celuch, 2006). An ideal numerical tool should integrate all the physical processes involved in the microwave heating problem, i.e., electromagnetics, thermodynamics, and fluid dynamics. There is no single modeling tool available to include all the effects as a package. However, there are separate software packages available to take care of each physical phenomenon (Kopyt & Gwarek, 2004).

There is still a lack of understanding and even more so predicting the nonuniform heating and complex interactions of microwaves with food involving electromagnetism, and heat and mass transfer. Knoerzer, Regier, & Schubert (2006) mentioned that it is very difficult to predict realistic temperatures or even electromagnetic field distributions inside a microwave oven when foods are involved. MW modeling requires the partial differential equations of electromagnetism and heat transfer to be solved in a coupled approach to minimize the variation in simulation results. In general, material properties have also a directional dependency, which has to be expressed by tensor properties. This is especially valid for inhomogeneous natural material, as a well-known example, the variation of mass transfer along and orthogonal orientation can be mentioned.

Chapman et al. (1993) developed a 3D FDTD-based electromagnetic model to calculate the resonant frequency and a Q factor (a dimensionless parameter, which is ratio of energy stored in the resonator to the energy supplied by the generator per cycle) in a cavity, and the specific absorption ratio (SAR) distribution in the sample. Xiaowei, Lin, and Yiqin (2010) recently developed a procedure for optimizing and calculating electromagnetic field distributions based on microwave oven cavity design and excitation of power. In the following section we discuss in more detail the modeling of microwave heating and the challenges in developing an accurate model.

Basak (1997) reported that the effective heat capacity (enthalpy) is the best suited method to analyze the thawing of heterogeneous material over the range of temperature. Swami (1982) used the finite difference method to describe microwave heating of cylindrical and rectangular shaped high moisture foods. The model predicted temperature distribution was in good agreement with experimental measurements for gel samples of high moisture content and different salt concentrations. Wei et al. (1985) conducted heat and mass transfer analysis during microwave heating in porous medium, water-filled sandstone. The model predicted temperatures and moisture profiles that compared favorably with experimental values. Chen, Singh, Haghighi, and Nelson (1993) derived and incorporated a heat generation term into an axisymmetric finite element model to analyze temperature distribution within a cylinder-shaped potato particulate.

In most modeling tools, space and time discrete numerical methods are used to solve for electromagnetic Maxwell's equations. Niziolek (2009) reviewed the available numerical methods for electromagnetic simulation with each of its strengths and weaknesses. The author mentioned that in real-world problems, it is difficult to find a method for all physical problems, thus hybrid methods (combination of two methods) often prove useful. Traditionally, Transmission Line Method (TLM) and Method of Moments (MOM) methods are used in microwave power engineering applications, but are less powerful and flexible for other applications (Yakovlev, 2001a). Table 18.1 summarizes the pro and cons of various numerical methods implemented by commercial softwares.

Table 18.1 **List of numerical methods available for modeling microwave power problems**

Numerical method	Advantages	Limitations	Commercial software	Applied research work in microwave power engineering
Finite Difference Time Domain (FDTD)	• Ability to obtain wideband results. • Quick implementation on parallel computers. • Good at modeling inhomogeneous and complex materials. • Computational memory required less.	• Errors are dispersive for larger time steps. • Difficult to model thin metal pieces. • Difficult to model materials with frequency dependent properties.	• Quickwave 3D 7.5 • Empire XCcel 5.40 • XFDTD 7.0 • FIDELITY 3.0	Ma et al. (1995) Mechenova and Yakovlev (2004). Celuch and Gwarek (2007) Kopyt and Celuch-Marcysiak (2003) Kopyt and Celuch-Marcysiak (2004) Tilford et al. (2007) Wäppling-Raaholt et al. (2002) Chen, Tang and Liu (2008) Wäppling-Raaholt, Risman, and Ohlsson (2006)
Finite Element Method (FEM)	• Good at handling complex geometries. • Can handle wide variety of engineering problem. • Easy to implement frequency, thermal, pressure, and force dependent material properties. • Best method for modeling resonant cavities.	• It obtains only approximate solution. • Large computational memory required. • Absorbing boundary required for radiation problem.	• HFSS 12.0 • Multiphysics 12.1(ANSYS) • COMSOL Multiphysics 5 • Abaqus 6.8 • LS-DYNA (also available in MOM) • NASTRAN • ALGOR • TWODEPEP (ANSYS)	Kopyt and Celuch (2004) Kopyt and Celuch (2006) Ehlers and Metaxas (2007) Zhang et al. (2001) Zhang and Datta (2003) Akarapu et al. (2004) Huo and Li (2005) Zhou et al. (1995)

Transmission Line Method (TLM)	• Analysis is performed in time domain. • Absorbing boundary condition can be easily modeled. • System explicitly solved.	• Requires more computer memory than FDTD method.	• MEFiSTo-3D Pro 2 • Micro-Stripes 5.6	–
Finite Volume Method (FVM)	• It is well suited for implementing in unstructured meshes. • Electromagnetic field values calculated for a small volume which represents more accurate value of the small domain. • It is well suited for flow dynamics equation solving.	• Dispersive errors more in surface integrals.	• Fluent 13.0 • EMC2000-VF • FLOW-3D 8.2 • CFX 4.3 • PHOENICS	Kopyt and Gwarek (2004) Kopyt and Celuch (2006) Kopyt and Celuch (2007) Verboven, Datta, Anh, Scheerlinck, and Nicolaï (2003) Dinčov et al. (2004)
Method of Moments (MOM)	• It can be used to solve a wide range of equations involving linear operations. • It is highly useful method to solve large and dense matrix equations.	• It is only best suited to solve for linear equations not for nonlinear problems.	• CONCEPT II 8.0	–
Boundary Element Method (BEM)	• Accurate modeling of infinite and semi-infinite domains. • Discretization required only for boundary not for entire domain.	• It does not model well for inhomogeneous or complex materials. • Not good for modeling the problems that combine small details geometries with larger objects. • Practical application not well established.	• LS-DYNA	Huo and Li (2005)

Yakovlev (2001a, 2001b, 2006) has extensively summarized the available software packages capable of modeling microwave heating problems and noted that conformal FDTD-based software packages, such as QuickWave-3D (QWED Sp. Z. o.o., Warsaw, Poland) and Microwave Studio (CST of America, Inc, Framingham, MA, USA), are effective for modeling microwave heating. Curet, Rouaud, and Boillereaux (2006) studied the microwave thawing of a piece of tylose using COMSOL Multiphysics software where the heating due to microwave is introduced in the heat equation solved by both Maxwell's law and Lambert's law separately. More recently, exchanging information between two different numerical methods in two different software packages has also been made possible. For example, one numerical software solves electromagnetic equations and another one solves heat transfer equations. This exchange of information between software packages enables the inclusion of all the physical phenomena in microwave heating. Knoerzer et al. (2006, 2008) developed a user-friendly MATLAB-based interface to couple two prominent software packages namely QuickWave-3D and COMSOL Multiphysics. Some customized software packages are also developed using object-oriented programming languages to couple the electromagnetic–heat transfer equations. It is also possible that commercial software packages can be coupled to custom-built software. Pandit and Prasad (2003) studied the possibility of linking finite element-based software with their own code written in C++.

18.3 Governing physics

Microwaves are often generated in a magnetron and are fed into the cavity of the microwave oven where they dissipate in the food and generate heat. Microwaves transport within the oven cavity and reflect back and forth on the metal boundary of the oven cavity, forming the standing electromagnetic field patterns in the oven.

The electromagnetic field (E), a vector quantity, at any point in the computational domain is governed by Maxwell's Equations. The combined wave form of Maxwell's Equations is expressed as:

$$\nabla \times \mu_r^{-1}(\nabla \times E) - \left(\frac{2 \cdot \pi \cdot f}{c}\right)^2 (\varepsilon' - i \cdot \varepsilon'')E = 0 \tag{18.1}$$

where f is the frequency (e.g., 2.45 GHz) of the microwave oven, c is the speed of light (3×10^8 m/s), ε', ε'', and μ_r are dielectric constant, dielectric loss factor, and magnetic permeability of the food product, respectively.

The solution of electromagnetic models requires discretization of the entire domain into small cells. A discretized cell forming the computational grid or "mesh," can be represented in Cartesian, spherical, or cylindrical coordinates. The typical case of the discretized grid in Cartesian coordinates is popular for FDTD and FEM methods. The main difference in implementing these two methods for solving electromagnetic and thermal fields is the location of the points where corresponding field equations are solved. The FDTD-based solver considers a rectangular structured mesh with the option of conformal boundary condition. The field is calculated at the center of each cell. To get the field value at nodal points of the mesh using the FDTD method, external interpolation functions have to be defined to find the nodal field values. However, the FEM method solves the field in the nodes of each cell. Electromagnetic field equations of (partial) differential forms are applied to the mesh for calculating electric field strength based on properties assigned for each cell and adjacent cells. The solution of electromagnetic equations requires a large number of iterations for each time step for meeting the criteria of convergence. In solving for field distributions in a small enclosed space like a microwave cavity, the metal walls of the model provide a boundary for the solution regime. The interactions at the conducting walls are typically accounted for boundary conditions where (parallel) electric field will not exist. Because metallic walls reflect all the incident microwave energy, there will be no temperature rise in the walls of the cavity (caused by microwave dissipation). The calculated electric field strength, E (V/m) is used as an input to determine the dissipated power density (W/m^3).

Applied microwave energy is converted into heat energy based on the electric field distribution at a particular location. The absorbed power term is considered as a source term in heat transfer equations to calculate transient temperature profile. The equation governing diffusion of heat into a (heat) conductor (i.e., food) is as follows:

$$\rho C_p \frac{\partial T}{\partial t} = k\nabla^2 T + P_v \tag{18.2}$$

where ρ is the density (kg/m^3), C_p is the specific heat at constant pressure (J/kg K), k is the thermal conductivity (W/m K), $\frac{\partial T}{\partial t}$ is the change in temperature over time (K/s), and P_v is the power density absorbed in the load (W/m^3). For the system under consideration, the boundary condition is relatively simple and is prescribed by the heat balance through radiation into the environment.

$$-k\hat{n}\cdot\nabla T = (T^4 - T_\infty^4)\sigma_s\varepsilon_s \tag{18.3}$$

where k is the thermal conductivity (W/m K) of the product, n is the unit vector normal to the surface, ∇T is the temperature gradient (K/m), ε_s is the surface emissivity, σ_s is the Stefan-Boltzmann constant (5.670×10^{-8} J s^{-1} m^{-2} K^{-4}), and T_∞ is the absolute ambient temperature (K). Conductive and convective effects on the gas side are ignored because inside the microwave cavity they are often negligible in rapid microwave heating processes (Akarapu et al., 2004). Otherwise corresponding terms (also, e.g., evaporation heat) have to be added on the right side of Eq. (18.3) leading to

$$-k\frac{\partial T}{\partial n} + h \cdot (T - T_a) = 0 \tag{18.4}$$

where k, $\frac{\partial T}{\partial n}$ is the change in temperature normal to the surface (K/m), h is the heat transfer coefficient (W/m^2 K), and T_a is the surrounding temperature (K).

18.4 Coupled electromagnetic and thermal models

Electromagnetic and thermal phenomena encountered in microwave heating should be solved in a coupled manner because the power dissipation calculated from the model describing the electromagnetic component induces a temperature changes and impacts other physical phenomenon, such as heat transfer and evaporation in heated foods (Bengtsson & Ohlsson, 1974), and phase changes during microwave thawing (Basak & Ayappa, 1997). These physical situations cause rapid changes in material properties, which in turn makes the problem highly nonlinear. Thus, the nonlinear problem of microwave heating needs to be approached as coupled electromagnetic and heat transfer equations in a cyclic manner. The heat generation rate strongly depends on the electric field amplitude, so the change in EM field distribution can increase or decrease the rate of temperature changes. At each time step of the heat transfer solver, updated information of electric field distribution is used as an input to calculate temperature rise in the material using heat transfer equations. The temperature rise in the product changes the thermophysical and dielectric properties. The process of coupling electromagnetic and heat transfer effects in microwave heating is shown in Fig. 18.1. Accurate modeling of a microwave heating process requires bringing together the most effective numerical methods for solving for both electromagnetics and thermodynamics. Zhang and Datta (2000) developed a methodology for coupling electromagnetics and heat transfer in the analysis of microwave heating using two finite element-based software packages and mentioned that coupling of electromagnetism and heat transfer should be needed when the dielectric properties of the product change significantly with temperature. Wäppling-Raaholt et al. (2002) and Knoerzer et al. (2006, 2008) successfully implemented coupled numerical solvers for electromagnetism and heat transfer with a custom developed MATLAB routine. In that study, the authors used the Maxwell's constitutive relations (equations describing the interaction of microwaves with the processed material) to predict the dissipated power distribution, followed by the Fourier heat conduction equation with a convective boundary condition on the food−air interface to predict the transient temperature fields. Akarapu et al. (2004) have developed algorithms to couple the electromagnetic and thermal equations mutually and one-way. Mutual coupling has been approached when the thermophysical properties are temperature dependent. One-way coupling has been applied when the properties are temperature independent. Real frozen cheese sauce heated in a microwave oven was modeled successfully using a coupled approach (Tilford et al., 2007). In this study, the authors accounted for the phase change effect with the placement of sauce in the center and at an offset in the rotating turntable.

18.5 Consideration for model development

18.5.1 Temperature-dependent material properties

Dielectric properties of foods are important parameters that influence the interactions between microwaves and food materials (see Chapter 2: Microwave heating and the

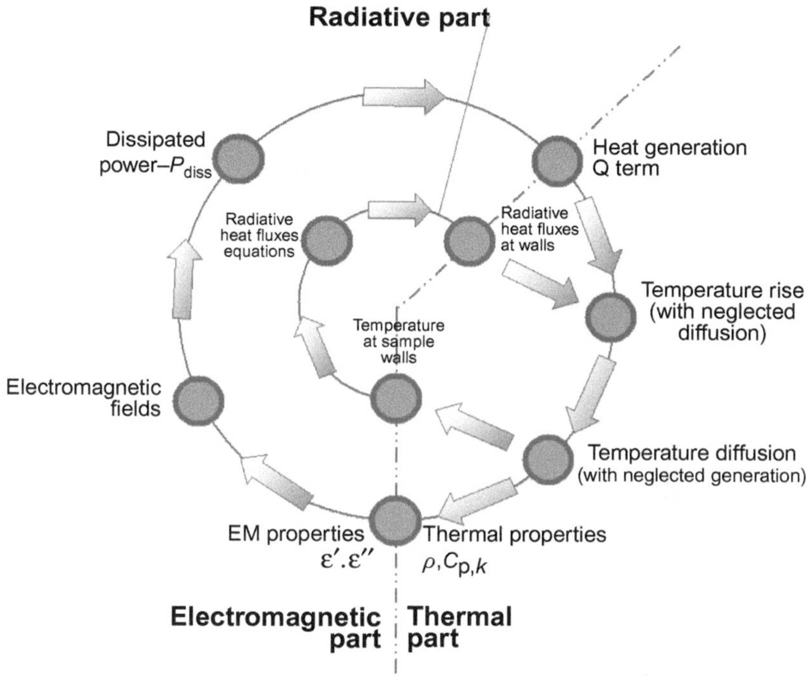

Figure 18.1 Multiphysics and their interaction in microwave heating (Celuch & Kopyt, 2009).

dielectric properties of foods). In particular in applications where phase changes occur during the process (e.g., from frozen to liquid), the dielectric properties change significantly and are, therefore, critical input parameters to the microwave heating model. Relative permittivity (ε_r), is a complex quantity, which is defined as:

$$\varepsilon_r = \varepsilon' - j.\varepsilon'' \qquad (18.6)$$

where the real component is the dielectric constant (ε'), the imaginary component is the dielectric loss factor (ε''), and $j = \sqrt{-1}$. The dielectric constant is related to the capacitance of a substance and its ability to store electric energy, while the loss factor determines the ability of the material to dissipate electric energy as heat. Generally, the dielectric properties of food material vary with composition, moisture, temperature, and frequency (Ryynänen & Ohlsson, 1996 and Chapter 2: Microwave heating and the dielectric properties of foods). During the microwave heating process, the electromagnetic energy dissipates in the material and is converted into heat, which is governed by:

$$Q = 2 \cdot \pi \cdot f \cdot \varepsilon_0 \cdot \varepsilon'' \cdot E^2 \qquad (18.7)$$

where ε_0 ($8.8 \times 10 - 12$ F/m) is the vacuum permittivity, f is the microwave frequency in Hz, ε'' is dielectric loss factor, and E is the electric field in V/m.

Mostly, researchers have measured the temperature-dependent dielectric properties after calibrating the dielectric systems using a short circuit, air, and water at room temperature (see Chapter 3: Determination of the dielectric properties of foods). However, for the dielectric measurement system, drift errors occur when a test system's temperature changes after performing calibration (Agilent, 2002). The measurement accuracy is compromised when the dielectric properties are measured at temperatures far from the calibration temperature. These errors, primarily attributed to temperature variation, can be eliminated by calibrating the dielectric properties system at multiple temperatures. Chen et al. (2015) developed a methodology to measure dielectric properties of lasagna ingredients using calibrating the system at multiple temperatures and quantify measurement errors to determine the optimal number of temperature calibration points from frozen to cooked temperature.

18.5.2 Geometric model development

It is critical to represent the microwave oven, food, and packages geometry accurately in the computational domain as it significantly affects the accuracy of the electric field distribution inside the cavity. Mostly, researchers have assumed that a microwave oven is a cavity in which the port is located on one of the cavity walls (e.g., Dinčov et al., 2004; Geedipalli et al. 2007). In modern microwave ovens, geometric features such as metal bumps, dimples, and turntable crevices have been introduced to improve the electric field distribution inside the cavity. Each of these features can dramatically change the electric field distribution inside the cavity. Therefore, a simulation model should include these features for accurate geometry and process representation. In industrial systems, an MW power source is usually located far away from the applicator and the MW power is transmitted through a long waveguide. In such systems, the computational domain may be limited to the applicator and a few wavelengths distance into the waveguide (Resurreccion, 2012).

Most of the models have assumed TE_{10} mode electromagnetic waves feed through a waveguide. However, the microwaves are fed to the domestic microwave oven cavity through a magnetron in very short, compact irregular shape waveguide. There is no such well-defined rectangular waveguide to assign TE_{10} model. To model these scenarios, Pitchai et al. (2012) assumed magnetron as a coaxial MW power source of transverse electromagnetic wave (TEM in the waveguide). They simulated the scattering parameter for TE_{10} and TEM electromagnetic mode in the frequency range of 2.4 to 2.5 GHz for a gellan gel cylinder (Ø 80 mm × 50 mm) heated in a 700 W rated power microwave oven.

Fig. 18.2 compares the simulated scattering parameters calculated using two input port conditions namely coaxial feed (TEM) and TE_{10} mode. In case of the TE_{10} mode, the reflected power is slightly higher than that of TEM mode throughout the frequency spectrum. It gives the indication that the electromagnetic energy couples better in TEM mode than TE_{10} mode with the load. Interestingly, the deep resonance (lowest value of S11) was found to occur at the same frequency, which

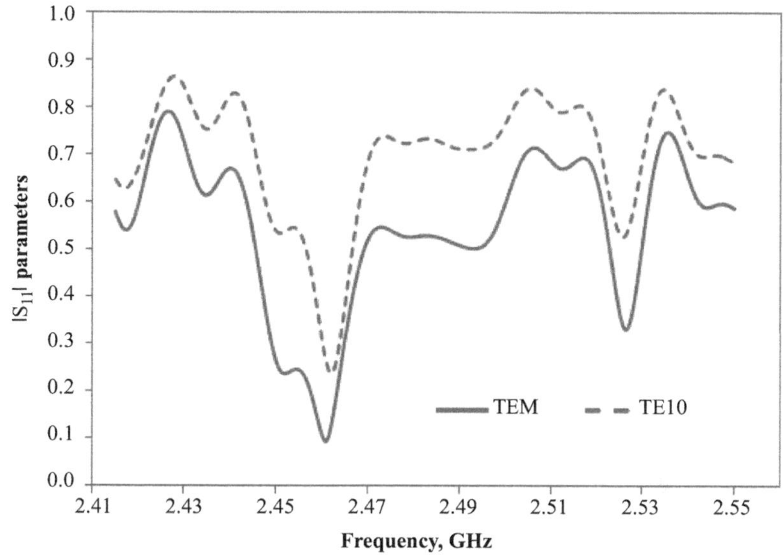

Figure 18.2 Scattering parameter of TEM and TE_{10} electromagnetic modes extracted from frequency spectrum (Pitchai et al., 2012).

is at 2.46 GHz in both types of feed conditions. Besides, the coaxial feeding in a microwave oven represents an actual scenario. Therefore, it is reasonable to assume the magnetron source as a coaxial feed in the microwave heat transfer model.

18.5.3 Optimizing model parameters

The focus of the modeling work was to simplify the models assumptions and to optimize electromagnetic and computational parameters. Electromagnetic models of microwave power problems are subjected to numerical error unless the parameters of electromagnetic fields and computational variables are optimized. Therefore, the complex interactions of electromagnetic fields with food need to be subjected to a numerical optimization process. This optimization process for each electromagnetic and thermal parameter can be easily done using a computational platform.

To be able to develop the electric field patterns and calculate accurate power absorption during each heating time step in electromagnetic simulation, electromagnetic field iterations need to reach steady state (Kopyt & Celuch-Marcysiak, 2003). It is important to choose the correct heating time step (Δt) at each thermal iteration to determine the convergence and the rate of convergence of the simulation (Celuch-Marcysiak, Gwarek, & Sypniewski, 2006). The choice of time step should be neither too big nor too small. In case of a too big heating time step, immediate divergence of the temperature fields would occur, whereas in case of a too small heating time step, the convergence of temperature fields would be reached but at the expense of a longer simulation time and possibly higher numerical errors. Mesh size optimization is important for obtaining better simulation results. Reducing the mesh size by half brings the space discretization errors down by a factor of four. However, it should also be mentioned that the computing time will increase 16-fold and memory occupation will increase eight-fold. Electromagnetic simulation in a dielectric material typically requires about 8 to 10 cells per wavelength (Akarapu et al., 2004).

The accuracy of the coupled MW heating model depends on the heating time step as well as the spatial and temporal discretization of the EM model. Meshing of the geometric model is a very important step in the model development. Due to different dielectric properties, wave length also differs in dielectric materials compared to air. As a rule of thumb, there should be at least 8 to 10 mesh elements per wavelength in each material for reasonable convergence of solutions for E-field results. It is highly recommended to run a mesh convergence study to ascertain that a chosen mesh scheme does not introduce a large error in E-field results. In FEM meshing, the quality of mesh should not be below 0.2 anywhere in the domain, otherwise the convergence rate can be very slow and simulated results could be erroneous.

The heating time step and mesh size are interrelated with each other in deciding the error bounds of the numerical computation. The level of dispersion depends on the direction of propagation, time step, and mesh size. To get more accurate simulation results, the numerical techniques should meet the criteria of Courant stability. The Courant stability factor is calculated by the following equation for the case of homogeneous, lossless, and equidistant meshes (Taflove & Hagness, 2000):

$$r = v\left(\frac{\Delta t}{a}\right) \tag{18.5}$$

where v is the phase velocity in the medium (m/s), Δt is the heating time step (s), r is the Courant stability criteria factor, and a is represents equal size of mesh in all three directions in Cartesian coordinates ($a = \Delta \mathrm{x} = \Delta \mathrm{y} = \Delta \mathrm{z}$). The Courant criteria dictates that factor $r \leq \frac{1}{\sqrt{3}}$ for stable operation of the numerical algorithm. Selection of time step in the time-dependent solver is also dictated by nonlinearity of thermal and dielectric properties of material. Time step size should be very small (less than 5 s) when material undergoes phase during heating of frozen foods (Celuch & Kopyt, 2009). One can set the time step manually or commercial software can decide the time step automatically based on encountered nonlinearity.

A more complete model of microwave heating requires that the load movement (rotation or linear movement) also be accounted for, as in many MW units load is moved to improve heating uniformity. The continuous load rotation can be discretized by defining a set of finite number of distinct positions of the load that it can assume within the cavity. Using such an approach, one full revolution of the turntable is represented by a finite number of steps with instantaneous transition between them. As expected, by increasing the number of angular steps, the accuracy of the result can be improved; however, increasing beyond a certain number of the steps, no further improvements of accuracy are achieved, yet the computation time is prolonged, as demonstrated by Kopyt and Celuch-Marcysiak (2003). They recommended 12 to 16 angular steps for reasonable accuracy.

The commercial software QuickWave-3D has in-built features to simulate moving loads. However FEM-based software such as HFSS and COMSOL Multiphysics does not have in-build features. COMSOL advanced user can use moving mesh feature of the software to get around the problem however, it takes excessive long computation time to simulate rotating objects. To overcome the issue, Pitchai et al. (2014) developed an approach to use a MATLAB (MathWorks Inc., Natick, MA, USA)-based algorithm to simulate microwave heating of rotating objects. They assumed that the load moves at a discrete angle instantaneously after being heated for a time step (Δt, s). At the end of each time step (Δt, s), the temperature field in the load domain was saved along with their coordinates in an external text file. In next angular step COMSOL Multiphysics software called the MATLAB routine to rotate the load by discrete angle (dθ) from previous position (Eqs. 18.7 and 18.8) and initialized the temperature from the previously saved file using interpolation function with space coordinate arguments, X', Y', and Z.

$$X' = \mathrm{x}\cos(\mathrm{d}\theta) - \mathrm{y}\sin(\mathrm{d}\theta) \tag{18.8}$$

$$Y' = \mathrm{x}\sin(\mathrm{d}\theta) + \mathrm{y}\cos(\mathrm{d}\theta) \tag{18.9}$$

Using the new temperature field, the electric field distribution is computed in COMSOL for the next time step (Fig. 18.3). This loop is continued, until the full heating time is reached. In a recent release of COSMOL software version 5.0, users can write JAVA methods to accomplish all pre-, real-time, and postprocessing within the software.

18.5.4 Electromagnetic frequency spectrum of magnetron

According to the Industrial, Scientific, and Medical (ISM) band allocated by Federal Communications Commission (FCC) for the food applications, microwave ovens are designed to be operated at 2450 MHz with a frequency tolerance limit of 50 MHz (Buffler, 1993). The domestic oven magnetron is complex in its operation due to various factors such as varying anode current and the cold-start process. Besides, a magnetron frequency spectrum depends on type of power supply, dielectric properties of food material, magnetron temperature, and magnetron construction. Due to these factors, a magnetron does not operate at a fixed single frequency but at a range of frequencies (Ghammaz, Lefeuvre, & Teissandier, 2003; Risman, 2009). Therefore, for accurate prediction of electric field density and temperature distribution in food, the magnetron frequency spectrum should be incorporated in the microwave heat transfer model to achieve greater accuracy.

Nonetheless, most microwave heat transfer simulations simplify that the magnetron generates a monochromatic electromagnetic wave of frequency of 2450 MHz to reduce the computational complexity. Pitchai et al. (2012) demonstrated through

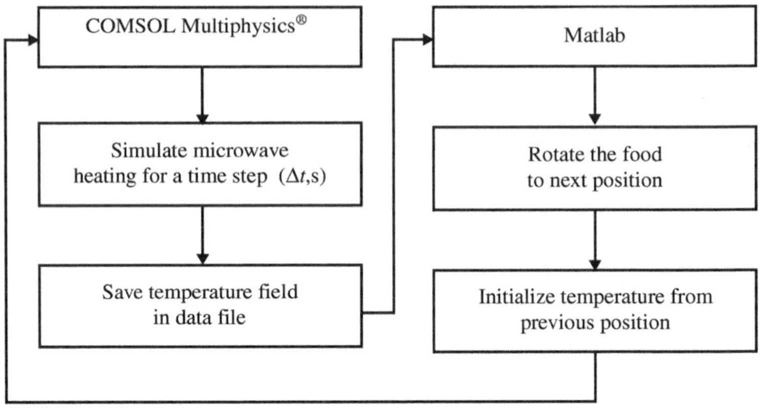

Figure 18.3 COMSOL–MATLAB interface depicting simulation strategy for rotating food in the turntable (Pitchai et al., 2014).

microwave heat transfer simulations that temperature field patterns are affected by the selected frequency (Fig. 18.4). Soltysiak, Erle, and Celuch (2010) demonstrated the effect of different monochromatic frequencies on temperature pattern with respect to different rotational position of the food on the turntable through modeling. The predicted temperature showed that temperature within the food varies over 10°C as the food is heated with different frequencies. Thus, it can be concluded that the magnetron frequency influences the temperature field patterns within the food (Schubert & Regier, 2005; Tang & Resurreccion, 2009).

Previously, researchers have not included the frequency spectrum in the electromagnetic and heat transfer simulation of domestic microwave heating of food. As a new paradigm in the microwave heat transfer simulations, Pitchai et al. (2016) developed a methodology to incorporate the electromagnetic frequency spectrum in the range of 2.4 to 2.5 GHz to the simulation in the form of Gaussian shape distribution. The electromagnetic frequency spectrum was assumed to be of Gaussian shape to cover the entire FCC allowable range of 0.05 GHz as a bell shape normal distribution curve with the central frequency of 2.45 GHz and variance of 0.05^2, 0.025^2, and 0.017^2 GHz^2 frequency (Fig. 18.5).

Three experimental surface temperature patterns at the top surface recorded through IR imaging were compared with the simulated temperature patterns using monochromatic frequency of 2.45 GHz and the three Gaussian frequency distribution after 6 min of microwave heating of mashed potato (Fig. 18.6). The 2.45 GHz frequency simulation showed the surface temperature pattern with location of cold middle spots consistent with all the three experimental surface temperature patterns. However, there was a location of cold spot of 40°C in the 2.45 GHz frequency simulation at the right-middle corner, which was not observed in the experimental temperature patterns. However, also all three Gaussian frequency distribution simulations showed a similar surface temperature pattern to that of the 2.45 GHz simulation; and there was no cold spot predicted in the right-middle corner in all three Gaussian frequency distribution simulation.

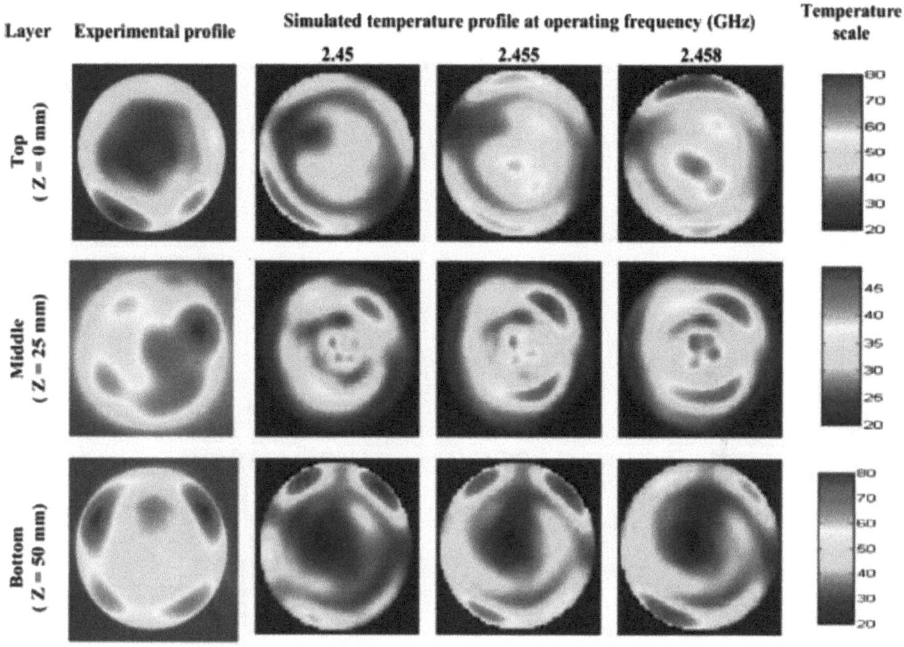

Figure 18.4 Effect of frequency on spatial heating profile of gellan gel cylinder (80 × 50 mm) subjected to 30 s heating in 700 W microwave oven (Pitchai et al., 2012).

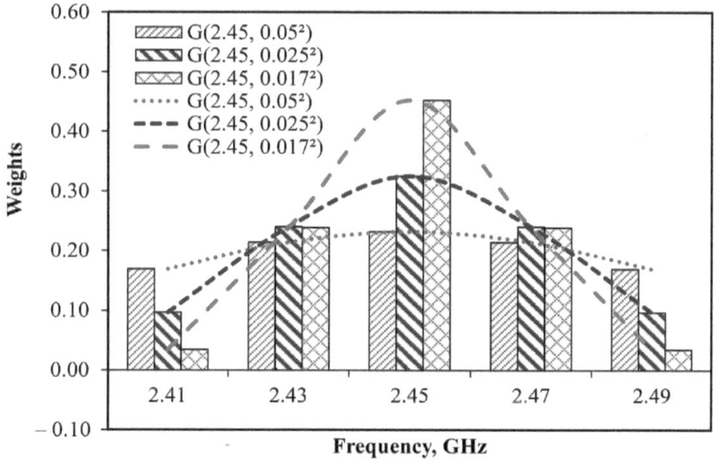

Figure 18.5 Electromagnetic Gaussian shape frequency spectrum with a central frequency of 2.45 GHz and three variances (Pitchai et al., 2016).

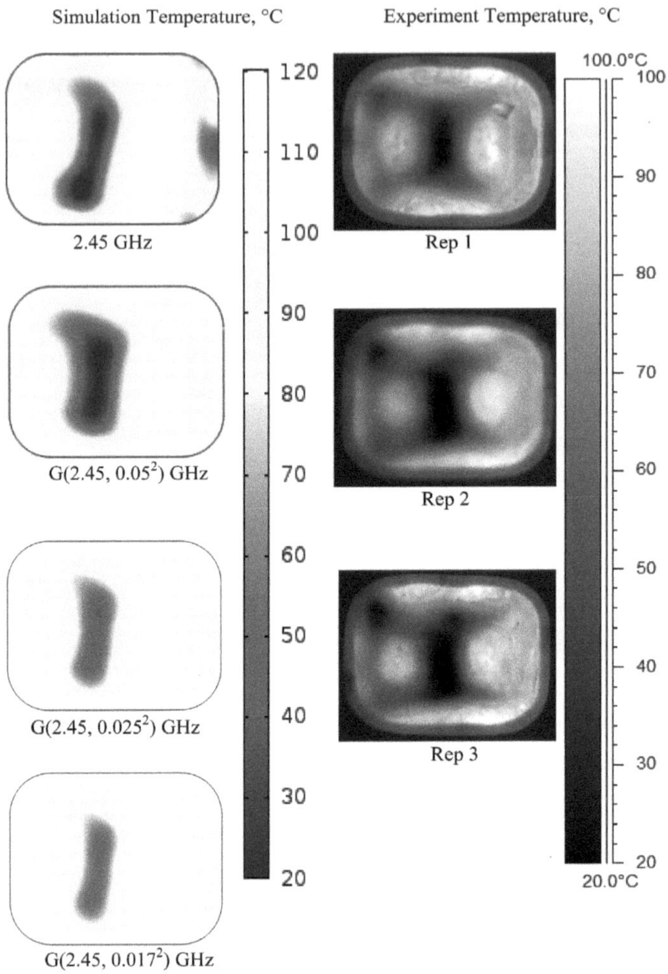

Figure 18.6 Comparison of spatial surface temperature profiles between experiment (3 replications) and simulations of different frequencies after microwave heating of frozen mashed potato for 6 min (Pitchai et al., 2016).

As shown in Fig. 18.6, as the variance of the Gaussian distribution decreases from 0.05^2 to 0.017^2 GHz^2, the regions of cold spots at the middle location shrink and do not match well with the temperature patterns observed in the experiments. It should be noted that as the variance decreases, the frequency spectrum operates close to 2.45 GHz and the contribution from 2.45 GHz was higher. Using a broader 0.05 GHz bandwidth, the averaged electromagnetic power density from all five frequencies was shared by similar weights and the average value worked the best. Therefore, it was important to include the contribution from the tail-end frequencies into the model. The computation time for simulating the 6 min of microwave heating took about 42 h for the 2.45 GHz frequency simulation (Dell Precision T7500 workstation with a 72 GB DDR3 RAM running on two quad-core Intel Xeon X5570 2.93 GHz processor), whereas the frequency spectrum simulation took 66 h. Although, the predicted surface temperature pattern of the frequency spectrum simulation yielded qualitatively a better match with the experimental temperature pattern than that of the single frequency model at 2.45 GHz, it is essential to evaluate the accuracy through transient point temperature measurements.

18.5.5 Active package components inclusion in model

In addition to heating load, there are other components in the MW system that influence EM field distribution in load. The use of active package components (see Chapter 13: Packaging for microwave foods) has been a preferred choice for overcoming some of the challenges associated with MW heating of complex food matrix. Susceptors, very thin metallic microwave absorbing films, are used in microwavable food packaging. They tend to heat up very rapidly during microwave heating and this effect is used to overcome major issues as the lack of browning and crisping. An ability of a thin metal layer to work as a susceptor is described by the surface resistance ($R_s = \frac{1}{\sigma d}$), where σ is electric conductivity of the metal used for the susceptor (S/m), d is the thickness of the metal layer (m). Note that a brute-force FDTD approach would require mesh refinement to the thickness of the susceptor and hence prohibitive computer resource.

In COMSOL Multiphysics, transition boundary condition is used on interior boundaries to model a sheet of active film that should be geometrically thin but not electrically thin. It represents a discontinuity in the tangential electric field. COMSOL software has transition boundary features allowing to take into account of thin metallic films. This approach completely eliminates the need for discretization of thin geometry. In QuickWave-3D software, the thin geometrical active features of package is modeled as a surrogate thin layer. Therefore, a thicker surrogate dielectric layer is defined instead of a thin metal film. The proper electric conductivity of the surrogate layer of thickness, d, is calculated using relation.

$$\sigma = \frac{1}{R_s d} \tag{18.10}$$

Electrical conductivity of susceptor film is a function of temperature. Due to rapid and uneven MW heating, the thin susceptor film starts to develop crack due to shrinking of plastic film.

A few researchers have attempted to develop susceptor-assisted microwave heat transfer models. Celuch et al. (2008) and Soltysiak et al. (2010) used finite-difference time domain (FDTD)-based software, QuickWave, to solve the susceptor-assisted microwave heating of a water container placed inside a cavity. For susceptor material, the authors created a 1-mm thick layer of a surrogate material whose properties were adjusted to mimic the susceptor film. Pitchai et al. (2011) developed a 3D Finite Element Method (FEM)-based computer model to simulate microwave interactions with a model food (1% gellan gel) and susceptor placed on top of a model food. In this study, they demonstrated that, for thin film a transition boundary can be used instead of geometrically creating a thin layer and discretizing it.

18.6 Examples of modeling of batch and continuous microwave heating processes

18.6.1 Batch processes

Modeling domestic microwave oven heating was pioneered by a group of scientists led by Dr. Ashim K. Datta at Cornell University. His research group was instrumental in developing modeling techniques for microwave oven heating. In his lab, research on microwave heating modeling evolved from analytical approaches to numerical approaches. The effect of geometric shape and applied power were systematically studied for model food (Chamchong & Datta, 1999a, 1999b). To improve the heating uniformity in microwave heating, the rotation of the load and combination of heating methods were described as important. During only 30 s heating of a potato slab, the heating uniformity of the rotated load increased by 37–43% while compared to the stationary load. It was observed that heating uniformity of the rotated load does not improve in different planes along the vertical axis of the potato. Thus, it was suggested that heating uniformity can be achieved better in different profiles when the system is designed to move the food at top and bottom (Geedipalli et al., 2007). Datta and Ni (2002) studied the possibility of incorporating infrared and hot air heating with microwave heating to reduce moisture build up in a surface profile, while at the same time increasing the surface temperature. When comparing the combination heating with microwave-only heating, it can be concluded that combination heating provides an opportunity to alleviate inherent problems of pressure-driven moisture buildup at the surface of high-moisture foods. Ni et al. (1999) studied the moisture loss due to evaporation and related that to heating uniformity. In this study, authors concluded that when the surface area increases for a given volume, the uniformity of heating increases but the total moisture loss reduces. Sterilization of solid foods using microwaves was modeled using numerical techniques (Zhang et al. 2001). Using the time–temperature history and first order thermal kinetics, spatial distributions of thermal death time (F_0 value), representing the sterilization process, were calculated.

A research group at the Institute of Radioelectronics, Warsaw University of Technology, Poland has developed a software platform for solving microwave power problems. The developed software (QuickWave 3D) is capable of modeling rotating and horizontally moving foods while heated in a microwave oven. Celuch and Gwarek (2000) summarized the advanced features of the FDTD method in microwave power applications. The authors have discussed the conformal geometry models, field excitation techniques, and parameter extraction procedure techniques.

A main advantage of the recent algorithm development is that conformal FDTD meshing helps to design curved sloped and boundary edges of complicated food shapes. As mentioned earlier, QuickWave-3D software can be coupled with external software to solve the multiphysics of microwave heating. Kopyt and Celuch (2006) have presented studies on coupling the FDTD-based software with FVTD- and FEM-based softwares. This hybrid technique of solving microwave power problems was necessary for accurate modeling.

Researchers from the University of Nebraska–Lincoln used COSMOL Multiphysics software to develop models for more realistic microwave ovens, food geometry and active packages to develop an holistic understanding of complex microwave interaction.

18.6.2 Continuous processes

The common objective of the design of a microwave system for thermal processing is to address the issue of uneven heat distribution and to optimize the energy efficiency by minimizing reflected power. Load rotation is the most popular temperature-equilibrating mechanism in domestic microwave oven (refer to Section 18.5.3). However, different types of load movements are also used in the engineering practice in industry, i.e., linear translation is widespread in industrial tunnel installations.

However, even minor changes in size, geometry, insertion of parts, and even orientation of any part in a microwave system can result in significant alteration of field distribution. Since the risk and cost is too high for the construction of a physical design and experimentation of the outcome, there is a need to predict a reasonable outcome of a design even before it is physically constructed to reduce capital costs and operating costs and time for experimentation.

Researchers at Washington State University (WSU, Washington, USA) started the development of a 915 MHz single-mode continuously operating microwave sterilization system using computer simulation as one of the tools for guiding their design and development process (Chen et al., 2008; Resurreccion et al., 2013).

Another continuously operating, i.e., flow through, MW system was simulated by Salvi, Boldor, Aita, and Sabliov (2011) using COMSOL Multiphysics. A numerical model was developed to simulate temperature profiles in Newtonian as well as non-Newtonian fluids during continuous flow microwave heating by simultaneously coupling of electromagnetism, fluid flow, and heat transport in COMSOL Multiphysics. Zhu et al. (2008) simulated microwave heating of a particulate flow containing spherical particles in flow through a microwave system. A 3D model that includes coupled Maxwell, continuity, Navier–Stokes, and energy equations is developed to describe transient temperature, electromagnetic, and fluid velocity fields. The hydrodynamic interaction between the solid particles and the carrier liquid is simulated by the force-coupling method. Lee et al. (2015) used computer simulation as a design tool for developing continuous flow hybrid microwave and pulsed ohmic heating systems for minimization of thermal lags in the processing of particulate foods.

Rogelio et al. (2015) used Microwave Studio to design and develop a batch type microwave vacuum dryer for low moisture food application. Cavity dimensions, microwave ports locations, and loading scheduling were optimized using commuter simulation. This is an excellent example of how the modeling-based approach in the design of a microwave applicator avoids pitfalls associated with MW-based processes.

Kurniawan et al. (2015) studied the effect of mode stirrers in a multimode microwave-heating applicator with the conveyor belt. A 3D simulation model was built and simulated using the COMSOL Multiphysics software. Since it is very difficult to simulate the mode stirrer's motion using COMSOL, eleven still angles of the mode stirrers are considered. The final temperature distribution result is obtained by taking the average of all eleven angles results. Also to get the final temperature of a heated material after it comes out of the cavity, the average of all temperature values in the direction of conveyor belt motion is taken as it is assumed that conveyor belt motion is much slower than the speed of the mode stirrers. The simulation results show that temperature distribution is more uniform for the case with mode stirrers compared to that obtained without mode stirrers.

18.7 Future outlooks

In the earlier days, due to the lack of powerful computers and computational techniques, it was impossible to model the coupled partial differential equations (Knoerzer et al., 2008). Remarkable progress has been achieved in recent years in the numerical simulation of microwave heating, which range from numerical modeling and computing to many applications in processing various kinds of materials. We believe that the development will continue based on more refined models, faster numerical methods, and more complete description of the complex permittivity of more materials.

With the help of numerical simulation, the complex process of interaction between microwave and materials will be further understood, the critical parameters influencing the process will be identified, and a rapid design of optimized and controllable applicators will be provided. Computer modeling and numerical simulation can provide valuable information involving various aspects of microwave process design and development, not accessible by any other means, including: microwave interaction with materials; effects of hybrid heating; heating multiple samples; the design and scale-up of a heating system, and process control parameters.

Recent advances in development of computational techniques and numerical algorithms have made it possible to develop predictive models, which will describe the process more accurately. Nowadays, in order to design thermal process, knowledge about the temperature distribution during heating treatment inside a microwave cavity is required. Government regulators require information on temperature distribution of the food product to regulate and approve the process design and scale-up the process from lab- or pilot-scale to full commercial scale operations.

Computer-aided modeling can be a tool to meet these requirements (Knoerzer et al., 2008). The challenge remains, however, for researchers in industry to learn and adopt the modeling theory and technology in a simple and holistic way to make an improvement in microwave food products and processes. Despite the recent advances in coupled simulation techniques and the quick rate of commercial softwares coming out with added new features, it is not easy to set up a model for coupled multiphysics problems. The ability to take advantage of recent development in coupled problems depends on the overall knowledge-based skills and proficiency of the model developer.

References

Agilent (2002). *Applying error correction to network analyzer measurements. Agilent Literature No. 5965-7709E*. Santa Clara, CA: Agilent Technologies.

Akarapu, R., Li, B. Q., Huo, Y., Tang, J., & Liu, F. (2004). Integrated modeling of microwave food processing and comparison with experimental measurements. *Journal of Microwave Power and Electromagnetic Energy*, *39*(3/4), 153.

Basak, T., & Ayappa, K. G. (1997). Analysis of microwave thawing of slabs with effective heat capacity method. *AICHE Journal*, *43*(7), 1662–1674.

Bengtsson, N. E., & Ohlsson, T. (1974). Microwave heating in the food industry. *Proceedings of the IEEE*, *62*(1), 44–55.

Buffler, C. R. (1993). *Microwave cooking and processing: Engineering fundamentals for the food scientist*. New York: Van Nostrand Reinhold.

Campanone, L. A., & Zaritzky, N. E. (2005). Mathematical analysis of microwave heating process. *Journal of Food Engineering*, *69*(3), 359–368.

Celuch, M., & Gwarek, W. (2000). Advanced features of FDTD modeling for microwave power applications. In *35th Microwave Power Symposium, Montreal, Canada*.

Celuch, M., Gwarek, W., & Soltysiak, M. (2008). Effective modeling of microwave heating scenarios including susceptors. In *International Conference on recent advances in microwave theory and applications, 2008. Microwave 2008, Jaipur, 2008* (pp. 404–405).

Celuch, M., & Gwarek, W. K. (2007). Properties of the FDTD method relevant to the analysis of microwave power problems. *Journal of Microwave Power and Electromagnetic Energy*, *41*(4), 62–80.

Celuch, M., & Kopyt, P. (2009). Modeling microwave heating of foods. In M. W. Lorence, & P. S. Pesheck (Eds.), *Development of packaging and products for use in microwave ovens*. Woodhead Publishing.

Celuch-Marcysiak, M., Gwarek, W. K., & Sypniewski, M. (2006). A novel FDTD system for microwave heating and thawing analysis with automatic time-variation of enthalpy-dependent media parameters. In *199–209. Ed. Anonymous*. Springer.

Chamchong, M., & Datta, A. K. (1999a). Thawing of foods in a microwave oven: I. Effect of power levels and power cycling. *Journal of Microwave Power Electromagnetic Energy*, *34*(1), 9–21.

Chamchong, M., & Datta, A. K. (1999b). Thawing of foods in a microwave oven: II. Effect of load geometry and dielectric properties. *Journal of Microwave Power and Electromagnetic Energy*, *34*(1), 22–32.

Chapman, B., Iskander, M. F., Smith, R. L., & Andrade, O. M. (1993). *Simulation of sintering experiments in single-mode cavities. Materials Research Society Symposium Proceedings*. Materials Research Society, 53.

Chen, D. S. D., Singh, R. K., Haghighi, K., & Nelson, P. E. (1993). Finite element analysis of temperature distribution in microwaved cylindrical potato tissue. *Journal of Food Engineering*, *18*(4), 351–368.

Chen, H., Tang, J., & Liu, F. (2008). Simulation model for moving food packages in microwave heating processes using conformal FDTD method. *Journal of Food Engineering*, *88*(3), 294–305.

Chen, J., Pitchai, K., Birla, S., Gonzalez, S. R., Jones, D., & Subbiah, J. (2015). Development of a multi-temperature calibration method for measuring dielectric properties of foods. *IEEE Transactions on Dielectrics and Electrical Insulation*, *22*(1), 626–634.

Curet, S., Rouaud, O., & Boillereaux, L. (2006). *Heat transfer models for microwave thawing applications*.

Dahl, C. A., Matthews, M. E., & Marth, E. H. (1981). Survival of *Streptococcus faecium* in beef loaf and potatoes after microwave-heating in a simulated cook/chill foodservice system. *Journal of Food Protection*, *44*, 128–133.

Datta, A. K., & Ni, H. (2002). Infrared and hot-air-assisted microwave heating of foods for control of surface moisture. *Journal of Food Engineering*, *51*(4), 355–364.

Dinčov, D. D., Parrott, K. A., & Pericleous, K. A. (2004). A new computational approach to microwave heating of two-phase porous materials. *International Journal of Numerical Methods for Heat and Fluid Flow*, *14*(6), 783–802.

Ehlers, R. A., & Metaxas, R. A. (2007). An investigation on the effect of varying the load, mesh and simulation parameters in microwave heating applications. *Journal of Microwave Power and Electromagnetic Energy*, *40*(4), 251–259.

Geedipalli, S., Rakesh, V., & Datta, A. (2007). Modeling the heating uniformity contributed by a rotating turntable in microwave ovens. *Journal of Food Engineering*, *82*(3), 359–368.

Ghammaz, A., Lefeuvre, S., & Teissandier, N. (2003). Spectral behavior of domestic microwave ovens and its effects on the ISM band. *Annals of Telecommunications*, *58*(7), 1178–1188.

Huo, Y., & Li, B. Q. (2005). Boundary/finite element modeling of three-dimensional electromagnetic heating during microwave food processing. *Journal of Heat Transfer*, *127*, 1159–1166.

Knoerzer, K., Regier, M., & Schubert, H. (2006). Microwave heating: A new approach of simulation and validation. *Chemical Engineering & Technology*, *29*(7), 796–801.

Knoerzer, K., Regier, M., & Schubert, H. (2008). A computational model for calculating temperature distributions in microwave food applications. *Innovative Food Science & Emerging Technologies*, *9*(3), 374–384.

Kopyt, P., & Celuch, M. (2006). *Towards a multiphysics simulation system for microwave power phenomena*. Microwave conference proceedings, 2005. APMC 2005. Asia-Pacific conference proceedings (p. 4). IEEE.

Kopyt, P., & Celuch, M. (2007). Coupled electromagnetic-thermodynamic simulations of microwave heating problems using the FDTD algorithm. *Journal of Microwave Power and Electromagnetic Energy*, *41*(4), 18–29.

Kopyt, P., & Celuch-Marcysiak, M. (2003). FDTD modelling and experimental verification of electromagnetic power dissipated in domestic microwave ovens. *Journal of Telecommunication and Information Technology*, *1*, 59–65.

Kopyt, P. & Celuch-Marcysiak, M. (2004). Coupled FDTD-FEM approach to modelling of microwave heating process. In *Fifth IEE international conference on computation in electromagnetics* (pp. 171–172).

Kopyt, P., & Gwarek, W. (2004). *A comparison of commercial CFD software capable of coupling to external electromagnetic software for modeling of microwave heating process. In Proceedings of the 6th seminar in computer modeling and microwave power engineering* (pp. 33–39). Austin, Texas, USA: *Worcester Polytechnic Institute, Worcester.*

Kurniawan, H., Alapati, S., & Che, W. S. (2015). Effect of mode stirrers in a multimode microwave-heating applicator with the conveyor belt. *International Journal of Precision Engineering and Manufacturing-Green Technology*, *2*(1), 31–36.

Lee, S. H., Choi, W., Yoon, S. K., & Jun, S. (2015). Development of a dual cylindrical microwave and ohmic combination heater for minimization of thermal lags in the processing of particulate foods. *LWT-Food Science and Technology*, *63*(2), 1220–1228.

Ma, L., Paul, D. L., Pothecary, N. M., Railton, C. J., Bows, J., Barratt, L., ... Simons, D. (1995). Experimental validation of a combined electromagnetic and thermal FDTD model of a microwave heating process. *IEEE Transactions on Microwave Theory and Techniques*, *47*.

Mechenova, V. A., & Yakovlev, V. V. (2004). Efficiency optimization for systems and components in microwave power engineering. *Journal of Microwave Power and Electromagnetic Energy*, *39*(1), 15–30.

Ni, H., Datta, A. K., & Parmeswar, R. (1999). Moisture loss as related to heating uniformity in microwave processing of solid foods. *Journal of Food Process Engineering*, *22*(5), 367–382.

Niziolek, M. (2009). Review of methods used for computational electromagnetics. In *Electrodynamic and mechatronics, 2009. SCE 11'09. 2nd international students conference on* (pp. 15–16). IEEE.

Ohlsson, T., & Bengtsson, N. E. (1971). Microwave heating profiles in foods: A comparison between heating experiments and computer simulation. *Microwave energy applications newsletter ed.* SIK Publication.

Pandit, R. B., & Prasad, S. (2003). Finite element analysis of microwave heating of potato-transient temperature profiles. *Journal of Food Engineering*, *60*(2), 193–202.

Pitchai, K., Birla, S., Jones, D., & Subbiah, J. (2011). Modeling of susceptor assisted microwave heating in domestic oven. In *Proceeding of COSMOL annual conference*, Oct 11–14, Boston.

Pitchai, K., Birla, S., Subbiah, J., Jones, D., & Thippareddi, H. (2012). Coupled electromagnetic and heat transfer model for microwave heating in domestic ovens. *Journal of Food Engineering*, *112*(1/2), 100–111.

Pitchai, K., Chen, J., Birla, S., Gonzalez, R., Jones, D., & Subbiah, J. (2014). A microwave heat transfer model for a rotating multi-component meal in a domestic oven: Development and validation. *Journal of Food Engineering*, *128*, 60–71.

Pitchai, K., Chen, J., Birla, S., Jones, D., & Subbiah, J. (2016). Modeling microwave heating of frozen mashed potato in a domestic oven incorporating electromagnetic frequency spectrum. *Journal of Food Engineering*, *173*, 124–131.

Resurreccion, F. P. (2012). *Microwave assisted thermal processing of homogeneous and heterogeneous food packed in a polymeric container, PhD dissertation.* Washington State University: Department of Biological Systems Engineering.

Resurreccion, F. P., Tang, J., Pedrow, P., Cavalieri, R., Liu, F., & Tang, Z. (2013). Development of a computer simulation model for processing food in a microwave assisted thermal sterilization (MATS) system. *Journal of Food Engineering, 118*, 406–416.

Risman, P. O. (2009). Advanced topics in microwave heating uniformity. In M. W. Lorence, & P. S. Pesheck (Eds.), *Development of packaging and products for use in microwave ovens* (pp. 66–104). Cambridge, UK: Woodhead Publishing Limited.

Risman, P. O., & Celuch-Marcysiak, M. (2002). Electromagnetic modelling for microwave heating applications. In: *13th IEEE international conference on microwaves, radar and wireless communications. 2000. MIKON-2000* (pp. 167–182).

Rogelio, J. P., Abulencia, G. L., & Baldovino, R. G. (2015). Optimization of microwave vacuum dryer design parameters using CST microwave studio for low-moisture food application. In: *Proceedings of the international multiconference of engineers and computer scientists 2015 Vol. I, IMECS 2015*, March 18–20, 2015, Hong Kong.

Ryynänen, S., & Ohlsson, T. (1996). Microwave heating uniformity of ready meals as affected by placement, composition, and geometry. *Journal of Food Science, 61*(3), 620–624.

Salvi, D., Boldor, D., Aita, G. M., & Sabliov, C. M. (2011). COMSOL multiphysics model for continuous flow microwave heating of liquids. *Journal of Food Engineering, 104*, 422–429.

Schubert, H., & Regier, M. (2005). *The microwave processing of foods*. Cambridge, UK: Woodhead Publishing Limited.

Shou-Zheng, Z., & Han-Kui, C. (1988). Power distribution analysis in rectangular microwave heating applicator with stratified load. *Journal of Microwave Power and Electromagnetic Energy, 23*(3), 139–143.

Swami, S. (1982). *Microwave heating characteristics of simulated high moisture foods. MS thesis ed.* Amherst, MA: University of Massachusetts.

Soltysiak, M., Erle, U., & Celuch, M. (2010). Influence of the magnetron operating frequency on the results of microwave heating. In *IEEE microwave symposium digest*.

Taflove, A., & Hagness, S. C. (2000). *Computational electrodynamics: The finite-difference time-domain method*. Norwood, MA: Artech House.

Tang, J., & Resurreccion, F. P., Jr. (2009). Electromagnetic basis of microwave heating. In M. W. Lorence, & P. S. Pesheck (Eds.), *Development of packaging and products for use in microwave ovens* (1st ed., pp. 3–36). Boca Raton, FL: CRC Press LLC.

Tilford, T., Baginski, E., Kelder, J., Parrott, K., & Pericleous, K. (2007). Microwave modeling and validation in food thawing applications. *Journal of Microwave Power and Electromagnetic Energy, 41*(4), 30–45.

Verboven, P., Datta, A. K., Anh, N. T., Scheerlinck, N., & Nicolaī, B. M. (2003). Computation of airflow effects on heat and mass transfer in a microwave oven. *Journal of Food Engineering, 59*(2–3), 181–190.

Wang, S., Luechapattanaporn, K., & Tang, J. (2008). Experimental methods for evaluating heating uniformity in radio frequency systems. *Biosystems Engineering, 100*(1), 58–65.

Wäppling-Raaholt, B., Risman, P. O., & Ohlsson, T. (2006). *Microwave heating of ready meals-FDTD simulation tools for improving the heating uniformity. Advances in Microwave and Radio Frequency Processing*. Springer Berlin Heidelberg.

Wäppling-Raaholt, B., Scheerlinck, N., Gait, S., Banga, J. R., Alonso, A., Balsa-Canto, E., ... Nicolai, B. M. (2002). A combined electromagnetic and heat transfer model for heating of foods in microwave combination ovens. *Journal of Microwave Power and Electromagnetic Energy, 37*(2), 97–111.

Watanabe, M., Suzuki, M., & Ohkawa, S. (1978). Analysis of power density distribution in microwave ovens. *Journal of Microwave Power and Electromagnetic Energy, 13*(2), 173–182.

Wei, C. K., Davis, H. T., Davis, E. A., & Gordon, J. (1985). Heat and mass transfer in water-laden sandstone: Microwave heating. *AICHE Journal, 31*(5), 842–848.

Xiaowei, G., Lin, M., & Yiqin, S. (2010). Electromagnetic field optimisation procedure for the microwave oven. *International Journal of Electronics, 97*(3), 339–347.

Yakovlev, V. V. (2001a). Comparative analysis of contemporary electromagnetic software for microwave power industry. *Ceramic Transactions., 111*, 551–558.

Yakovlev, V. V. (2001b). *Commercial EM codes suitable for modeling of microwave heating – A comparative review. Scientific computing in electrical engineering: proceedings of the 3rd international workshop,* August 20–23, 2000, Warnemünde, Germany (p. 87). Springer Verlag.

Yakovlev, V. V. (2006). Examination of contemporary electromagnetic software capable of modeling problems of microwave heating. *Advances in Microwave and Radio Frequency Processing, Part III*, 178–190.

Zhang, H., & Datta, A. K. (2000). Coupled electromagnetic and thermal modeling of microwave oven heating of foods. *Journal of Microwave Power and Electromagnetic Energy, 35*(2), 71–85.

Zhang, H., & Datta, A. K. (2003). Microwave power absorption in single-and multiple-item foods. *Food and Bioproducts Processing, 81*(3), 257–265.

Zhang, H., Datta, A. K., Taub, I. A., & Doona, C. (2001). Electromagnetics, heat transfer, and thermokinetics in microwave sterilization. *AICHE Journal, 47*(9), 1957–1968.

Zhou, L., Puri, V. M., Anantheswaran, R. C., & Yeh, G. (1995). Finite element modeling of heat and mass transfer in food materials during microwave heating – Model development and validation. *Journal of Food Engineering, 25*(4), 509–529.

Zhu, J., Kuznetsov, A. V., & Sandeep, K. P. (2008). Investigation of a particulate flow containing spherical particles subjected to microwave heating. *Heat Mass Transfer, 44*, 481–493.

Index

Note: Page numbers followed by "*f*" and "*t*" refer to figures and tables, respectively.

I

U

V

Kolibri 360 GmbH
Ferdinand-Jühlke-Straße 7
99095 Erfurt, Deutschland
produktsicherheit@kolibri360.de